T0305972

ATMOSPHERIC LIDAR FUNDAMENTALS

Laser Light Scattering from Atoms and Linear Molecules

Lidar is a remote sensing technique that employs laser beams to produce a high-resolution, four-dimensional probe, with important applications in atmospheric science. Suitable as a detailed reference or an advanced textbook for interdisciplinary courses, this book discusses the underlying principles of light-scattering theory and describes widely used lidar systems in current research, exploring how they can be employed effectively for atmospheric profiling. This self-contained text provides a solid grounding in the essential physics of light–matter interactions and the fundamentals of atmospheric lidars through a discussion of the principles that govern light–matter interactions and an exploration of both historical and recent scientific developments in lidar technology. This is an essential resource for physicists, optical engineers, and other researchers in atmospheric science and remote sensing.

CHIAO-YAO SHE is Emeritus Professor of Physics at Colorado State University and a Fellow of the Optical Society of America. Dr. She earned his Ph.D. from Stanford University in 1964. He has decades of experience in areas of laser light scattering spectroscopy and atmospheric lidar that enabled important scientific investigations. He also pioneered two narrowband atmospheric lidars that have been duplicated internationally for atmospheric research and observations. He has an outstanding research and publication record including more than two hundred papers on related topics and two book chapters.

JONATHAN S. FRIEDMAN received his Ph.D. from Colorado State University in 1991 and is currently the Director of the Puerto Rico Photonics Institute, which he founded in 2011 at the Universidad Ana G. Mendez. For twenty-six years, he was a research associate at the Arecibo Observatory for Cornell University, and at SRI International. In addition to developing and employing lidar technology to study the atmosphere, he has also published extensively in the field of lidar science and spectroscopic physics.

ATMOSPHERIC LIDAR FUNDAMENTALS

Laser Light Scattering from Atoms and Linear Molecules

CHIAO-YAO SHE
Colorado State University

JONATHAN S. FRIEDMAN
Ana G. Mendez University

CAMBRIDGE
UNIVERSITY PRESS

University Printing House, Cambridge CB2 8BS, United Kingdom

One Liberty Plaza, 20th Floor, New York, NY 10006, USA

477 Williamstown Road, Port Melbourne, VIC 3207, Australia

314–321, 3rd Floor, Plot 3, Splendor Forum, Jasola District Centre, New Delhi – 110025, India

103 Penang Road, #05–06/07, Visioncrest Commercial, Singapore 238467

Cambridge University Press is part of the University of Cambridge.

It furthers the University's mission by disseminating knowledge in the pursuit of education, learning, and research at the highest international levels of excellence.

www.cambridge.org
Information on this title: www.cambridge.org/9781316518236
DOI: 10.1017/9781108968713

First published 2022

A catalogue record for this publication is available from the British Library.

ISBN 978-1-316-51823-6 Hardback

To
Lucy, Colleen, and Camille
Ivonne, Coralis, Bart, and Sheila

Contents

Foreword

In studying the several chapter drafts, I can see that this book will be a valuable addition to the literature, especially in that the historical evolution of the field of atmospheric lidar is so nicely laid out. This is hardly surprising, as the author, Professor Chiao-Yao She (formerly of the Colorado State University in Fort Collins, Colorado), has been a pioneer in the field over the decades. Importantly, the book also covers recent advances in atmospheric lidars from selected literature between 2005 and 2021, not contained in previous books of this type. Additionally, there is ample material to get the new researcher "up to speed" in the latest topics, such as the laser guide star idea and approaches.

One fundamental issue is the dynamic range requirements that must be met for any real success in this atmospheric lidar business: the transmitted pulses are measured in joules, while the returns are measurable as time-dispersed single photon detection events. These signal scales are separated by the (detection) sensitivity factor ~1E19, which is attenuation incurred in these experiments thanks to the unfavorable geometrical factors. While we have all experienced some "dark days" when sunlight is diminished by clouds and raindrops scatter sunlight before it arrives to warm us, this attenuation factor is rarely more than 30×, not the 10^{19}-fold encountered in upper atmospheric lidar. On the other hand, we know that nearby scattered outbound laser light will be far too intense for the sensitive receivers. A first idea is to use the time delay of the interesting echo, since it is (just) possible to design detectors that have fast-enough recovery from overload.

Another and more powerful approach is to use notch-like narrowband spectral filters in the return detection system, as light scattered by dust and aerosols will be substantially unchanged in wavelength. But atmospheric molecules have their own frequency shifts resulting from different modes of motion, so the receiver can be off-tuned the proper amount to receive the shifted light and mostly block the transmitter's pulse. But to get deep-enough absorption of the scattered primary beam is not trivial, so the community was glad to have molecules such as iodine, which offer strong absorption bands near 532 nm, the wavelength of powerful frequency-doubled Nd-YAG pulse lasers. Under proper conditions these absorbing cells can be essentially transparent for the desired, shifted wavelengths resulting from Cabannes, pure rotational Raman, and vibrational Raman scattering, corresponding to their (random) translational, rotational, and vibrational motions respectively. In the upper atmosphere, where aerosol scattering is not a concern, the community uses very narrowband Fabry–Perot étalon (narrower than 2 GHz) and Faraday filters to diminish skylight background.

I was particularly rewarded by the author's historical coverage of the LIDAR science and technology. Professor She introduced atomic and molecular (iodine) filters to the tropospheric lidar community in 1983. After hearing of the atmospheric Na layer in 1987, he and his students developed Na laser–induced fluorescence lidar in 1989, then pioneered atmospheric temperature measurements in the sodium layer near 90 km, at the border between the atmosphere and space, studying gravity waves, tides, and secular (long-term) changes. His group continually advanced the technology, adding dual two-pass acousto-optic modulators to enable atmospheric wind measurement, Na vapor Faraday filters for daylight measurement, and sum-frequency generation to reach 589 nm as an attempt to improve the lidar transmitter. Most of these contributions can be found in Chapter 7. This book includes stories about how his group got the first results; for example, the excitement of receiving Na signal in 2000 for the first time in Arctic Norway under 24-hour daylight. As the reader may know, this reflected light from the 90-km-altitude Na layer is now routinely used to represent a "point-source" of astronomical light. Its return image is used to dynamically adjust the multi-segment optics of many modern telescopes to largely remove the degrading of obtained images by atmospheric scintillation and shearing.

For this reader, Section 8.5 is delicious – it incorporates some analysis of signal levels and expected turbulence scaling, but the presentation about the LGS work is the home run. I had known about and even visited Roger Angel's rotating furnace for making large parabolic mirror blanks. But I didn't know the where and who about the guide-star development. It looks like Mr Fugate's Albuquerque team is one of the main heroes.

This book project seems to have morphed from a history of stuff that happened and new lidars that were developed, with many interesting observations and results, into a book for every optics student to enjoy and be inspired by. I was astonished to find a reference to the Fugate lecture on YouTube. Perhaps there is some public information about one of the current frontier telescopes (Hawaii and/or Chile), where this spin-off from the field's work is leading to applications in astronomy.

Also, I wonder if there is something interesting about atmospheric chemistry that could be diagnosed using the spatial resolution possibilities. Would that occur only if the reactions had a nonlinear density dependence?

Professor She and his coauthor have been active for decades in training researchers in the atmospheric LIDAR business. His research program had its home base in Colorado, with many Ph.D.-level researchers being educated in the LIDAR arts. He also was involved in many collaborations at other labs at geophysically interesting places, such as northern Norway. Such experiences for students definitely will involve broad learning not usually available in a regular course of study. You will like this book.

Dr. John L. "Jan" Hall
Senior Fellow of NIST, Emeritus
Adjoint Professor of Physics, University of Colorado
Nobel Laureate in Physics 2005

Preface

As its title, *Atmospheric Lidar Fundamentals: Laser Light Scattering from Atoms and Linear Molecules*, suggests, this book treats the fundamentals of atmospheric lidar from first physics principles. The introductory chapter gives a brief description of the relation between the optical processes involved and atmospheric lidars. The relevant optical processes are first discussed classically in Chapter 2. Supplemented by Appendix A, Chapter 3 discusses the processes of light absorption, scattering, and laser-induced fluorescence (LIF) for atomic systems. Here, we employ semiclassical (quantum) theory along with spectroscopy of atoms relevant to atmospheric lidar and Na laser guide stars (LGS). This is followed in Chapter 4 by quasi-elastic (Rayleigh) and inelastic (Raman) laser light scattering from linear molecules. Utilizing the results developed in Chapters 2–4, we use the following three chapters to discuss atmospheric lidars: in Chapter 5, the lidar equation; in Chapter 6, lidars using broadband pulsed lasers, including Rayleigh–Mie, and conventional and Stokes-vector-based polarization lidars; and in Chapter 7, lidars using narrowband pulsed lasers for the measurement of aerosol optical properties and atmospheric state parameters (density, temperature, and wind) via Cabannes scattering, pure rotational Raman scattering, and laser-induced fluorescence. Finally, in Chapter 8 we give an overview of transmitting and receiving optics and telescopes along with a brief account on atmospheric turbulence and Laser Guide Star (LGS) for ground-based astronomical telescopes.

The monograph is written in a textbook format. As such, it follows, as is typical, a logical order of information dissemination, starting with an introduction, followed by basic principles and then applications. If the research career of one author (CYS) of this book (briefly described in Appendix C) might inspire many scientists interested in applications, the chronology of events is less orderly than one might anticipate. Such an evolution often begins with a practical problem requiring a creative solution. This is followed by one or more ideas and preferably, if possible, some preliminary investigations. Following the good fortune to secure funding, integration of a team of graduate students in a university project leads to mining the often scattered and difficult literature and, ultimately, to deeper understanding of the associated basics. In the process, dissertations are written and advanced degrees are awarded. With luck, this deeper basic understanding inspires more exciting methods for solving deeper and broader basic and applied problems. This is our roundabout way of stating our strong belief that, in addition to necessary and timely technology, a sound physical understanding of the basics and quantitative knowledge of various light scattering processes will enhance the creativity of students and researchers in the field of laser remote sensing, or LIDAR (**LI**ght **D**etection **A**nd **R**anging). Our limited experience in working with young students and researchers in applied sciences worldwide has led us to know that they are typically highly motivated to succeed, and, thanks to the Internet, they are increasingly skilled at computer programming and knowledgeable about openly available technical information. At the same time, they are generally less proficient with relevant basic physics principles and historical developments. A common element is that their lives are a lot busier than those of their counterparts decades ago. Our motivation for this book is then primarily to support these lovely young researchers by presenting a concise, single source for the principles of relevant optical processes coupled with showcase examples of modern atmospheric lidars.

This is an interdisciplinary book in the field of applied optics, overlapping the disciplines of atmospheric lidars and optical spectroscopy. It is intended for advanced undergraduate and graduate students in physics, chemistry, optical engineering, and atmospheric sciences, as well as researchers in research institutes and optics industries who develop and/or use lidars for remote sensing and atmospheric studies. The book could be a text or major reference for an interdisciplinary special topics–type course for advanced undergraduate and graduate students in a physics or atmospheric sciences department. It could also be a text for a semester course at the same level in light scattering and spectroscopy of atoms and simple molecules in departments of physics, chemistry, or optical engineering. Such a course, though it may be considered old fashioned nowadays, is in our view very useful for the scientific well-being of most students; it may be constructed from the materials in Chapters 1–4 along with Appendices A and B. This text could also be used for a course in atmospheric lidars in engineering or atmospheric/environmental sciences, in which case the course would most benefit from the materials in Chapters 1, 5, 6, 7, and 8 along with Appendix B.

We acknowledge the contributions of our many research collaborators, colleagues, visitors, and graduate students, too numerous to name here (see Appendix C), who have sustained and stimulated our learning process and have directly or indirectly influenced the writing of this book and the content therein. We are grateful for many discussions with Jay Yu on the coherent detection technique for wind measurement, and with Raul Alvarez, John Hair, and Dave Krueger on Cabannes scattering and associated temperature measurements, which we pursued more than two decades ago. We thank Chet Gardner for suggesting the inclusion of a brief discussion on laser guide stars, and Paul Hickson for providing relevant expert references on this exciting development. We also thank Jan Hall, who kindly authored the foreword to this book and made suggestions to improve the presentation of Chapter 7 while reading draft chapters. We are deeply grateful for the assistance of Zhao-Ai Yan, who prepared all the figure presentations for this book. JSF gratefully acknowledges the patient mentorship and leadership of the late Craig Tepley, who transformed him from a crude postdoc to a more accomplished researcher.

1
Introduction

Since the comprehensive textbook entitled *Laser Remote Sensing: Fundamentals and Applications*, written by R. M. Measures, was published by Wiley-Interscience in 1984 [1.1], there have been tremendous advances both in laser technology – especially in the realm of spectral purity and high power – and in information/digital technology. These advances in the past three-plus decades have transformed narrowband lidars from being demonstrations of remote sensing concepts to working instruments capable of probing the atmosphere on regular and quasi-continuous bases. Nevertheless, until 2005 there was no substantial textbook development about atmospheric lidars to keep up with these advances. At that time, two extensive collections of book chapters, written by experts and researchers specializing in different aspects and types of lidars, were published. They are *Laser Remote Sensing*, published by CRC Press with Takashi Fujii and Tetsuo Fukuchi [1.2] as editors, and *Lidar: Range-Resolved Optical Remote Sensing of the Atmosphere*, published by Springer with Claus Weitkamp [1.3] as the editor. These books brought up to date all important advances and practices in lidar research. The book chapters in these collections, though already more than a decade old at the time of this writing, contain sufficient information for beginners to learn and to initiate a chosen type of lidar for their remote sensing and research needs. When coupled with manufacturers' websites that now provide up-to-date information on the availability of lidar hardware, these resources represent sufficient information for state-of-the-art system development.

What is lacking is a book on the fundamental treatment of Rayleigh and Raman scattering as well as laser-induced fluorescence (LIF) relating to the atmospheric components and sensing relevant to lidar-enabled research. Since we strongly believe that, in addition to technology, a sound physical understanding and consistent knowledge of various scattering cross sections and associated frequency distribution will enhance the creativity of students and researchers in the field of laser remote sensing, we decided to write this book. In order to maintain physical clarity, we shall treat single scattering of atoms and linear molecules in some detail, as they are the dominant scattering species within different layers (heights) of the neutral atmosphere. At the same time, light interaction with these species can be understood by manageable quantum mechanical manipulations. With this aim in mind, this book first presents classical light scattering theory (Chapter 2), followed by semiclassical (quantum) treatment of light absorption and scattering from atoms (Chapter 3), and Rayleigh and Raman scattering from linear molecules (Chapter 4). These treatments are supplemented by an overview of electric dipole interactions and structures of atoms and of linear molecules (Appendix A), and by a discussion of coordinate systems used and their transformations (Appendix B). The second half of the book treats the fundamentals of atmospheric lidars, starting with the lidar equation in Chapter 5. We move on to lidars using broadband pulsed lasers in Chapter 6, including Rayleigh–Mie, conventional, and Stokes-vector–based polarization lidars. In Chapter 7, we treat lidars employing narrowband pulsed laser transmitters, considering measurements of aerosol optical properties and atmospheric state parameters (density, temperature, and wind) via Cabannes scattering, pure rotational Raman scattering, and laser-induced fluorescence. Finally, Chapter 8 begins with a brief account on transmitting and receiving optics and telescopes. It then moves to an overview of atmospheric turbulence and the basics of laser guide stars (LGS). Much of the material in these chapters has already been documented in the book chapters edited by Weitkamp [1.3] and by Fujii and Fukuchi [1.2] in 2005.

Research literature, since 2005, has guided much of the new material presented in this book, particularly that on Stokes-vector–based polarization lidars (detailed in Chapter 6) and on atmospheric parameter measurements with narrowband laser-induced resonance fluorescence (LIF) lidars (detailed in Chapter 7). The new materials also include relevant literature as recent as 2020 and 2021, noticeably, the proposal of maintenance-free Cabannes lidar at 770 nm (in Sections 7.3.4 and 7.3.5) and the use of large power-aperture product lidar at 589 nm for the challenging investigations of atmospheric turbulence in "real time." The latter research could benefit both atmospheric and astronomical communities (see Section 7.5.3), as it has the potential to (1) characterize the turbulent heating of a transient event, such as wave breaking, and (2) to follow the Na layer centroid, as a LGS, in a timescale of seconds in order to negate the wavefront distortions caused by atmospheric turbulence using adaptive optics for modern telescopes. Brief discussions on atmospheric turbulence and LGS can be found in Sections 3.5, 7.5.3, and 8.5.

The basic relevant light scattering processes are depicted in Fig. 1.1, in which light scattering from a particle with a simplified energy level structure is shown.

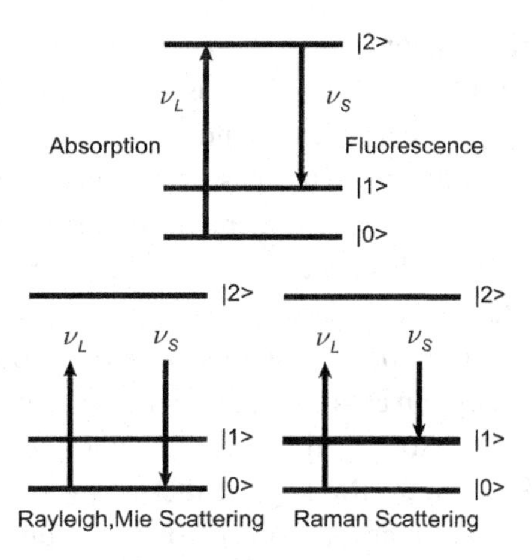

Fig. 1.1 Simple optical scattering processes relevant to atmospheric lidars.

When the incident laser optical frequency matches that of an electronic transition (of a metal atom, for example), the incident photon may be absorbed by the atom, which in turn is excited. The excited atom then emits a photon a brief time (ca. 10^{-8} s) later by the process of fluorescence. This is depicted in the top panel of Fig. 1.1. If the incident photon at the frequency ν_L is from an incoherent source, the direction of the fluorescence emission will be random and isotropic. If the incident light is coherent (as if from a laser), it will induce a dipole moment, which then emits fluorescence with an angular distribution (i.e., an antenna pattern) – often referred to as the Hanle effect in the lidar literature and detailed in Chapter 3. This angular pattern depends on the polarization of the incident light and on the atomic transition in question. The combined processes of absorption and fluorescence are called laser-induced fluorescence (LIF). In this book, we will treat only resonant scattering, or LIF, from atoms (not molecules). The simplest case is a system with two states, |0> and |2>, as depicted in the upper panel in Fig. 1.1, or a system with a single resonance frequency. Such a system may be modeled by a classical damped harmonic oscillator, as described in Chapter 2. For realistic atoms, like Na, K, or Fe, in which ground and excited electronic states have many levels – for the example shown in Fig. 1.1, the electronic ground state is split into |0> and |1> – quantum treatment must be invoked; this is done in Chapter 3.

When the incident photon energy, $h\nu_L$, is much less than that needed to cause absorption by an atom or molecule (e.g., nitrogen and oxygen molecules), non-resonance optical scattering can occur. The frequency of the scattered photon can be nearly the same as that of the incident photon or quite different. In the former case, the molecule returns to its original state, designated as $|0\rangle$. The process is called Rayleigh scattering if the dimension of the scattering particle, a, is much smaller than the wavelength of the light λ ($a \ll \lambda$), and it is called Mie scattering if a is comparable to, or larger than λ ($a \gtrsim \lambda$). This is depicted in the lower-left panel of Fig. 1.1. In molecules, the electronic ground state is split into sublevels due to rotational and vibrational motions. In this case, the scattered photon frequency may be less than the incident photon frequency, and the molecule is promoted to a higher rotational or vibrational level, designated as $|1\rangle$ in the lower-right panel of Fig. 1.1. This process is called Raman scattering.

Besides its angular distribution, the fluorescence (thus LIF) cross section is on the same order as the absorption cross section for a given atom, about 5×10^{-16} m^2 for the Na D transition. The cross section for rotational Raman scattering in visible wavelengths is ~ 2.5% of that for Rayleigh scattering, which in turn is 15 orders of magnitude smaller than that for absorption or LIF, and about 3 orders of magnitude larger than that of vibrational Raman scattering.

At higher spectral resolution, Rayleigh scattering of diatomic molecules has a fine structure that was unresolved in the days of Lord Rayleigh. It consists of a center peak called the Cabannes line, plus two sidebands of (pure) rotational Raman lines [1.4; 1.5; 1.6]. Due to the absence of rotational motions, quasi-elastic scattering from atoms, such as those of noble gases, has only the Cabannes line. The nonresonant scattering spectrum of a diatomic molecule, such as N_2 or O_2, in a relative frequency scale, $\nu_S - \nu_L$, with ν_S and ν_L being, respectively, the scattering frequency and incident laser frequency, is shown schematically in the upper panel of Fig. 1.2. In addition to Rayleigh scattering (Cabannes scattering plus rotational Raman scattering), Fig. 1.2 includes vibrational Raman scattering resulting from molecular vibration. Positive frequency shifts correspond to the "anti-Stokes" line (transition from $|1\rangle$ to $|0\rangle$, not included in Fig. 1.1), and negative frequency shifts correspond to the "Stokes" line (in the lower-right panel of Fig. 1.1), which results in lower frequency emission and the molecule gaining energy (transition from $|0\rangle$ to $|1\rangle$).

Cabannes spectra along with their Gaussian estimates are shown on an expanded frequency scale in the lower panel of Fig. 1.2. It is well known that at thermal equilibrium the scattering spectrum from a gas is Doppler broadened, giving rise to a Gaussian spectrum with full width at half-maximum (FWHM) depending on the gas temperature. These are shown as black thin-dashed and solid lines in the lower part of Fig. 1.2, respectively corresponding to atmosphere at 288.15 K (mean temperature at sea level) and at 226.15 K (mean temperature at 30 km altitude). Comparing the Cabannes spectra, gray thick solid and dashed lines, of atmospheric scattering at altitudes of 0 km (288.15 K, 1 atm) and 30 km (226.15 K, 0.012 atm), respectively, we see that the sea-level Cabannes spectrum is not Gaussian, whereas at 30 km it is virtually indistinguishable from a Gaussian. The reason that the Cabannes scattering at sea level is not Gaussian is that in addition to temperature (entropy) fluctuations, the molecular ensemble is subject to pressure fluctuations

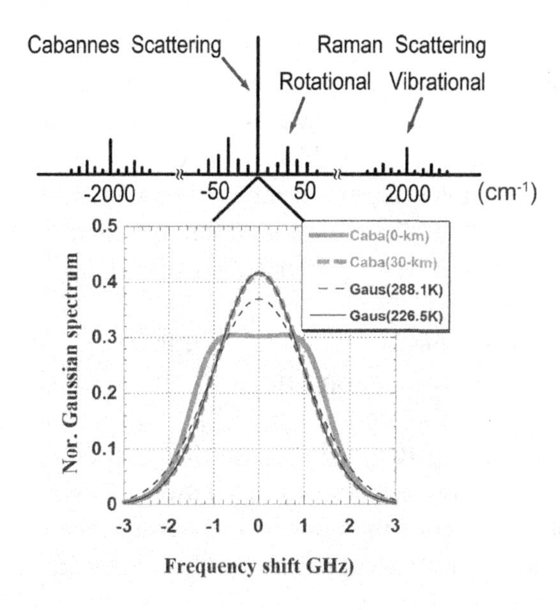

Fig. 1.2 Rayleigh and vibrational Raman scattering spectrum of a diatomic molecule in frequency shift, $\nu_S - \nu_L$ (1 cm^{-1} = 30 GHz).

due to collisions. Atmospheric pressure fluctuations, responsible for sound wave propagation, give rise to Brillouin scattering, resulting in a scattering angle-dependent frequency shift.

For backscattering, which is of interest here, the magnitude of the mean scattering wave vector is twice that of the incident wave vector. This means the wavelength of the backward propagating (sound or perturbation) wave in the atmosphere equals half of the incident optical wavelength (i.e., $\Lambda = \lambda/2$). Thus, the mean Brillouin frequency shift is $\Delta\nu_B = c_s/\Lambda$, where c_s is the speed of sound. For atmosphere at 288 K at sea level, $c_s = 340$ m/s, which for $\lambda = 532$ nm results in a frequency shift of $\Delta\nu_B = 1.28$ GHz.

While a more thorough description of the Cabannes spectrum will be given in Chapter 4, an oversimplified account describes the atmospheric Cabannes spectrum as the result of translational motion of the molecules. In the absence of collisions, molecules in thermal equilibrium move independently and randomly, giving rise to a Doppler-broadened Cabannes line. As pressure increases, the chance of colliding with other molecules increases. This allows the molecules to move correlatively and sound waves to propagate, which diverts part of the scattering power to the Brillouin wings, with two sidebands centered at $\sim \pm 1.28$ GHz for 1 atm, in the backward direction. The relative contribution of the sidebands (Brillouin scattering) is determined by a comparison between the scattering wavelength, $\Lambda = 266$ nm, and the mean free path between collisions $\bar{\ell}$. A simple calculation yields a mean free path of 66 nm at sea level and 44,000 nm at 30 km. Thus, at sea level many collisions occur within one scattering wavelength, which makes the Cabannes spectrum much broader (shown as the gray solid curve), while at 30 km $\bar{\ell} \gg \Lambda$, the Cabannes spectrum (thin black solid curve) appears to be identical to the associated Doppler-broadened spectrum (thick gray dashed curve). The details of nonresonant light scattering from linear molecules, their spectral distribution and relative line strengths, will be treated quantum mechanically in more detail in Chapter 4.

Lidar remote sensing of the atmosphere began in 1954 with the use of searchlights for the study of atmospheric temperature, density, and pressure in the altitude range between 10 and 68 km [1.7]. Soon after the invention of the ruby laser in 1960, the first Rayleigh–Mie, differential absorption, Raman, and fluorescence lidar observations were reported between 1969 and 1970. Since that time, lidar remote sensing of the atmosphere has become increasingly active. A selected collection of milestone papers on the achievements during the first three decades was published in 1997 [1.8]. We hope with this book to cover the fundamentals that made those achievements, as well as those in the years that followed, possible. For this objective, we have made no attempt to include an exhaustive list of relevant publications in the literature; rather, we refer to selective publications that allow us to explain the physics fundamentals for the atmospheric lidars in question.

References

1.1 Measures, R. M. (1984). *Laser Remote Sensing: Fundamentals and Applications.* Wiley-Interscience.
1.2 Takashi Fujii, T. and T. Fukuchi, eds. (2005). *Laser Remote Sensing*. CRC Press – Technology & Engineering.
1.3 Weitkamp, C., ed. (2005). *Lidar: Range-Resolved Optical Remote Sensing of the Atmosphere*. Springer-Verlag.
1.4 Inaba, H. (1976). Detection of atoms and molecules by Raman scattering and resonance fluorescence. In *Laser Monitoring of the Atmosphere*, E. D. Hinkley, ed., Springer-Verlag, 153–236.
1.5 She, C. Y. (2005). On atmospheric lidar performance comparison: from power–aperture product to power–aperture–mixing ratio–scattering cross-section product, *J. Mod. Optics*, **52**, 2723–2729.
1.6 Young, A. T. (1982). Rayleigh scattering. *Phys. Today*, January issue, pp. 42–48.
1.7 Elterman, L. (1954). Seasonal trends of temperature, density and pressure to 67.6 km obtained with the search light probing technique. *J. Geophys. Res.* **59**, 351–358.
1.8 Grant, W. B., E. V. Browell, R. T. Menzies, K. Sassen, and C.-Y. She, eds. (1997). *Selected Papers on Laser Applications in Remote Sensing*. SPIE Milestone Series, Vol. MS 141. SPIE Optical Engineering Press.

2

Classical Light Scattering Theory

When a monochromatic laser beam at an optical frequency (angular frequency), $\nu\,(\omega = 2\pi\nu)$, illuminates a neutral atom or molecule, the electrons oscillate relative to the nucleus at the optical frequency, which induces multipole electric and magnetic moments on the atom or molecule. Since the speed of electrons is much less than that of light, the electric dipole interaction energy is by far the largest, so we only need to consider its effect on light scattering. Therefore, we can conceptually picture an induced electric dipole moment, $\vec{\mathrm{p}}(t) = 0.5(\vec{p}e^{-i\omega t} + \vec{p}{*}e^{i\omega t}) = 0.5|p|(\hat{e}e^{-i\omega t} + \hat{e}{*}e^{i\omega t})$, oscillating at the same (angular) frequency ω on the atom or molecule, resulting from this dominating interaction. The induced oscillating electric dipole moment in turn radiates electromagnetic waves. The intensity of the radiation at a point $\vec{r}$ from this electric dipole moment is given by the Poynting vector, $\overrightarrow{\mathscr{P}}$, along its wave-propagating direction $\hat{k} = \hat{r}$. Classical electromagnetic theory – in chapter 9 of Griffiths [2.1], or in chapter 2 of Corney [2.2] – shows that the cycle-averaged Poynting vector $< \overrightarrow{\mathscr{P}} >$ is the radiated power per unit area, giving rise to radiated power per unit solid angle, $d\mathrm{P}/d\Omega = r^2[\hat{r}{\bullet} < \overrightarrow{\mathscr{P}} >]$. Since only the transverse dipole component radiates, it is convenient to project the dipole moment onto the scattering coordinates $(\hat{e}_a{}', \hat{e}_b{}', \hat{r})$; see Fig. B.1, as $\vec{p} = \overrightarrow{p_\perp} + p_{\|}\hat{r} = p_a\hat{e}_a{}' + p_b\hat{e}_b{}' + p_{\|}\hat{r}$. The directions of the radiation electric and magnetic fields are then, respectively, parallel to $(\hat{r} \times \vec{p}) \times \hat{r} = p_a\hat{e}_a{}' + p_b\hat{e}_b{}'$ and to $(\hat{r} \times \vec{p})$. The resulting $< \overrightarrow{\mathscr{P}} >$ and $d\mathrm{P}/d\Omega$ can then be derived in terms of the absolute square of the transverse components of the induced dipole moments as $|p_a|^2 + |p_b|^2$, yielding respectively,

$$< \overrightarrow{\mathscr{P}} > = \frac{\omega^4}{32\pi^2 \mathrm{e}_0 c^3 r^2} |\overrightarrow{p_\perp}|^2 \hat{r}; \text{ with } |\overrightarrow{p_\perp}|^2 = |p_a|^2 + |p_b|^2 \text{ and } |p_{a,b}|^2 = |\vec{p}{\bullet}\hat{e}_{a,b}{}'|^2, \text{ and} \tag{2.1.a}$$

$$\frac{d\mathrm{P}}{d\Omega} = r^2[\hat{r}{\bullet} < \overrightarrow{\mathscr{P}} >] = \frac{\omega^4}{32\pi^2 \varepsilon_0 c^3} |\overrightarrow{p_\perp}|^2 = \frac{\omega^4 |p|^2}{32\pi^2 \mathrm{e}_0 c^3} \left[(\hat{e}{\bullet}\hat{e}_a{}')^2 + (\hat{e}{\bullet}\hat{e}_b{}')^2 \right], \tag{2.1.b}$$

where ω, e_0, and c are, respectively, angular frequency, vacuum permittivity, and the speed of light. We note that the last expression of (2.1.b) is identical to (2) of [2.3], where the unit vector $\hat{e}\,'$ represents the polarization of the scattering radiation, which lies on the $\hat{e}_a{}' - \hat{e}_b{}'$ plane.

The dipole moment in question is induced by the incident laser with electric field, $\vec{\mathrm{E}}(t) = \frac{1}{2}(\vec{\mathscr{E}}e^{-i\omega t} + \vec{\mathscr{E}}{*}e^{i\omega t})$ and intensity $I = \frac{\mathrm{e}_0 c}{2}|\mathscr{E}|^2$. For light scattering from an isotropic medium like atmospheric atoms, the polarizability is a scalar $\alpha(\omega)$, leading to $\alpha(\omega)\vec{\mathscr{E}}$, or $\vec{p} = |p|\hat{e} = \alpha(\omega)\mathscr{E}\hat{e}$, where $|p|$ and $\hat{e}$ are, respectively, the magnitudes of the induced dipole moment and its polarization direction. Thus, the absolute square of the transverse components of the induced dipole moment and light scattering differential cross section summed over both scattering polarizations may be written as

$$|p_a|^2 = |\alpha(\omega)\mathscr{E}|^2(\hat{e}\bullet\hat{e}_a{}')^2; \text{ and } |p_b|^2 = |\alpha(\omega)\mathscr{E}|^2(\hat{e}\bullet\hat{e}_b{}')^2, \text{ and} \tag{2.2.a}$$

$$\frac{d\sigma}{d\Omega} = \frac{d\mathrm{P}}{d\Omega}/I = \frac{\omega^4}{32\pi^2\mathrm{e}_0c^3}\left(|p_a|^2 + |p_b|^2\right)/\left(\frac{\mathrm{e}_0c}{2}|\mathscr{E}|^2\right) = \frac{\omega^4|\alpha(\omega)\mathscr{E}|^2}{16\pi^2\mathrm{e}_0^2c^4}(\hat{e}\bullet\hat{e}')^2, \tag{2.2.b}$$

where $(\hat{e}\bullet\hat{e}')^2 = (\hat{e}\bullet\hat{e}_a{}')^2 + (\hat{e}\bullet\hat{e}_b{}')^2$ is the sum of received scattered light from the two independent polarizations. We note that (2.2.b) is identical to Eq. (3) of She et al. [2.3].

In general, the induced dipole moment is related to the second rank polarizability tensor $\overleftrightarrow{\alpha}$ of the atom or molecule in question as $\vec{p} = \overleftrightarrow{\alpha}\bullet\vec{\mathscr{E}}$ – see, for example, chapter 3 of Long [2.4] – or in the Cartesian components as $p_i = \alpha_{ij}(\omega)\mathscr{E}_j$, with $i, j = (1,2,3)$ representing (x,y,z) axes. For scattering from an isotropic medium, which we assume for the treatments in Chapters 2 and 3 of this book, the directions of $\vec{p}, \vec{\mathscr{E}}$, and $\hat{e}$ are parallel to one another. This assumption will not be valid for a medium with linear molecules, where the polarizabilities parallel and perpendicular to the molecular axis are different, necessitating the use of tensor polarizability, $\overleftrightarrow{\alpha}$; this situation will be treated in Chapter 4.

2.1 A Simple Model of Atomic Polarizability and Differential Scattering Cross Section

To reveal the frequency structure of $\alpha(\omega)$ and $|\alpha(\omega)|^2$, we need a physical model for the atom, thus its polarizability. A simple (weakly) damped harmonic electron oscillator with one resonance frequency ω_0 is a suitable model in which a dipole moment $p(t) = -ex(t)$ may be induced under the illumination by an optical field, with a linear equation of motion for electronic displacement x as

$$\frac{d^2x(t)}{dt^2} + \gamma\frac{dx(t)}{dt} + \omega_0^2 x(t) = -\frac{e}{m_e}\mathrm{E}(t), \tag{2.3.a}$$

where m_e, γ, ω_0, and e are respectively the mass, damping rate, resonance frequency, and charge of the electron oscillator, which may be thought of as driving the superposition of external, restoring, and damping forces, as given in (2.3.a); see chapter 2 of [2.5]. Without an externally applied force (i.e., setting $E(t)$ to zero), a perturbation in displacement $x(0)$ initiated at $t = 0$ will oscillate at the frequency ω_0' and decay at its amplitude rate of $(\gamma/2)$, resulting in the following solution to (2.3.a):

$$x(t) = 0.5x(0)e^{-(\gamma/2)t}\left[e^{-i\omega_0' t} + e^{i\omega_0' t}\right];$$
$$\omega_0' = \sqrt{\omega_0^2 - (\gamma/2)^2} \approx \omega_0, \text{ since } \gamma \ll \omega_0. \tag{2.3.b}$$

The decay rate of the individual dipole energy will then be γ. With this model, we can relate the induced dipole moment $p(t) = 0.5(pe^{-i\omega t} + p^* e^{i\omega t})$ and the applied electric field $E(t) = 0.5(\mathscr{E}e^{-i\omega t} + \mathscr{E}^* e^{i\omega t})$, and easily determine the associated frequency-dependent complex polarizability, $\alpha(\omega)$, as, for example, in Chapter 8 of Griffiths [2.1]:

$$(-\omega^2 - i\gamma + \omega_0^2)x = -\frac{e}{m_e}E \to p = -ex$$
$$= \frac{e^2/m_e}{\omega_0^2 - \omega^2 - i\omega\gamma}E \to \alpha(\omega) = \frac{e^2/m_e}{\omega_0^2 - \omega^2 - i\omega\gamma}. \tag{2.3.c}$$

Because the damping rate $\gamma \ll \omega, \omega_0$, we can decompose the polarizability into two contributions, corresponding to the rotating and counterrotating wave terms as

$$\alpha(\omega) = \frac{e^2/m_e}{\omega_0^2 - \omega^2 - i\omega\gamma} \approx \frac{e^2}{2\omega_0 m_e}\left[-\frac{1}{(\omega - \omega_0) + i(\gamma/2)} + \frac{1}{(\omega + \omega_0) + i(\gamma/2)}\right]. \tag{2.3.d}$$

Notice that when $\omega \approx \omega_0$, the first term is much larger; we only need to keep it. This is the case for resonant scattering. We can then make the rotating-wave approximation leading to resonant polarizability, $\alpha^R(\omega)$. When $\omega, \gamma \ll \omega_0$, we need to keep both (rotating and counterrotating) terms, and they may be combined to yield the nonresonant polarizability, $\alpha^{NR}(\omega)$. These are explicitly given in (2.3.e):

$$\alpha^R(\omega) = -\frac{e^2/2m_e\omega_0}{(\omega-\omega_0)+i(\gamma/2)};\ \alpha^{NR}(\omega) = \frac{e^2}{m_e\omega_0^2} = \left(\frac{e^2}{4\pi\mathrm{e}_0 m_e c^2}\right)\frac{4\pi\mathrm{e}_0 c^2}{\omega_0^2} = r_e^2\frac{4\pi\mathrm{e}_0 c^2}{\omega_0^2}, \tag{2.3.e}$$

where $r_e = e^2/4\pi\mathrm{e}_0 m_e c^2 = 2.82\times10^{-15}$m is the classical electron radius. Since both rotating and counterrotating terms make equal contributions to $\alpha^{NR}(\omega)$, it is a factor of two larger than $\alpha^R(\omega)$ in the limit of $\omega, \gamma \ll \omega_0$. Substituting (2.3.e) into (2.2.b), we obtain

$$\frac{d\sigma^{RS}}{d\Omega} = r_e^2\frac{\pi\omega_0^2}{2\gamma}\mathrm{g}(\omega-\omega_0)(\hat{e}\bullet\hat{e}')^2;\ \text{with } \mathrm{g}(\omega-\omega_0) = \frac{\gamma/2\pi}{(\omega-\omega_0)^2+(\gamma/2)^2},\ \text{and} \tag{2.3.f}$$

$$\frac{d\sigma^{NR}}{d\Omega} = r_e^2\left(\frac{\omega}{\omega_0}\right)^4(\hat{e}\bullet\hat{e}')^2 = \sigma_\pi^C(\hat{e}\bullet\hat{e}')^2;\ \text{with } r_e = \frac{e^2}{4\pi\mathrm{e}_0 m_e c^2}. \tag{2.3.g}$$

Here, $\mathrm{g}(\omega-\omega_0)$ is the Lorentzian lineshape function with full width at half maximum of $\Delta\omega_L = \gamma = 1/\tau$, where γ and τ are, respectively, the energy damping rate and the radiative lifetime of an electron oscillator.

The quantity σ_π^C is known as the differential Cabannes scattering (CS) cross section; for scattering from air molecules at 532 nm, $\sigma_\pi^C = 5.96\times10^{-32}\mathrm{m}^2/\mathrm{sr}$ (see Table 1, She [2.6]). This leads to an estimate of the ratio of resonant wavelength λ_0 to the laser wavelength λ by equating $(\omega/\omega_0)^4 = (\lambda_0/\lambda)^4$ to $\sigma_\pi^C/r_e^2 = 5.96\times10^{-32}/(2.82\times10^{-15})^2 \approx 7.5\times10^{-3}$, resulting in $\lambda_0 \approx 157$ nm, or about 7.9 eV, which represents the effective dipole-allowed excited states of atmospheric molecules. This value is interesting, as the ionization potentials (wavelengths) of O_2 and N_2 are, respectively, 12.2 (101.6) and 15.6 eV (79.5 nm); see Table 37, Herzberg [2.7]. Of practical interest is to compare the resonant differential cross section at 589 nm from a Na atom to the nonresonant cross section from an air molecule at 532 nm, because these are the wavelengths of a Na lidar and a Cabannes (often termed a Rayleigh) lidar. Using (2.3.e), we see that the peak backward resonant differential cross section is $d\sigma^{RS}/d\Omega = r_e^2(\omega_0/\gamma)^2$. If we substitute $(\hat{e}\bullet\hat{e}')^2 = 1$ and $\omega_0/2\Gamma \approx 1/(2\times10^{-8})$ into (2.3.e), we obtain $d\sigma^{RS}/d\Omega = 2\times10^{-14}\mathrm{m}^2/\mathrm{sr}$, roughly 3×10^{17} larger than σ_π^C and in reasonable agreement with quantum mechanical results to be discussed at the end of Section 3.2.

2.2 Resonant and Nonresonant Scattering from the Atmosphere

To apply the above treatment to a system of atmospheric metal atoms and air molecules, we must account for their motion due to the background wind as well as random motions relative to one another with a speed variation dependent on background or ambient temperature. At thermal equilibrium (ignoring the effects of pressure for simplicity), the probability of finding an air molecule with line of sight (LOS) thermal velocity u is given by the well-known Maxwellian distribution:

$$\mathcal{G}(u, T) = \sqrt{\frac{m}{2\pi k_B T}} \exp\left[-\frac{mu^2}{2k_B T}\right] \equiv \frac{1}{\sqrt{2\pi}\sigma_u} \exp\left[-\frac{u^2}{2\sigma_u^2}\right]. \tag{2.4}$$

Here, the variance of this velocity distribution is $\sigma_u^2 = k_B T/m$, with k_B, m, and T being, respectively, Boltzmann constant, mass of the molecule (or metal atom), and temperature. As we pointed out in Chapter 1, for atmospheric molecules above 30 km altitude, their correlated motions (Brillouin scattering) are negligible; we need only consider the Doppler broadening due to their random motion and Doppler shift due to background wind. For air molecules moving (in dynamic equilibrium) with the background LOS wind, V, the instantaneous LOS velocity is the sum of thermal velocity and background LOS wind (i.e., $\mathcal{V} = u + V$).

For nonresonant scattering (or Cabannes scattering) from a single atom or molecule (ignoring correlated motion for simplicity here), the differential cross section has a mild 4th-power frequency dependence, as can be seen in (2.3.g); σ_π^C is nearly constant. The Doppler-shifted frequency of a particle moving with the LOS (thermal) velocity, u, in the background LOS wind V becomes $\nu_s = \nu - 2(u + V)/\lambda$, where λ is the wavelength of the incident light. For an ensemble of atmospheric atoms or molecules, the velocity distribution, $\mathcal{G}(u, T)$, transformed into the frequency domain becomes the normalized Gaussian spectrum for CS, $G^{CS}(\nu - \nu_s - 2V/\lambda, T)$; it is obtained from (2.4) by substituting $0.5(\nu - \nu_s - 2V/\lambda)$ for u/λ, and $\lambda d\nu/2$ for du. The spectral differential resonant backscattering cross section is then simply the product of $d\sigma_\pi^{CS}/d\Omega$ and the backward-scattering spectrum, $G^{CS}(\nu - \nu_s - 2V/\lambda, T)$, where the backscattering Doppler shift of $-2V/\lambda$ from the incident frequency, $\nu = c/\lambda$, is well known, and it will be further treated in Chapter 3. Thus, with $G^{CS}(\nu - \nu_s - 2V/\lambda, T)$, the spectral backscattering differential cross section is peaked at $\nu_s = \nu - 2V/\lambda$ as

$$\frac{d\sigma_\pi^{CS}}{d\Omega} = \sigma_\pi^C(\nu) \rightarrow \frac{d^2\sigma_\pi^{CS}(\nu - \nu_s, T, V)}{d\nu_s d\Omega} = \sigma_\pi^C G^{CS}(\nu - \nu_s - 2V/\lambda,\ T),\ \text{where}$$
$$G^{CS}(\nu - \nu_s\ - 2V/\lambda,\ T) = \frac{\lambda}{2}\frac{1}{\sqrt{(2\pi)}\sigma_u}\exp\left[-\frac{u^2}{2\sigma_u^2}\right]$$
$$\equiv \frac{1}{\sqrt{(2\pi)}\sigma_\nu^{CS}}\exp\left[-\frac{(\nu - \nu_s - 2V/\lambda)^2}{2(\sigma_\nu^{CS})^2}\right]. \qquad (2.5)$$

When integrating over the entire scattering spectrum, we retrieve $d\sigma_\pi^{CS}(\nu)/d\Omega = \sigma_\pi^C$. Here, $\sigma_\nu^{CS} = (2/\lambda)\sqrt{k_B T/M}$ is the standard deviation (RMS variation) of the CS frequency spectrum; its corresponding full width at half maximum (FWHM) is then

$$\Delta\nu_D^{CS} = 2\sqrt{2\ln 2}\sigma_\nu^{CS} = \frac{1}{\lambda}\sqrt{\frac{32(\ln 2)k_B T}{m}} \rightarrow \Delta\nu_D^{CS}(\text{GHz}) = \frac{1}{\lambda(\text{nm})}\sqrt{\frac{1.84\times 10^5 T(\text{K})}{M(\text{amu})}}, \qquad (2.5.\text{a})$$

where k_B is the Boltzmann constant, and m and M are the mass of air molecule in kg and amu. For air molecules (M = 28.97 amu) at 300 K, the FWHM Doppler width from (2.5.a), $\Delta\nu_D^{CS}$ is 2.59 GHz at 532 nm. Similarly, it is 2.15 GHz and 1.26 GHz, respectively, for Na (M = 22.99 amu) at 589 nm and for K (M = 39.1 amu) at 770 nm, both at 200 K.

For absorption and resonant backscattering (or laser induced fluorescence, LIF), we note that a metal atom can absorb a photon only when the incident laser frequency in its rest frame (i.e., $\nu - \upsilon/\lambda$) falls within the natural linewidth, $\Delta\nu$, of an atomic transition, where υ is the LOS velocity of the atom in question. Thus, the peak absorption occurs when $\nu - \upsilon/\lambda = \nu_0$. In the presence of background LOS wind, V, the LOS individual velocity is $\upsilon = u + V$. Thus, the frequency in the resonant differential scattering cross section (2.3.f) should be its Doppler-shifted frequency (i.e., $\nu - (u+V)/\lambda$) and

$$\frac{d\sigma^{RS}}{d\Omega} = \frac{\pi^2\nu^4 r_e{}^2}{\nu_0^2\Delta\nu}(\hat{e}\bullet\hat{e}')^2 g\left(\nu - \frac{u+V}{\lambda} - \nu_0\right),\ \text{with} \qquad (2.6.\text{a})$$

$$g\left(\nu - \frac{u+V}{\lambda} - \nu_0\right) = \frac{\Delta\nu}{2\pi\left\{\left(\nu - \frac{u+V}{\lambda} - \nu_0\right)^2 + (\Delta\nu/2)^2\right\}};\ \Delta\nu = \frac{1}{2\pi\tau} = \frac{\gamma}{2\pi} = \frac{\Delta\omega_L}{2\pi}. \qquad (2.6.\text{b})$$

Since frequency is the quantity that is measured, it is often desirable to express the differential cross section as a function of frequency, as given above. In (2.6.a), we have related the natural lineshape function $g(\omega - \omega_0)$ in angular frequency, (2.3.f), to that in frequency, $g(\nu - \nu_0)$, with $\nu_0 = \omega_0/2\pi$. Here, $\Delta\nu = \gamma/2\pi$ is the full width at half maximum (FWHM) linewidth due to lifetime (natural) broadening, and $\gamma = 1/\tau = \Delta\omega_L$ is the energy damping rate of the oscillator, or the inverse of lifetime.

For an ensemble of atoms in thermal equilibrium at temperature T, (2.6.a) should be integrated over its thermal distribution $\mathscr{G}(u, T)$ in the frequency domain. This is a Gaussian function $G^{RS}[(\nu - V/\lambda - \nu_0), T]$, which may be obtained from (2.4) by substituting $(\nu - \nu_0 - V/\lambda)$ for u/λ, and $\lambda d\nu$ for du, leading to the resonant differential scattering cross section:

$$\frac{d\sigma^{RS}}{d\Omega} = \frac{\pi^2 \nu^4 r_e{}^2}{\nu_0^2 \Delta\nu} (\hat{e} \bullet \hat{e}')^2 \mathbb{G}[(\nu - V/\lambda - \nu_0), T] \text{ , with} \tag{2.7.a}$$

$$\mathbb{G}[(\nu - V/\lambda - \nu_0), T] = \int_{-\infty}^{\infty} g\left(\nu - \frac{u+V}{\lambda} - \nu_0\right) \mathscr{G}(u, T) du \text{ , and}$$

$$\xrightarrow{\nu' = u/\lambda} \mathbb{G}[(\nu - V/\lambda - \nu_0), T] = \int_{-\infty}^{\infty} g\left(\nu - \frac{V}{\lambda} - \nu' - \nu_0\right) G^{RS}\left(\nu', \mathrm{T}\right) d\nu'. \tag{2.7.b}$$

The spread of Gaussian function, $G^{RS}[(\nu - V/\lambda - \nu_0), T]$, results from absorption rather than backward scattering; thus, its standard deviation (RMS variation) is a factor of 2 smaller (i.e., $\sigma_\nu^{RS} = \sigma_\nu^{CS}/2 = (1/\lambda)\sqrt{k_B T/M}$); the same for the full width at half maximum (FWHM),

$$\Delta\nu_D^{RS} = 2\sqrt{2\ln 2}\sigma_\nu^{RS} = \frac{1}{\lambda}\sqrt{\frac{8(\ln 2)k_B T}{M}} \rightarrow \Delta\nu_D^{RS}(\mathrm{GHz}) = \frac{1}{\lambda(\mathrm{nm})}\sqrt{\frac{4.61 \times 10^4 T(\mathrm{K})}{M(\mathrm{amu})}}. \tag{2.7.c}$$

For Na (M = 22.99 amu) at 589 nm and K (M = 39.1 amu) at 770 nm, the Doppler-broadened widths at 200 K are 1.08 GHz and 0.63 GHz, respectively. Since typically, $\Delta\nu \approx 10$ MHz and $\Delta\nu_D^{RS} \approx 1{,}000$ MHz at 200 K, the Voigt function, $\mathbb{G}[(\nu - \nu_0),\ T]$ in (2.7.b) may generally be approximated by the Gaussian function $G^{RS}[(\nu - \nu_0), T]$.

Equation (2.7.a) represents the magnitude of the resonant backscattering scattering cross section, which is dictated by the absorption process in LIF; it does not reveal the fluorescence emission frequency or spectrum. The fluorescence frequency is determined by the Doppler effect to be further analyzed in Chapter 3; it turns out to be $\nu_e = 2\nu_0 - \nu$, within the natural linewidth of the respective spontaneous emission. This can be briefly understood by considering resonant scattering as a combination of absorption followed by emission. As noted, the peak absorption occurs when $\nu - \mathcal{V}/\lambda = \nu_0$, that is, the resonant frequency and the incident laser frequency select atoms with correct LOS velocity $\mathcal{V}$, and these are excited by absorbing the incident photon. Once excited (coherently, say by a laser), the atom could emit a photon at its transition frequency ν_0 in its rest frame. Since the moving atom carries much higher momentum than that of the emitted photon, its velocity will remain essentially the same to satisfy linear momentum conservation. The emitted frequency in the laboratory frame (or to a detector on ground) will be Doppler shifted again by the same amount i.e., $\nu_e = \nu_s = \nu_0 - \mathcal{V}/\lambda = \nu_0 - (\nu - \nu_0) = 2\nu_0 - \nu$, leading to an emission (or resonantly backscattered) frequency depending only on the laser and atomic transition frequencies (i.e., independent of atom's motion), in other words the background conditions (temperature or LOS wind). Therefore, the spectral differential resonant backscattering cross section, or simply, LIF spectrum of atmospheric metal atoms, is the product of $d\sigma^{RS}/d\Omega$ and the natural-broadened emission lineshape function centered at ν_e:

$$\frac{d^2\sigma_\pi^{RS}}{d\nu_s d\Omega} = \frac{2\pi^3\nu^4 {r_e}^2}{\nu_0^2\gamma}(\hat{e}\bullet\hat{e}')^2\mathfrak{G}[(\nu - V/\lambda - \nu_0), T]\mathrm{g}(\nu_s - 2\nu_0 + \nu). \tag{2.8}$$

Thus, the spectral backward resonant scattering cross section is a Lorentzian line centered at $\nu_s = 2\nu_0 - \nu$ with magnitude dependent on the resonant cross-section weighted by the propagation factor, $(\hat{e}\bullet\hat{e}')^2$, and the thermal distribution as a function of resonance frequency and background LOS wind, $\mathfrak{G}[(\nu - V/\lambda - \nu_0)$, via the absorption process.

2.3 Physical Causes of Broadened Light Scattering Spectra

We have introduced the Doppler-broadened Lorentzian-scattered light spectra for an atmosphere free of collisions. Doppler shift is the frequency difference between scattered and incident light, $\Delta\omega = \omega_S - \omega_L = -(\mathcal{V}/c)\,\omega_L$, when a monochromatic light beam at frequency ω_L impinges on and is scattered by a particle moving away with the LOS speed $\mathcal{V}$. For atmosphere in thermal equilibrium at temperature T, the Maxwellian speed distribution of the molecular population divides molecules into different speed groups, leading to a Gaussian-shaped, inhomogeneous Doppler-broadened frequency spectrum. Lorentzian broadening, on the other hand, is due to the decay of individual dipole moments at a rate of γ, which corresponds to the Einstein A coefficient of spontaneous emission from an excited atom/molecule. Therefore, both broadening mechanisms could be associated with individual molecules without collisions with other molecules during the scattering event. In nonresonant scattering, the individual dipole moment is not supposed to decay, and the resulting CS scattering spectrum is purely Doppler-broadened, as given in (2.5). On the other hand, in resonant scattering, not only does the individual dipole decay via spontaneous emission, the effect of the Maxwellian distribution also broadens the spectrum, leading to a LIF spectrum as the convolution of Doppler-broadened and Lorentzian lineshapes, or the Voigt function as given in (2.7).

As molecular density in gas increases, since the impact duration is much shorter than the time between collisions, the collisional perturbation causes a sudden and random change in the phase of the scattered light. Like the decay of individual dipoles, the effect of dephasing collisions also leads to a Lorentzian lineshape; an excellent discussion of this mechanism is given in Chapter 2 of Siegman [2.5]. Briefly, suppose the external electric field initially (at $t = 0$) induces a macroscopic dipole moment of $\mathcal{N}(0)$ molecules per unit volume (each with dipole moment p), with all individual dipoles oscillating in phase. As time passes, some dipoles suffer dephasing collisions, and the density of dipoles that continue to oscillate in phase is reduced to $\mathcal{N}(t) = \mathcal{N}(0)e^{-t/T_2}$, where T_2 is the mean time of dephasing collisions, giving a dipole energy decay rate of $2/T_2$. This leads to a Lorentzian distribution with FWHM of $2/T_2$. The lineshape resulting from dephasing collisions was attributed to Lorentz's 1906 paper, and the dephasing time T_2 is the inverse of the collision frequency λ_c, with $\lambda_c = \mathcal{N}\sigma\overline{v}$ [2.8]. Here, σ and $\overline{v}$ are, respectively, the collision cross section and average molecular speed. Since the dephasing collision could occur with or without internal energy transfer, we include the dephasing broadening in both nonresonant scattering and resonant scattering spectra in the form of a Voigt function, that is, the convolution between the respective Doppler-broadened function $G^{CS}(v, T)$ or $G^{RS}(v, T)$ and the Lorentzian function as

$$g(v) = \frac{\Delta v}{2\pi\left\{(v)^2 + (\Delta v/2)^2\right\}}; \text{ with } \Delta v = \frac{1}{\pi T_2} \text{ for } CS, \;\; \Delta v = \frac{\gamma}{2\pi} + \frac{1}{\pi T_2} \text{ for } RS. \tag{2.9}$$

The time between collisions gets shorter as gas density increases, leading to a change in the velocities of the scattering molecules. In this case, the gas can no longer be unambiguously divided into different speed groups during a scattering event. The Maxwellian speed distribution then collapses and the collisions become diffusive, and substantial narrowing of the Doppler-broadened line occurs, as was first considered by Dicke [2.9]. The simple Voigt-type folding can no longer describe the resulting lineshape function. Galatry [2.10] has proposed a spectral lineshape model to account for simultaneous Doppler broadening and both collisional effects (dephasing and Dicke narrowing). Though the Galatry model has been used to successfully fit the lineshape of the Q-branch vibrational-rotational Raman spectrum in nitrogen [2.11], given in Section 4.3.4, it is inadequate for describing the Cabannes scattering spectrum. In fact, the dynamics of the two-body collisions become much more complex as the gas density increases, since in addition to Doppler and pressure-induced broadening and narrowing effects, gas density perturbations associated with the translational degrees of freedom can also propagate in a gas. The shape of the Cabannes scattering spectrum for pressures larger than 5 kPa requires solving the kinetic equations and figuring out the density–density correlation functions for molecular gas, an area of intensive research in the 1960s and 1970s. The shape and dependence of the Cabannes spectrum will be discussed in some detail in Section 4.4.

References

2.1 Griffith, D. J. (1998). *Introduction to Electrodynamics*. 2nd ed. Prentice Hall.
2.2 Corney, A. (1977). *Atomic and Laser Spectroscopy.* Oxford Press.
2.3 She, C.-Y., H. Chen, and D. A. Krueger. (2015). Optical processes for middle atmospheric Doppler lidars: Cabannes scattering and laser induced resonance fluorescence. *Jour. Opt. Soc. Am.* **B32**(9), 1575–1592.
2.4 Long, D. A. (2002). *The Raman Effect: A Unified Treatment of the Theory of Raman Scattering by Molecules*. John Wiley & Sons, Ltd.
2.5 Siegman, A. E. (1986). *Lasers*. University Science Books.
2.6 She, C.-Y. (2001). Spectral structure of laser light scattering revisited: Bandwidths of non-resonant scattering lidars. *Appl. Optics* **40**(27), 4675–4884.
2.7 Herzberg, G. (1950). *Molecular Spectra and Molecular Structure: I. Spectra of Diatomic Molecules*, 2nd ed. Van Nostrand Reinhold Company.
2.8 Foley, H. M. (1946). The pressure broadening of spectral lines. *Phys. Rev.* **69**(1–12), 616–628.
2.9 Dicke, R. H. (1953). The effect of collisions upon the Doppler width of spectral lines. *Phys. Rev.* **89**(2), 472–473.
2.10 Galatry, L. (1961). Simultaneous effect of Doppler and foreign gas broadening on spectral lines. *Phys. Rev.* **122**(4), 1218–1223, DOI: https://doi.org/10.1103/PhysRev.122.1218.
2.11 Rahn, L. A., and R. E. Palmer. (1986). Studies of nitrogen self-broadening at high temperature with inverse Raman spectroscopy. JOSA-B, **3**(9), 1164–1169.

3

Semiclassical Treatment of Light Absorption and Scattering from Atoms

Atoms and molecules are sub-nanoscale objects with dimensions in the order of 10^{-10} m; their internal structures can only be described by quantum mechanics. The wavelength of interacting visible light is around 5×10^{-7} m; being substantially larger than atomic dimensions, it may be treated classically. In this book we adopt semiclassical (quantum) theory for treating light–matter interactions, in which atoms (and molecules) are treated quantum mechanically and the impinging and scattered light classically.

3.1 Doppler Shift in Backscattering: Nonresonant versus Resonant Processes

Consider a monochromatic laser beam at frequency ν (wavelength, λ) impinging onto an atmospheric layer at a range r, which contains air molecules and metal atoms, using sodium (Na) as our example. If the laser wavelength can excite the metal atoms (for Na, $\lambda = 589$ nm) but not the air molecules (the air is transparent), finite probabilities exist that an incident photon will be backscattered via nonresonant Cabannes scattering (CS) from air molecules or via resonant scattering or laser-induced fluorescence (LIF) from metal atoms. Here, for simplicity, we ignore other weaker processes (such as rotational and vibrational Raman effects) that may also take part in the scattering processes. We also ignore CS from metal atoms, since it is much weaker than LIF from these same atoms. We then consider CS from air molecules and LIF from metal atoms by the same laser beam separately.

Cabannes scattering from air molecules is a nonresonant, two-photon process that can occur with any incident frequency at which air is transparent. Here we consider the use of the frequency that can induce LIF from the metal atoms under consideration. Since the two-photon CS process occurs instantly (a two-photon, one-step process), one can mentally conceptualize it as two simultaneously occurring one-photon steps. Classically, the process is shown in Fig. 3.1(a) with solid arrows in the laboratory frame of reference. The incident laser beam (long solid upward arrow) at frequency (wavelength) ν (λ) induces a dipole moment, shown as a filled ellipse, on a molecule moving with radial velocity, υ. In its rest frame, the molecule experiences a Doppler-shifted frequency, $\nu - \upsilon/\lambda$. The molecule then "reradiates" at this same frequency in its rest frame and returns to its original unpolarized state, shown as a filled circle. The result is a twice Doppler-shifted scattering frequency, ν_s, viewed from the laboratory frame of reference (long solid downward arrow) as:

$$\nu_s = \nu - 2\upsilon/\lambda. \tag{3.1.a}$$

The modern (quantum) view describes CS as a superposition of two components (terms) [3.1; 3.2], representing the two contributions to the polarizability as depicted in (2.3.c) and (3.3). The rotating wave component consists of absorption of a Doppler-shifted photon, which promotes the moving atom at its rest frame from ground state |1> to virtual state |vir>, as shown by the upward dashed arrow in Fig. 3.1(a). This is followed immediately by emission of a photon at the same frequency in the rest frame, as shown by the downward dashed arrow, and returns the molecule to its ground state. The net frequency shift as viewed from the laboratory system is again given in (3.1.a). Since the two conceptualized steps occur simultaneously in a nonresonant scattering, energy conservation is not required in the intermediate (virtual) step, according to the uncertainty principle. The counterrotating component (term), a reversed process (not shown) with emission from the virtual state followed by absorption from ground state to the virtual state, is equally likely. Both contributions (terms) must be included in the calculation of polarizability, (2.3.c) and (3.3), and of the associated differential cross sections.

As shown in Appendix A of She et al. [3.3], the frequency shift may be derived from nonrelativistic energy and momentum conservation of the scattering process, and $\Delta\nu_s = -2\upsilon/\lambda$ is valid only under the assumption that the linear momentum of the molecule is much larger than that of a photon, h/λ – with h being the Planck constant, so that the velocity of the molecule before and after the scattering remains approximately the same.

The process of LIF consists of two sequential one-photon processes: induced excitation/absorption followed by fluorescence emission – thus, energy conservation is required in both steps. Consider a two-level atom with transition frequency ν_{21} moving away and along the laser beam with LOS velocity, υ. It sees the incident light at the Doppler-shifted frequency, $\nu - \upsilon/\lambda$ where ν is the frequency of the incident light in the lab frame, and it can be excited (or absorb the photon) if $\nu - \upsilon/\lambda = \nu_{21}$ (i.e., the laser frequency selects atoms with a specific radial velocity for excitation). Once excited (coherently by a laser, say), the atom could later emit a photon at the transition frequency, ν_{21}, in its rest frame. Since the moving atom carries much

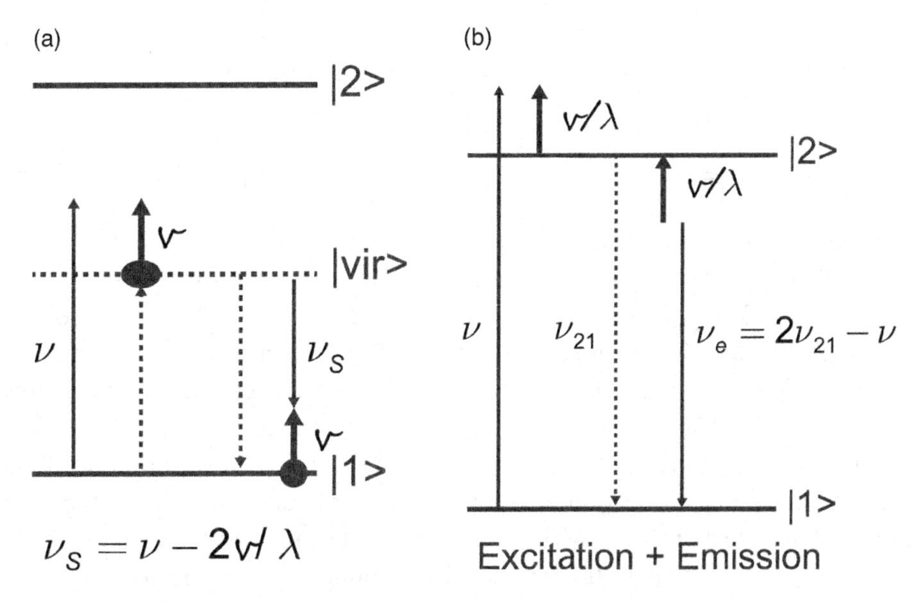

Fig. 3.1 (a) The process of nonresonant Cabannes scattering from a molecule moving with LOS velocity v. The frequencies of the incident laser and backscattered light in the laboratory frame, ν and ν_s, are represented by long solid arrows; the corresponding transitions in the molecule's rest frame are shown by dashed arrows. The energy levels, |1>, |vir>, and |2> are, respectively, the ground, virtual, and excited states of the molecule. See the text for a detailed discussion of the relevance of both rotating and counterrotating (not shown) contributions. (b) The process of laser resonance scattering (or LIF) by a photon at frequency ν and wavelength λ on a moving atom with radial velocity v along the laser beam. For absorption to take place, the laser frequency in the lab frame must be $\nu = \nu_{21} + v/\lambda$, as shown on the left. This atom then emits a photon at ν_{21} in its rest frame, or at $\nu_e = \nu_{21} - v/\lambda = 2\nu_{21} - \nu$ in the lab frame, as shown on the right, which is seen to be independent of v. The dashed downward arrow indicates the emission frequency in the atom's rest frame. Adapted from Figs. 1 and 2 of [3.3]. © The Optical Society. Used with Permission.

higher momentum than the emitted photon, its velocity will remain essentially the same, thus satisfying linear momentum conservation. The backward emitted frequency in the laboratory frame (or to a detector on ground) will be Doppler shifted again by the same amount (i.e., $\nu_e = \nu_{21} - v/\lambda = \nu_{21} - (\nu - \nu_{21}) = 2\nu_{21} - \nu$), leading to an emission frequency depending only on the laser and atomic transition frequencies (i.e., independent of atom's background condition [temperature or LOS wind]). The absorption/excitation and emission process associated with LIF shown in Fig. 3.1(b) represents the rotating wave component. In resonance, the counter-rotating component is not only negligibly small, but it also cannot occur since it is unable to conserve energy in either of the one-photon processes.

If the excited atom is also allowed to transition to a different ground state, |1'>, as occurs in a real atom such as Na, then the emission frequency may be written as (3.1.b):

$$\nu_e = \nu'_{21} - \nu/\lambda = \nu'_{21} - (\nu - \nu_{21}) = \nu_{21} + \nu'_{21} - \nu, \tag{3.1.b}$$

where ν'_{21} may or may not be equal to ν_{21}. The emission is significant only when the difference between the scattered frequency, ν_s, and the emitted frequency, ν_e, as determined by (3.1.b), is within the natural linewidth of |1> to |2> and |2> to |1'> transitions, so that both sequential one-photon processes can occur. Though the LIF emission frequency is independent of atomic velocity, υ, because only the atoms with correct υ may absorb photons of frequency ν, the associated scattering cross section, or the magnitude (not the frequency), of the emission spectrum resulting from the absorption process does depend on the environment (temperature and wind) in which the atom moves. Thus, the apparent Doppler shift of emission from atmospheric metal atoms in the backward direction, as seen from the scattering cross-section expression, is therefore a factor of two smaller than that for the CS process. Appendix A of She et al. [3.3] gives an exact derivation of the emission frequency in terms of laser and atomic transition frequencies as well as conservation of energy and linear momentum. There, instead of dealing with the algebraic complexity in a nonrelativistic treatment, the authors found conceptual clarity and simplicity in the relativistic treatment. They found the relationship between various frequencies, as given in Appendix A (A3) of their paper, to be $\nu_e \nu = \nu'_{21}\nu_{21}$. This reduces to (3.1.b) under the assumption that the energy transitions in question are larger than the separation between the two ground states. It is interesting to note that (3.1.b) may also be written as $(\nu_e - \nu'_{21}) + (\nu - \nu_{21}) = 0$, suggesting that the detuning (offset) of the emission frequency is precisely balanced by that of the absorption frequency. Thus, the frequency of LIF emission, ν_e, as given in (3.1.b), depends only on laser and atomic transition frequencies. It is independent of background atmospheric parameters (temperature or winds), a fact not well appreciated in the literature.

3.2 Quantum Polarizability and Absorption Cross Section

Since the system in question is an ensemble of independent atoms, its quantum state is represented by a density matrix $\rho(t)$, whose time development is governed by the Liouville equation with built-in damping, given as (A.2) in Appendix A. Its system Hamiltonian $\mathcal{H}(t) = \mathcal{H}_0 + \mathcal{H}_I(t)$ is the sum of the unperturbed Hamiltonian $\mathcal{H}_0$ and the interaction Hamiltonian $\mathcal{H}_I(t)$, the latter being time dependent because it represents the interaction of the quantum atoms with the classical external field.

For the case of interest (e.g., the electric dipole interaction, $\mathcal{H}_I(t) \ll \mathcal{H}_0$), the Liouville equation may be solved to different orders by perturbation theory from a hierarchy of equations given in (A.3).

For the electric dipole interaction in question, the interaction Hamiltonian is $\mathcal{H}_I(t) = 0.5[\mathcal{H}_I e^{-i\omega t} + \mathcal{H}_I{}^\dagger e^{i\omega t}]$, with $\mathcal{H}_I = -(\vec{p}\bullet\vec{\mathscr{E}})$ and $\mathcal{H}_I{}^\dagger = -(\vec{p}^\dagger\bullet\vec{\mathscr{E}}^*)$, where $\vec{p}^\dagger$ is the Hermitian adjoint of the dipole operator $\vec{p}$, since the electric dipole operator, $\vec{p}$, is Hermitian, $\vec{p}^\dagger = \vec{p}$. For the evaluation of the induced dipole moment and associated electronic polarizability and absorption cross section, we need only to solve the first-order equation in (A.3) by setting $n = 1$. The solution for the density matrix element can be shown (see (B1) of [3.3]) to be:

$$\rho_{\mu m}{}^{(1)}(t) = \frac{1}{2}\left[\rho_{\mu m}^{(1)}(\omega)e^{-i\omega t} + \rho_{\mu m}^{(1)}(-\omega)e^{i\omega t}\right], \text{ with}$$
$$\rho_{\mu m}^{(1)}(\omega) = \frac{1}{i\hbar}\frac{<f\mu|\left(\vec{p}\bullet\vec{\mathscr{E}}\right)|Fm>}{\Gamma_{Ff} - i(\omega + \omega_{Ff})}\frac{1}{g_f} \text{ and } \rho_{\mu m}^{(1)}(-\omega) = \rho_{m\mu}^{(1)*}(\omega), \tag{3.2}$$

where $|f,\mu>$ and $|F,m>$, respectively, denote the degenerate initial and the final states of the associated transition, with μ and m as the respective magnetic quantum numbers having degeneracies $g_f = 2f + 1$ and $g_F = 2F + 1$. The factor Γ_{Ff} represents the damping rate for the density matrix element, resulting from the interactions with other atoms as well as the background radiation that leads to spontaneous emission.

Assuming the polarization of the induced dipole moment is $\hat{e}$, we can write the expectation value of the induced dipole moment operator, $\vec{p} = p\hat{e}$, as $\langle\vec{p}(t)\rangle = 0.5(\langle p\rangle\hat{e}e^{-i\omega t} + \langle p^\dagger\rangle\hat{e}^* e^{i\omega t}) = 0.5\left(\vec{\mathcal{P}}(\omega)e^{-i\omega t} + \vec{\mathcal{P}}^*(\omega)e^{i\omega t}\right)$, where the angle bracket indicates the average over a quantum state. In terms of the density matrix, $\langle\vec{p}(t)\rangle$ may be evaluated by the trace $\text{Tr}\left[\rho^{(1)}(t)(\vec{p}\bullet\hat{e})\right]$ with (3.2). The induced dipole moment can also be expressed in terms of the polarizability $\alpha(\omega)$ as $\langle\vec{p}(t)\rangle = \frac{1}{2}\left(\alpha(\omega)\mathscr{E}\hat{e}e^{-i\omega t} + \alpha^*(\omega)\mathscr{E}^*\hat{e}^* e^{i\omega t}\right)$ with $\vec{\mathscr{E}} = \mathscr{E}\hat{e}$. From the induced dipole moment so evaluated, the associated polarizability may be deduced as

$$\alpha(\omega) = \frac{1}{\hbar}\sum_{m,\mu}\left[-\frac{<Fm|(\vec{p}\bullet\hat{e})|f\mu><f\mu|(\vec{p}\bullet\hat{e})|Fm>}{(\omega - \omega_{Ff}) + i\Gamma_{Ff}} + \frac{<f\mu|(\vec{p}\bullet\hat{e})|Fm><Fm|(\vec{p}\bullet\hat{e})|f\mu>}{(\omega + \omega_{Ff}) + i\Gamma_{Ff}}\right]\frac{1}{g_f}, \tag{3.3}$$

where the first and second terms are, respectively, the rotating and counterrotating wave contributions to the polarizability. Equations (3.2) and (3.3) are not difficult to derive, and they were given as (B1) and (B2) in [3.3]. Using the Wigner–Eckart theorem, as given in (5.4.1) of Edmonds [3.4], we define the reduced matrix element, $<F||p||f>$, which one may relate to the linestrength as defined in (4.35) of [3.5], as $S_{Ff} = \sum_{m,\mu} |<Fm|\,\vec{p}\,|f\mu>|^2$. With the help of the symmetry of the 3-j coefficients given in (3.7.3)–(3.7.8) of [3.4], we can show

$$S_{Ff} = \sum_{m,\mu} |<Fm|\vec{p}|f\mu>|^2 = \sum_{m} <Fm|\vec{p}\bullet\vec{p}^{\dagger}|Fm> = |<F||p||f>|^2, \text{ and} \tag{3.4.a}$$

$$\sum_{m,\mu} \langle Fm|(\vec{p}\bullet\hat{e})|f\mu\rangle\langle f\mu|(\vec{p}\bullet\hat{e})|Fm\rangle = \left|\langle F||p||f\rangle\right|^2/3 = S_{Ff}/3, \tag{3.4.b}$$

where (3.4.b) is an identity analogous to $x^2 = r^2/3$ for the vector $\vec{r} = x\hat{i} + y\hat{j} + z\hat{k}$. Since the denominators of (3.3) are independent of m and μ, they may be moved out of the summation sign. With the substitution of (3.4.b), (3.3) may be expressed separately for the resonant and nonresonant cases as

$$\alpha^R(\omega) = -\frac{1}{\hbar}\sum_{m\mu}\frac{<Fm|(\vec{p}\bullet\hat{e})|f\mu><f\mu|(\vec{p}\bullet\hat{e})|Fm>}{(\omega-\omega_{Ff})+i\Gamma_{Ff}}\frac{1}{g_f}$$
$$\approx -S_{Ff}\Big/\left\{\left(3\hbar g_f\right)\left[(\omega-\omega_{Ff})+i\Gamma_{Ff}\right]\right\}, \text{ and} \tag{3.5.a}$$

$$\alpha^{NR}(\omega) = \frac{1}{\hbar}\sum_{m\mu}\frac{2<Fm|(\vec{p}\bullet\hat{e})|f\mu><f\mu|(\vec{p}\bullet\hat{e})|Fm>}{\omega_{Ff}}\frac{1}{g_f} \approx 2S_{Ff}\Big/\left(3\hbar g_f\omega_{Ff}\right), \tag{3.5.b}$$

where S_{Ff} and $\langle F||p||f\rangle$ are the linestrength and reduced matrix element for the transition and are respectively defined in (4.35) of [3.5] and in (5.4.1) of [3.4]. Since the absorption oscillator strength for a classical oscillator is $f_{fF} = 1$, the equivalence between (3.5) and its classical counterpart (2.3.e) is evident when we

(1) let $\omega_{Ff} = \omega_0$ and identify $\left(2m_e\omega_{Ff}S_{Ff}/3\hbar e^2 g_f\right)$ as the f_{fF} of the transition, and
(2) set Γ_{Ff} to $\gamma/2$.

Here, the former is the damping rate of the associated off-diagonal density matrix element, and the latter is the amplitude damping rate of a classical oscillator. Since γ is the energy damping rate, and spontaneous emission is the only damping mechanism for the present system, we equate γ with the Einstein A coefficient, A_{Ff}, leading to $\Gamma_{Ff} = A_{Ff}/2$. Does the equivalency between quantum and classical expressions in polarizability imply that we can substitute $\alpha^R(\omega)$ and $\alpha^{NR}(\omega)$ into (2.2.b) and obtain the resonant and nonresonant differential cross sections, as in (2.3.f) and (2.3.g), respectively? The answer turns out to be yes for nonresonant scattering, where a stationary dipole moment is induced with S_{Ff} and $\omega_{Ff} \gg \omega$ as, respectively, the effective linestrength and frequency of the excited virtual state. Since the Cabannes scattering of interest comes from atmospheric molecules, we shall defer this discussion to Chapter 4. For resonance scattering, since the excited state is a real state with degeneracy g_F, it is more complex. Further, as will be explained in Subsection 3.3.1, the phases of the excited substates tend to be randomized in time, degrading the coherence of the induced dipole moment to a degree dependent on the structure of the atom. The effect of this decoherence is to affect the differential scattering cross section and as such demands a different treatment, which is presented in Section 3.3.

Considering a light beam propagating through a medium with absorption cross section per atom, σ^A, we can relate σ^A to the imaginary part of the resonant polarizability, $\alpha^R(\omega)$. This relation may be derived by expressing the induced dipole moment as $\vec{p}(t) = \frac{1}{2}(\alpha(\omega)\mathscr{E}\hat{e}e^{-i\omega t} + \alpha^*(\omega)\mathscr{E}^*\hat{e}^*e^{i\omega t})$, and the polarization of its associated medium as $\vec{P}(t) = \frac{1}{2}(P(\omega)e^{-i\omega t} + P^*(\omega)e^{i\omega t}) = \mathcal{N}\vec{p}(t)$, with $P(\omega) = \mathcal{N}\alpha(\omega)\mathscr{E}$ and $P(\omega) \equiv \mathsf{e}_0\chi(\omega)\mathscr{E} = \mathcal{N}\alpha(\omega)\mathscr{E}$, where $\mathcal{N}$ and $\chi(\omega)$ are, respectively, the number of atoms per unit volume and medium susceptibility. The laser beam is attenuated exponentially with a rate of $\mathcal{N}\sigma^A$ per unit length of travel or of $\mathcal{N}\sigma^A c$ per unit time. Alternatively, in terms of the imaginary part of resonant susceptibility, $\chi''(\omega)$, or of resonant polarizability, $\alpha''(\omega)$, the exponential decay rate of the beam intensity per unit length is $\omega\chi''(\omega)/c = \omega\mathcal{N}\alpha''(\omega)/\mathsf{e}_0 c$, resulting in $\sigma^A(\omega) = \omega\alpha''(\omega)/\mathsf{e}_0 c$, with e_0 and c being, respectively, the vacuum permittivity and the speed of light. Using (3.5.a) to determine the imaginary part of $\alpha^R(\omega)$ (i.e., $\alpha''(\omega)$), along with the relationship between linestrength and Einstein A coefficient, given in (A.8) (i.e., $A_{Ff} = \omega_{Ff}^3 S_{Ff}/3\pi\mathsf{e}_0 c^3\hbar\mathsf{g}_F$), and replacing $\pi\omega_{Ff}S_{Ff}/3\hbar c\mathsf{e}_0$ by $\lambda_{Ff}^2 A_{Ff}\mathsf{g}_F/4$, with λ_{Ff} being the transition wavelength, we can express σ^A in terms of the lineshape function as

$$\sigma^A(\omega) = \frac{\omega}{c\mathsf{e}_0}\alpha''(\omega) = \frac{S_{Ff}}{\left(3\hbar\mathsf{g}_f\right)}\frac{\Gamma_{Ff}}{(\omega-\omega_{Ff})^2+\Gamma_{Ff}{}^2}\frac{\omega}{c\mathsf{e}_0} = \frac{S_{Ff}}{\left(3\hbar\mathsf{g}_f\right)}\frac{\pi\omega}{c\mathsf{e}_0}\,\mathrm{g}(\omega-\omega_{Ff})\,;$$

$$\sigma^A(\omega) = \frac{\lambda_{Ff}^2}{4}\frac{\mathsf{g}_F}{\mathsf{g}_f}A_{Ff}\mathrm{g}(\omega-\omega_{Ff});\ \mathrm{g}(\omega-\omega_{Ff}) = \frac{\Gamma_{Ff}/\pi}{(\omega-\omega_{Ff})^2+(\Gamma_{Ff})^2}. \tag{3.6}$$

Again, the FWHM of the lineshape function in angular frequency is $\Delta\omega = 2\Gamma_{Ff} = A_{Ff}$, with $\tau = 1/A_{Ff}$ being the lifetime of the excited state. Equation (3.6) is equivalent to that derived from transition probabilities between two degenerate dipole-allowed states as, for example, given in Eq. (A.13).

Applying this to the D_2 transition of the Na atom (λ = 589.158 nm) without hyperfine interactions (i.e., ${}^2S_{1/2} \rightarrow {}^2P_{3/2}$), with $g_2/g_1 = 2, A_{21} \sim A_0 = 6.15\times10^7$ s^{-1} ($\tau \sim$ 16 ns), or $\Delta\nu = A_0/(2\pi) \sim$ 10 MHz and the peak absorption cross section is $\sigma^A(\nu_{21}) = (g_2/g_1)(\lambda^2/2\pi) = 1.11 \times 10^{-13}$m^2. The corresponding differential cross section is then 8.79×10^{-15} m^2/sr, 1.5×10^{17} times larger than the differential backscattering Cabannes cross section $\sigma_\pi^C = 5.96 \times 10^{-32}$ m^2/sr (as compared to the classical result of 3×10^{17} times larger; see Section 2.1). In reality, atmospheric Na atoms are in thermal equilibrium with the atmosphere, and unlike nonresonant scattering, only a fraction of the atoms may be in resonance at a given laser frequency. At 200 K, the Doppler broadened width is about 1 GHz with a peak absorption cross section of 8.0×10^{-17} m^2. Thus, the ratio of LIF to Rayleigh/Cabannes scattering signals is $\sim 1.3\times10^{15}$ [3.6].

3.3 Differential Resonance Scattering Cross Sections

We first outline a derivation of the differential resonance scattering cross sections. In this book, we consider the resonance scattering process as synonymous with the laser-induced fluorescence process. In this process, atoms in a degenerate initial state $|f>$ are coherently and resonantly excited (by a laser) to a degenerate (or nearly degenerate) higher state $|F>$, before they return via spontaneous emission to a lower final degenerate state $|f'>$, which may or may not be the same as $|f>$. Since, depending on the atom in question, there may be more than one allowed emission final state $|f'>$, we will deal with two related resonant scattering cross sections in Subsection 3.3.1:

(1) The differential LIF cross section, $d\sigma^R_{fFf'}/d\Omega$, for the fFf' pathway (i.e., excitation from an initial state $|f>$ to the excited state $|F>$) followed by emission from $|F>$ to a specific final state $|f'>$.
(2) The (total) differential LIF cross section summing over spontaneous emissions to all possible final states; we term this the differential LIFΣ cross section, $d\sigma^R_{fF\Sigma}/d\Omega = \sum_{f'} d\sigma^R_{fFf'}/d\Omega$.

We then provide physical insight and understanding on the impact of coherent excitation, often referred to as the Hanle Effect, in Subsection 3.3.2. Since most of these materials are taken from [3.3], we refer for mathematical details to that reference. We apply those results to Na and Fe atoms in Section 3.4 and illustrate the angular distribution of radiation patterns for realistic atoms, ultimately obtaining the associated backward differential LIFΣ cross sections of interest to respective resonance scattering (LIF) lidars.

3.3.1 Derivation of Differential Cross Sections

First-order perturbation accounts for excitation of the system from a ground state to the excited substates with different magnetic quantum numbers m. If an incoherent source is used to illuminate the atomic ensemble, the relative phases of different substates will be random, and the emission from this excitation will be isotropic (independent of the polarization of the exciting field) with the corresponding differential cross section $\sigma^A/4\pi$. On the other hand, if the system is excited by a coherent source, like a laser, then the excited atoms are prepared in a way that the phases of the excited substates $|m\rangle$ and $|m'\rangle$ are the same for all atoms in the ensemble. These excited atoms are described by wave functions that are a coherent superposition (mixing) of the magnetic substates, giving rise to nonzero second-order off-diagonal matrix elements $\rho^{(2)}_{mm'}(t)$. These decay spontaneously at a rate $\Gamma_{mm'}$, and the coherence is diminished accordingly. We call $\Gamma_{mm'}$ the decoherence rate, which quantifies the damping of coherent superposition of excited substates. As in the classical case, the emission of a coherently excited atomic dipole will have specific angular dependence (an antenna pattern, if you will) that depends on the polarization of the exciting field and the quantum structure of the atom in question. If the substates are Zeeman split, this coherence is degraded. The state of polarization of fluorescence light may change from almost perfect polarization at zero magnetic field to complete depolarization at high-field of 5–10 G, depending on the decoherence rate $\Gamma_{mm'}$, which is related but not necessarily equal to the lifetime of the excited state. Since the angular frequency between the Zeeman substates, $(E_m - E_{m'})/\hbar$, lies in the radio frequency band, the ensemble of atoms is said to possess Hertzian coherence. A technique using magnetic depolarization of resonance radiation, commonly referred to as a Hanle Effect experiment, has been used to measure the lifetime of atomic excited states – see for example, chapters 15 and 16 of Corney [3.5]. To prepare such a coherently excited emitter, we need to go to the second-order perturbation and determine the associated density matrix, $\rho^{(2)}(t)$ at radio (or near DC) frequencies in terms of the first-order density matrix, $\rho^{(1)}(t)$, by letting $n = 2$ in (A.3). The second-order equation is then:

$$\frac{\partial \rho^{(2)}_{mm'}(t)}{\partial t} = \frac{1}{i\hbar}\left\{\hbar\omega_{mm'}\rho^{(2)}_{mm'}(t) + [\mathcal{H}_I(t), \rho^{(1)}(t)]_{mm'}\right\} - \Gamma_{mm'}\rho^{(2)}_{mm'}(t),$$

with $\omega_{mm'} = \omega_B(m - m')$, where $\omega_B = g_F\mu_B B/\hbar$, and

$$\mathcal{H}_I(t) = -0.5\left[\left(\vec{p}\bullet\vec{\mathcal{E}}\right)e^{-i\omega t} + \left(\vec{p}^{\dagger}\bullet\vec{\mathcal{E}}^{*}\right)e^{i\omega t}\right], \tag{3.7}$$

where g_F (not to beconfused with $\mathcal{g}_F$), μ_B, B, and ω_B are the hyperfine Landé g-factor, Bohr magneton, magnetic field (or induction) in Wb·m^{-2} and Larmor frequency respectively. The interaction Hamiltonian, $\mathcal{H}_I$ in (3.7), results from the electric dipole interaction. Although the substates in both the excited and ground states are split in the presence of a weak magnetic field, we ignore the Zeeman splitting in comparison to the energy difference between $|f>$ and $|F>$ in carrying out the first-order perturbation calculation. On the other hand, since the frequency difference between the Zeeman-split excited substates, $(E_m - E_{m'})/h$, is comparable to the lifetime broadening of the excited state and the spectral width of the Hertzian coherence, we explicitly include the Zeeman splitting in the unperturbed Hamiltonian in the second-order treatment. The steady-state solution to the second-order equation, yielding the off-diagonal matrix element, $\rho^{(2)}_{mm'}(t)$, is given in (A.17).

The quantum expression for differential scattering power, or the Poynting vector of the radiating dipole, in analogy to (2.1.b) or (2) of [3.3], may be expressed as $dP/d\Omega$, with the factor $|p|^2(\hat{e} \bullet \hat{e}')^2$ now given by the expectation value of $(\vec{p} \bullet \hat{e}')(\vec{p} \bullet \hat{e}'^*)$, or $\mathrm{Tr}\left[\rho^{(2)}(\vec{p} \bullet \hat{e}')(\vec{p} \bullet \hat{e}'^*)\right]$, the trace of the operator in the square bracket. Following (B3d) of [3.3], we can derive the associated differential resonance scattering cross section resulting from excitation ($|f\rangle$ to $|F\rangle$) followed by fluorescence emission ($|F>$ to $|f'>$) as $d\sigma^{RS}_{fFf'}/d\Omega$:

$$\frac{dP}{d\Omega} = \left(\frac{\omega^4}{32\pi^2 \mathrm{e}_0 c^3}\right) \mathrm{Tr}\left[\rho^{(2)}(\vec{p} \bullet \hat{e}')(\vec{p} \bullet \hat{e}'^*)\right] \rightarrow$$

$$\frac{d\sigma^R_{fFf'}}{d\Omega} = \left(\frac{\omega^4}{16\pi\hbar^2 \mathrm{e}_0^2 c^4}\right) g(\omega - \omega_{Ff}) \frac{1}{g_f} \sum_{mm'} \frac{F_{mm'} G_{m'm}}{[\Gamma_{mm'} + i\omega_B(m - m')]}, \tag{3.8}$$

with $F_{mm'} \equiv \sum_{\mu} <Fm|(\vec{p} \bullet \hat{e})|f\mu><f\mu|(\vec{p} \bullet \hat{e}^*)|Fm'>$ and $G_{m'm} \equiv \sum_{\mu'} \langle Fm'|(\vec{p} \bullet \hat{e}') |f\mu'><f\mu'|(\vec{p} \bullet \hat{e}'^*)|Fm>$ being, respectively, the excitation and emission matrix elements (see Chapter 15 of Corney [3.5]) summed over, respectively, the initial substates $|\mu>$ and the final substates $|\mu'>$. Here, $\hat{e}$ and $\hat{e}'$ are, respectively, the polarizations of incident and scattered beams.

Following (B5b) of [3.3], we define the polarization (tensor) function, $\Phi^\kappa_q(\hat{e})$, in terms of polarization vectors and the 3-j coefficients as

$$\Phi^\kappa_q(\hat{e}) = \sum_{q_1,q_2=-1}^{1} (-1)^{q_1+q_2} e_{q_1} e_{q_2} \begin{pmatrix} 1 & 1 & \kappa \\ q_1 & q_2 & -q \end{pmatrix}; \; \hat{e} = \sum_{q=-1}^{1} e_q \hat{\varepsilon}_{-q}. \tag{3.9}$$

Here, e_q are components of our chosen polarization vectors $\hat{e}$ projected on the unit vectors $(\hat{\varepsilon}_{+1}, \hat{\varepsilon}_0, \hat{\varepsilon}_{-1})$ of the complex coordinate system; see (B.2.a) in Appendix B. The products $e_{q_1} e_{q_2}$ of (3.9) for the real unit vectors $\hat{e}_a \hat{e}_b$ were worked out in [3.3] and tabulated in its Table 4. We point out that the polarization function (3.9) is a property of the polarization vector, and it is independent of the quantum structure of the atom in question. The property of the 3-j coefficients in (3.9) spells out the selection rules of the electric dipole transition (i.e., $q = q_1 + q_2$). Since $\Phi_q^\kappa(\hat{e})$ is a second-rank tensor formed by the (direct) product of two unit vectors, it can be reduced to the sum of three irreducible tensors with $\kappa = 0, 1$, and 2. For our choice of coordinate system for incident and scattered polarizations depicted in Fig. B.1, the polarization function is zero for $\kappa = 1$ (i.e., $\Phi_q^1(\hat{e}) = 0$) as shown in p. 1590 of [3.3]. With considerable effort exerted in manipulating 3-j and 6-j coefficients, detailed in Appendix B of [3.3], and by keeping only $\kappa = 0$ and $\kappa = 2$ terms in (B7b), we can derive (B7c) of [3.3] (see typos corrected in Erratum, p. 1954, ibid.). The differential resonance scattering (LIF) cross section and its associated strength factor $C_{fFf'}(\hat{e};\hat{e}')$ can then be expressed as

$$\frac{d\sigma_{fFf'}^R}{d\Omega} = \left(\frac{\omega^4}{16\pi\hbar^2 \mathsf{e}_0^2 c^4}\right) \frac{|<F||p||f>|^2 |<F||p||f'>|^2}{g_f \Gamma_{mm'} g_F} g(\omega - \omega_{Ff}) C_{fFf'}(\hat{e};\hat{e}')$$

$$= \frac{9}{4\pi} \frac{\lambda_{Ff}^2}{16} \frac{A_{Ff}}{\Gamma_{mm'}} \frac{g_F}{g_f} A_{Ff'} g(\omega - \omega_{Ff}) C_{fFf'}(\hat{e};\hat{e}'),$$

with $C_{fFf'}(\hat{e};\hat{e}') = g_F B_{fFf'}^{(0)} \Phi_0^0(\hat{e}) \Phi_0^0(\hat{e}') + g_F B_{fFf'}^{(2)} \sum_q (-1)^q \frac{\Phi_q^2(\hat{e}) \Phi_{-q}^2(\hat{e}')}{1 - iq\omega_B/\Gamma_{mm'}}$, where

$$B_{fFf'}^{(0)} = (-1)^{f+f'} \begin{bmatrix} F & F & 0 \\ 1 & 1 & f \end{bmatrix} \begin{bmatrix} F & F & 0 \\ 1 & 1 & f' \end{bmatrix} \text{ and}$$

$$B_{fFf'}^{(2)} = 5(-1)^{f+f'} \begin{bmatrix} F & F & 2 \\ 1 & 1 & f \end{bmatrix} \begin{bmatrix} F & F & 2 \\ 1 & 1 & f' \end{bmatrix}. \tag{3.10}$$

In order to get the second equality of (3.10) we have employed $|<F||p||f>|^2 = S_{Ff}$ and $|<F||p||f'>|^2 = S_{Ff'}$, and substituted $A_{Ff} = \omega_{Ff}^3 S_{Ff}/3\pi \mathsf{e}_0 c^3 \hbar g_F$ and $A_{Ff'} = \omega_{Ff}^3 S_{Ff'}/3\pi \mathsf{e}_0 c^3 \hbar g_F$. Equation (3.10) is identical to (16) of [3.3]. As defined in (3.9), $\Phi_0^0(\hat{e})$ and $\Phi_q^2(\hat{e})$ are the zeroth- and second-order polarization tensor elements (for the latter, $q = -2, \ldots, 2$), and $\hat{e}$ indicates the direction of the incident beam polarization. Similarly, $\Phi_0^0(\hat{e}')$ and $\Phi_q^2(\hat{e}')$ are the corresponding tensor elements for the scattering polarization, $\hat{e}'$. We also point out that via $\hat{e}$ and $\hat{e}'$, the differential scattering cross section in (3.10) depends on the propagation and polarization directions of the incident and scattered beams; thus it is a function of θ and ϕ. The existence of this angular distribution is the result of coherent excitation (i.e., the Hanle Effect [3.5]).

By projecting $\hat{e}$ onto the unit vectors, $\hat{e}_a$ or $\hat{e}_b$, and $\hat{e}'$ to $\hat{e}_a'$ or $\hat{e}_b'$, we take advantage of the simplicity of the real basis unit vectors and evaluate independently the strength factors of (3.10) for the four cases. These are denoted as: $C_{fFf'}(\hat{e}_a;\hat{e}_a{}')$, $C_{fFf'}(\hat{e}_a;\hat{e}_b{}')$, $C_{fFf'}(\hat{e}_b;\hat{e}_a{}')$ and $C_{fFf'}(\hat{e}_b;\hat{e}_b{}')$, each of which may be realized in laboratory experiments by the use of a polarizer and analyzer. However, for atmospheric applications, we consider only the cases where the radiated power of both $\hat{e}'_a$ and $\hat{e}'_b$ polarizations is received and evaluate $C_{fFf'}(\hat{e}_a;\hat{e}') \equiv C_{fFf'}(\hat{e}_a;\hat{e}'_a) + C_{fFf'}(\hat{e}_a;\hat{e}'_b)$ and $C_{fFf'}(\hat{e}_b;\hat{e}') \equiv C_{fFf'}(\hat{e}_b;\hat{e}'_a) + C_{fFf'}(\hat{e}_b;\hat{e}'_b)$ as in (B8a) and (B8c) of [3.3]. Due to the invariance of $g_F B^{(0)}_{fFf'} = 1/3$ (see p. 1591 of [3.3]), these strength factors may be given as

$$C_{fFf'}(\hat{e}_a;\hat{e}') = \frac{2}{9}\left\{1 + \frac{9}{2} g_F B^{(2)}_{fFf'} \mathcal{F}_a(\theta,\phi;\phi_0;\omega_B/\Gamma_{mm'})\right\}, \text{ where}$$

$$\mathcal{F}_a(\theta,\phi;\phi_0;\omega_B/\Gamma_{mm'}) = \left[\frac{2 - 3\sin^2\theta}{30} + \frac{\sin^2\theta}{10}\left(\frac{\cos2(\phi_0 - \phi) - (2\omega_B/\Gamma_{mm'})\sin2(\phi_0 - \phi)}{1 + (2\omega_B/\Gamma_{mm'})^2}\right)\right]; \tag{3.11.a}$$

$$C_{fFf'}(\hat{e}_b;\hat{e}') = \frac{2}{9}\left\{1 + \frac{9}{2} g_F B^{(2)}_{fFf'} \mathcal{F}_b(\theta,\phi;\theta_0\phi_0;\omega_B/\Gamma_{mm'})\right\}, \text{ where}$$

$$\mathcal{F}_b(\theta,\phi;\theta_0\phi_0;\omega_B/\Gamma_{mm'}) = \left[\begin{array}{l} \frac{1 - 3\sin^2\theta_0}{30} + \frac{(1 - 3\sin^2\theta_0)(1 - 3\sin^2\theta)}{30} + \frac{\sin2\theta\sin2\theta_0}{10}\left(\frac{\cos(\phi_0 - \phi) - (\omega_B/\Gamma_{mm'})\sin(\phi_0 - \phi)}{1 + (\omega_B/\Gamma_{mm'})^2}\right) \\ - \frac{\cos^2\theta_0\sin^2\theta}{10}\left(\frac{\cos2(\phi_0 - \phi) - (2\omega_B/\Gamma_{mm'})\sin2(\phi_0 - \phi)}{1 + (2\omega_B/\Gamma_{mm'})^2}\right) \end{array}\right]. \tag{3.11.b}$$

Note that (3.11.a) and (3.11.b) here are identical to (B8a)–(B8d) of [3.3]. With $\mathcal{F}_a(\theta,\phi;\phi_0;\omega_B/\Gamma_{mm'})$ and $\mathcal{F}_b(\theta,\phi;\theta_0\phi_0;\omega_B/\Gamma_{mm'})$ explicitly given, the fact that the differential scattering cross section depends on the propagation and polarization directions of the incident and scattering beams is clear. Notice that $\mathcal{F}_a(\theta,\phi;\phi_0;\omega_B/\Gamma_{mm'})$ is independent of incident polar angle, θ_0, because $\hat{e}_a$ is on the x-y plane. After the external magnetic field B ($\omega_B = g_F\mu_B B/\hbar$, with μ_B being the Bohr magneton) and the quantum structure coefficient $B^{(2)}_{fFf'}$ (calculated from the 6-j coefficients provided by (3.10)) are given, the radiation pattern of a coherently excited atomic dipole may be evaluated numerically.

To allow for the possibility of more than one emission final state, we define the differential LIFΣ cross section, summing all allowed emission pathways in (3.10), as

$$\frac{d\sigma^R_{fF\Sigma}}{d\Omega} = \sum_{f'} \frac{d\sigma^R_{fFf'}}{d\Omega} = \frac{1}{4\pi}\frac{\lambda^2_{Ff}}{4}\frac{3}{2}\frac{g_F}{g_f} A_{Ff}\, g(\omega - \omega_{Ff})\left[3\sum_{f'} \alpha_{Ff'} C_{fFf'}(\hat{e};\hat{e}')\right] \tag{3.12}$$

where $\alpha_{Ff'} = A_{Ff'}/\sum_{f'} A_{Ff'}$. We point out that, unlike the excitation process, the emission is spontaneous. Thus, we can simply sum over the emissions to all allowed final-states, giving rise to (3.12). Also note the lineshape function $g(\omega - \omega_{Ff})$ depends only on the absorption (excitation) process; it is independent of the emission frequency, $\omega_{Ff'}$. To arrive at (3.12), we have introduced the fractional transition rate, $\alpha_{Ff'}$. We have also employed the identification of $\Gamma_{mm'} = \frac{1}{2}\sum_{f'} A_{Ff'}$, which requires a bit of explanation. We point out that, unlike $\Gamma_{Ff} = A_{Ff}/2$, the physical meaning of the damping rate for $\rho_{mm'}{}^{(2)}(t)$ between excited substates, $\Gamma_{mm'}$, termed the decoherence rate here, appears to be more subtle and fuzzy. We set it to $\Gamma_{mm'} = \frac{1}{2}\sum_{f'} A_{Ff'}$ since emission in general ends up in more than one $|f'>$ state, and any of the allowed $|f>$ states could terminate the excited state coherence. The equating of $\Gamma_{mm'}$ to $\frac{1}{2}\sum_{f'} A_{Ff'}$ is not only consistent with what was done with the damping of $\rho_{m\mu}{}^{(1)}(t)$ between the excited and ground states (i.e., $\Gamma_{Ff} = A_{Ff}/2$), but perhaps more importantly, it leads to the equality between the integrated differential cross section over all solid angles and the absorption cross-section σ^A (as shown on p. 1582 of [3.3]) to comply with conservation of energy. By doing so, we also deduce/define $\Gamma_{mm'}$, the decoherence rate, quantitatively.

3.3.2 Impacts of Coherent Excitation and the Hanle Effect

The main consequence of coherent excitation is that it leads to a specific radiation pattern from an induced dipole polarization, or the Hanle Effect [3.5; 3.7]. To ascertain that the radiation pattern resulting from coherent excitation of a quantum dipole system is conceptually similar to that of a classical electric dipole, we consider two simple cases of incident polarizations, $\hat{e}_a = \hat{x}$ and $\hat{e}_b = -\hat{z}$, as is often done in the treatment of classical electromagnetic radiation [3.8]. To excite these dipoles, the incident wave is pointing along $\hat{k}_0$ in the direction with $\theta_0 = 90°$ and $\phi_0 = 90°$. For simplicity, we consider the case with $B = 0$, so the strength factors $C_{fFf'}(\hat{e}_a;\hat{e}')$, given in (3.11.a), and $C_{fFf'}(\hat{e}_b;\hat{e}')$ in (3.11.b), become, respectively,

$$C_{fFf'}(\hat{e}_a;\hat{e}') = \frac{2}{9}\left\{1 + \frac{9}{2}\mathrm{g}_F B^{(2)}_{fFf'}\mathscr{F}_a(\theta,\phi;\omega_B = 0)\right\}, \text{ with}$$
$$\mathscr{F}_a(\theta,\phi;\omega_B = 0) = \left[\frac{2 - 3\sin^2\theta(1 + \cos 2\phi)}{30}\right] \tag{3.13.a}$$

$$C_{fFf'}(\hat{e}_b;\hat{e}') = \frac{2}{9}\left\{1 + \frac{9}{2}\mathrm{g}_F B^{(2)}_{fFf'}\mathscr{F}_b(\theta,\phi;\omega_B = 0)\right\}, \text{ with}$$
$$\mathscr{F}_b(\theta,\phi;\omega_B = 0) = \left[\frac{-2(2 - 3\sin^2\theta)}{30}\right]. \tag{3.13.b}$$

We now further apply these formulae to the simplest quantum structure without spin-orbit or hyperfine interactions, with the fractional transition rate $\alpha_{10} = 1$ and substitution of $f = f' = \ell_g = 0$ and $F = \ell_e = 1$ into the 6-j coefficients for $B^{(2)}_{fFf'}$. As a result, we obtain $\mathrm{g}_1 = 3$ and $B^{(2)}_{010} = 5/9$, leading to the distribution function,

$3C_{010}(-\hat{z};\hat{e}') = \sin^2\theta = (\hat{z} \bullet \hat{e}'_a)^2 + (\hat{z} \bullet \hat{e}'_b)^2$ and $3C_{010}(\hat{x};\hat{e}') = \sin^2\phi + \cos^2\theta \cos^2\phi = (\hat{x} \bullet \hat{e}'_a)^2 + (\hat{x} \bullet \hat{e}'_b)^2$, and the associated differential scattering cross section (3.12) becomes

$$\frac{d\sigma^{RS}_{010}}{d\Omega} = \frac{1}{4\pi} \frac{\lambda^2_{Ff}}{4} \frac{9}{2} A_{10} g(\omega - \omega_{m\mu}) \left[3C_{010}(\hat{e};\hat{e}')\right]. \tag{3.14}$$

That the angular distribution of radiation for this simplest quantum system agrees with that of the classical dipole, $(\hat{e} \bullet \hat{e}')^2 = (\hat{e} \bullet \hat{e}'_a)^2 + (\hat{e} \bullet \hat{e}'_b)^2$, clearly demonstrates that the modification of the spatial intensity distribution from that of an isotropic pattern is not of quantum origin. Rather, it is due to the excitation of the atomic system by a laser-like source, which coherently prepares the mixed excited substates before fluorescence emission. Indeed, the Hanle Effect has a classical explanation and is most prominent when the excited substates are degenerate (i.e., no Zeeman splitting or when $B = 0$); thereby, it is an effective tool for measuring the lifetime of atomic excited states; see Chapter 15 of Corney [3.5]. On the other hand, for useful and complex atoms, like Na, K, or Fe, whose structures do not have a classical analog, quantum calculations must be invoked to properly handle the atom's quantum structure and to numerically evaluate the differential cross section and its associated angular distribution from (3.10) via 6-*j* coefficients.

Since the system under consideration, like the classical electric dipole, is lossless, the differential scattering cross section averaged over all solid angles, called an isotropic average, should reduce to the absorption cross section divided by 4π (i.e., $\sigma^A/(4\pi)$). This is certainly the case for the two instances considered in (3.13) since the isotropic averages, denoted by angular brackets, $\langle\cos^2\theta \cos^2\phi\rangle + \langle\sin^2\phi\rangle$ and $\langle\sin^2\theta\rangle$, are, respectively, $1/6 + 1/2 = 2/3$, and simply $2/3$. This equality, $\langle d\sigma^R_{fF\Sigma}/d\Omega\rangle = \sigma^A/(4\pi)$, should also be true generally, including in a system with more than one emission final state, $|f'\rangle$. If we average over all angles θ and ϕ by direct integration, we find $\langle\mathscr{F}_a(\theta,\phi;\phi_0;\omega_B/\Gamma_{mm'})\rangle = \langle\mathscr{F}_b(\theta,\phi;\theta_0\phi_0;\omega_B/\Gamma_{mm'})\rangle = 0$ in (3.11), even if $B \neq 0$. Thus, the isotropic average of either $C_{fFf'}(\hat{e}_a;\hat{e}')$ or $C_{fFf'}(\hat{e}_b;\hat{e}')$ is reduced to its first term (i.e., 2/9), leading to the isotropic average of $\langle d\sigma^{RS}_{fF\Sigma}/d\Omega\rangle = \langle\sum_{f'} d\sigma^{RS}_{fFf'}/d\Omega\rangle = \sigma^A/(4\pi)$. Had the excitation been incoherent (i.e., $\rho^{(2)}_{mm'}(t) = 0$ except where $m = m'$), the scattering would be isotropic, and the absorption cross section σ^A obtained from 4π times the value from (3.12) and (3.14) by replacing, respectively, $[3\sum_{f'} \alpha_{Ff'} C_{fFf'}(\hat{e};\hat{e}')]$ and $[3C_{010}(\hat{e};\hat{e}')]$ with their isotropic average, 2/3.

Employing (3.11.a) and (3.11.b) for incident polarizations, $\hat{e}_a$ and $\hat{e}_b$, respectively, the distribution function, the last factor in (3.12), takes the following form:

$$\mathscr{D}_{fF\Sigma}(\hat{e}_{a,b};\hat{e}') = \left[3\sum_{f'} \alpha_{Ff'} C_{fFf'}(\hat{e}_{a.b};\hat{e}')\right]$$
$$= \frac{2}{3}\left\{1 + \frac{9}{2} g_F \sum_{f'} \alpha_{Ff'} B^{(2)}_{fFf'} \mathscr{F}_{a,b}(\theta,\phi;\phi_0;\omega_B/\Gamma_{mm'})\right\}. \tag{3.15}$$

For the $B = 0$ case, $\mathscr{F}_a(\theta,\phi;0)$ and $\mathscr{F}_b(\theta,\phi;0)$ are explicitly given in (3.13.a) and (3.13.b). For backscattering, we can substitute $\theta = 90°$ and $\phi = 270°$ into these expressions and obtain the same backward scattering distribution function as

$$\mathscr{D}^{\pi}_{fF\Sigma}(\hat{e};\hat{e}') = \left[3\sum_{f'} \alpha_{Ff'} C^{\pi}_{fFf'}(\hat{e}_{a,b};\hat{e}')\right]$$
$$= \frac{2}{3}\left\{1 + \frac{3}{10} g_F \sum_{f'} \alpha_{Ff'} B^{(2)}_{fFf'}\right\} \equiv \frac{2}{3} q^{\pi}_{fF\Sigma}(\omega_B = 0, \theta_0). \tag{3.16}$$

Here, as in (19e) and Table 3 of [3.3], we also denote the factor in the curly brackets as the backward q-factor, $q^{\pi}_{Ff}(0,\theta_0)$, which is independent of the incident beam direction, θ_0. The first term of (3.16) is its spatial or isotropic average, corresponding to incoherent fluorescence emission. Including the second term gives rise to an antenna pattern, whose existence depends on the product of the fractional transition rate, $\alpha_{Ff'}$, and the coherence associated with the pathway, fFf', with the 6-j coefficient $B^{(2)}_{fFf'}$, as given in (3.10). This second term can be zero in two ways. If there is only one final state, the 6-j coefficient $B^{(2)}_{fFf'}$ may be zero. For the case with more than one final state, since $B^{(2)}_{fFf'}$ may be positive (constructive) or negative (destructive), the sum over all states, $\sum_{f'} \alpha_{Ff'} B^{(2)}_{fFf'}$, could be zero.

3.4 Application to Real Atoms with Interest to Atmospheric Lidars

The angular dependence in (3.15) lies in the strength factor $C_{fFf'}(\hat{e};\hat{e}')$. We define the quantity in the square bracket in (3.12) as the distribution function of the fluorescence intensity, $\mathscr{D}_{fF\Sigma}(\hat{e};\hat{e}') \equiv [3\sum_{f'} \alpha_{Ff'} C_{fFf'}(\hat{e};\hat{e}')]$; it plays the same role as $(\hat{e} \bullet \hat{e}')^2$ in the differential scattering for classical dipole radiation, see (2.2.b); its isotropic average is invariant and has a value of 2/3. As pointed out for the simplest spinless system, for incident polarization $\hat{e} = -\hat{z}$, the factor $3C_{010}(-\hat{z};\ \hat{e}')$, which may be obtained by substituting $\theta_0 = 90°$ into (3.11.b), was shown to be $\sin^2\theta$, a distribution identical to that from the classical z-polarized electric dipole. One can also show, in general, that the isotropic average of (3.12) is $\sigma^A/(4\pi)$ with σ^A given in (3.6).

3.4.1 Illustration of Radiation Patterns of Transition in Real Atoms

For a realistic atom, the radiation pattern of a coherently excited dipole would depend on the incident polarization as well as the structure of the atom. Therefore, for simplicity we discuss the Hanle Effect by considering only the distribution function, which emulates the radiation patterns of $\hat{e}_b = -\hat{z}$ polarization, from the relevant atomic transitions in ^{23}Na, ^{39} K (93%), and ^{56}Fe (91.8%). We plot the distribution function of the fluorescence intensity $\mathscr{D}_{fF\Sigma} = \left[3\sum_{f'} \alpha_{Ff'} C_{fFf'} \; (\hat{e}_b = -\hat{z}; \hat{e}')\right]$ for these cases and compare them to the "antenna pattern" of a classical z-polarized electric dipole, $\sin^2\theta$. Due to a judicious choice of the incident beam polarization parallel to the atomic magnetic axis $\hat{z}$, which leads to the setting of $\theta_0 = 90°$ in (3.11.b), the relevant (3 times) strength factor function from (3.11.b), summed over allowed final states $|f'\rangle$, is (for this incident polarization) independent of the external B field and derived to be

$$\mathscr{D}_{fF\Sigma}(\hat{e}_b;\hat{e}') \equiv \left[3\sum_{f'} \alpha_{Ff'} C_{fFf'}(\hat{e}_b;\hat{e}')\right]$$
$$= \frac{2}{3}\left\{1 + \frac{9}{2}\left(g_F \sum_{f'} \alpha_{Ff'} B^{(2)}_{fFf'}\right)\left[-\frac{2}{30}\left(2 - 3\sin^2\theta\right)\right]\right\}. \tag{3.17}$$

The value of (3.17) in the backward direction, that is, $\theta = 90°$ (note $\theta_0 = 90°$ has already been applied to (3.11.b)) – which reduces to $\mathscr{D}^{\pi}_{ffF\Sigma}(-\hat{z};\hat{e}') \equiv \left[3\sum_{f'} \alpha_{Ff'} C^{\pi}_{fFf'}(-\hat{z};\hat{e}')\right] = \frac{2}{3}\left\{1 + \frac{3}{10} g_F \sum_{f'} \alpha_{Ff'} B^{(2)}_{fFf'}\right\}$, as seen in (3.16) – is of particular interest. This backscattering expression is valid for arbitrary incident polarization with $B = 0$. Because the spin-orbit and hyperfine splitting are small and thus may be neglected compared to the energy difference between $3p$ and $3s$ levels of ^{23}Na and $4p$ and $4s$ levels of ^{39}K atoms, the Einstein coefficient $A_{Ff'}$, as explained in Section A.2.2 in Appendix A, is related to the spontaneous emission rate, A_0, of the s-to-p transition. The values of $\alpha_{Ff'}$ for the Na D_2 transition can then be directly calculated from (A.18.c), with the numerical results given in the bracket above "Na D_2 HFS" in Fig. A.1. Using these values and the values of $B^{(2)}_{fFf'}$ tabulated in Table 5 of [3.3], we calculate the transition-dependent factor in (3.16), $\left(g_F \sum_{f'} \alpha_{Ff'} B^{(2)}_{fFf'}\right)$, for the Na D_2 (or ^{39} K D_2) transitions as given in Table 3.2.a. The reason that we do not tabulate this factor for the Na D_1 (or ^{39} K D_1) transitions is because the fortuitous combination of $\alpha_{Ff'}$ and $B^{(2)}_{fFf'}$ makes the two terms in $\sum_{f'} \alpha_{Ff'} B^{(2)}_{fFf'}$ in all

D_1 transitions mutually cancel – see discussions below (22b) of [3.3]. Such cancellations, or *null effect*, as to be discussed below, exist in the D_2 transitions only in the cases of $F = 2$ and $F = 0$ excited states; see Table 3.2.a.

Of course, in general the polarization of the incident beam in the geometric z-axis is not parallel to the atom's magnetic axis, which lies parallel to the Earth's magnetic field (Z-axis); that is, the incident angle between the polarization and the Earth's magnetic field, θ_0, is no longer 90°. In this case, one must consider the transformation between X-Y-Z coordinates and the geometric coordinates x-y-z as given in (B.6.b) and shown in Fig. B.4 of Appendix B. Further complications occur with multiple incident beams associated with multiple observation directions. Though complicated, all can be handled mathematically by multiple transformations of coordinates. One such example has been already given in Appendix A of [3.9]. Fortunately, the Earth's magnetic field is typically weak compared to the energy separation, and setting $\omega_B = B = 0$ yields a good approximation in practice [3.3; 3.9].

There are two differences when working with ^{56}Fe transitions. First, the states are labeled by fine-structure angular momenta, J and j. Though the formulae for the differential cross section (3.12) and the distribution function (3.14) remain the same, the quantum numbers (f, F, f') therein should be replaced by (j, J, j'). Second, because spin-orbit coupling in ^{56}Fe is not negligible, it affects the A coefficients of the split states. We must calculate the $\alpha_{Jj'}$ directly from the measured values of the relevant A_{Jj} – listed in Table A.3, which is taken from the NIST atomic database [3.10] – using the expression in (3.12). Similarly, we calculate the $B^{(2)}_{fFf'}$ with the formula given by (3.10) and using the 6-j coefficient calculator [www-stone.ch.cam.ac.uk/wigner.shtml]. We note that the numerical results of Table 3.1 were published in [3.11], where the value of $\alpha_{43} = 0.100$ in the upper-right entry was misprinted as –0.509. For ^{56}Fe we are interested in five transitions: at 368 nm $(a\,^5D_4 \rightarrow z^5F^0_4)$, 372 nm $(a\,^5D_4 \rightarrow z^5F^0_5)$, 374 nm $(a\,^5D_3 \rightarrow z^5F^0_4)$, 386 nm $(a\,^5D_4 \rightarrow z^5D_4)$, and 392 nm $(a\,^5D_3 \rightarrow z^5D_4)$. Values of both $\alpha_{Jj'}$ and $B^{(2)}_{jJj'}$ for these transitions are tabulated in Table 3.1. With these values, we calculate $\left(g_J \sum_{j'} \alpha_{Jj'} B^{(2)}_{jJj'}\right)$; the results are tabulated in Table 3.2.b. Notice there is no fortuitous cancellation in ^{56}Fe transitions.

With the tabulation values in Tables 3.2.a and 3.2.b we plot the distribution functions of fluorescence intensity, (3.17), of Na D_2 (same for ^{39}K D_2) transitions in Fig. 3.1a, and of ^{56}Fe transitions in Fig. 3.1b, taken from Fig. 4 of the recently published [3.11]. As can be seen in these figures, laser excitation of both Na and ^{56}Fe transitions are not as coherent as a classic dipole radiator, as they show a considerably smaller Hanle Effect. Due to the fortuitous cancellations for D_1 transitions in Na and K atoms, these transitions exhibit no coherence and thus present spontaneous emission patterns (i.e., the intensity is independent of scattering direction). This is the case only for the (2,2), (1,2), and (1,0) excitations of the D_2 transition, as shown in thin black-solid line in Fig. 3.2.a.

Table 3.1 *The values of $\alpha_{Jj'}$ and $B^{(2)}_{jJj'}$ in ^{56}Fe transitions of interest*

$\alpha_{Jj'}$; $B^{(2)}_{jJj'}$	$z^5F, J=4$	$z^5F, J=5$	$z^5D, J=4$
$a^5D, j'=(3,3)$	$\frac{14.1}{14.1+1.38}=0.911$; $5\left(\frac{1}{6}\sqrt{\frac{11}{42}}\right)^2=0.0364$		$\frac{1.08}{9.69+1.08}=0.100$; $5\left(\frac{1}{6}\sqrt{\frac{11}{42}}\right)^2=0.0364$
a^5D, $j'=(3,4)$	$\frac{1.38}{14.1+1.38}=0.089$; $-5\left(\frac{1}{6}\sqrt{\frac{11}{42}}\right)\left(\frac{1}{30}\sqrt{\frac{77}{6}}\right)=-0.0509$		$\frac{9.69}{9.69+1.08}=0.900$; $-5\left(\frac{1}{6}\sqrt{\frac{11}{42}}\right)\left(\frac{1}{30}\sqrt{\frac{77}{6}}\right)=-0.0509$
a^5D, $j'=(4,3)$	$\frac{14.1}{14.1+1.38}=0.911$; $-5\left(\frac{1}{6}\sqrt{\frac{11}{42}}\right)\left(\frac{1}{30}\sqrt{\frac{77}{6}}\right)=-0.0509$		$\frac{1.08}{9.69+1.08}=0.100$; $-5\left(\frac{1}{6}\sqrt{\frac{11}{42}}\right)\left(\frac{1}{30}\sqrt{\frac{77}{6}}\right)=-0.0509$
a^5D, $j'=(4,4)$	$\frac{1.38}{14.1+1.38}=0.089$; $5\left(\frac{1}{30}\sqrt{\frac{77}{6}}\right)^2=0.0713$	1.000; $5\left(\frac{1}{15}\sqrt{\frac{13}{11}}\right)^2=0.0263$	$\frac{9.69}{9.69+1.08}=0.900$; $5\left(\frac{1}{30}\sqrt{\frac{77}{6}}\right)^2=0.0713$

Table 3.2.a $g_F \sum_{f'} \alpha_{Ff'} B^{(2)}_{fFf'}$ *for Na* D_2 *transition*

$g_F \sum_{f'} \alpha_{Ff'} B^{(2)}_{fFf'}$	$F = 3$	$F = 2$	$F = 1$	$F = 0$
$\left(2, \sum_{f'}\right)$	2/5	0	-1/15	
$\left(1, \sum_{f'}\right)$		0	1/3	0

Table 3.2.b $g_J \sum_{j'} \alpha_{Jj'} B^{(2)}_{jJj'}$ *for Fe transitions*

$g_J \sum_{j'} \alpha_{Jj'} B^{(2)}_{jJj'}$	$z^5F^0_4$	$z^5F^0_5$	$z^5D^0_4$
$a\,^5D_3, \left(\sum_{j'}\right)$	0.257		−0.380
$\left(a\,^5D_4, \sum_{j'}\right)$	−0.360	0.289	0.532

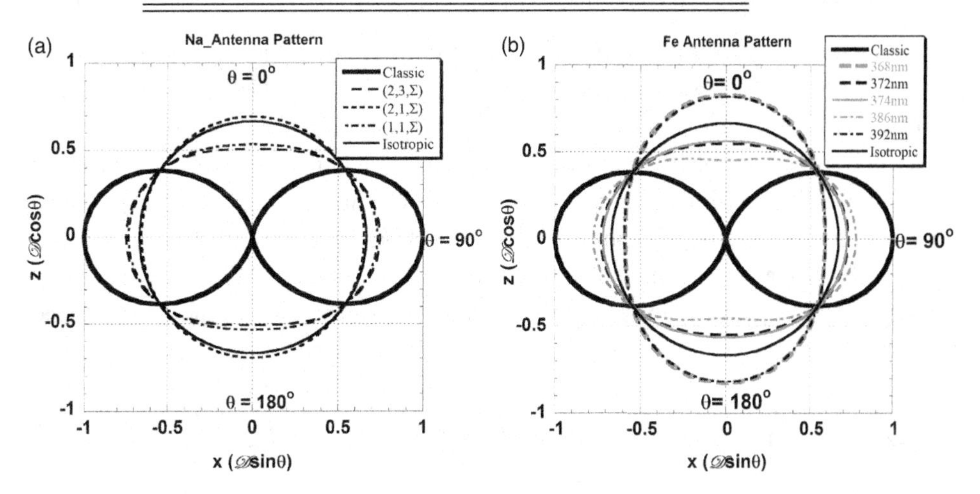

Fig. 3.2 Radiation patterns of the z-polarized, dipole $\hat{e} = -\hat{z}$ with fluorescence intensity distribution $\mathscr{D} = \left[3\sum_{f'} \alpha_{Ff'} C_{fFf'}(-\hat{z};\hat{e}')\right]$ for (a) Na D_2 and (b) ^{56}Fe transitions. Reprinted with permission from [3.11]. © The Optical Society.

More precisely, for the D_2 transitions, since $B^{(2)}_{101} = 0$, the (1,0) excitation shows no Hanle Effect. For (2,2) and (1,2) excitations, $\alpha_{22} = \alpha_{21} = 1/2$, see Fig. A.1, along with $B^{(2)}_{221} = B^{(2)}_{121} = 7/60$ and $B^{(2)}_{222} = B^{(2)}_{122} = -7/60$; see Table 5 of [3.3]. Thus, the enhancement in one fluorescence pathway cancels exactly the reduction in the other pathway in the fluorescence intensity (3.14). The same is true for the (1,1) and (2,1) excitations of the D_1 transition. In other words, though $B^{(2)}_{111} = 1/180$ and

$B_{112}^{(2)} = -1/36$, fortuitously $\alpha_{12} = 5/6$ and $\alpha_{11} = 1/6$, giving rise to exact cancellation for the (1,1) excitation. Likewise, since $B_{211}^{(2)} = 5/36$ and $B_{212}^{(2)} = -1/36$, the same cancellation exists in the (2,1) excitation. Such complete cancellations between hyperfine states are possible only when the spin-orbit and hyperfine interactions are negligible, so that the Einstein A coefficients from a hyperfine state may be simply related to that of the basic orbital transition ($3s$-$3p$ in the case of Na) A_0 via the structure of the 6-j coefficients, independent of spin-orbit and hyperfine interactions. Due to the presence of significant spin-orbit interaction, no fortuitous cancellation exists in ^{56}Fe transitions, which all exhibit the Hanle Effect to a different degree. Interestingly, we note the radiation patterns of the 372 and 374 nm emissions are nearly identical; see Table 3.2.b, a fact relevant to the iron Boltzmann lidar data processing discussed in Chapter 7.

In Fig. 3.2(a), the Na (2,2), (1,2), and (1,0) excitations exhibit spontaneous emission behavior (thin solid, labeled in the legend as "isotropic") with a backward scattering value of interest equal to the isotropic value, $3C_{fFf'}^{\pi}(-\hat{z};\hat{e}') = \langle 3C_{fFf'}(-\hat{z};\hat{e}')\rangle = 2/3$; this is quite different from the classic dipole radiation pattern in the thick solid line. The backward values, $\left[3\sum_{f'}\alpha_{Ff'}C_{fFf'}^{\pi}(-\hat{z};\hat{e}')\right]$, for excitations (2,3), (2,1), and (1,1), shown respectively in long dash, short dash, and dash-dot lines, are 0.747, 0.653, and 0.733, corresponding to, respectively, 112%, 98%, and 110% of its isotropic average. In Fig. 3.2(b), none of the Fe transitions exhibits spontaneous emission behavior (shown in black thin solid lines and labeled as "isotropic" only for reference). All transitions exhibit the Hanle Effect with the backscattering values for 368 nm (gray-dash), 372 nm (black dash), 374 nm (gray thin solid), 386 nm (gray dash-dot), and 392 nm (black dash-dot) of, respectively, 0.880, 1.087, 1.077, 1.160, and 0.886 times their isotropic average of 2/3. In both figures, the classic dipole pattern is shown in the black solid line for comparison.

3.4.2 *Differential Backscattering and Absorption Cross Sections of Na, ^{39}K, and ^{56}Fe Atoms*

We now evaluate the strength factors and scattering cross sections for useful cases of backscattering from atomic systems of Na, ^{39}K, and ^{56}Fe with arbitrary incident beam direction. We do so by substituting $(\pi - \theta_0,\ \phi_0 + \pi)$ for $(\theta,\ \phi)$ into (3.11.a) and (3.11.b), and obtain, respectively

$$C_{fFf'}^{\pi}(\hat{e}_a;\hat{e}') = \frac{2}{9}\left\{1 + \frac{9g_F}{2}B_{fFf'}^{(2)}\mathcal{F}_{Ffa}^{\pi}(\omega_B,\theta_0)\right\};$$

$$\mathcal{F}_{Ffa}^{\pi}(\omega_B,\theta_0) = \frac{1}{15}\left[1 - \frac{3}{2}\sin^2\theta_0\left(\frac{(2\omega_B/\Gamma_{mm'})^2}{1+(2\omega_B/\Gamma_{mm'})^2}\right)\right] \tag{3.18.a}$$

$$C^{\pi}_{fFf'}(\hat{e}_b;\hat{e}') = \frac{2}{9}\left\{1 + \frac{9g_F}{2}B^{(2)}_{fFf'}\mathscr{F}^{\pi}_{Ffb}(\omega_B,\theta_0)\right\};$$

$$\mathscr{F}^{\pi}_{Ffb}(\omega_B,\theta_0) = \frac{1}{15}\left[1 - \frac{9\,\sin^2 2\theta_0}{2}\frac{(\omega_B/\Gamma_{mm'})^4}{[1+(\omega_B/\Gamma_{mm'})^2][1+(2\omega_B/\Gamma_{mm'})^2]}\right]. \tag{3.18.b}$$

We point out the existence of notation discrepancies in the literature. Although $\mathscr{F}_a$ and $\mathscr{F}_b$ in (3.13) here and in (A4) of (3.11) represent the same quantities as F_a and F_b in (A8) of [3.3], the expressions $\mathscr{F}^{\pi}_{Ffa}(\omega_B,\theta_0)$ and $\mathscr{F}^{\pi}_{Ffb}(\omega_B,\theta_0)$ above are not the same as those for $\boldsymbol{\mathcal{F}}^{\pi}_{Ffa}$ and $\boldsymbol{\mathcal{F}}^{\pi}_{Ffb}$ in (18a) and (18b) of [3.3], F^{π}_{Ffa} and F^{π}_{Ffb} in (A1) of [3.9], or $\mathscr{F}^{\pi}_{Ffa}$ and $\mathscr{F}^{\pi}_{Ffb}$ in (A5) of [3.11]. They are, in fact, related as $\boldsymbol{\mathcal{F}}^{\pi}_{Ffe}(\omega_B,\theta_0) = 15\mathscr{F}^{\pi}_{Ffe}(\omega_B,\theta_0)$, with the subscript *Ffe* representing either *Ffa* or *Ffb*. The definition of the q-factors, $q^{\pi}_{Ffe}(\omega_B = 0,\theta_0)$ in (3.16) and $q^{\pi}_{Ffe}(\omega_B,\theta_0)$ in (3.19), are the same in all publications mentioned.

Substituting (3.18), the backscattering differential LIFΣ cross section may be derived from (3.12) as

$$\frac{d\sigma^{\pi}_{f F\Sigma}}{d\Omega} = \frac{1}{4\pi}\frac{\lambda^2_{Ff}}{4}\frac{g_F}{g_f}A_{Ff}g(\omega-\omega_{Ff})q^{\pi}_{Ffe}(\omega_B,\theta_0), \text{ with}$$

$$q^{\pi}_{Ffe}(\omega_B,\theta_0) = \left[1 + \frac{9g_F}{2}\sum_{f'}\alpha_{Ff'}B^{(2)}_{fFf'}\mathscr{F}^{\pi}_{Ffe}(\omega_B,\theta_0)\right] \tag{3.19}$$

Here, we have defined as in [3.3] the dimensionless q-factor for backscattering. The total differential cross section (or scattered power), and indeed the q-factor $q^{\pi}_{Ffe}(\omega_B,\theta_0)$ in (3.19), consist of two contributions to emission to the final states $|f>$ and $|f'>$. For zero magnetic field, we can write them explicitly, since $\mathscr{F}^{\pi}_{Ffe}(0,\theta_0) = \frac{1}{15}$ and $\alpha_{Ff} + \alpha_{Ff'} = 1$, as

$$q^{\pi}_{Ffe}(0,\theta_0) = \left[\alpha_{Ff}\left(1 + \frac{3g_F}{10}B^{(2)}_{fFf}\right) + \alpha_{Ff'}\left(1 + \frac{3g_F}{10}B^{(2)}_{fFf'}\right)\right]. \tag{3.20.a}$$

Though the q-factor in (3.19) and (3.20.a) does not reveal the frequencies of the emission lines (fluorescence) for the relevant excitation, from $|f>$ to $|F>$ (for Na and ^{39}K) or from $|j>$ to $|J>$ (for ^{56}Fe), the transition results in two naturally broadened emission lines, respectively centered at angular frequencies $\omega_{fFf} = 2\omega_{Ff} - \omega$ and $\omega_{fFf'} = \omega_{Ff} + \omega_{Ff'} - \omega$. The spectral backward differential LIFΣ cross section or, simply, the LIF emission spectrum as a function scattering frequency ω_s, which reveals emission frequencies, may be deduced from (3.19) and (3.20.a) as:

$$\frac{d\sigma^{\pi}_{fF\Sigma}}{d\omega_s d\Omega} = \frac{\lambda^2_{Ff}}{16\pi}\frac{g_F}{g_f}A_{Ff}g(\omega-\omega_{Ff})$$
$$\times\left[\alpha_{Ff}\left(1+\frac{3g_F}{10}B^{(2)}_{fFf}\right)g(\omega_s-\omega_{fFf})+\alpha_{Ff'}\left(1+\frac{3g_F}{10}B^{(2)}_{fFf'}\right)g(\omega_s-\omega_{fFf'})\right].$$
(3.20.b)

Note that if we integrate the above expression over ω_s, we recover (3.20.a). For a broadband receiving system typically used, the individual emission lines cannot be resolved. To apply (3.19) to atmospheric Na, ^{39}K, or ^{56}Fe atoms, we make several modifications:

(1) Since the Earth's magnetic field is typically small, we may set $2\omega_B/\Gamma_{mm'}$ in (3.18) to zero; this leads to a simplification, with $\mathcal{F}^{\pi}_{Ff'a}(0,\theta_0) = \mathcal{F}^{\pi}_{Ff'b}(0,\theta_0) = 1/15$ (or $\mathcal{F}^{\pi}_{Ffe}(0,\theta_0) = 1/15$). Then, we have $q^{\pi}_{Ffe}(0,\theta_0) = \left[\alpha_{Ff}\left(1+0.3g_F B^{(2)}_{fFf}\right)\right.$ $\left.+\alpha_{Ff'}\left(1+0.3g_F B^{(2)}_{fFf'}\right)\right]$ or $q^{\pi}_{fF\Sigma}(0) = \sum_{f'}\alpha^{\pi}_{fFf'}(0)$, which can be similarly numerically evaluated as the quantities in Tables 3.2.a and 3.2.b. For Na and K D-transitions, these quantities, $\alpha_{Ff}(1+0.3g_F B^{(2)}_{fFf})$ and $\alpha_{Ff'}(1+0.3g_F B^{(2)}_{fFf'})$, are denoted respectively as $\alpha^{\pi}_{fFf}(0)$ and $\alpha^{\pi}_{fFf'}(0)$, or $\alpha^{\pi}_i(0)$, as in Eq. (3) of [3.9].

(2) To write the resulting expression in the frequency domain and to account for atomic motions in ambient temperature and wind, we replace the lineshape function $g(\omega-\omega_{Ff})$ by $\mathbb{G}[(\nu - V/\lambda - \nu_{Ff}), T]/(2\pi)$, where T and V are temperature and LOS wind, and $\mathbb{G}[(\nu - V/\lambda - \nu_{Ff}),\ T]$ is a Voigt function, as given in (2.7.b), which results from the convolution of natural linewidth with a Gaussian function depicting Doppler broadening and Doppler shift.

(3) Since both atomic lower (ground) states are occupied at ambient temperature, the differential cross section should be multiplied by the population of the lower state $|f\rangle$, $\boldsymbol{n}_f$ with $f = 1$ or 2 given in (3.22.c).

Implementing these modifications to (3.19) yields

$$\frac{d\sigma^{\pi}_{fF\Sigma}}{d\Omega} = \frac{\lambda^2_{Ff}}{32\pi^2}\frac{g_F}{g_f}A_{Ff}\boldsymbol{n}_f\mathbb{G}[(\nu - V/\lambda - \nu_{Ff}),\ T]\sum_{f'}\alpha^{\pi}_{fFf'}(0). \tag{3.21}$$

Here, we note that for $\mathbf{B} = 0$, the backward q-factor is not only independent of θ_0 but also independent of incident polarization, $\hat{e}$. The fractional Einstein A coefficients, $\alpha_{Ff'}$, for atomic iron are defined above in terms of two measured coefficients, A_{Ff} and $A_{Ff'}$; see Table A.3. For Na and K atoms, since the spin-orbit coupling and hyperfine splitting are relatively small compared to the s-p transition energy, as explained in Appendix A, the fractional coefficients can all be expressed in terms of the Einstein A coefficient for the s-p transition in question, A_0, as given by (A.18.c) and Fig. A.1.

To apply (3.20) to Na, ^{39}K, and ^{56}Fe, we consider their respective Doppler FWHM at ambient temperature, $\Delta\nu_{FWHM} = 2\sqrt{8\ln(2)k_BT/mc^2}$, to their respective energy structures. At 200 K, $\Delta\nu_{FWHM}$ is calculated to be 1.055 GHz for Na, 0.810 GHz for ^{39}K, and 0.676 GHz for ^{56}Fe. Since the level spread of the excited states in Na D_2 transition is about 0.109 GHz, much smaller than its Doppler width, a laser beam can excite atoms moving at appropriate velocities to all excited states. The same is true for K D_1 transitions, as the level spread of their excited states is 0.189 GHz, again notably smaller than their Doppler width. However, the situation is different for ^{56}Fe transitions. As can be seen in Fig. A.2, the closest excited states that can be reached from ground state $a\,^2D_4$ by optical absorption are $z\,^5F_4^0$ and $z\,^5F_5^0$; their level difference is about 292 cm^{-1}, or 8768 GHz, much larger than their Doppler width. With these in mind, we need to include all excited states (and both ground states) in computing the differential LIFΣ cross sections for K and Na D transitions, while only one relevant excited state (with one of the two ground states: a^2D_4 or a^2D_3) is needed for a specific Fe transition.

For the D_1 and D_2 transitions of atmospheric ^{39}K and Na atoms, we replace A_{Ff} by $\alpha_{Ff}A_0$, and substitute $\alpha_{Ff}A_0$ and $g_F\sum\limits_{f'}\alpha_{Ff'}B^{(2)}_{fFf'}$ (which is zero for all D_1 transitions) with appropriate numerical values and state designations from, respectively, Fig. A.1 and Table 3.2.a. In so doing, we obtain (22a) and (22b) of [3.3]:

$$\frac{d\sigma^{\pi}_{D_1}}{d\Omega}(\nu) = \frac{\lambda^2_{D_1}}{32\pi^2}\left\{\begin{array}{l}\frac{n_1}{6}[\mathrm{G}_{31}(\nu - V/\lambda - \nu_{31}, \mathrm{T}) + 5\mathrm{G}_{41}(\nu - V/\lambda - \nu_{41}, \mathrm{T})] \\ \quad + \frac{n_2}{10}[5\mathrm{G}_{32}(\nu - V/\lambda - \nu_{32}, \mathrm{T}) + 5\mathrm{G}_{42}(\nu - V/\lambda - \nu_{42}, \mathrm{T})]\end{array}\right\}A_0; \quad (3.22.\mathrm{a})$$

$$\frac{d\sigma^{\pi}_{D_2}}{d\Omega}(\nu) = \frac{\lambda^2_{D_2}}{32\pi^2}\left\{\begin{array}{l}\frac{n_1}{6}[2\mathrm{G}_{51}(\nu - V/\lambda - \nu_{51}, T) + 5.5\mathrm{G}_{61}(\nu - V/\lambda - \nu_{61}, T) \\ +5\mathrm{G}_{71}(\nu - V/\lambda - \nu_{71}, T)] + \frac{n_2}{10}[0.98\mathrm{G}_{62}(\nu - V/\lambda - \nu_{62}, T) \\ +5\mathrm{G}_{72}(\nu - V/\lambda - \nu_{72}, T) + 15.68\mathrm{G}_{82}(\nu - V/\lambda - \nu_{82}, T)]\end{array}\right\}A_0, \text{ where}$$

(3.22.b)

$$n_1 = \frac{g_1}{g_1 + g_2 \times \exp[-0.04798 \times (\Delta\nu(\mathrm{GHz})/T)]};$$
$$n_2 = \frac{g_2 \times \exp[-0.04798 \times (\Delta\nu(\mathrm{GHz})/T)]}{g_1 + g_2 \times \exp[-0.04798 \times (\Delta\nu(\mathrm{GHz})/T)]}. \quad (3.22.\mathrm{c})$$

In deriving (3.22.c), we have used the fact that $\Delta E/k_BT = h\Delta\nu(\mathrm{GHz})/k_BT = 4.798 \times 10^{-2}\ (\Delta\nu(\mathrm{GHz})/T)$, where $\Delta E = E_2 - E_1$ is the energy difference between the two ground states, and $\Delta\nu$ its corresponding frequency. Here, A_0 is 6.29×10^7 s^{-1} for Na and 3.85×10^7 s^{-1} for K, λ_{D_1} is 589.76 nm for Na and 770.11 nm for K, and λ_{D_2} is 589.16 nm for Na and 766.70 nm for K.

For ^{56}Fe transitions, substituting into (3.21) the tabulated values for A_{Ff} and $g_F \sum_{f'} \alpha_{Ff'} B^{(2)}_{fFf'}$, respectively from Table A.3 and Table 3.2.b, we can write the differential LIFΣ cross sections for each of the five transitions of interest. We do this for the three employed by Fe fluorescence lidars below:

$$\text{For the 372 nm line}: \frac{d\sigma^{\pi}_{fF\Sigma}}{d\Omega} = \frac{(372\ \text{nm})^2}{32\pi^2}\frac{11}{9}\boldsymbol{n}_1(1.62\times 10^7 \text{s}^{-1})\,\mathbb{G}[(\nu - V/\lambda - \nu_{372}),\ T][1.087] \tag{3.23.a}$$

$$\text{For the 374 nm line}: \frac{d\sigma^{\pi}_{fF\Sigma}}{d\Omega} = \frac{(374\ \text{nm})^2}{32\pi^2}\frac{9}{7}\boldsymbol{n}_2(1.41\times 10^7 \text{s}^{-1})\,\mathbb{G}[(\nu - V/\lambda - \nu_{374}),\ T][1.077] \tag{3.23.b}$$

$$\text{For the 386 nm line}: \frac{d\sigma^{\pi}_{fF\Sigma}}{d\Omega} = \frac{(386\ \text{nm})^2}{32\pi^2}\frac{9}{9}\boldsymbol{n}_1(9.69\times 10^6 \text{s}^{-1})\,\mathbb{G}[(\nu - V/\lambda - \nu_{386}),\ T][1.160] \tag{3.23.c}$$

Here, $\boldsymbol{n}_1$ and $\boldsymbol{n}_2$ are the populations of the two ground (lowest) states; they can be deduced from (3.22.c) by using $g_1 = 9$, $g_2 = 7$, and $\Delta\nu = 12,480$ GHz. Though there are higher-level states in the ground term, due to their higher energies they have negligible populations at ambient temperature. We also note that, though there are two emission pathways (see Fig. A.2) for the excited states $z^5F^0_4$ and $z^5D^0_4$, due to their much narrower Doppler width as compared to $\Delta\nu$, the populations of a^2D_4 and a^2D_3 cannot contribute to fluorescence for, respectively, the 374 nm and 386 nm excitations of ^{56}Fe atoms.

Since, as pointed out in Section 3.3.2, the absorption cross section σ^A may be obtained from 4π times the isotopic average of (3.12), that is, replacing $[3\sum_{f'} \alpha_{Ff'} C_{fFf'}(\hat{e};\hat{e}')]$ by $\langle 3\sum_{f'} \alpha_{Ff'} C_{fFf'}(\hat{e};\hat{e}')\rangle = 2/3$, the LIFΣ cross sections can either be constructed from the isotropic average of (3.12) (by noticing the difference in formulation between ω and ν space) or by ignoring the Hanle Effect – that is, setting $g_F \sum_{f'} \alpha_{Ff'} B^{(2)}_{fFf'}$ in (3.20.a) to its null isotropic average. By doing so, we would obtain $\sigma^A_{D_1}$ and $\sigma^A_{D_2}$, as She et al. demonstrated in [3.3] (21a) and (21b) for Na and K. For absorption cross section of iron transitions, we simply multiply (3.23) by 4π and set the q-factor to unity.

3.5 Rudimentary Physics of Na Laser Guide Stars

As is well known, atmospheric turbulence limits the performance of ground-based optical telescopes for astronomy and space surveillance. One way to undo its adverse effects is to measure the phase front distortion of light emitted from a reference star and to opto-mechanically compensate for the blurring effect of atmospheric turbulence in real time. Tremendous progress has been made in the past two decades using an artificial guide star created in the earth's Na layer by a laser beam tuned to the NaD_2 resonance. A brief account on the relevant effects of atmospheric turbulence are given in Chapter 8. The objective of a lidar transmitter in a Na laser guide star (LGS) system, which we discuss here, is totally different from that used to measure atmospheric temperature and wind. The latter, as described in [3.9] and in Chapter 7, probes the dynamics of the atmosphere to the extent possible without saturating or perturbing it, while the former creates a bright point source (a "star") by fully saturating the NaD_2 transition and its associated fluorescence emission. As this opens new applications of LIF, we briefly discuss the saturation of a two-level system and optical pumping of Na atoms, making them behave like a two-level system coherently interacting with a resonant laser beam and at the same time emitting fluorescence. We will also consider the associate obstacles and remedies in practice.

3.5.1 Saturation of a Two-Level System

Consider a population of atoms with density N_0 that are described by a two-level system with an electric dipole–allowed transition between excited state $|F, m\rangle$ and ground state $|F, \mu\rangle$ and with degeneracies g_F and g_f, under continuous illumination with coherent or incoherent resonant light of intensity (irradiance) I. In steady state, the number densities of the two levels N_F and N_f ($N_f + N_F = N_0$) of this system of atoms may be evaluated by employing the rate equations, considering the energy balance between the atoms via absorption plus stimulated and spontaneous emission of radiation. The rate of fluorescence (spontaneous) emission from this system is $R = (N_F/N_0)A_{Ff}$, where A_{Ff} is the Einstein A coefficient of the system. The steady-state solution for the fluorescence rate R as a function of resonant laser beam intensity I is well known [3.12, 3.13]:

$$R = R_s \frac{I/I_s}{1 + I/I_s}, \text{ with } R_s = \frac{A_{Ff} g_F}{g_f + g_F} \text{ and } I_s = \frac{8\pi hc}{\left(1 + g_F/g_f\right)\lambda_{Ff}^3 g(\nu - \nu_{Ff})}, \tag{3.24}$$

where R_s, I_s, and $g(\nu - \nu_{Ff}) = (\Delta\nu/2\pi)/[(\nu - \nu_{Ff})^2 + (\Delta\nu/2)^2]$ are respectively, the saturated emission rate, saturating intensity, and the Lorentzian lineshape function of the transition with FWHM $\Delta\nu = A_{Ff}/2\pi$. Due to random motion of atoms in an ambient temperature, only a portion of atoms (such as those moving perpendicular to the laser beam) can interact with the laser beam on resonance. Under strong illumination, the population of this portion of atoms in the ground state will be partially depleted, leading to the phenomenon of “hole-burning” and spontaneous emission at the saturated rate, along with “power broadening” of the FWHM linewidth by a factor $\sqrt{1+(I/I_s)}$, which in turn increases the share of atoms in the Doppler distribution able to participate in the interaction. For the Na D_2 transition, the Einstein coefficient is $A_{Ff} = 6.15 \times 10^7\ \mathrm{s}^{-1}$; see Table A.2. Equation (3.24) introduces saturation and the spontaneous emission rate deduced from the energy balance rate equations, or density matrix treatment with off-diagonal elements set to zero. For LGS, since the light source is a highly coherent narrowband laser, the off-diagonal elements are no longer zero. For a simple two-level system, instead of the full density matrix treatment, one often uses the optical Bloch equation; see Chapter 2 of [3.14], for example, to handle this coherent interaction, including damping and magnetic field effects. Though we will not discuss the mathematical treatment of the optical Bloch equation here, its results are described conceptually in the subsections below.

3.5.2 Optical Pumping of the NaD_2 Transition

A two-level system not only is the simplest in structure but is an effective system to evaluate the effect of interaction with high-intensity saturating radiation. The earth’s sodium layer is nearly ideal for the creation of an LGS, since the cross section of the NaD_2 transition is very large and the Na layer is situated at a distance sufficiently distant to be starlike (not suffer too much “cone effect”), and close enough for a telescope to receive a large fluorescence photon flux. However, the NaD_2 transition is not quite a two-level system. The question is then how to turn it into a workable two-level system to increase the LGS efficiency. The well-studied process of optical pumping [3.15, 3.16] offers a solution under ideal conditions, thus it is a starting point for our LGS discussion. To proceed, we consider the NaD_{2a} transition (i.e., between $|f, \mu\rangle$ and $|F, m\rangle$ states); its linestrength S_{Ff}, given in (A.8.c), is repeated as (3.25):

$$S_{Ff} = \sum_{m,\mu} |\langle Fm|\vec{p}|f\mu\rangle|^2 = \sum_{m,\mu,q} |\langle Fm|p_q^{\kappa}|f\mu\rangle|^2$$
$$= \left\{ \sum_{m,q,\mu} \begin{pmatrix} f & \kappa & F \\ -\mu & q & m \end{pmatrix}^2 \right\} |\langle F\|p\|f\rangle|^2. \tag{3.25}$$

Written in this form, the linestrength is the sum of all allowed electric dipole transitions between magnetic substates $|f, \mu\rangle$ and $|F, m\rangle$, with each term in the curly brackets representing the associated share (or probability) of the transition. We consider the $|f = 2\rangle$ to $|F = 3\rangle$ D_{2a} transition because it is the only Na D_2 transition that can be optically pumped into a two-level structure by a circularly

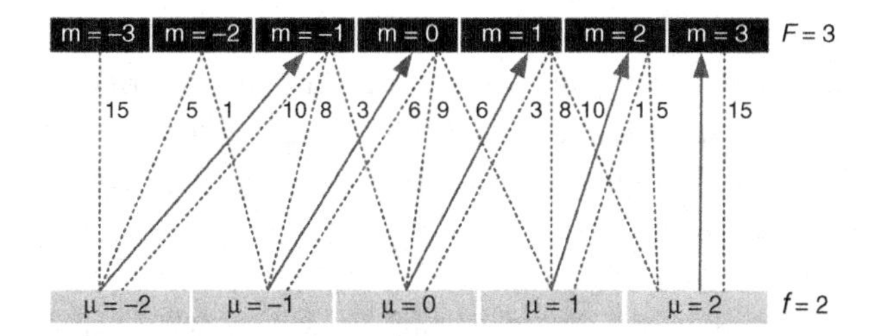

Fig. 3.3 The sharing of the linestrength $|\langle F = 3\|p\|f = 2\rangle|^2$ (short-dash lines), along with the absorption of a left-hand circularly polarized laser beam (solid lines with arrow). The share of each linestrength in question is the number shown divided by 105, each allowed absorption between a pair of nondegenerate magnetic states is connected by an arrow-headed solid line. See further discussions in the text.

polarized beam. In Fig. 3.3, the fractional share of each line to the overall linestrength, $S_{Ff} = |<F\|\, p\, \|f>|^2$, is the number accompanying each dashed line divided by 105, which is calculated from the square of the associated 3-*j* coefficient in (3.25). Also shown explicitly by arrow-headed solid lines are the paths of absorption of a left-hand circularly polarized (LHC, σ_+ or $q = -1$) photon, which promotes the atom from each $|2, \mu>$ to a connected $|3, m>$ state.

Several interesting and important messages are contained in Fig. 3.3:

(1) The sum of all numbers by the short-dashed lines shown is 105, as expected. Similar to the fact that $|\langle p_x\rangle|^2 = |\langle p_y\rangle|^2 = |\langle p_z\rangle|^2 = |\langle p\rangle|^2/3$, the sum of the squared magnitudes of the allowed transitions for each $q = \mu - m$ value, $|\langle p^1_{-1}\rangle|^2 = |\langle p^1_0\rangle|^2 = |\langle p^1_1\rangle|^2 = |\langle p^1\rangle|^2/3$ is also $35 = 105/3$. It is also interesting to note that the share for a particular circularly polarized excitation in question (i.e., $|f = 2,\ \mu = \pm 2>$ to $|F = 3, m = \pm 3>$) is 15/105; comparing this to the mean share of the five linearly polarized (or circularly polarized, for that matter) excitations ($m = \mu$, $(35/5)/105 = 7/105$) suggests that the absorption cross section for $|f = 2> \leftrightarrow |F = 3>$ by a circularly polarized resonant beam is a tad larger than the factor of 2 observed for the linearly polarized beam; see Fig. 2 of [3.17].

(2) With the absorption of an LHC-polarized photon, the atom at the ground state $|2, \mu>$ gains one unit of angular momentum and is promoted to the excited state $|3, \mu + 1>$, as shown by the connecting line with arrow in the figure. Once excited, it can spontaneously emit a photon of different polarization and return to $|2, \mu>$, $|2, \mu + 1>$, or $|2, \mu + 2>$ to comply with the selection rule. As an example, consider $\mu = -1$: the atom starting in the $|2, -1>$ state, after being excited to the $|3, 0>$ state, relaxes to one of $|2, -1>$, $|2, 0>$, or $|2, 1>$, with respective probabilities 6/21, 9/21, or 6/21 (as shown in the connecting dashed lines in Fig. 3.3). The atom has only 6/21 chance of returning to the original ground state and 15/21 chance of relaxing to a state with higher μ value. Similarly, for atoms starting in $|2, 1>$, the chance of

returning to $|2,1>$ is 1/6 with 5/6 chance of relaxing to $|2,2>$. Thus, after absorbing an LHC photon, the chance for the atom to return to its original state is less than that to a ground state with higher μ value. The only exception is the ground state with highest angular momentum, $|2,2\rangle$, which returns to the same state with certainty.

(3) Once in the $|2,2>$ state, the atoms are locked in a two-level system transition between $|2,2>$ and $|3,3>$. This is achieved under continuous illumination by a resonant LHC laser beam, which, after a few Rabi cycles (explained in Section 3.5.3) of absorption and emission, forces the interacting atom into this two-level system. That is, as the result of the "optical pumping" just described, after several cycles of interaction with a circularly polarized laser beam, we have a nondegenerate two-level system with $g_F = g_f = 1$ interacting with a laser beam with intensity I. For a more thorough discussion, see [3.15, 3.16]. For the Na D_2 transition, with I, the resonant saturation intensity for the $|2,2> \leftrightarrow |3,3>$ transition may be calculated from the third equation in (3.24), as $I_{s0} = \pi hcA_{3,3;2,2}\lambda^3 = 26.8\ \mathrm{Wm}^{-2}$ with $A_{3,3;2,2} = A_{32}/7 = 8.79 \times 10^6\ \mathrm{s}^{-1}$. As part of the atmospheric Na layer at an ambient temperature of ~ 200 K, only the atoms with velocity within the natural linewidth of about 10 MHz (roughly 1–2 % of the thermal population) can interact with the resonant laser beam and participate in the steady-state emission. If there were no spin-exchange collisions (explained below), depending on laser intensity, the thermal population of the ground state, $|f = 2>$, would be considerably depleted, resulting in a hole-burned population.

3.5.3 The "Three Devils," Down-Pumping and Repumping of Sodium LGS

As stated, under the continuous illumination of a powerful circularly polarized laser beam at resonance – for example, the single-frequency (15-kHz linewidth) 50 W CW laser [3.18] used for a Na LGS system at the Starfire Optical Range (SOR) – the Na atoms are locked in the two-level system between states $|2,2>$ and $|3,3>$, coherently and dynamically interacting with the CW laser beam described by the two-level optical Bloch equation; see Chapter 2 of [3.14]. Its analytic solution (assuming no damping) is well-known in that the quantum state of the atom is in a coherent mixture of the two states with amplitude oscillating/cycling from one state to the other at the Rabi frequency ω_{Ra}, which is the ratio of the electric dipole interaction energy $\vec{p}\bullet\vec{E}$ divided by $\hbar$, the Planck constant over 2π. Since the electric dipole moment is induced by the impinging electric field of the laser in an isotropic medium, they are parallel to each other (i.e., $\vec{p}\bullet\vec{E} = |p||\mathscr{E}|$, the product of the amplitudes of the dipole moment and laser electric field). The Rabi frequency and its associated period τ_{Ra} are then given as

$$\omega_{Ra} = \frac{\vec{p} \bullet \vec{E}}{\hbar} = \frac{|p||\mathscr{E}|}{\hbar} \to \tau_{Ra} = \frac{h}{|p||\mathscr{E}|}. \tag{3.26.a}$$

To provide an example, we calculate the Rabi period for a two-level system under illumination from a CW single-frequency laser with intensity $I = 50\ \text{Wm}^{-2}$, or $|\mathscr{E}| = 194\ \text{Vm}^{-1}$. The magnitude of the dipole moment for the $|2,2\rangle \leftrightarrow |3,3\rangle$ transition is $|\langle 3,3||p||2,2\rangle|$, which may be estimated as follows. From Fig. A.1, the linestrength of the $|f = 2\rangle \leftrightarrow |F = 3\rangle$ transition, $|\langle F = 3||p||f = 2\rangle|^2$, is 14/6 of the 3$s-3p$ transition linestrength S_0; for Na it is 19 $a.u.$ according to Table A.2 with 1 $a.u.$ = $(ea_0)^2$ and the product of electron charge and Bohr radius $(ea_0) = 8.47{\times}10^{-30}$ C-m. For the two-level system in question, $|2,2\rangle \leftrightarrow |3,3\rangle$, its linestrength, $|\langle 3,3||p||2,2\rangle|^2$ is (1/7 = 15/105) of $|\langle F = 3||p||f = 2\rangle|^2$ $= (14/6)S_0 \approx 44(ea_0)^2$ as shown in Fig. 3.3, leading to $|p| = \sqrt{44/7}(ea_0)$ $= 21.2 \times 10^{-30}$C-m, and the Rabi period $\tau_{Ra} = 0.16\ \mu s$ via (3.26.a).

Of course, the reality is much more complex. There exist many mechanisms in the interacting system that work against optical pumping; these reduce the effectiveness of a Na LGS. Holzlöhner et al. [3.18] called the most important ones the "three devils" of sodium LGS, in order of importance: Larmor precession, radiation pressure, and stimulated emission. We discuss these briefly along with the associated problem of down-pumping and potential solution of repumping below.

It is well known that, when a magnetic moment $\vec{\mu}$ is placed in a magnetic field $\vec{B}$, it experiences a torque $\vec{\mu} \times \vec{B}$, causing it to precess around the magnetic field at the Larmor frequency $\omega_B = \gamma B = eB/2m_e$, where γ, e, and m_e are the gyromagnetic ratio, electron charge, and electron mass. The same is true for a Na atom in a quantum state in Earth's magnetic field: its angular momentum resulting from atomic electrons rotating around the nucleus gives rise to an intrinsic magnetic moment, which precesses around the external magnetic field. The associated magnetic moment, $\vec{\mu}$, magnitude of the torque, Larmor frequency, ω_B, and Larmor period, τ_{La}, are:

$$\vec{\mu} = -g_F \mu_B \vec{F},\ |Torque| = \mu B\ \sin\theta,\ \omega_B = \frac{g_F \mu_B B}{\hbar},\ \tau_{La} = \frac{h}{|g_F|\mu_B B}, \tag{3.26.b}$$

where θ is the angle between the laser beam (nominally vertical) and earth's magnetic field, and $\mu_B = \gamma\hbar$ is the Bohr magneton, $\vec{F}$ the total atomic angular momentum in a hyperfine state in units of $\hbar$, and g_F the Landé g-factor given in Eqs. (3.86) and

(18.24) of [3.5] as $g_F = g_J \frac{F(F+1)+J(J+1)-I(I+1)}{2F(F+1)}$, with $g_J = \frac{3J(J+1)+S(S+1)-L(L+1)}{2J(J+1)}$. The Landé g-factor for $| f = 1>$ and $| f = 2>$ states are respectively –1/2 and 1/2. For Na atoms in these states in a magnetic field of 50 μT, the Larmor period is calculated from (3.26.b) to be 2.8 μs. Since the magnetic torque is proportional to $\sin\theta$, the effect of Larmor precession increases as θ increases; for most observatories, θ is larger than 60° [3.19].

Unlike the electric dipole interaction, Larmor precession imparts no energy to the system. However, it works against optical pumping. Consider a Na atom in $|2, 2>$, the ground state of the two-level system resulting from optical pumping. Larmor precession mixes states with different magnetic quantum numbers of the same angular momentum periodically and returns to the original $|2, 2>$ state at the end of a Larmor period (i.e., 2.8 μs for Na in Earth's magnetic field of 50 μT). In the presence of laser illumination, the atom could be excited to the $|3, 3>$ state, the excited state of the two-level system, from time to time and emit photons spontaneously if the cycling time (0.16 μs for laser intensity of 50 Wm^{-2}) is shorter than the Larmor period. If, however, the cycling time (Rabi period) is much longer, the magnetic quantum number of the ground state will be scrambled, decreasing the effectiveness of optical pumping.

An atom in the $|2, 2>$ ground state may move out of its natural broadened transition linewidth in the presence of radiation pressure. This is because an atom recoils when it spontaneously emits a photon. A backward emitting atom thus moves forward along the laser beam and redshifts the laser frequency by about 50 kHz [3.18, 3.19]. After recycling absorption/emission about 100 times, the atom can no longer be excited by the laser beam (i.e., it has redshifted out of the natural linewidth and left the two-level system), representing a loss of LGS atoms. Thus, under strong illumination, along with high photon flux returns, the population of the $|f, 2>$ ground state is quickly and inevitably reduced from its initial thermal equilibrium value. However, before this happens significantly, the Na atom would have experienced several spin-exchange collisions, particularly with O_2, converting atoms in $|f = 1>$ to the $|f = 2>$ state, thus restoring the thermal equilibrium population [3.20]. The rate of spin-exchange collisions is difficult to measure, resulting in different estimates [3.18, 3.19]; we take it to be around 1/(250 μs). Thus, the ground state of the participating population (within the power-broadened natural linewidth, ~10 – 15 MHz) is continually but slowly replenished via spin-exchange thermalizing collisions.

In general, increasing the laser intensity and reducing the cycling time will increase the photon flux return and the brightness of LGS. However, as laser intensity increases, stimulated emission becomes more significant. Since stimulated emission emits forward into space, it cannot be utilized for a

LGS. Thus, though the photon flux increases as the intensity of the illuminating laser increases, the fraction lost to stimulated emission also increases, contributing to saturation.

Both Rabi oscillations and Larmor precession are damped in the Na layer due to both velocity-changing collisions at a rate about 1/(35 μs) [3.18, 3.19] and spin-exchange collisions, mentioned above. To understand the complicated interplay between the three devils as well as to account for the damping effects to determine quantitative values of the LGS return photon flux, numerical simulations are necessary. For example, a numerical model called Exciter, used in [3.18], has shown that at $I = 46\ \mathrm{Wm}^{-2}$, the simulated output consists of 10.2% stimulated emission and 89.8% spontaneous emission, with one spontaneous emission every 9.5 μs on average. Of these, 70.6% occur from the $|2,2> \leftrightarrow |3,3>$ transition, indicating effective optical pumping. Further increasing laser power has implications. For example, for $I = 100\ \mathrm{Wm}^{-2}$ and $I = 1,000\ \mathrm{Wm}^{-2}$ the time for one spontaneous emission to occur decreases to 5.1 μs and 1.1 μs, and the percentage of stimulated emission increases to 17.9% and 55.2%, respectively. The resulting specific photon rate in $\mathrm{s}^{-1}\mathrm{sr}^{-1}/(\mathrm{Wm}^{-2})$ as a function of laser intensity in Wm^{-2} under different conditions is shown in Fig. 3 of [3.18]. For example, the magenta curve in Fig. 3(a) of [3.18] exhibits typical behavior of saturation, starting from about 230 $\mathrm{s}^{-1}\mathrm{sr}^{-1}/(\mathrm{Wm}^{-2})$ at low intensity and falling steadily in the semilog plot.

In addition to the three devils, the process of down-pumping also reduces LGS efficiency. While near-stationary Na atoms and those moving perpendicular to the laser beam can interact with the resonant laser beam coherently, resulting in Rabi oscillations, atoms moving against the laser beam with speed corresponding to 60 and 34 MHz Doppler-shifts can also interact with the same laser beam via $|f = 2> \leftrightarrow |F = 2>$ and $|f = 2> \leftrightarrow |F = 1>$ transitions, respectively. Once excited to the $|F = 2>$ or $|F = 1>$ state, the atom can emit a photon spontaneously and return to either the $|f = 2>$ or $|f = 1>$ ground state. Those atoms that end up in the $|f = 1>$ state (1.772 GHz below $|f = 2>$) are lost, as they can no longer interact with the laser beam, giving rise to the phenomenon of down-pumping. To minimize the loss of these atoms, in addition to the CW excitation of the D_{2a} line, which we have only considered thus far, we pump the D_{2b} line (termed repumping) either with another CW laser with 10–20% power or by widening the laser line to ≥ 2 GHz. The model in [3.18] showed that “a narrow-band laser with 12% repumping beats the return flux of a single-line 2 GHz-linewidth laser by a factor of 3.7 under standard conditions.” Another method to improve LGS brightness is to switch the exciting CW laser beam between left and right circular polarization at the Larmor frequency, as proposed in 2015 [3.19]. Their model simulation shows that “the return flux is increased when the angle between geomagnetic field and laser beam is larger than 60°, as much as 50% at 90°.”

As can be appreciated, the physics of LGS is quite involved. Within the scope of this book, we can only present the prerequisite physical concepts and some associated simple calculations. To gain a conceptual understanding, we recommend the technical report (PDF) published online as part of the 2009 Advanced Maui Optical and Space Surveillance Technologies Conference [3.20]. To understand much of the intricate interplay between Rabi oscillation and Larmor precession, radiation pressure, damping, and collisions (both velocity-changing and spin-exchange), as well as stimulated emission, down-pumping, and repumping, numerical analysis is necessary. For these as well as more substantive discussions, we recommend [3.18].

References

3.1 Wang, D.-W., A.-J. Li, L.-G. Wang, S.-Y. Zhu, and M.-S. Zubairy. (2004). Effect of counterpropagating terms on polarizability in atom-field interactions, *Phys. Rev. A*, **80** (6), 063826.

3.2 Loudon, R. and S. M. Barnett. (2006). Theory of linear polarizability of a two-level atom. *J. Phys. B*, **39**(15), S555–S563.

3.3 She, C.-Y., H. Chen, and D. A. Krueger. (2015). Optical processes for middle atmospheric Doppler lidars: Cabannes scattering and laser induced resonance fluorescence. *J. Opt. Soc. Am.*, **B32**(9), 1575–1592, and Erratum, ibid., p. 1954.

3.4 Edmonds, A. R. (1957). *Angular Momentum in Quantum Mechanics*. Princeton University Press.

3.5 Corney, A. (1977). *Atomic and Laser Spectroscopy.* Oxford University Press.

3.6 She, C.-Y. (2005). On atmospheric lidar performance comparison: from power-aperture product to power-aperture-mixing ratio scattering cross section product. *J. Mod. Opt.*, **52**(18), 2723–2729.

3.7 Fricke, K. H. and U. von Zahn. (1985). Mesopause temperatures derived from probing the hyperfine structure of the D_2 resonance line of sodium by lidar. *J. Atmos. Terr. Phys.*, **47**, 499–512.

3.8 Griffiths, D. J. (1981). *Introduction to Electrodynamics*. 2nd ed. Prentice Hall.

3.9 Krueger, D. A., C.-Y. She, and T. Yuan. (2015). Retrieving mesopause temperature and line-of-sight wind from full-diurnal-cycle Na lidar observations. *Appl. Opt.*, **54** (32), 9469–9489.

3.10 https://physics.nist.gov/PhysRefData/ASD/lines_form.html: Enter Fe I 365 nm to 395 nm.

3.11 She, C. Y. (2019). The Hanle effect in laser-induced fluorescence and Na and Fe resonance scattering lidars. *Appl. Opt.*, **58**(31), 8354–8361.

3.12 She, C. Y., K. Billman, and W. M. Fairbank, Jr. (1978). Measuring the velocity of individual atoms in real-time. *Opt. Lett.*, **2**(2), 30–32.

3.13 She, C. Y. and J. R. Yu. (1995). Doppler-free saturation fluorescence spectroscopy of Na atoms for atmospheric applications. *Appl. Opt.*, **34**(6), 1063–1075.

3.14 Loudon, R. (1983). *The Quantum Theory of Light*. Oxford.

3.15 McClelland, J. J. and M. H. Kelley. (1985). Detailed look at aspects of optical pumping in sodium. *Physical Rev. A*, **31**(6), 3704–3710.

3.16 Ungar, P. J., D. S. Weiss, E. Riis, and S. Chu. (1989). Optical molasses and multilevel atoms: theory. *J. Opt. Soc. Am. B*, **6**(11), 2058.

3.17 Jeys, T. H., R. M. Heinrichs, K. F Wall, J. Korn, and T. C. Hotaling. (1992). Observation of optical pumping of mesospheric sodium. *Opt. Letters*, **17**(16), 1143–1147.

3.18 Holzlöhner, R. et al. (2010). Optimization of CW sodium laser guide star efficiency. *Astron. Astrophys.*, **510**, A20. doi: https://doi.org/10.1051/0004-6361/200913108.

3.19 Fan, T., T. Zhou, and Y. Feng. (2016). Improving sodium laser guide star brightness by polarization switching. *Nature-Scientific Reports*, **6**(1): 19859. doi: https://doi.org/10.1038/srep19859.

3.20 Kibblewhite, E. (2009). The physics of the sodium laser guide star: Predicting and enhancing the photon returns of sodium guide stars for different laser technologies. *Proceedings of the Advanced Maui Optical and Space Surveillance Technologies Conference*, Wailea, Maui, HI, ed. S. Ryan. https://amostech.com/2009-technical-papers.

4

Rayleigh and Raman Scattering from Linear Molecules

We now entertain the question of light scattering from an ensemble of linear molecules. The semiclassical (quantum) treatment and its formalism are the same as that for coherently excited light scattering from an ensemble of atoms presented in Chapter 3, with two complications and one simplification. First, though under most circumstances linear molecules are randomly orientated, the medium is not exactly isotropic; there may exist an induced dipole moment that is perpendicular to the applied electric field. Second, in addition to the motion of electrons around the nuclei, there exists vibrational motion along the symmetry axis of the molecule (called the figure axis) and rotation about its center of mass. In other words, associated with each electronic state there exist a number of vibrational and rotational energy levels, with magnitudes $E_{ele} \gg E_{vib} \gg E_{rot}$, where E_{ele} is the electronic, E_{vib} the vibrational, and E_{rot} the rotational energies. The one simplification is that we shall limit ourselves to nonresonant light scattering only. This is because for most lidar applications, the energy difference between any molecular electronic transition and a laser photon is much larger than the width of the associated electronic transition in the molecule. In other words, we will be dealing with Rayleigh and ordinary vibrational and rotational Raman scattering from an ensemble of linear molecules. Not only do we exclude ourselves from resonant Raman scattering, but we also limit ourselves to those scattering processes having the ground electronic state of the molecule as both initial and final electronic states.

Following the developments in Chapters 2 and 3, we may start here by generalizing (2.1.b) and (2.2.b) to an anisotropic medium, and express the classical differential scattering light power and the associated cross section in the $\hat{e}'$-polarization of a radiating dipole moment $\vec{p}(t) = 0.5\left(\vec{\mathcal{P}}(\omega)e^{-i\omega t} + \vec{\mathcal{P}}^*(\omega)e^{i\omega t}\right)$ induced by an $\hat{e}$-polarized electric field of incident light $\vec{E}(t) = 0.5(\mathcal{E}\hat{e}\, e^{-i\omega t} + \mathcal{E}^*\hat{e}^* e^{i\omega t})$, respectively, as:

$$\frac{dP}{d\Omega} = \frac{\omega^4}{32\pi^2 \mathrm{e}_0 c^3}\left|\vec{p} \bullet \hat{e}'\right|^2; \quad \frac{d\sigma}{d\Omega} = \frac{\omega^4\left|\overline{\overline{\alpha_{e'e}(\omega)}}\right|^2}{16\pi^2 \mathrm{e}_0^2 c^4}, \tag{4.1}$$

where the complex polarizability tensor $\alpha_{e'e}(\omega)$ in (4.1) is a generalization of $\alpha(\omega)$ in (2.2.b). To apply (4.1) to the ensemble of linear molecules in question, an isotropic average over all possible molecular orientations should be taken, as discussed below; this is denoted by a double bar in the term $\overline{\overline{\alpha_{e'e}(\omega)}}$. For the (isotropic) medium of an atomic ensemble, this polarizability is a scalar (i.e., $\overline{\overline{\alpha_{e'e}(\omega)}} = \alpha(\omega)(\hat{e} \bullet \hat{e}')$) and (4.1) reduces to (2.2.b).

In the semiclassical (quantum) treatment, we need to evaluate the isotropic average of the expectation value of the induced dipole moment operator and the associated differential scattering power and cross section. The expectation value (denoted by angle brackets) of the induced dipole moment and its components may be expressed in terms of the associated polarizability tensor elements as

$$\begin{aligned}\langle \vec{p}\,(t)\rangle &= \frac{1}{2}\left(\langle \vec{p}\rangle e^{-i\omega t} + \langle \vec{p}^{\dagger}\rangle e^{i\omega t}\right) = \frac{1}{2}\left(\vec{\mathcal{P}}(\omega)e^{-i\omega t} + \vec{\mathcal{P}}^{*}(\omega)e^{i\omega t}\right)\\ \rightarrow \langle \vec{p}(t)\rangle_{\hat{e}'} &= \frac{1}{2}\left(\alpha_{e'e}(\omega)\mathcal{E}\, e^{-i\omega t} + \alpha^{*}_{e'e}(\omega)\mathcal{E}^{*} e^{i\omega t}\right), \text{ with } \hat{e}' = \hat{e_a}' \text{ or } \hat{e_b}'.\end{aligned} \tag{4.2.a}$$

Here, the polarizability tensor $\alpha_{e'e}(\omega)$ – see (4.2.b) below – is a generalization of $\alpha(\omega)$ in (3.3), which may be derived similarly. It is consistent with the rotating and counterrotating terms depicted in Fig. 3.1(a) and in agreement with the expression on p. 17 of Shen [4.1], or (4.2.11) of Long [4.2]:

$$\begin{aligned}\alpha_{\rho\sigma}(\omega) = \frac{1}{\hbar}\sum_{m,\mu}\Bigg[&\frac{<f\mu|(\vec{p} \bullet \hat{\sigma})|Fm><Fm|(\vec{p} \bullet \hat{\rho})|f\mu>}{(\omega_{Ff} - \omega) - i\Gamma_{Ff}}\\ &+\frac{<Fm|(\vec{p} \bullet \hat{\rho})|f\mu><f\mu|(\vec{p} \bullet \hat{\sigma})|Fm>}{(\omega + \omega_{Ff}) + i\Gamma_{Ff}}\Bigg]\frac{1}{g_f}\end{aligned} \tag{4.2.b}$$

where the matrix elements, $\rho, \sigma = x, y, z$ or X, Y, Z, respectively, denote the space-fixed or rotating (molecular) Cartesian coordinates. For the scattering problems under consideration, we are interested in $\hat{e}'$-polarized scattering induced by an $\hat{e}$-polarized field (i.e., $(\rho, \sigma) \rightarrow (e', e)$). Note by inspection, except for the sign in front of $i\Gamma_{Ff}$, consistent with damping, $\alpha_{\rho\sigma}(\omega) = \alpha_{\sigma\rho}^{*}(\omega)$. Thus, when damping is ignored or negligible, as appropriate for nonresonant scattering, the polarizability tensor (second-rank) matrix $\overleftrightarrow{\alpha}$ is Hermitian. With appropriate choice of the phase of quantum states, the polarizability tensor in real Cartesian coordinates is real and symmetric (i.e., $\alpha_{\rho\sigma}(\omega) = \alpha_{\sigma\rho}(\omega)$).

To apply the above expression to the ensemble of linear molecules of interest here, we need to include the vibrational and rotational manifold of the ground state explicitly and compute its isotropic average, $\overline{\overline{\alpha_{e'e}(\omega)}}$. Once this is done, for the nonresonant scattering considered in this chapter, we can simply substitute $\overline{\overline{\alpha_{e'e}(\omega)}}$ into (4.1) and compute the differential cross section in question. Recall that $\alpha_{e'e}(\omega)$ was computed from the expectation value of the induced dipole moment via the first-order perturbation density matrix, $\rho^{(1)}(t)$, oscillating at the laser frequency, ω, as shown in Chapter 3. The calculated induced dipole moment in this case is stationary and is fully coherent, as in the classical case. Since the difference between incident laser frequency, ω, and the effective transition frequency, ω_{Ff}, is much larger than its energy damping rate, Γ_{Ff}, there can be no decoherence effect among the phases of the degenerate excited substates. The quantity $\left|\vec{p} \bullet \hat{e}'\right|^2$ is then calculated as $\left|\mathrm{Tr}\left[\rho^{(1)}\left(\vec{p} \bullet \hat{e}'\right)\right]\right|^2$. In contrast, for the resonant case $|\omega - \omega_{Ff}| \leq \Gamma_{Ff}$, the phases of the coherently excited degenerate substates will be randomized in time, unless the coherent excitation corresponds to the simple s-to-p transition or is nondegenerate, such as in a classical system, see Section 3.3.2. To handle the decoherence properly, the quantity $\left|\vec{p} \bullet \hat{e}'\right|^2$ should be calculated by $\mathrm{Tr}\left[\rho^{(2)}(\vec{p} \bullet \hat{e})(\vec{p} \bullet \hat{e}'^{*})\right]$; see (3.8), where $\rho^{(2)}$ is quasi-d.c. or oscillating at radio frequency. For the topics of interest to this chapter, we need not worry about decoherence, only evaluate $\mathrm{Tr}\left[\rho^{(1)}(\vec{p} \bullet \hat{e}')\right]$ for an ensemble of atmospheric molecules.

4.1 Formulation and Evaluation of Polarizability Tensor

We note that the polarizability is a material property that relates an applied electric field, a vector, to the induced dipole moment, another vector, and as such it is then a second-rank tensor $\overleftrightarrow{\alpha}$. The result of the quantum calculation (4.2.b) shows the matrix to be Hermitian in complex spherical coordinates; it is real and symmetric in real Cartesian coordinates. Here, we consider more generally a tensor ${}^{C}\overleftrightarrow{\alpha}$ of nine elements, α_{ij}, arranged in a 3×3 matrix, constructed by the tensor (direct) product of two real Cartesian vectors in space-fixed Cartesian coordinates with unit vectors $(\hat{x}, \hat{y}, \hat{z})$, as shown in (4.3.a). By inspection, it may be decomposed into a sum of three matrices, as in (4.3.b):

$$
{}^{C}\overleftrightarrow{\alpha} = \begin{bmatrix} \alpha_{xx} & \alpha_{xy} & \alpha_{xz} \\ \alpha_{yx} & \alpha_{yy} & \alpha_{yz} \\ \alpha_{zx} & \alpha_{zy} & \alpha_{zz} \end{bmatrix} = {}^{C}\overleftrightarrow{\alpha}^{(0)} + {}^{C}\overleftrightarrow{\alpha}^{(1)} + {}^{C}\overleftrightarrow{\alpha}^{(2)},
$$

$$
\text{with Trace}\left[{}^{C}\overleftrightarrow{\alpha}\right] = \alpha_{xx} + \alpha_{yy} + \alpha_{zz} \equiv 3a, \tag{4.3.a}
$$

$$
{}^{C}\overleftrightarrow{\alpha}^{(0)} = \begin{bmatrix} a & 0 & 0 \\ 0 & a & 0 \\ 0 & 0 & a \end{bmatrix}; \quad {}^{C}\overleftrightarrow{\alpha}^{(1)} = \begin{bmatrix} 0 & \frac{1}{2}\left(\alpha_{xy} - \alpha_{yx}\right) & \frac{1}{2}\left(\alpha_{xz} - \alpha_{zx}\right) \\ \frac{1}{2}\left(\alpha_{yx} - \alpha_{xy}\right) & 0 & \frac{1}{2}\left(\alpha_{yz} - \alpha_{zy}\right) \\ \frac{1}{2}\left(\alpha_{zx} - \alpha_{xz}\right) & \frac{1}{2}\left(\alpha_{zy} - \alpha_{yz}\right) & 0 \end{bmatrix};
$$

$$
{}^{C}\overleftrightarrow{\alpha}^{(2)} = \begin{bmatrix} \alpha_{xx} - a & \frac{1}{2}\left(\alpha_{xy} + \alpha_{yx}\right) & \frac{1}{2}\left(\alpha_{xz} + \alpha_{zx}\right) \\ \frac{1}{2}\left(\alpha_{yx} + \alpha_{xy}\right) & \alpha_{yy} - a & \frac{1}{2}\left(\alpha_{yz} - \alpha_{zy}\right) \\ \frac{1}{2}\left(\alpha_{zx} + \alpha_{xz}\right) & \frac{1}{2}\left(\alpha_{zy} + \alpha_{yz}\right) & \alpha_{zz} - a \end{bmatrix}. \tag{4.3.b}
$$

We note that there is only one independent element in ${}^{C}\overleftrightarrow{\alpha}^{(0)}$, and the trace of this matrix is $3a$. There are three independent elements in ${}^{C}\overleftrightarrow{\alpha}^{(1)}$, and $\mathrm{Tr}\left[{}^{C}\overleftrightarrow{\alpha}^{(1)}\right] = 0$. Finally, there are five independent elements in ${}^{C}\overleftrightarrow{\alpha}^{(2)}$, and its trace is also 0. This means that the second-rank Cartesian tensor ${}^{C}\overleftrightarrow{\alpha}$, with nine elements, is reducible, which, with an appropriate transformation, may be transformed into three independent (or irreducible) subspaces, respectively, with one, three, and five elements. This is similar to decomposition of quantum mechanical angular momentum, in which the sum of two $j = 1$ momenta yield three momenta with $j = 0$, 1, and 2, each in an irreducible subspace. Symbolically, we could describe this decomposition of the tensor product as $1 \otimes 1 = 0 \oplus 1 \oplus 2$, where $\otimes$ and $\oplus$ respectively refer to the tensor product and the sum of irreducible tensors in respective subspaces. Following the notations of quantum mechanics, we write these irreducible tensors as ${}^{IR}\overleftrightarrow{\alpha}^{(j)}$ with $j = 0$, 1, and 2 denoting irreducible tensors of rank 0, 1, and 2, each with elements denoted by integer m, having values 0; –1, 0, 1; and –2, –1, 0, 1, 2.

4.1.1 Transformation to Irreducible Representation and Placzek Invariants

The mathematics for the transformation from the Cartesian second-rank tensor ${}^{C}\overleftrightarrow{\alpha}$ to its irreducible components ${}^{IR}\overleftrightarrow{\alpha}^{(j)}$ is tedious, but it has been worked out in Appendix 14 of Long [4.2], among others. Conceptually, we first transform ${}^{C}\overleftrightarrow{\alpha}$ into

$^{S}\overleftrightarrow{\alpha}$ by going from the Cartesian coordinates to the well-known (complex) spherical coordinates with the unit vectors $(\hat{\varepsilon}_1, \hat{\varepsilon}_0, \hat{\varepsilon}_{-1})$; the result is given in (A14.3.16) of [4.2]. Since the nine complex spherical tensor elements $\alpha_{\lambda,-\mu}$ with $\lambda, \mu = -1, 0, 1$ transform as the product of two angular momentum states $|1, \lambda>$ and $|1, -\mu>$ under rotation, the procedure for converting combined quantum states into an allowed irreducible set of states under rotation may be used to reduce the tensor elements into the associated irreducible tensors. With the help of Clebsch–Gordan or Wigner 3-j coefficients, this transformation relating $^{S}\overleftrightarrow{\alpha}$ and $^{IR}\overleftrightarrow{\alpha}^{(j)}$ is given in (A19.2.23), p. 542 of [4.2] and grouped in j = 0, 1, and 2 irreducible subspaces in Table A14.2, p. 484 of [4.2], in which the elements (or states) with different j-values are orthogonal to one another. Here, the 0th rank tensor and its only element are seen from the combination of $\alpha_{\lambda,-\mu}$ as $\alpha_0^{(0)} = -(\alpha_{1-1} + \alpha_{00} + \alpha_{-11})/\sqrt{3}$, leading to $\left|\alpha_0^{(0)}\right|^2 = \frac{1}{3}\left[\mathrm{Tr}\left(^{S}\overleftrightarrow{\alpha}\right)\right]^2$.

To summarize below for comparison, we write down the three representations of the second-rank tensor, $^{C}\overleftrightarrow{\alpha}$, $^{S}\overleftrightarrow{\alpha}$, and $^{IR}\overleftrightarrow{\alpha}^{(j)}$ by representing the nine elements of the first two as matrices, while we write out the nine elements for the irreducible tensor components in a row as

$$^{C}\overleftrightarrow{\alpha} = \begin{bmatrix} \alpha_{xx} & \alpha_{xy} & \alpha_{xz} \\ \alpha_{yx} & \alpha_{yy} & \alpha_{yz} \\ \alpha_{zx} & \alpha_{zy} & \alpha_{zz} \end{bmatrix} \to {}^{S}\overleftrightarrow{\alpha} = \begin{bmatrix} \alpha_{1-1} & \alpha_{10} & \alpha_{11} \\ \alpha_{0-1} & \alpha_{00} & \alpha_{01} \\ \alpha_{-1-1} & \alpha_{-10} & \alpha_{-11} \end{bmatrix} \to$$

$$^{IR}\overleftrightarrow{\alpha} = \left[\alpha_0^{(0)}; \alpha_{-1}^{(1)}, \alpha_0^{(1)}, \alpha_1^{(1)}; \alpha_{-2}^{(2)}, \alpha_{-1}^{(2)}, \alpha_0^{(2)}, \alpha_1^{(2)}, \alpha_2^{(2)}\right].$$

We already notice that the magnitude square of the 0th-rank irreducible tensor is an invariant, because the trace of a 3 × 3 matrix is independent of its representations. Placzek in 1934 recognized that this is true generally for an irreducible subspace of any rank: that is, the sum of the square of all elements in an irreducible subspace, like the square of the magnitude of a vector, should be invariant under rotation. These Placzek invariants [4.3], one for each irreducible tensor, are then

$$\mathscr{G}^{(j)} = \sum_{m=-j}^{m=j} \left|\alpha_m^{(j)}\right|^2, \text{ with } j = 0,\ 1,\ \text{and } 2. \tag{4.4}$$

The mathematical proof for the invariance of $\mathscr{G}^{(j)}$ can be found in (A14.7.2) to (A14.7.5) of [4.2], and it will not be repeated here. Using the transformation coefficients given in Table A14.3 of [4.2], one can relate $\alpha_m^{(j)}$ in the irreducible representation to α_{ij} in the Cartesian representation, and with that information evaluate the Placzek invariants $\mathscr{G}^{(j)}$ with j = 0, 1, and 2 in terms of Cartesian tensor elements in (4.4.a) to (4.4.c) below – as given in (A14.7.19) to (A14.7.21) in [4.2]:

$$\mathscr{G}^{(0)} = \frac{1}{3}\left\{\left|\alpha_{xx} + \alpha_{yy} + \alpha_{zz}\right|^2\right\}, \tag{4.4.a}$$

$$\mathscr{G}^{(1)} = \frac{1}{2}\left\{\left|\alpha_{xy} - \alpha_{yx}\right|^2 + \left|\alpha_{yz} - \alpha_{zy}\right|^2 + \left|\alpha_{zx} - \alpha_{xz}\right|^2\right\}, \text{ and} \tag{4.4.b}$$

$$\mathscr{G}^{(2)} = \frac{1}{2}\left\{\left|\alpha_{xy} + \alpha_{yx}\right|^2 + \left|\alpha_{yz} + \alpha_{zy}\right|^2 + \left|\alpha_{zx} + \alpha_{xz}\right|^2\right\}$$
$$+\frac{1}{3}\left\{\left|\alpha_{xx} - \alpha_{yy}\right|^2 + \left|\alpha_{yy} - \alpha_{zz}\right|^2 + \left|\alpha_{zz} - \alpha_{xx}\right|^2\right\}. \tag{4.4.c}$$

Again, $\mathscr{G}^{(0)}$ in (4.4.a) is as expected $\frac{1}{3}\left[\mathrm{Trace}\left({}^{C}\overleftrightarrow{\alpha}\right)\right]^2$. For the polarizability tensor in question, which is real and symmetric (i.e., $\mathscr{G}^{(1)} = 0$), thus only $\mathscr{G}^{(0)}$ and $\mathscr{G}^{(2)}$ are relevant to us.

4.1.2 Transformation to Principal Axes and Isotropic Averages of a Linear Molecule

Since the polarizability tensor of interest is a real and symmetric tensor, its six matrix elements can be used to depict a quadric (surface). In this case, there exists a geometrical interpretation, and the tensor may be represented by an ellipsoid. As such, one can rotate the fixed coordinates (x, y, z) to the principal axes (X, Y, Z) via appropriate Euler angles (α, β, γ), reducing the number of nonzero matrix elements to three; see Fig. A14.1 in [4.2], for example. The equations of the quadric (ellipsoidal surface) in the two different representations are respectively

In (x, y, z): $\alpha_{xx}x^2 + \alpha_{yy}y^2 + \alpha_{zz}z^2 + 2\alpha_{yz}yz + 2\alpha_{zx}zx + 2\alpha_{xy}xy = 1$, and

In (X, Y, Z): $\alpha_{XX}X^2 + \alpha_{YY}Y^2 + \alpha_{ZZ}Z^2 = 1$. (4.5.a)

We say the matrix $\overleftrightarrow{\alpha}$ is diagonalized and transformed into $\overleftrightarrow{A}$ by the rotation matrix $D(\alpha, \beta, \gamma)$ given in (B.5.b), that is,

$$\begin{bmatrix} p_x \\ p_y \\ p_z \end{bmatrix} = \begin{bmatrix} \alpha_{xx} & \alpha_{xy} & \alpha_{xz} \\ \alpha_{yx} & \alpha_{yy} & \alpha_{yz} \\ \alpha_{zx} & \alpha_{zy} & \alpha_{zz} \end{bmatrix} \begin{bmatrix} E_x \\ E_y \\ E_z \end{bmatrix} \xrightarrow{D(\alpha,\beta,\gamma)} \begin{bmatrix} p_X \\ p_Y \\ p_Z \end{bmatrix} = \begin{bmatrix} \alpha_{XX} & 0 & 0 \\ 0 & \alpha_{YY} & 0 \\ 0 & 0 & \alpha_{ZZ} \end{bmatrix} \begin{bmatrix} E_X \\ E_Y \\ E_Z \end{bmatrix}, \text{ with}$$

$$\overleftrightarrow{\alpha} = \begin{bmatrix} \alpha_{xx} & \alpha_{xy} & \alpha_{xz} \\ \alpha_{yx} & \alpha_{yy} & \alpha_{yz} \\ \alpha_{zx} & \alpha_{zy} & \alpha_{zz} \end{bmatrix}, \overleftrightarrow{A} = \begin{bmatrix} \alpha_{XX} & 0 & 0 \\ 0 & \alpha_{YY} & 0 \\ 0 & 0 & \alpha_{ZZ} \end{bmatrix} \text{ and } \overleftrightarrow{\alpha} = D^T(\alpha, \beta, \gamma)\, \overleftrightarrow{A}\, D(\alpha, \beta, \gamma).$$

(4.5.b)

Here, $D^T(\alpha,\beta,\gamma)$ is the transpose of $D(\alpha,\beta,\gamma)$ given in (B.5.b) in matrix form. Though tedious, we can work out the matrix elements of $\overleftrightarrow{\alpha}$ in terms of $(\alpha_{XX},\alpha_{YY},\alpha_{ZZ})$ and deduce the properties of a 2nd-rank tensor with geometrical interpretation. Without going into details (an exercise for the reader), one can show $\alpha_{xy}=\alpha_{yx}$, $\alpha_{yz}=\alpha_{zy}$, $\alpha_{zx}=\alpha_{xz}$ (i.e., the matrix is symmetric independent of representation), and that the Placzek invariants are indeed valid, giving

$$\begin{aligned}\mathscr{G}^{(0)} &= \frac{1}{3}\left\{\left|\alpha_{XX}+\alpha_{YY}+\alpha_{ZZ}\right|^2\right\},\ \mathscr{G}^{(1)}=0,\\ \mathscr{G}^{(2)} &= \frac{1}{3}\left\{\left|\alpha_{XX}-\alpha_{YY}\right|^2+\left|\alpha_{YY}-\alpha_{ZZ}\right|^2+\left|\alpha_{ZZ}-\alpha_{XX}\right|^2\right\}\end{aligned} \tag{4.6}$$

If the tensor is expressed in its irreducible representation, then it will transform under rotation as a standard angular momentum basis ket; that is, an element with a given j-value in one coordinate system will transform into a superposition of elements (substates) in another coordinate system with the same j-value. Thus, the space-fixed irreducible element $\alpha_m^{(j)}$ is related to the molecule-fixed irreducible element $\alpha_{m'}^{(j)}$ as

$$\alpha_m^{(j)} = [D^T(\alpha,\beta,\gamma)\overleftrightarrow{A}D(\alpha,\beta,\gamma)]_m^{(j)} = \sum_{m'=-j}^{m'=j}\alpha_{m'}^{(j)}D_{m'm}^{(j)}, \tag{4.7}$$

where $D_{m'\,m}^{(j)}$ is a shorthand of the irreducible matrix element of finite rotation, $D_{m'm}^{(j)}(\alpha,\beta,\gamma) = \langle jm'|D(\alpha,\beta,\gamma)|jm\rangle$. Equation (4.7) is the same as (6.2.5) of [4.2], and (5.2.1) of [4.4].

To evaluate a physical quantity of interest relating to light scattering from a molecular ensemble, we need to take an isotropic average of that quantity. This is because individual molecules (and the associated molecular system) in a molecular ensemble are randomly orientated with respect to the fixed coordinate system where the light scattering experiment is monitored. Since a physical quantity of interest $F(\alpha,\beta,\gamma)$ is described by its molecular coordinates (principal axes), which are related to the fixed coordinates via Euler angle rotations, its isotropic average, denoted by double over bars $\overline{\overline{F(\alpha,\beta,\gamma)}}$, amounts to the evaluation of the following integral:

$$\overline{\overline{F(\alpha,\beta,\gamma)}} = \frac{1}{8\pi^2}\int_0^{2\pi}\int_0^{\pi}\int_0^{2\pi}F(\alpha,\beta,\gamma)d\alpha(\sin\beta)d\beta d\gamma. \tag{4.8.a}$$

As a simple but useful example, we calculate the isotropic averages of products of selected Cartesian polarizability tensor elements, since the light scattering cross section is proportional to them. Instead of expressing the quantity of interest in

Euler angles using formulae invoking transformation functions $D^{(j)}_{m'm}(\alpha, \beta, \gamma)$, as in the more general treatment in terms of Placzek invariants given in (A14.6.1) of [4.2], resulting in Table A14.9 of [4.2], here we calculate the isotropic averages for an ensemble of linear molecules by direct integration and present selected results in Table 4.1. To this end, we choose the figure axis of a linear molecule as the Z-axis and note that the transformation from the fixed to molecular coordinates can be achieved with two rotations by substituting $(\alpha = \Phi,\ \beta = \Theta,\ \gamma = 0\ )$ into (B.5.b). Asserting cylindrical symmetry of a linear molecule, we substitute $\alpha_{ZZ} = \alpha_{\parallel}$ and $\alpha_{XX} = \alpha_{YY} = \alpha_{\perp}$ into $\overleftrightarrow{A}$ and can evaluate $\overleftrightarrow{\alpha}$ from $D^T(\Phi, \Theta, 0)\, \overleftrightarrow{A} D(\Phi, \Theta, 0)$, obtaining:

$$\overleftrightarrow{\alpha} = \begin{bmatrix} \alpha_{\perp} + (\alpha_{\parallel} - \alpha_{\perp})\cos^2\Phi \sin^2\Theta & (\alpha_{\parallel} - \alpha_{\perp})\sin\Phi \cos\Phi \sin^2\Theta & (\alpha_{\parallel} - \alpha_{\perp})\sin\Theta \cos\Theta \cos\Phi \\ (\alpha_{\parallel} - \alpha_{\perp})\sin\Phi \cos\Phi \sin^2\Theta & \alpha_{\perp} + (\alpha_{\parallel} - \alpha_{\perp})\sin^2\Phi \sin^2\Theta & (\alpha_{\parallel} - \alpha_{\perp})\sin\Phi \sin\Theta \cos\Theta \\ (\alpha_{\parallel} - \alpha_{\perp})\sin\Theta \cos\Theta \cos\Phi & (\alpha_{\parallel} - \alpha_{\perp})\sin\Phi \sin\Theta \cos\Theta & \alpha_{\perp}\sin^2\Theta + \alpha_{\parallel}\cos^2\Theta \end{bmatrix}, \tag{4.8.b}$$

where the matrix is symmetric, as expected. Substituting $\alpha_{ZZ} = \alpha_{\parallel}$ and $\alpha_{XX} = \alpha_{YY} = \alpha_{\perp}$ into (4.6) and defining the invariant mean $a = (\alpha_{\parallel} + 2\alpha_{\perp})/3$ and anisotropy $\gamma = |\alpha_{\parallel} - \alpha_{\perp}|$, we obtain $\mathscr{G}^{(0)} = 3a^2$; $\mathscr{G}^{(1)} = 0$; $\mathscr{G}^{(2)} = 2\gamma^2/3$, providing a physical (geometrical) interpretation of the Placzek invariant. Here, γ is the anisotropy of the ellipsoid, not to be confused with the third Euler angle γ in (4.8.a). Using (4.8.b), we express the isotropic averages of products of selected polarizability tensor elements of the molecular ensemble in question as

$$\begin{aligned} \overline{\overline{\alpha_{xx}{}^2}} &= \alpha_{\perp}{}^2 + 2\alpha_{\perp}(\alpha_{\parallel} - \alpha_{\perp})\overline{\overline{\cos^2\Phi \sin^2\Theta}} + (\alpha_{\parallel} - \alpha_{\perp})^2\,\overline{\overline{\cos^4\Phi \sin^4\Theta}} = a^2 + \frac{4}{45}\gamma^2, \\ \overline{\overline{\alpha_{xy}{}^2}} &= (\alpha_{\parallel} - \alpha_{\perp})^2\,\overline{\overline{\sin^2\Phi \cos^2\Phi \sin^4\Theta}} = \frac{1}{15}\gamma^2, \text{ and} \\ \overline{\overline{\alpha_{xx}\alpha_{yy}}} &= \alpha_{\perp}{}^2 + \alpha_{\perp}(\alpha_{\parallel} - \alpha_{\perp})\overline{\overline{\sin^2\Theta}} + (\alpha_{\parallel} - \alpha_{\perp})^2\,\overline{\overline{\sin^2\Phi \cos^2\Phi \sin^4\Theta}} = a^2 - \frac{2}{15}\gamma^2, \end{aligned} \tag{4.8.c}$$

where $\overline{\overline{xxx}}$ indicates isotropic average, i.e., $\overline{\overline{F(\Phi, \Theta)}} = \frac{1}{4\pi}\int_0^{2\pi}\int_0^{\pi} F(\Phi, \Theta)\sin\Theta d\Theta d\Phi$,

$$\tag{4.8.d}$$

and the mean $a = (\alpha_{\parallel} + 2\alpha_{\perp})/3$ and anisotropy $\gamma = |\alpha_{\parallel} - \alpha_{\perp}|$ are obviously invariants. The isotropic averages of relevant trigonometry functions may be calculated with (4.8.d), from which various products (and squares) of tensor elements may be evaluated as given in Table 4.1.

Table 4.1 *Isotropic averages of trigonometry functions and polarizability tensor products*

Isotropic averages of $F(\Phi, \Theta)$	Isotropic averages of $\alpha_{\rho\sigma}\alpha_{\rho'\sigma'}$
$\overline{\overline{\sin^2\Theta}} = \frac{2}{3}$; $\overline{\overline{\cos^2\Theta}} = \frac{1}{3}$;	$\overline{\overline{\alpha_{xx}}}^2 = \overline{\overline{\alpha_{yy}}}^2 = \overline{\overline{\alpha_{zz}}}^2 = a^2 + \frac{4}{45}\gamma^2$;
$\overline{\overline{\sin^2\Phi \sin^2\Theta}} = \overline{\overline{\cos^2\Phi \sin^2\Theta}} = \frac{1}{3}$;	$\overline{\overline{\alpha_{xy}}}^2 = \overline{\overline{\alpha_{yz}}}^2 = \overline{\overline{\alpha_{zx}}}^2 = \frac{1}{15}\gamma^2$;
$\overline{\overline{\cos^2\Phi \sin^2\Theta}} = \overline{\overline{\cos^2\Phi \cos^2\Theta}} = \frac{1}{6}$;	$\overline{\overline{\alpha_{xx}\alpha_{yy}}} = \overline{\overline{\alpha_{yy}\alpha_{zz}}} = \overline{\overline{\alpha_{zz}\alpha_{xx}}} = a^2 - \frac{2}{45}\gamma^2$;
$\overline{\overline{\cos^4\Phi \sin^4\Theta}} = \overline{\overline{\sin^4\Phi \sin^4\Theta}} = \frac{1}{5}$;	The above relations are consistent with the symmetry of a symmetric top. The isotropic averages of all other products of two tensor elements are zero.
$\overline{\overline{\sin^2\Phi \cos^2\Phi \sin^4\Theta}} = \frac{1}{15}$	

The isotopic averages of the tensor element products given in the right column of Table 4.1 are consistent with those listed in Tables A14.5 and A14.9 of [4.2]. They were used in the expression for differential cross section in (A2) of [4.5]. We point out that though these results account for the random orientations of linear molecules, the effects of molecular vibration and rotation have not yet been accounted for, a topic we treat next.

4.2 Differential Light Scattering Cross Section from an Ensemble of Linear Molecules

Without losing generality, we consider an x-polarized laser beam impinging vertically (or z-pointing) onto an ensemble of linear molecules, and calculate the nonresonant differential scattering cross section for scattered light in the direction $\hat{k}$ with polarizations $\hat{e}_a{}'$ and $\hat{e}_b{}'$. As explained in Appendix B, $\left(\hat{e}_a{}', \hat{e}_b{}', \hat{k}\right)$ form a Cartesian coordinate system; the receiving polarization unit vectors, in terms of polar and azimuthal angles as in (B.1.c) and (B.1.d), are $\hat{e}_u{}' = (\sin\phi \quad -\cos\phi \quad 0)$ and $\hat{e}_b{}' = (\cos\theta\cos\phi \quad \cos\theta\sin\phi \quad -\sin\theta)$, respectively. Notating the induced dipole moment from the x-polarized field as $\vec{p} = (\alpha_{xx}(\omega),\ \alpha_{yx}(\omega),\ \alpha_{zx}(\omega))\mathscr{E}$, we consider the two scattering polarizations separately by calculating $|\vec{p} \bullet \hat{e}_a{}'|^2$ and $|\vec{p} \bullet \hat{e}_b{}'|^2$ for (4.1); the results are given respectively as (4.9.a) and (4.9.b). Invoking the isotropic averages given in the right columns of Table 4.1, we derive the respective differential cross sections; the results are given in (4.9.c) and (4.9.d). If we collect both scattered polarizations in the receiver (as in most lidar applications), the differential cross section becomes (4.9.e), the sum of (4.9.c) and (4.9.d).

$$|\vec{p} \bullet \hat{e}_a{}'|^2 = \left[\alpha_{xx}{}^2(\omega)\sin^2\phi + \alpha_{yx}{}^2(\omega)\cos^2\phi\right]|\mathscr{E}|^2, \text{ and} \tag{4.9.a}$$

$$|\vec{p} \bullet \hat{e}_b{}'|^2 = \left[\alpha_{xx}{}^2(\omega)\cos^2\theta\cos^2\phi + \alpha_{yx}{}^2(\omega)\cos^2\theta\sin^2\phi + \alpha_{zx}{}^2(\omega)\sin^2\theta\right]|\mathscr{E}|^2 \tag{4.9.b}$$

$$\frac{d\sigma_a}{d\Omega}=\frac{\omega^4}{16\pi^2\mathrm{e}_0^2c^4}\left\{\left(a^2+\frac{4}{45}\gamma^2\right)\sin^2\phi+\frac{\gamma^2}{15}\cos^2\phi\right\},\tag{4.9.c}$$

$$\frac{d\sigma_b}{d\Omega}=\frac{\omega^4}{16\pi^2\mathrm{e}_0^2c^4}\left\{\left(a^2+\frac{4}{45}\gamma^2\right)\cos^2\theta\cos^2\phi+\frac{\gamma^2}{15}(\cos^2\theta\sin^2\phi+\sin^2\theta)\right\},\text{ and}\tag{4.9.d}$$

$$\frac{d\sigma_{a+b}}{d\Omega}=\frac{\omega^4}{16\pi^2\mathrm{e}_0^2c^4}\left\{\left(a^2+\frac{4}{45}\gamma^2\right)(1-\sin^2\theta\cos^2\phi)+\frac{1}{15}\gamma^2(1+\sin^2\theta\cos^2\phi)\right\}.\tag{4.9.e}$$

A special case of particular interest to lidar applications is backscattering; the differential backscattering cross sections may be obtained from (4.9.c) and (4.9.d) by substituting $(\pi-\theta_0,\ \phi_0+\pi)$ for $(\theta,\ \phi)$ with $(\theta_0=0$, and $\phi_0=0)$ for a zenith pointing beam with the incident polarization $\hat{e}=\hat{x}$. From (B.1.c) and (B.1.d), we obtain $\hat{e}_a{}'=\hat{y}$ and $\hat{e}_b{}'=-\hat{x}$ with $(\theta=\pi,\ \phi=\pi)$ corresponding to perpendicular and parallel scattering polarizations; their corresponding differential backscattering cross sections from (4.9.c), (4.9.d), and (4.9.e) are

$$\frac{d\sigma_a^\pi}{d\Omega}=\frac{\omega^4}{16\pi^2\mathrm{e}_0^2c^4}\left(\frac{4}{45}\gamma^2\right);\ \frac{d\sigma_b^\pi}{d\Omega}=\frac{\omega^4}{16\pi^2\mathrm{e}_0^2c^4}\left(a^2+\frac{1}{15}\gamma^2\right);$$
$$\frac{d\sigma_{a+b}^\pi}{d\Omega}=\frac{\omega^4}{16\pi^2\mathrm{e}_0^2c^4}\left(a^2+\frac{7}{45}\gamma^2\right),\tag{4.9.f}$$

where $d\sigma_a^\pi/d\Omega$ and $d\sigma_b^\pi/d\Omega$ are, respectively, perpendicular and parallel components (to the incident polarization) of the backscattering cross section. It is of interest to note, had we considered 90° scattering, by setting $(\theta_0=\pi/2$, and $\phi_0=0)$, leading to $(\theta=\pi/2,\ \phi=\pi)$ or $\hat{e}_a{}'=\hat{y}$ and $\hat{e}_b{}'=-\hat{z}$, the perpendicular and parallel components of the differential backscattering cross sections are the same as given in (4.9.f). We now compare the isotropic average of $d\sigma_{a+b}/d\Omega$ in (4.9.e) and its backscatter value $d\sigma_{a+b}^\pi/d\Omega$ in (4.9.f). The ratio of the latter to the former is termed the q-factor for backward Rayleigh scattering, $q^R(\pi)$. The isotropic average of the former $\overline{\overline{d\sigma_{a+b}/d\Omega}}$ can be readily calculated by using the isotropic averages of the relevant trigonometric functions on the left column of Table 4.1. The results are

$$\overline{\overline{\frac{d\sigma_{a+b}}{d\Omega}}}=\frac{\omega^4}{16\pi^2\mathrm{e}_0^2c^4}\left(a^2+\frac{10}{45}\gamma^2\right)\frac{2}{3};\ \text{and}\ q^R(\pi)=\frac{3}{2}\frac{45a^2+7\gamma^2}{45a^2+10\gamma^2}=\frac{3}{2}\frac{45+7R_A}{45+10R_A},\tag{4.9.g}$$

where $q^R(\pi)$ agrees with (A9) of [5.3]. For dry atmosphere, the relative anisotropy $R_A = (\gamma/a)^2 = 0.22$ and $q^R(\pi) = 1.497$.

Though we consider only a vertical pointing incident beam, the above procedures may be straightforwardly applied to an incident beam pointing at $\hat{k}_0 = (\sin\theta_0 \cos\phi_0 \quad \sin\theta_0 \sin\phi_0 \quad \cos\theta_0)$, as given in (B.1.a), with its polarization vector $\hat{e} = (\cos\theta_0 \cos\phi_0 \quad \cos\theta_0 \sin\phi_0 \quad -\sin\theta_0)$, as given in (B.1.d), being the rotated x-polarization with Euler angles $(\phi_0, \theta_0, 0)$. This gives rise to an induced dipole moment of

$$\begin{bmatrix} p_x \\ p_y \\ p_z \end{bmatrix} = \begin{bmatrix} \alpha_{xx} \cos\phi_0 \cos\theta_0 + \alpha_{xy} \sin\phi_0 \cos\theta_0 - \alpha_{xz} \sin\theta_0 \\ \alpha_{yx} \cos\phi_0 \cos\theta_0 + \alpha_{yy} \sin\phi_0 \cos\theta_0 - \alpha_{yz} \sin\theta_0 \\ \alpha_{zx} \cos\phi_0 \cos\theta_0 + \alpha_{zy} \sin\phi_0 \cos\theta_0 - \alpha_{zz} \sin\theta_0 \end{bmatrix}. \tag{4.10}$$

This leads to expressions for $|\vec{p} \bullet \hat{e}_a{}'|^2$ and $|\vec{p} \bullet \hat{e}_b{}'|^2$, and the nonzero isotropic averages of tensor products associated with differential cross sections depending on both (ϕ, θ) and (ϕ_0, θ_0), not considered further.

The differential cross sections of interest, as given in (4.9.c) and (4.9.d) where $a^2 = \mathscr{G}^{(0)}/3$ and $\gamma^2 = 3\mathscr{G}^{(2)}/2$ are independent of the values of (ϕ, θ) and can be calculated from invariants $\mathscr{G}^{(j)} = \sum_m |\alpha_m^{(j)}|^2$ given in (4.4) with $j = 0$ and $j = 2$. To relate the Placzek invariants to molecular scattering cross sections, we need to understand the dependence of $\mathscr{G}^{(j)}$ for a specified transition between rotational and vibrational states, to be derived in (4.14) and (4.15). To this end, before we produce expressions for nonresonant light scattering cross sections of linear molecules, we revisit the eigenstates of a symmetric top in the Born–Oppenheimer approximation. These result from the product of electronic, vibrational, and rotational states, such as given in Section A.3.3, and cast the polarizability tensor element (4.2.b) in terms of the initial, higher, and final quantum states

4.2.1 The Eigenstates of a Symmetric Top

Following (A.21.a) and (A.22.a) in Section A.3.3, we express the eigenenergies and eigenstates in notations similar to chapter 4 of Long [4.2] as

$$E_{n\Lambda,\mathrm{v},J\Lambda} = \hbar\omega_{e^\eta} + \hbar\omega_{\mathrm{v}^\eta} + \hbar\omega_{J^\eta} \quad \text{and}$$
$$|\Psi_{n\Lambda,\mathrm{v},J\Lambda M}\rangle = |ele\rangle|vib\rangle|rot\rangle = |e^\eta\rangle|\mathrm{v}^\eta\rangle|J^\eta\rangle, \tag{4.11.a}$$

where the total angular momentum of the molecule is $\hbar\vec{J}$. This includes the contribution from the rotations of the nuclear frame and of electrons in the electronic state $|n, \Lambda\rangle$ (i.e., $\vec{J} = \vec{L} + \vec{N}$), with its component on the figure axis,

$J_Z = \Lambda$, which is the Z-projection of electronic angular momentum of electrons in the state $|n, \Lambda>$. The parameters in (4.11.a) are related to those in (A.21.a), (A.21.b), and (A.21.c); these are defined in Section A.3 and tabulated in Table A.4 for molecular oxygen and nitrogen. The leading terms of the associated electronic, vibrational, and rotational energies from (A.21.b) are given in (4.11.b), and the rotational states in (4.11.c) from (A.22.c):

$$\hbar\omega_{e^\eta} = E_{n,\Lambda},\ \hbar\omega_{v^\eta} = \hbar\omega_e(\mathrm{v}+\frac{1}{2}),\ \hbar\omega_{J^\eta} = hcB_{\mathrm{v}}J(J+1);\ B_{\mathrm{v}} = B_e - \alpha_e(\mathrm{v}+\frac{1}{2}),\ \text{and} \tag{4.11.b}$$

$$\left|e^\eta\right> = \left|n,\Lambda\right>,\ \left|\mathrm{v}^\eta\right> = \left|\mathrm{v}\right>,\ \left|J^\eta\right> = \left|J,K,M\right> = \sqrt{\frac{2J+1}{8\pi^2}}D^J_{K,M}. \tag{4.11.c}$$

Here, the index η refers to initial, higher, and final states, $\eta = i, h, f$, each consisting of an electronic, vibrational, and rotational component.

Since the first excited electronic states for oxygen and nitrogen molecules, according to Table A.4, are about 8,000 cm^{-1} and 70,000 cm^{-1}, respectively, above the ground state, for nonresonant light scattering the electronic terms of both initial and final states refer to the same (ground) state, and we denote them as $|e^i> = |e^f> = |e^g>$. Vibrational and rotational levels being far less energetic, their initial and final levels could be different (i.e. $|\mathrm{v}^i> \neq |\mathrm{v}^f>$ and $|J^i> \neq |J^f>$). The transition frequency between initial and higher and that between higher and final states are then expressed as $\omega_{hi} = \omega_{e^h e^g} + \omega_{\mathrm{v}^h\mathrm{v}^i} + \omega_{R^h R^i}$ and $\omega_{hf} = \omega_{e^h e^g} + \omega_{\mathrm{v}^h\mathrm{v}^f} + \omega_{R^h R^f}$.

4.2.2 Placzek Invariants and Placzek-Teller Coefficients for the Polarizability Tensor

Armed with the expressions for eigenstates of a symmetric top, and the associated transition energies/frequencies, we can employ the information in (4.2.b) and derive the polarizability tensor elements for linear molecules. We also note that the projection of ground state angular momentum for nitrogen and oxygen molecules on the figure axis is $\Lambda = 0$ with degeneracy $g_i = g_f = 1$. The results for the polarizability tensor elements (4.2.b) in the fixed Cartesian coordinates are

$$(\alpha_{\rho\sigma})_{e^f\mathrm{v}^f J^f; e^i\mathrm{v}^i J^i} = \frac{1}{\hbar}\sum_{e^h\neq e^g,\, \mathrm{v}^h,\, J^h}$$

$$\left\{\begin{array}{l} \dfrac{<J^i|<\mathrm{v}^i|<e^g|(\vec{p}\bullet\hat{\rho})|e^h>|\mathrm{v}^h>|J^h><J^h|<\mathrm{v}^h|<e^h|(\vec{p}\bullet\hat{\sigma})|e^g>|\mathrm{v}^f>|R^f>}{\omega_{e^h e^g}+\omega_{\mathrm{v}^h\mathrm{v}^i}+\omega_{R^h R^i}-\omega-i\Gamma_{e^h\mathrm{v}^h R^h}} \\ +\dfrac{<J^f|<\mathrm{v}^f|<e^g|(\vec{p}\bullet\hat{\sigma})|e^h>|\mathrm{v}^h>|J^h><J^h|<\mathrm{v}^h|<e^h|(\vec{p}\bullet\hat{\rho})|e^g>|\mathrm{v}^i>|J^i>}{\omega_{e^h e^g}+\omega_{\mathrm{v}^h\mathrm{v}^f}+\omega_{R^h R^f}+\omega+i\Gamma_{e^h\mathrm{v}^h R^h}} \end{array}\right\}.$$

This polarization tensor matrix element may be cast into a simpler matrix element between the initial and final states, as given in (4.12.a), if we apply the completeness relationship for higher vibrational and rotational quantum states (i.e., $\sum_{J^h}|J^h><J^h| = \sum_{v^h}|v^h><v^h| = 1$) with the associated operator defined in (4.12.b):

$$\left(\alpha_{\rho\sigma}\right)_{e^g v^f R^f; e^g v^i R^i} = <J^f|<v^f|<e^g|\hat{\alpha}_{\rho\sigma}(\sum\{\ldots\}, R)|e^g>|v^i>|J^i>; \tag{4.12.a}$$

$$\hat{\alpha}_{\rho\sigma}(\sum\{\ldots\}, R) = \frac{1}{\hbar}\sum_{e^h \neq e^g}\left\{\frac{(\vec{p}\bullet\hat{\rho})|e^h><e^h|(\vec{p}\bullet\hat{\sigma})}{\omega_{e^h e^i} + \omega_{v^h v^i} + \omega_{R^h R^i} - \omega - i\Gamma_{e^h v^h R^h}}\right.$$
$$\left. + \frac{(\vec{p}\bullet\hat{\sigma})|e^h><e^h|(\vec{p}\bullet\hat{\rho})}{\omega_{e^h e^g} + \omega_{v^h v^f} + \omega_{R^h R^f} + \omega_1 + i\Gamma_{e^h v^h R^h}}\right\}. \tag{4.12.b}$$

These expressions are the same as (4.5.5) and (4.5.6) of [4.2]. Here, $\sum\{\ldots\}$ and R represent summation over higher electronic and associated vibration and rotational states and distance between nuclei, respectively. Note that $\sum_{e^h \neq e^g}|e^h><e^h| \neq 1$, since the summation does not include the ground electronic state. The matrix element, $\alpha_{\rho\sigma}(R) \equiv <e^g|\hat{\alpha}_{\rho\sigma}(\sum e^h, R)|e^g>$, is the electronic polarization tensor matrix element in Cartesian coordinates, similar to that in (4.1); in general, it depends on internuclear spacing but is independent of the vibration and rotation states of the molecule of interest. As shown in Table 4.1, the isotropic average of the product of two such tensor elements, $\overline{\overline{\alpha_{\rho\sigma}\alpha_{\rho'\sigma'}}}$, can be expressed in terms of quantities a^2 and γ^2, which represent polarized and depolarized scattering [4.5], respectively. These quantities, $a^2 = \mathscr{G}^{(0)}/3$ and $\gamma^2 = 3\mathscr{G}^{(2)}/2$, though they generally depend on R, are invariants with respect to coordinate systems. Their associated Placzek invariants $\mathscr{G}^{(0)}$ and $\mathscr{G}^{(2)}$, given in (4.4), may be evaluated in the irreducible representation in either fixed or rotating coordinates. In short, instead of evaluating the tensor elements $\alpha_{\rho\sigma}(R)$ in the fixed Cartesian (reducible) representation, we first calculate them in the irreducible representation $\alpha_m^{(j)}(R)$ with $j = 0$ and 2 and $m = -j, -j+1, \ldots, j-1, j$. This facilitates the calculations of the relevant tensor element products $\overline{\overline{\alpha_{\rho\sigma}\alpha_{\rho'\sigma'}}}$, as needed for scattering cross section; the results are listed in Table 4.1. When the vibrational and rotational motions with the associated transitions are included, the same procedure can be followed; we carry out the following in sequence as

$$\left(\alpha_{\rho\sigma}\right)_{v^f R^f; v^i R^i} = <R^f|<v^f|\alpha_{\rho\sigma}|v^i>|R^i> \rightarrow \left(\alpha_m^{(j)}\right)_{v^f R^f; v^i R^i} = <R^f|<v^f|\alpha_m^{(j)}|v^i>|R^i>$$
$$\rightarrow \sum_{m=-j}^{m=j}\left|\left(\alpha_m^{(j)}\right)_{v^f R^f; v^i R^i}\right|^2 \rightarrow \overline{\overline{\left(\alpha_{\rho\sigma}\right)_{v^f R^f; v^i R^i}\left(\alpha_{\rho'\sigma'}\right)_{v^f R^f; v^i R^i}}}.$$

Here, though not explicitly shown, the matrix elements of the polarizability tensor are generally functions of the nuclear separation R, in either the reducible or irreducible representation. Notice that $|v^i>|J^i>$ and $|v^f>|J^f>$ are, respectively, the initial and final vibration and rotation states in the fixed coordinates. We are reminded again that the advantage of using quantities in the irreducible representation is that their components, denoted by index m, transform under coordinate rotations into themselves in the same subspace with the same value of j, as given in (4.7).

We now proceed to calculate $\alpha_m^{(j)}$, the irreducible tensor elements in the fixed coordinates, by writing out $|J>=|J,K,M>$ and invoking (4.7) as

$$\left(\alpha_m^{(j)}\right)_{v^fJ^fK^fM^f;v^iJ^iK^iM^i} = <J^fK^fM^f|<v^f|\sum_{m'=-j}^{m'=j} \alpha_{m'}^{(j)} D_{m'm}^{(j)}(\alpha,\beta,\gamma)|v^i>|J^iK^iM^i>, \tag{4.13.a}$$

where $\alpha_{m'}^{(j)}$ are the associated irreducible tensor elements in the molecular coordinates. Since $D_{m'm}^{(j)}(\alpha,\beta,\gamma)$ is the irreducible representation of the rotation operator between the fixed and molecular coordinates, we can separate the rotation and vibration transitions into products of matrix elements associated with rotational or vibrational transitions. These are given by

$$\left(\alpha_m^{(j)}\right)_{v^fJ^fK^fM^f;v^iJ^iK^iM^i} = \sum_{m'=-j}^{m'=j} <v^f|\alpha_{m'}^{(j)}|v^i><J^fK^fM^f|D_{m'm}^{(j)}(\alpha,\beta,\gamma)|J^iK^iM^i>. \tag{4.13.b}$$

This separation is possible because the $\alpha_{m'}^{(j)}$ in molecular coordinates (fixed with the molecule) are independent of molecular rotations. We note that (4.13.b) is the same as (6.2.6) of [4.2]. By considering $D_{m'm}^{(j)}(\alpha,\beta,\gamma)$ as the rotational wave function, as given in (4.11.c), we first calculate the second matrix element in (4.13.b) by the formula for the following matrix element, derived in (A.22.d) as

$$<J^fK^fM^f|D_{m'm}^{(j)}(\alpha,\beta,\gamma)|J^iK^iM^i> = \sqrt{(2J^f+1)(2J^i+1)}\begin{pmatrix} J^f & j & J^i \\ -M^f & m & M^i \end{pmatrix}\begin{pmatrix} J^f & j & J^i \\ -K^f & m' & K^i \end{pmatrix}.$$

We substitute the above identity into (4.13.b), noting that only the second 3-j coefficient is under the summation of m' in (4.13.b). Realizing that, in any coordinate system, a matrix in its irreducible representation with a given j-value is diagonal, we can express the Placzek invariants for the polarization tensor, given in (4.4), of a specified transition between rotational and vibrational states (in the ground electronic state) as

$$\mathscr{G}^{(j)} = \sum_m \left| \left(\alpha_m^{(j)}\right)_{v^f J^f K^f M^f; v^i J^i K^i M^i} \right|^2 = (2J^f+1)(2J^i+1)\sum_{m=-j}^{m=j}\begin{pmatrix} J^f & j & J^i \\ -M^f & m & M^i \end{pmatrix}^2$$

$$\times \left[\sum_{m'=-j}^{m'=j} <v^f|\alpha_{m'}^{(j)}|v^i> \begin{pmatrix} J^f & j & J^i \\ -K^f & m' & K^i \end{pmatrix} \right] \left[\sum_{m''=-j}^{m''=j} <v^f|\alpha_{m''}^{(j)}|v^i> \begin{pmatrix} J^f & j & J^i \\ -K^f & m'' & K^i \end{pmatrix} \right]$$

$$= (2J^i+1)(2J^f+1)\sum_{m=-j}^{m=j}\begin{pmatrix} J^f & j & J^i \\ -M^f & m & M^i \end{pmatrix}^2 \sum_{m'=-J}^{m'=J}\begin{pmatrix} J^f & j & J^i \\ -K^f & m' & K^i \end{pmatrix}^2 \left| <v^f|\alpha_{m'}^{(j)}|v^i> \right|^2. \tag{4.14.a}$$

Selection rules demand that for given M^i and M^f or K^i and K^f, there is only one value allowed for m' and m'', suggesting $m' = m'' = K^f - K^i$. This compresses the product of the two associated summations into one. The same rule allows us to set $m = \Delta M$ $\left(\text{where } \Delta M = M^f - M^i\right)$ and $m' = \Delta K$ $\left(\text{where } \Delta K = K^f - K^i\right)$, thus reducing (4.14.a) to (4.14.b):

$$\left(\mathscr{G}^{(j)}\right)_{v^f J^f K^f M^f : v^i J^i K^i M^i} = (2J^i+1) b^{(j)}_{J^f K^f M^f; J^i K^i M^i} \left| <v^f|\alpha_{\Delta K}^{(j)}|v^i> \right|^2, \text{ with}$$

$$b^{(j)}_{J^f K^f M^f; J^i K^i M^i} \equiv (2J^f+1)\begin{pmatrix} J^f & j & J^i \\ -M^f & \Delta M & M^i \end{pmatrix}^2 \begin{pmatrix} J^f & j & J^i \\ -K^f & \Delta K & K^i \end{pmatrix}^2. \tag{4.14.b}$$

This is the expression for the invariants between substates $|J^i M^i>$ and $|J^f M^f>$, and it is identical to (6.3.13) of [4.2]. Since the invariants we seek are proportional to the scattering cross sections (or intensity) of interest, we define a new form of Placzek invariants $\left(\mathscr{G}^{(j)}\right)_{v^f J^f K^f : v^i J^i K^i}$ and a new factor $b^{(j)}_{J^f K^f; J^i K^i}$ by summing (4.14.b) over the degeneracies of both initial and final rotational states and dividing by $(2J^i+1)$, which gives (4.15.a)

$$\left(\mathscr{G}^{(j)}\right)_{v^f J^f K^f : v^i J^i K^i} = \frac{1}{(2J^i+1)} \sum_{M^i, \Delta M, M^f} \left(\mathscr{G}^{(j)}\right)_{v^f J^f K^f M^f : v^i J^i K^i M^i} \equiv b^{(j)}_{J^f K^f; J^i K^i} \left| <v^f|\alpha_{\Delta K}^{(j)}|v^i> \right|^2. \tag{4.15.a}$$

For a justification of this definition, we note that for a given M^i, we include all allowed ΔM and M^f. We then evaluate the mean value of the invariance by summing over M^i and dividing by $(2J^i+1)$. Since $|<v^f|\alpha_{\Delta K}^{(j)}|v^i>|^2$ is independent of the rotational quantum numbers, this leads to the new rotational transition factors $b^{(j)}_{J^f K^f; J^i K^i}$, first derived in 1933 and now termed Placzek–Teller coefficients [4.6]. These are given in (4.15.b):

$$b^{(j)}_{J^fK^f;J^iK^i} \equiv \sum_{M^i,\Delta M,M^f} b^{(j)}_{J^fK^fM^f;J^iK^iM^i} = (2J^f+1)\begin{pmatrix} J^f & j & J^i \\ -K^f & \Delta K & K^i \end{pmatrix}^2. \quad (4.15.b)$$

In deriving (4.15.a) and (4.15.b) from (4.14.b), we have used a summation formula of 3-j coefficients given in (A.22.e) and reproduced in (4.15.c):

$$\sum_{m_1,m_2,m_3} \begin{pmatrix} j_1 & j_2 & j_3 \\ m_1 & m_2 & m_3 \end{pmatrix}^2 = 1 \rightarrow \sum_{M^i,\Delta M,M^f} \begin{pmatrix} J^f & j & J^i \\ -M^f & \Delta M & M^i \end{pmatrix}^2 = 1. \quad (4.15.c)$$

The formula (4.15.b) has also been rederived using modern notation (see (16) in [4.7]). We can then express the square of the mean and anisotropy polarizabilities, a^2 and γ^2, in terms of the Placzek invariants, calculable from (4.15.a) for specified quantum states, respectively, as $a^2 = \mathscr{G}^{(0)}/3$ and $\gamma^2 = 3\mathscr{G}^{(2)}/2$, and substitute them into (4.9.c)–(4.9.e) for the calculation of differential cross sections.

4.2.3 Rotational and Vibrational Contributions to Placzek Invariants

The Placzek invariants of interest given in (4.15.a) are expressed as the product of $b^{(j)}_{J^fK^f;J^iK^i}$ for a rotational transition and $|<\mathrm{v}^f|\alpha^{(j)}_{\Delta K}|\mathrm{v}^i>|^2$ for a vibrational transition with $j = 0$ and $j = 2$. As noted in Section A.3.2, the ground state electronic term of the molecules of interest, such as N_2, O_2, and CO_2, is Σ (i.e., $\Lambda = 0$), suggesting that we are interested only in invariants with $\Delta K = K^i = K^f = 0$. We first calculate the strength relating rotational transitions between the initial state $J^i = J$ and the final state J^f, $b^{(0)}_{J^f0;J^i0}$ and $b^{(2)}_{J^f0;J^i0}$ (step A), before we evaluate the strengths of the vibrational transitions $|<\mathrm{v}^f|\alpha^{(0)}_0|\mathrm{v}^i>|^2$ and $|<\mathrm{v}^f|\alpha^{(2)}_0|\mathrm{v}^i>|^2$ (step B).

(Step A) $b^{(0)}_{J^f0;J^i0}$ and $b^{(2)}_{J^f0;J^i0}$: For $j = 0$, the selection rules for 3-j coefficients demand that $J^f = J^i = J$ with $b^{(0)}_{J^f0;J^i0} = 1$, as expected. For $j = 2$, taking $J^i = J$, the final state angular momenta may be $J^f = J-2$, J, or $J+2$, corresponding to O-branch, Q-branch, and S-branch scattering. Using (3.7.17) of [4.4], we can compute all four relevant invariants as

$$b^{(0)}_{J,0;J,0} = (2J+1)\begin{pmatrix} J & 0 & J \\ 0 & 0 & 0 \end{pmatrix}^2 = (2J+1)\frac{2J!}{(2J+1)!} = 1, \quad (4.16.a)$$

$$b^{(2)}_{J-2,0;J,0} = (2J-3)\begin{pmatrix} J-2 & 2 & J \\ 0 & 0 & 0 \end{pmatrix}^2 = \frac{3J(J-1)}{2(2J+1)(2J-1)}, \quad (4.16.b)$$

$$b^{(2)}_{J,0;J,0} = (2J+1)\begin{pmatrix} J-2 & 2 & J \\ 0 & 0 & 0 \end{pmatrix}^2 = \frac{J(J+1)}{(2J+3)(2J-1)}, \text{ and} \tag{4.16.c}$$

$$b^{(2)}_{J+2,0;J,0} = (2J+5)\begin{pmatrix} J+2 & 2 & J \\ 0 & 0 & 0 \end{pmatrix}^2 = \frac{3(J+1)(J+2)}{2(2J+3)(2J+1)}. \tag{4.16.d}$$

The three invariants for $j = 2$, $b^{(2)}_{J-2,0;J,0}$ (O-branch), $b^{(2)}_{J,0;J,0}$(Q-branch), and $b^{(2)}_{J+2,0;J,0}$(S-branch) are now termed Placzek–Teller coefficients for the $\sum$-term (i.e., $K = 0$); they may also be obtained from Table 6.2 of [4.2] by setting $\Delta K = K^i = K^f = 0$. The coefficients $b^{(2)}_{J^f,0;J,0}$ for a given J sum (over three possible J^f) to unity, so they may be thought of as shares of the transitions among three groups. The fractional sharing among the three branches depends on the rotational quantum number of the molecule, J, as shown in Fig. 4.1. At $J = 0$, the only branch that can conserve energy is the S-branch; at $J = 1$, the S-branch shares with the Q-branch (40%), and its share decreases to 60%. When $J > 2$, we note the following: (1) the Q-branch gradually decreases from 40% to 25% as J increases, and (2) all O-, Q-, and S-branches are possible and respectively approach their classical limits of 37.5%, 25.0%, and 37.5%, respectively, at large J (> 30) values. We also note from (4.15.a) that the invariant associated with the coefficients $b^{(2)}_{J^f,0;J,0}$ is $\alpha^{(2)}_0 = 2\gamma^2/3$, while the invariant associated with $b^{(0)}_{J,0;J,0}$ is $\alpha^{(0)}_0 = 3a^2$. Their respective relations to the mean and anisotropy of polarizability, and the associated significance to the light scattering spectrum are discussed below.

(Step B) $|\langle v^f|\alpha^{(0)}_0(R)|v^i\rangle|^2$ and $|\langle v^f|\alpha^{(2)}_0(R)|v^i\rangle|^2$: These invariants between initial and final vibrational states depict, respectively, the electronic polarizability of the linear

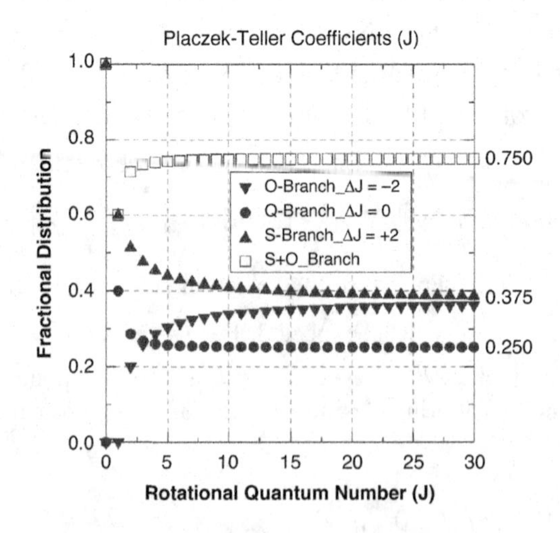

Fig. 4.1 The Placzek–Teller coefficients as a function of rotational quantum number J, showing the fractional distribution of the O-, Q-, and S-branches of the pure rotational Raman scattering. Notice that the distribution between the Q-branch and the O+S branches of 0.25 and 0.75 is valid for $J \geq 5$.

molecule and its change due to a vibrational transition. Here, we have restored their explicit dependence on R. Since we are considering homogeneous diatomic molecules, nitrogen and oxygen, we have only one vibrational mode to deal with. Physically, the electronic wave functions of a diatomic molecule are functions of the internuclear distance, R, displacing it from its equilibrium separation R_e in a normal mode vibration, $q = R - R_e$. Since $q \ll R_e$, we can expand the tensor elements in a Taylor series to first order as

$$\alpha_0^{(j)}(R) = \alpha_0^{(j)}(R_e, q) = \alpha_0^{(j)}(R_e, 0) + \left(\frac{\partial \alpha_0^{(j)}}{\partial q}\right)_0 q; \; j = 0, \, 2. \tag{4.17.a}$$

The normal mode vibration can be modeled by a simple harmonic oscillator, employing (4.17.a). The relevant matrix element square may be found in elementary texts or in (5.7.6) to (5.7.8) of [4.2]:

$$\text{For } \mathrm{v}^f = \mathrm{v}^i, \; \left| < \mathrm{v}^f | \alpha_0^{(j)} | \mathrm{v}^i > \right|^2 = \left| < \mathrm{v}^f | \alpha_0^{(j)}(R_e, 0) | \mathrm{v}^i > \right|^2 = \alpha_0^{(j)}(R_e, 0); \tag{4.17.b}$$

$$\text{For } \mathrm{v}^f = \mathrm{v}^i + 1, \; \left| < \mathrm{v}^f | \alpha_0^{(j)} | \mathrm{v}^i > \right|^2 = \left(\frac{\partial \alpha_0^{(j)}}{\partial q}\right)_0^2 \frac{\hbar(\mathrm{v}^i + 1)}{2\omega_0} = \frac{\hbar}{2\omega_0} \left(\frac{\partial \alpha_0^{(j)}}{\partial q}\right)_0^2 (\mathrm{v}^i + 1); \tag{4.17.c}$$

$$\text{For } \mathrm{v}^f = \mathrm{v}^i - 1, \; \left| < \mathrm{v}^f | \alpha_0^{(j)} | \mathrm{v}^i > \right|^2 = \left(\frac{\partial \alpha_0^{(j)}}{\partial q}\right)_0^2 \frac{\hbar(\mathrm{v}^i)}{2\omega_0} = \frac{\hbar}{2\omega_0} \left(\frac{\partial \alpha_0^{(j)}}{\partial q}\right)_0^2 (\mathrm{v}^i). \tag{4.17.d}$$

Therefore, the Rayleigh and Raman mean polarizabilities, a and a', and their associated anisotropies, γ and γ', may be expressed in terms of the parameters given in (14.17) as

$$a^2 = \frac{1}{3}\alpha_0^{(0)}(R, 0), \; \gamma^2 = \frac{3}{2}\alpha_0^{(2)}(R, 0); \; a'^2 = \frac{\hbar}{6\omega_0} \left(\frac{\partial \alpha_0^{(0)}}{\partial q}\right)_0^2, \; \gamma'^2 = \frac{3\hbar}{4\omega_0} \left(\frac{\partial \alpha_0^{(2)}}{\partial q}\right)_0^2. \tag{4.17.e}$$

We note that, since at temperatures of interest for atmospheric lidar only the ground (v = 0) and the first excited (v = 1) vibrational levels are occupied, the strength of anti-Stokes vibrational Raman scattering – by setting $\mathrm{v}^i = 1$ in (4.17.d) – is the same as Stokes vibrational Raman scattering from the v = 0 state – by setting $\mathrm{v}^i = 0$ in (4.17.c). To summarize, for Rayleigh and Raman scattering, we only need the results for $j = 0$ and $j = 2$. For $v^f = v^i$, (4.17.b) leads to a^2 with $j = 0$ for Rayleigh scattering, and to γ^2 with $j = 2$ for pure rotational Raman scattering. For $v^f = v^i \pm 1$, (4.17.c) and (4.17.d) lead to a'^2 with $j = 0$ and γ'^2 with $j = 2$, respectively, these being responsible for polarized (resulting from vibrations) and depolarized (from rotations) vibrational-rotational Raman scattering.

4.2.4 Differential Rayleigh and Raman Scattering Cross Section of Nitrogen and Oxygen

With the knowledge of mean and anisotropy-induced polarizabilities, we can express the differential backscattering cross sections for Rayleigh and Raman processes in terms of a^2 and γ^2, and a'^2 and γ'^2, respectively, as given in (4.17.e). In this connection, we invoke (4.9.f) and write the differential backscattering cross sections for Rayleigh and (vibrational) Raman scattering when detecting both scattering polarizations as

$$\frac{d\sigma_{a+b}^{\pi,Ray}}{d\Omega} = \frac{\omega^4}{16\pi^2 \mathrm{e}_0^2 c^4}\left(a^2 + \frac{7}{45}\gamma^2\right); \quad \frac{d\sigma_{a+b}^{\pi,Ram}}{d\Omega} = \frac{\omega^4}{16\pi^2 \mathrm{e}_0^2 c^4}\left(a'^2 + \frac{7}{45}\gamma'^2\right). \tag{4.18}$$

Here, the term with parameter a^2 is the differential cross section of the polarized (sometimes referred to as Cabannes) backscattering, denoted as σ_π^P in [4.5], and the term with γ^2 is that for the depolarized pure rotational Raman (PRR) backscattering from the sum of O-, Q-, and S-branches, denoted as σ_π^{DP} in [4.5]. Likewise, a'^2 and γ'^2, which are given in (4.14.e), refer, respectively, to pure vibrational Raman scattering and to the sum of O-, Q-, and S-branches of the associated rotational-vibrational Raman scattering (RVR). These parameters, once provided, may be used to calculate and compare the differential cross sections of interest. To provide such a comparison with a 2001 publication [4.5], we point out that other than the factor (ν/ν_s), (4.18) differs from (A6) of [4.5] by a factor $\left(1/\mathrm{e}_0^2\right)$, due only to a difference in the definition of polarizability. Also, while the parameters a^2 and γ^2 given in (4.17.e) and (4.18) are the same as those in Table 9.3 in the book chapter by Wandinger [4.8], the parameters a'^2 and γ'^2 differ, respectively, by $\hbar/(6\omega_0)$ and $3\hbar/(4\omega_0)$. These parameters are repeated in Table 4.2 with our definition.

As an example, we utilize the tabulated values of $(a/4\pi\mathrm{e}_0)^2$ and $(\gamma/4\pi\mathrm{e}_0)^2$ in Table 4.2 and calculate the differential scattering cross sections for Cabannes and rotational Raman scattering at 532 nm for molecular nitrogen. Here, since the Q-branch of rotational Raman scattering – or about 25% of PRR $(0.25\sigma_\pi^{DP})$ – cannot be spectrally separated from the polarized scattering (σ_π^P), it is included in Cabannes

Table 4.2 *The square of mean and anisotropy parameters of nitrogen and oxygen for Rayleigh and vibrational* Raman scattering.*

Gas	$(a/4\pi e_0)^2$ in m^6	$(\gamma/4\pi e_0)^2$ in m^6	$(a'/4\pi e_0)^2$ in m^6	$(\gamma'/4\pi e_0)^2$ in m^6
N_2	3.17×10^{-60}	0.52×10^{-60}	1.05×10^{-63}	7.61×10^{-63}
O_2	2.66×10^{-60}	1.26×10^{-60}	9.78×10^{-64}	1.74×10^{-62}

*Vibrational frequency is 2,331 cm^{-1} for N_2 and 1,556 cm^{-1} for O_2.

scattering (i.e., $\sigma_\pi^C = \sigma_\pi^P + 0.25\sigma_\pi^{DP}$). For the same reason, the rotational Raman scattering in the temperature range of interest is about 75% of PRR (i.e., $\sigma_\pi^{RR} = 0.75\sigma_\pi^{DP}$). The calculated σ_π^C and σ_π^{RR} for molecular nitrogen are then $6.21 \times 10^{-32}\text{m}^2/\text{sr}$ and $1.18 \times 10^{-33}\text{m}^2/\text{sr}$, respectively; these values compare reasonably well to those of $6.02 \times 10^{-32}\text{m}^2/\text{sr}$ and $1.08 \times 10^{-33}\text{m}^2/\text{sr}$ in Table 1 of [4.5]. Similar comparison calculations can be made for dry atmosphere, which consists of 79.05% nitrogen and 20.95% oxygen.

Likewise, using the values of $(a'/4\pi e_0)^2$ and $(\gamma'/4\pi e_0)^2$ in Table 4.2, we calculate $\sigma_\pi^{QRVR} = 2.62 \times 10^{-35}\text{m}^2/\text{sr}$ for pure vibrational Raman scattering (including Q-branch RVR, as the Q-branch rotation-vibration coupling is relatively small; see Section 4.3.4) and $\sigma_\pi^{OSRVR} = 1.73 \times 10^{-35}\text{m}^2/\text{sr}$ for O+S-branch RVR for molecular nitrogen. Compared to the values in Table 2 of [4.5], $4.05 \times 10^{-35}\text{m}^2/\text{sr}$ and $7.67 \times 10^{-36}\text{m}^2/\text{sr}$, these are, respectively, too small and too big. However, their sum, the total differential vibrational Raman scattering cross section, $4.35 \times 10^{-35}\text{m}^2/\text{sr}$, compares well to the value $4.81 \times 10^{-35}\text{m}^2/\text{sr}$ listed in [4.5]. These parameters could, of course, be calculated if the electronic, vibrational, and rotational wave functions of nitrogen and oxygen were known to the accuracies required. At present, calculations rely on the measured scattering cross sections, and the experimental values of these parameters are continuously being improved for practical applications.

4.2.5 Relating the Mean and Anisotropy Polarizabilities to Macroscopic Quantities

Before we move on, a comparison of the various expressions of differential cross section as appears in the literature is in order. Here, we compare the expressions in [4.5] with (4.9), which follows the 2015 paper by She et al. [4.9], and use the MKS system with $p_i \equiv \alpha_{ij}E_j$ (with α in units of Cm^2V^{-1}). Notice that other than the factor (ν/ν_s) in (A3) of [4.5] and $(1/e_0^2)$ in (4.9.c) and (4.9.d), these are in complete agreement. This minor discrepancy owes to the fact that [4.5] defines the differential cross section on a photon basis as opposed to on the basis of power, as presented here, and that [4.5] uses $p_i \equiv e_0\alpha_{ij}E_j$ to define polarizability, in contrast to $p_i \equiv \alpha_{ij}E_j$

here. Equation (4.9.f) parallels that obtained by combining (9.5) plus (9.6) in the book chapter by Wandinger [4.8], with the distinction that (9.5) of [4.8] includes only the Q-branch pure rotational Raman scattering (PRR) in the classical limit (large J), while the O- and S-branches are represented in (9.6) of [4.8]. In contrast, we include all three branches of PRR for any J value in (4.9.f).

Also following [4.5], we introduce useful terminology for Rayleigh scattering cross sections (per molecule) and the associated macroscopic parameter, the refractive index. These include simplified notation of polarized (coherent), depolarized, and Rayleigh backscattering differential cross sections, respectively given by $\sigma_\pi^P, \sigma_\pi^{DP}$, and σ_π^R, as they relate to (4.9.f), as follows:

$$\sigma_\pi^R = \frac{d\sigma_{a+b}^\pi}{d\Omega} = \frac{\omega^4}{16\pi^2 \mathrm{e}_0^2 c^4}\left(a^2 + \frac{7}{45}\gamma^2\right); \quad \sigma_\pi^P = \frac{\omega^4}{16\pi^2 \mathrm{e}_0^2 c^4}\left(a^2\right);$$

$$\sigma_\pi^{DP} = \frac{\omega^4}{16\pi^2 \mathrm{e}_0^2 c^4}\left(\frac{7}{45}\gamma^2\right). \tag{4.19.a}$$

The polarized and depolarized scatterings are termed Cabannes and pure rotational Raman scattering, respectively, here. Since unlike O- and S-branches, the Q-branch PRR cannot be separated from the Cabannes scattering (discussed in more detail below), we sometimes lump, as done in [4.8], the Q-branch PRR with the Cabannes scattering and refer to the sum of O- and S-branches as PRR. This is shown in [4.5] for the classical limit, respectively, as $\sigma_\pi^C = \sigma_\pi^P + 0.25\sigma_\pi^{DP}$, and $\sigma_\pi^C = \sigma_\pi^P + 0.75\sigma_\pi^{DP}$. In addition, we are also interested in the average differential cross section over all directions, that is, the isotropic average of (4.9.e), σ_T^R, as well as the Rayleigh extinction cross section $\sigma^R = 4\pi\sigma_T^R$; these are

$$\sigma_T^R = \frac{d\sigma_{a+b}}{d\Omega} = \frac{\omega^4}{16\pi^2 \mathrm{e}_0^2 c^4}\left\{\left(a^2 + \frac{4}{45}\gamma^2\right)\left(1 - \overline{\overline{\sin^2\theta\cos^2\phi}}\right) + \frac{1}{15}\gamma^2\left(1 + \overline{\overline{\sin^2\theta\cos^2\phi}}\right)\right\}, \text{ leading to}$$

$$\sigma_T^R = \frac{\omega^4}{16\pi^2 \mathrm{e}_0^2 c^4}\left(\frac{2}{3}\right)\left(a^2 + \frac{10}{45}\gamma^2\right); \; \sigma^R = 4\pi\sigma_T^R = \frac{\omega^4}{16\pi^2 \mathrm{e}_0^2 c^4}\left(\frac{8\pi}{3}\right)a^2 F_K; \; F_K = \left(1 + \frac{2}{9}\frac{\gamma^2}{a^2}\right), \tag{4.19.b}$$

where $F_K = 1 + 2R_A/9$ is the King factor, with relative anisotropy (anisotropy factor), $R_A = \gamma^2/a^2$. Recall that $a = (2\alpha_\perp + \alpha_\parallel)/3$ is the average polarizability of the molecule in question, responsible for trace scattering, which is isotropic and polarized. We denote the polarized part of the Rayleigh cross section (extinction cross section of Cabannes scattering) as σ^C; it is σ^R when F_K is set to unity (zero anisotropy). For symmetric top molecules, the Rayleigh extinction cross section is larger than the corresponding isotropic "atom" by the factor F_K (i.e., $\sigma^R = \sigma^C F_K$). It is possible to relate the King factor to depolarization ratios, and thus it is separately measurable. A number of depolarization ratios for

Rayleigh scattering may be defined (scattered intensity of perpendicular polarization component to that of the parallel component), each corresponding to a different measurement geometry. Two polarization ratios are particularly useful; they are defined as

$$\delta^t = \delta = \left(\frac{\sigma_\perp{}^R}{\sigma_\parallel{}^R}\right)_{lin\ pol} = \frac{3R_A}{45 + 4R_A}; \ \delta_0^t = \left(\frac{\sigma_\perp{}^R}{\sigma_\parallel{}^R}\right)_{natural} = \frac{6R_A}{45 + 7R_A}. \quad (4.19.c)$$

Here, the superscript "t" indicates total scattering [4.10], that is, Rayleigh = Cabannes + pure rotational Raman scattering. The polarization ratio δ^t is conceived for a laser light scattering experiment with linearly polarized incident light; it makes no difference whether one measures the scattering in the backward or in the 90° direction (see [4.5]). The ratio δ_0^t is for natural, unpolarized incident light [4.10]. Both δ^t and δ_0^t are measures of the anisotropy of the medium and can be related to R_A, as given in (4.19.c).

The scattering cross section (per molecule) discussed above is a microscopic property, while the refractive index n is a useful and measurable macroscopic property of the gas. According to Young [4.10], it was Lord Rayleigh who showed that compared to the primary (incident) radiation, the phase of the secondary radiation in the forward direction (forward scattering) is not random. Nevertheless, the coherence in forward scattering contains a phase delay related to the refractive index of the gas. It is thus the polarized part of the Rayleigh extinction σ^C that is related to the refractive index. To obtain this relationship, we consider the molecular ensemble in question as a dielectric medium and recall the relationship of its displacement, D, electric field, E, and polarization, P (the electric dipole moment per unit volume, not to be confused with the polarization direction of the electric field, $\hat{e}_a$ or $\hat{e}_b$). These macroscopic field quantities and the medium's electric susceptibility χ_e are related; see p. 176 of [4.11], as given in (4.20.a):

$$D = \mathsf{e}_0 E + P; \ P = \mathsf{e}_0 \chi_\mathsf{e} E \rightarrow D = \mathsf{e}_0(1 + \chi_e)E \rightarrow \mathsf{e} \equiv \mathsf{e}_0(1 + \chi_e), \quad (4.20.a)$$

where e and e_0 are permittivity of the medium and vacuum, respectively, and their ratio is the dielectric constant $\mathsf{K} = \mathsf{e}/\mathsf{e}_0$. Since by definition, the polarization $P = \mathsf{e}_0 \chi_e E$ equals the product of the molecular number density $\mathcal{N}$ times the induced dipole moment per molecule, p, which in turn is the product of polarizability (per molecule), α, and the inducing electric field, E, we need to point out the difference between the electric field in $p = \alpha E$ and that in $P = \mathsf{e}_0 \chi_e E$. The former is the electric field on a given "test" molecule that results from the induced dipole moments of all other molecules and is denoted as E_{else}, following the description on p.191 of [4.11]. The latter is the macroscopic electric field at the location of the test molecule, resulting from the induced dipole moments of all molecules including the test molecule (i.e., $E = E_{self} + E_{else}$), where E_{self} is the electric field produced by the induced dipole moment of the test molecule. Since E_{self} is related to E_{else} and in turn to E, as given in (4.20.b), we have the necessary elements to derive the electric displacement D and, in turn, the macroscopic susceptibility χ_e, dielectric constant $\mathsf{K} = (1 + \chi_e)$, and refractive index $n = \sqrt{\mathsf{K}}$. All relate back to the microscopic polarizability α, or the average polarizability a given in (4.20.c):

$$P = \mathsf{e}_0 \chi_e E;\ E = E_{self} + E_{else};\ P = \mathcal{N} p = \mathcal{N} \alpha E_{else},\ \text{and}$$
$$E_{self} = -\frac{P}{3\mathsf{e}_0} \rightarrow E_{self} = -\frac{\mathcal{N}\alpha}{3\mathsf{e}_0} E_{else} \rightarrow E_{else} = \left(1 - \frac{\mathcal{N}\alpha}{3\mathsf{e}_0}\right)^{-1} E; \tag{4.20.b}$$

$$D = \mathsf{e}_0 \left[1 + \mathcal{N}\alpha \left(1 - \frac{\mathcal{N}\alpha}{3\mathsf{e}_0}\right)^{-1}\right] E = \mathsf{e}_0 (1 + \chi_e) E;\ \mathsf{K} \equiv \frac{\mathsf{e}}{\mathsf{e}_0}.$$
$$\therefore\ \mathsf{K} = n^2 = \left[1 + \mathcal{N}\alpha \left(1 - \frac{\mathcal{N}\alpha}{3\mathsf{e}_0}\right)^{-1}\right] \rightarrow \alpha \equiv a = \frac{3\mathsf{e}_0}{\mathcal{N}} \frac{\mathsf{K} - 1}{\mathsf{K} + 2} = \frac{3\mathsf{e}_0}{\mathcal{N}} \frac{n^2 - 1}{n^2 + 2}. \tag{4.20.c}$$

The last relation of (4.20.c) is referred to as the Clausius–Mossotti formula (in terms of K), or the Lorentz–Lorentz equation (in terms of n); see p. 192 of [4.11]. Using the Lorentz–Lorentz equation, we can now relate σ^C and $\sigma^R = \sigma^C F_K$ to refractive index as

$$\sigma^C = \frac{\omega^4}{16\pi^2 \mathsf{e}_0^2 c^4} \left(\frac{8\pi}{3}\right) a^2 = \frac{24\pi^3}{\mathcal{N}^2 \lambda^4} \left(\frac{n^2 - 1}{n^2 + 2}\right)^2 \xrightarrow{n \approx 1} \frac{32\pi^3}{3\mathcal{N}^2 \lambda^4} (n - 1)^2;$$
$$\sigma^R = \sigma^C F_K = \frac{24\pi^3}{\mathcal{N}^2 \lambda^4} \left(\frac{n^2 - 1}{n^2 + 2}\right)^2 F_K;\ F_K = 1 + \frac{2}{9} R_A = \left(\frac{6 + 3\delta_0^t}{6 - 7\delta_0^t}\right). \tag{4.20.d}$$

We note that the approximated expression in the first row of (4.20.d) is in agreement with (7) of [4.5], and the expression in the second row is identical to (96) of the classic article by King [4.12]. It also agrees with (A8) of [4.5] and (4) and (5) in a 2005 article [4.13] that employed this equation to determine σ^R at different wavelengths from measured/calculated refractive indices and depolarization ratios δ_0^t of the atmosphere under a variety of conditions, including moisture content.

4.3 Rayleigh and Vibrational Raman Spectra of Nitrogen and Oxygen Molecules

As pointed out in Chapter 1 and illustrated in Fig. 1.2, we follow [4.5] and [4.10] and use modern terminology to classify the light scattering spectrum.

Rayleigh scattering consists of Cabannes scattering and pure rotational Raman (PRR) scattering. The Cabannes scattering spectrum results from fluctuations in molecular translational motions. Its scattering intensity is proportional to the normalized Cabannes frequency spectrum $\mathscr{R}(\nu_S - \nu_L; T, P)$, with ν_S and ν_L being, respectively, the frequencies of scattered and incident radiation. The shape of the spectrum depends on temperature and pressure and will be discussed in Section 4.4. The PRR scattering is the result of molecular rotational motion without vibrations (i.e., no change in vibrational quantum number). Since Q-branch PRR scattering (depolarized, proportional to γ^2) transitions do not change the rotational quantum number either, their spectra are centered around and not resolvable from the Cabannes scattering spectrum (polarized and proportional to a^2), that is, they are presumably buried in it. The O- and S-branches of PRR scattering are frequency resolvable, with their centers shifted away from the Cabannes scattering spectrum. The shifts of PRR lines are seen both toward higher frequency – termed anti-Stokes ($\nu_s > \nu_L$) – and toward lower frequency – termed Stokes ($\nu_s < \nu_L$).

There are two groups of vibrational Raman scattering lines, resulting from vibrational motion at frequency ν_0 (pure vibrational Raman, or PVR) along with associated vibrational-rotational motion (vibrational-rotational Raman, or VRR) of the molecule. The Stokes group (centered at $\nu_s - \nu_L = -\nu_0$) results from the $\mathrm{v} = 0$ to $\mathrm{v} = 1$ transition, and the anti-Stokes group (centered at $\nu_s - \nu_L = \nu_0$) results from $\mathrm{v} = 1$ to $\mathrm{v} = 0$. In each group, there are three branches, O-, Q-, and S-branches, respectively corresponding to the simultaneous rotational transitions from J to J–2, J, and J+2. Equation (4.9.e) gives the differential scattering cross section into the (θ, ϕ) direction from an isotropically oriented ensemble of linear molecules when both scattering polarizations are received. For backscattering, the differential cross section is given in (4.18) with Cabannes and PRR components in separate terms, respectively proportional to a^2 and γ^2; the differential cross section is expressed similarly for PVR and VRR, respectively proportional to a'^2 and γ'^2.

For a linear molecular ensemble at thermal equilibrium at temperature T, the scattering cross sections should be weighted by the appropriate population product of initial vibrational and rotational states, $N(\mathrm{v}) \cdot N(J)$, and (4.18) can then be generalized to include VRR as (4.21):

$$\frac{d\sigma_{a+b}^{\pi,Ray}}{d\Omega} = \frac{\omega^4}{16\pi^2 \mathrm{e}_0^2 c^4} N(\mathrm{v}) N(J) \left(a^2 + \frac{7}{45}\gamma^2 \right);$$
$$\frac{d\sigma_{a+b}^{\pi,Ram}}{d\Omega} = \frac{\omega^4}{16\pi^2 \mathrm{e}_0^2 c^4} N(\mathrm{v}) N(J) \left(a'^2 + \frac{7}{45}\gamma'^2 \right), \tag{4.21}$$

where a^2 and γ^2, and a'^2 and γ'^2 are, respectively, the squares of the mean (third of the trace) and anisotropy polarizabilities for polarized and depolarized Rayleigh and vibrational Raman scattering. Before we discuss the frequency spectra, we need to review various relevant transition energies as well as the population factors, $N(v)$ and $N(J)$.

4.3.1 Rayleigh and Vibrational Raman Transition Energies

Since we are interested in nonresonant scattering resulting from transitions between rotational and vibrational states within the same ground electronic state, the electronic term of the ground electronic state may be taken as zero. In the quasi-harmonic approximation discussed in Section A.3, we have defined the vibrational frequency ν_0 and rotational constant B_{v} and we now also set the energy of the ground vibrational level $\nu_0/2$ to zero as reference; the eigenstate $|e^g>|\mathrm{v},J>$ energy, $E_{e^g,\mathrm{v},J}$, expressed in units of cm^{-1} – see (A.21.d) and Table A.4 – is then

$$E_{e^g,\mathrm{v},J} = \mathrm{v}\nu_0 + B_{\mathrm{v}}J(J+1); \quad \nu_0 = \nu_e(1-0.5x_e),\ B_{\mathrm{v}} = B_0 - \alpha_e\mathrm{v}. \qquad (4.22)$$

The molecular energy change of the vibrational-rotational transition $\Delta E_{0,\Delta\mathrm{v},\Delta J} = E_{e^g,\mathrm{v}^f,J^f} - E_{e^g,\mathrm{v}^i=\mathrm{v},J^i=J}$ is precisely the difference between the incident light frequency ν_L and the scattering frequency ν_S, $\nu_s = \nu_L - \Delta E_{0,\Delta\mathrm{v},\Delta J}$ (in cm^{-1}). For Stokes vibrational Raman (SVR) scattering, we are concerned with the vibrational transition $|\mathrm{v}^i = \mathrm{v} = 0> \rightarrow |\mathrm{v}^f = 1>$ (molecule gaining vibrational energy) with $\Delta\mathrm{v} = \mathrm{v}_f - \mathrm{v}_i = 1$, and $\Delta J = J^f - J^i = -2,\ 0,\ 2$. The energy differences between final and initial states $\Delta E_{0,\Delta\mathrm{v}=1,\Delta J}$ are as follows:

In the O-branch, $\Delta J = -2$ (i.e., $J^i = J \rightarrow J^f = J-2$), the molecule loses rotational energy, so

$$\begin{aligned}\Delta E_{0,1,-2} &= \nu_0 + B_1(J-2)(J-1) - B_0J(J+1)\\ &= \nu_0 + (B_1 - B_0)J^2 - (3B_1 + B_0)J + (2B_1). \qquad (4.23.a)\end{aligned}$$

In the Q-branch, $\Delta J = 0$ (i.e., $J^i = J \rightarrow J^f = J$), the molecule experiences no rotational energy change, so

$$\Delta E_{0,1,0} = \nu_0 + (B_1 - B_0)J(J+1) = \nu_0 + (B_1 - B_0)J^2 + (B_1 - B_0)J. \quad (4.23.b)$$

In the S-branch, $\Delta J = +2$ (i.e., $J^i = J \rightarrow J^f = J+2$), the molecule gains rotation energy, so

$$\Delta E_{0,1,+2} = \nu_0 + B_1(J+2)(J+3) - B_0 J(J+1)$$
$$= \nu_0 + (B_1 - B_0)J^2 + (5B_1 - B_0)J + (6B_1). \quad (4.23.c)$$

For depolarized PRR scattering, in which there is no vibrational energy change, the transition energies for the O- ($\Delta J = -2$), Q- ($\Delta J = 0$), and S- ($\Delta J = +2$) branches, $\Delta E_{0,0,\Delta J} = \nu_L - \nu_s$, may be obtained directly from (4.23.a), (4.23.b), and (4.23.c) by setting $\nu_0 = 0$, and $B_1 = B_0$. Thus, the Q-branch PRR scattering spectrum coincides with the Cabannes scattering spectrum.

Table 4.3 *Vibration and rotation constants for molecular energies**

Gas	ν_0 in cm^{-1}	B_0 in cm^{-1}	B_1 in cm^{-1}	D_0 in cm^{-1}	D_1 in cm^{-1}	Comment
N_2	2329.9	1.98957	1.97219	5.7606×10^{-6}	5.7610×10^{-6}	From [4.15]
O_2	1556.4	1.43768	1.42188	4.85×10^{-6}	4.864×10^{-6}	From [4.16]

*Values taken from [4.14], [4.15], and [4.16]

In the case of anti-Stokes vibrational Raman (AVR) scattering, we are concerned with the vibrational transition $|\mathrm{v}=1\rangle \to |\mathrm{v}^f=0\rangle$. The scattering frequency, ν_S, relative to the incident frequency, ν_L, is $\nu_S = \nu_L + \Delta E_{0,-1,\Delta J}$ (in cm^{-1}), in which the molecule loses vibrational energy, and the relevant transitions are $\Delta\mathrm{v} = -1$, and $\Delta J = -2,\ 0,\ 2$. Thus, the transition energies for the O-branch ($\Delta J = -2$), Q-branch ($\Delta J = 0$), and S-branch ($\Delta J = +2$), $\Delta E_{0,-1,\Delta J}$, may be obtained respectively from (4.23.a), (4.23.b), and (4.23.c) by changing ν_0 to $-\nu_0$ and interchanging B_1 and B_0. Since $B_1 \approx B_0$, as can be seen in Table 4.3, we set $B_1 \approx B_0$ for most of the discussion in this book except in Section 4.3.4 when we discuss rotation-vibration coupling.

In contrast to those of Table A.4, the constants in Table 4.3 were deduced from more recent measurements, and with higher resolution. For example, the value of $\nu_0 = 2329.9$ cm^{-1} for the N_2 molecule in Table 4.3 is different from the value $\nu_0 = 2345.5$ cm^{-1}, calculated from constants in Table A.4. For our presentation of Raman spectra, the values in Table 4.3 will be used.

4.3.2 Population of a Rotational and Vibrational State $|\mathrm{v}, J\rangle$

Up to this point, we have discussed various invariants of the polarizability tensor associated with an initial rotation-vibration state $|\mathrm{v}, J\rangle$ and its transition to a final state $|\mathrm{v}^f, J^f\rangle$, both in the ground electronic state $|e^g\rangle$. The typical vibrational energy, ν_0 (rotational energy, B_v), is 2,000 cm^{-1} (2 cm^{-1}) (see Table A.4 for N_2 and O_2 molecules). Since 1 cm^{-1} is 1.44 K, 200–300 K may be considered high-temperature for rotational motion, while it is regarded as low-temperature for vibrational motion. Each occupied level/state contributes to the differential scattering cross section (or to the total polarizability invariants) weighted by its population.

To determine the probability of occupation or population of a level with the same v but different J, in addition to the rotational degeneracy $(2J+1)$, we must consider nuclear spin I and its degeneracy, $(2I+1)$. The population, $N(J)$, and its associated partition function, $Z(T)$, as a function of rotational energy, B_0 (in cm^{-1}) and temperature, T (in K), respectively, are

$$N(J) = \frac{(2I+1)(2J+1)}{Z(T)} \mathrm{e}^{-hcB_0 J(J+1)/k_B T};$$

$$Z(T) = \sum_{I,J} (2I+1)(2J+1)\mathrm{e}^{-hcB_0 J(J+1)/k_B T}. \qquad (4.24.a)$$

For homogeneous diatomic molecules, the association of angular momentum, J, to the nuclear spin, I, is dictated by the Pauli exclusion principle. The fact that nuclear spins are 1 and 0 for N and O atoms, respectively, gives rise to molecular nuclear spins of $I = 0$, 1, and 2 for nitrogen and $I = 0$ for oxygen molecules. Since both oxygen and nitrogen molecules have integral spin, according to the Pauli exclusion principle their total wave function, $|\Psi_{total}>$, which is the product of $|I>$ and $|\mathrm{v}>|e^g>|J>$, with $|I>$ being the nuclear spin wave function, must be symmetric (or even) with respect to the exchange of (identical) nuclei. However, as discussed in Appendix A, the ground-state orbital electronic wave functions for nitrogen and oxygen are, respectively, symmetric [*singlet* (electron spin), ${}^1\Sigma_g{}^+$] and antisymmetric [*triplet* (electron spin), ${}^3\Sigma_g{}^-$] relative to the exchange of nuclei and the parity of nuclear spin states is even (+) for $I = 0$ and 2, and odd for $I = 1$. We can then determine the exchange symmetry of J from Table 4.4 by demanding that the total wave function $|\Psi_{total}>$ be symmetric with respect to nuclear exchange, leading to a specific association between J and I. This restricts the available rotational states. Following Table 4.4, we conclude that for oxygen molecules, only levels with odd J can be populated, and for nitrogen molecules, levels with even J values are doubly populated compared to those with odd J.

With the restrictions resulting from the above discussions, we may derive the population $N(J)$ and its associated partition function $Z(T)$ from (4.24.a) for molecular nitrogen and oxygen, respectively, as (4.24.b) and (4.24.c):

$$N(J = \text{even}) = \frac{6(2J+1)}{Z_{N_2}(T)} \mathrm{e}^{-hcB_{N_2} J(J+1)/k_B T};$$

$$N(J = \text{odd}) = \frac{3(2J+1)}{Z_{N_2}(T)} \mathrm{e}^{-hcB_{N_2} J(J+1)/k_B T}, \quad \text{where}$$

$$Z_{N_2}(T) = 6\sum_{J=\text{even}} (2J+1)e^{-hcB_{N_2}J(J+1)/k_BT} + 3\sum_{J=\text{odd}} (2J+1)e^{-hcB_{N_2}J(J+1)/k_BT}, \text{ and} \qquad (4.24.b)$$

$$N(J=\text{odd}) = \frac{(2J+1)}{Z_{O_2}(T)} e^{-hcB_{O_2}J(J+1)/k_BT}, \text{ where}$$

$$Z_{O_2}(T) = \sum_{J=\text{odd}} (2J+1)e^{-hcB_{O_2}J(J+1)/k_BT};$$

$$N(J=\text{even}) = 0. \qquad (4.24.c)$$

Table 4.4 *Exchange symmetry, components of wave function and associated rotations*

Molecule	I	$g_I = 2I+1$	$\|I>$	$\|e^g>$	$\|v>$	$\|J>$	Allowed $\|J>$ for even $\|\Psi_{total}>$
N_2, $^1\Sigma_g^+$	0, 2	1+5 = 6	(+)	(+)	(+)	(+)	Even J level populated ($g_I = 6$)
N_2, $^1\Sigma_g^+$	1	3	(−)	(+)	(+)	(−)	Odd J level populated ($g_I = 3$)
O_2, $^3\Sigma_g^-$	0	1	(+)	(−)	(+)	(−)	Only odd J level populated

Since atmospheric thermal energy is much higher than the molecular rotational energy (i.e., $hcB_0 \ll k_BT$), the summations in the partition functions over either odd or even J values may be approximated by their associated integrals, resulting in $k_BT/(2hcB_0)$. Multiplying by the molecular nuclear spin degeneracy (9 for N_2 and 1 for O_2), we have $Z_{N_2}(T) \approx 9k_BT/(2hcB_{N_2})$ and $Z_{O_2}(T) \approx k_BT/(2hcB_{O_2})$. Since, at temperatures of interest, the thermal energy is negligible compared to the vibrational energies, most molecules occupy the ground vibrational state (v = 0). Nonetheless, theoretically, thermal statistics allow us to calculate the population of the ground and the first excited vibrational states), $N(0)$ and $N(1)$, as

$$N(0) = \frac{1}{Z(\text{v})}, \quad N(1) = \frac{e^{-h\nu_0/kT}}{Z(\text{v})}, \quad \text{and } Z(\text{v}) = \sum_{\text{v}=0}^{\text{v}=1} e^{-hc\text{v}\nu_0/kT} \approx (1 + e^{-h\nu_0/kT}). \qquad (4.24.d)$$

Using the data from Table 4.3 with $\nu_0 = 2331\ \text{cm}^{-1}$ for nitrogen and $\nu_0 = 1556\ \text{cm}^{-1}$ for oxygen, we calculate the populations at 250 K to be $N(1) = 1.5\times10^{-6}$ for N_2 and 1.3×10^{-4} for O_2, and $N(0) = 0.999998$ for N_2 and 0.9998 for O_2. Practically speaking, only the ground vibrational state is occupied.

4.3.3 Rayleigh and Vibrational Raman Scattering Spectra

As an example, we show the molecular transitions and the associated spectral lines for Raman scattering from oxygen molecules at 250 K in Fig. 4.2. We show the allowed transitions from the $J = 2$ initial state in the upper panel, and the spectral

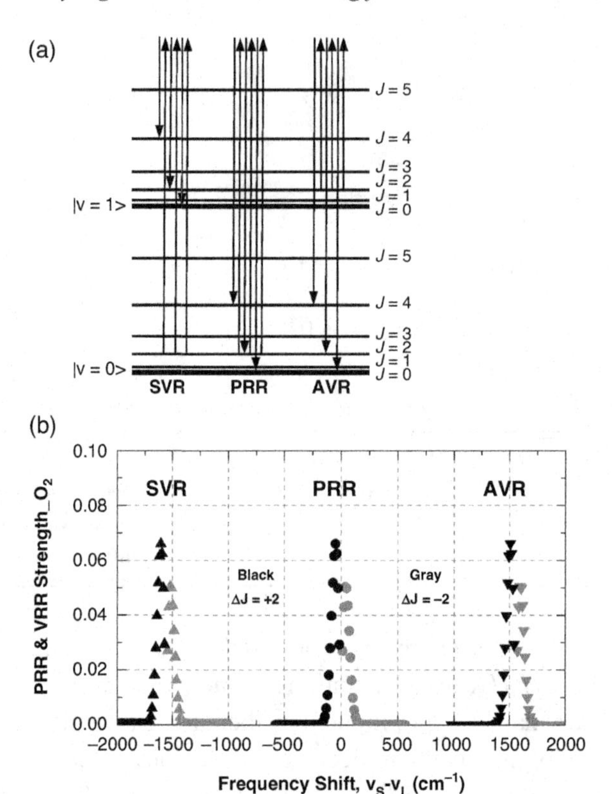

Fig.4.2 A schematic of spectral lines of Raman scattering from a homogeneous diatomic molecule in three groups: SVR ($\Delta J = +2$, S-branch), PRR, and ARR ($\Delta J = -2$, O-branch) with N(v) set to unity in lower panel, see text for details.

strength of all lines in the lower panel – normalized within each group of SVR, PRR, and AVR. The scale of the strength for PRR and VRR (i.e., SVR and AVR) lines are, respectively, in units of $(\gamma/4\pi e_0)^2$ and $(\gamma'/4\pi e_0)^2$. The Q-branch lines (not shown) of the PRR and VRR spectra are not resolved, but their shared strengths summed over all J's represent ~ 25% of the whole for the respective group/scale. In these frequency locations (i.e., at 0 cm^{-1}, and – ν_0 cm^{-1} and + ν_0 cm^{-1}), there is, in addition, polarized scattering with strengths of Cabannes and pure vibrational scattering, $(a/4\pi e_0)^2$ and $(a'/4\pi e_0)^2$, respectively; see Table 4.2. In these plots, the vibrational state population N(v) was set to unity. Since at 250 K, $N(1) = 1.5\times10^{-6}$ for N_2 and 1.3×10^{-4} for O_2, as opposed to $N(0) = 0.999998$ and 0.9998, respectively, anti-Stokes VRR (AVR) scattering is negligible compared to Stokes VRR (SVR) were we to account for the effect of N(v).

In Fig. 4.2, the Stokes vibrational-rotational Raman (SVR) scattering (group centered at $\nu_S - \nu_L = -\nu_0$, left), pure rotational Raman (PRR) scattering (group centered at $\nu_S - \nu_L = 0$, center), and anti-Stokes vibrational-rotational Raman scattering (AVR) (group centered at $\nu_S - \nu_L = \nu_0$, right) are shown. In the upper panel, within each of the three vibrational groups, we select $J = 2$ as the initial state as an example, and show all allowed transitions in order O-, Q-, and S-branch. Note that the energy scale between the two thick horizontal lines marked with $|v = 0>$ and $|v = 1>$ is much larger (1556 cm^{-1} for oxygen and 2331 cm^{-1} for nitrogen, for example) than the scale between the thin lines, say about 25 cm^{-1} between $J = 0$ and $J = 5$. Though for PRR, the $\Delta J = -2$ (O-branch, gray) and +2 (S-branch, black) lines are often referred to, respectively, as anti-Stokes and Stokes lines, we prefer to use the value of ΔJ to distinguish the subgrouping within each group of SVR, PRR, and AVR: $\Delta J = -2$ depicts the molecule losing rotational energy, and the reverse for $\Delta J = +2$. In the lower panel, the spectral strengths of lines normalized within each group of SVR, PRR, and AVR for molecular oxygen are shown, including the effects of both the population of the rotational states $N(J)$ at 250 K and the Placzek–Teller coefficients with N(v) set to unity.

To understand the spectral structure of Rayleigh and vibrational Raman scattering, we first consider ensembles of nitrogen and oxygen molecules at a constant temperature, say 250 K, and plot the fractional population of rotational levels $N(J)$, (4.24.b) and (4.24.c), at the ground vibrational state as a function of energy (in cm^{-1}), that is, setting v = 0 in (4.22). The results are shown in Fig. 4.3, where only levels with odd J values for oxygen molecules are occupied (black circles). For molecular nitrogen, however, the levels with even J values (black triangles) have twice the population of those with odd J values (black inverted triangles). The sum of the fractional populations for all allowed J values of a given molecule is unity, as

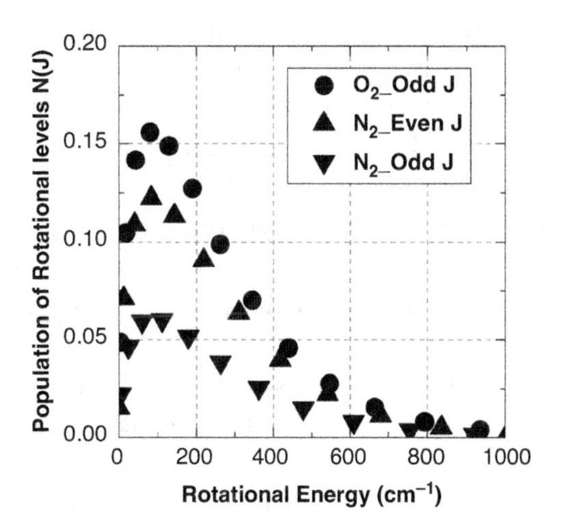

Fig. 4.3 Population of rotational levels at the ground vibrational state as a function of rotational energy.

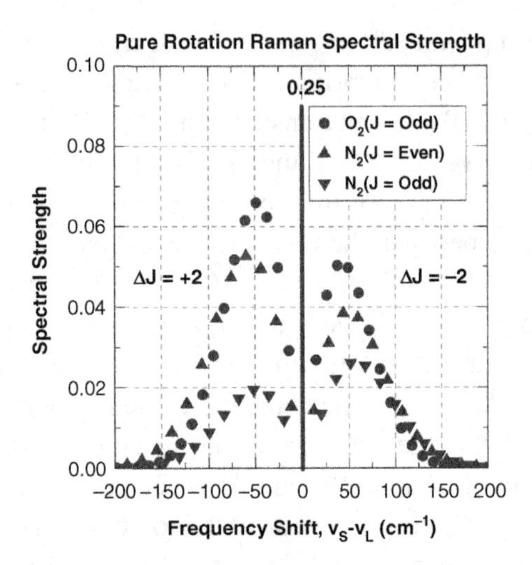

Fig. 4.4 PRR spectra of oxygen (circles) and nitrogen (J even: triangles + J odd: inverted triangles). The anti-Stokes ($\nu_S > \nu_L$) and Stokes ($\nu_S < \nu_L$) scatterings are due to $\Delta J = -2$ and $\Delta J = +2$ transitions, respectively. The sum of the strengths of all N_2 (or O_2) lines is unity, thus these may be referred to as normalized PRR spectra. Notice that the range of frequency shifts is much smaller than that of rotational energy in Fig. 4.3.

one would expect – for both oxygen and nitrogen. Note that the horizontal axis of Fig. 4.3 is the rotational energy for all allowed J values; at 250 K, the peak populations occur for oxygen at $J = 7$, while for nitrogen they occur at $J = 7$ for odd J and $J = 6$ for even J.

Now, if we take the oxygen PRR spectrum (black circles in Fig. 4.4) along with the vertical line indicating the unshifted component and move them to the right (left), centered on $+1556\ \mathrm{cm}^{-1}$ ($-1556\ \mathrm{cm}^{-1}$), we then obtain the anti-Stokes (Stokes) vibrational-rotational Raman scattering spectrum of oxygen. The same procedure with the nitrogen PRR spectrum (triangles and inverted triangles in Fig. 4.4), but shifting by $+2331\ \mathrm{cm}^{-1}$ ($-2331\ \mathrm{cm}^{-1}$), gives us the anti-Stokes (Stokes) vibrational-rotational Raman scattering spectrum for nitrogen. If the thermal population of the vibrational state is then accounted for, the strengths of PRR and Stokes VRR spectra are nearly the same, but strength of the anti-Stokes VRR spectra would be substantially diminished. If the vibrational Raman spectrum of the atmosphere at 250 K is of interest, one then needs to combine the lines indicated by the circles and the upright and inverted triangles with weighting factors of 0.21 for oxygen and 0.79 for nitrogen to account for their relative abundances, as well as scaling to the relative line strengths $(\gamma'/4\pi e_0)^2$ between the two species (see Table 4.2). Due to rotational-vibrational coupling, to be discussed in Section 4.3.4, there is a nonzero difference between rotational constants B_0 and B_1, $B_1 - B_0 \approx 0.017\ \text{cm}^{-1} \neq 0$, according to Table 4.3; thus, the scattering frequency of the Q-branch VRR depends on rotational quantum number J, as seen from (4.23.b).

4.3.4 Rotation-Vibration Coupling in the Q-Branch VRR Spectra

Since the rotational constants for the ground vibrational state, B_0 and D_0, are different from those of the first excited vibrational state, B_1 and D_1, the spacings between the pure rotational Raman (PRR) lines are, strictly speaking, different from those between the vibrational rotation Raman (VRR) lines. However, their differences are small, and both O-branch and S-branch Raman lines of the two spectral groups in fact look quite similar. In contrast, the high-resolution Q-branch Raman lines of PRR look quite different from those of VRR, since there is vibration-rotation coupling for VRR lines but not for PRR lines, and the Q-branch PRR lines are, in fact, degenerate, with $\nu_S = \nu_L$, independent of J. The scattering frequencies of the Q-branch VRR lines are given by $\nu_S = \nu_L - [\nu_0 + (B_1 - B_0)J(J+1)]$ when the much smaller D_0 and D_1 are neglected, and this corresponds to a Raman shift of $[\nu_0 + (B_1 - B_0)J(J+1)]$ that depends on the value of J. Since $B_1 - B_0$ is quite small, the typical resolution of optical spectrometers cannot distinguish the rotational components. In 2000, Bendtsen and Rasmussen [4.15] measured the vibrational Raman spectrum of $^{14}N_2$, using a high-resolution incoherent Fourier transform spectrograph with a precision of 0.015 cm^{-1} and resolved the Q-branch fundamental VRR lines, deducing values of $\nu_0 = 2329.9\ \text{cm}^{-1}$ and $B_1 - B_0 = -0.0174\ \text{cm}^{-1}$, which we have included in Table 4.3. Using these values, we calculate the Q-branch of VRR lines for different J values as follows:

J	1	2	3	4	5	6	7
$\Delta\nu$ (cm^{-1})	2329.9	2329.8	2329.7	2329.6	2329.4	2329.2	2328.9
J	8	9	10	11	12	13	14
$\Delta\nu$ (cm^{-1})	2328.7	2328.3	2328.0	2327.6	2327.2	2326.7	2326.3

In the late 1980s, there was strong interest in coherent nonlinear laser spectroscopy for gas diagnostics. These pump/probe laser techniques include coherent anti-Stokes Raman scattering (CARS), stimulated Raman gain spectroscopy (SRS), and inverse Raman spectroscopy (IRS); see, for example, [4.17] and chapter 1 in [4.2]. The spectral resolution of these similar techniques is essentially limited by the linewidth of the pump laser, which has been steadily improving. In their 1986 experiment, Rahn and Palmer [4.18] utilized the IRS technique and measured the Q-branch VRR spectrum of N_2 at various temperatures and pressures with a resolution of 0.0015 cm^{-1}, an order of magnitude

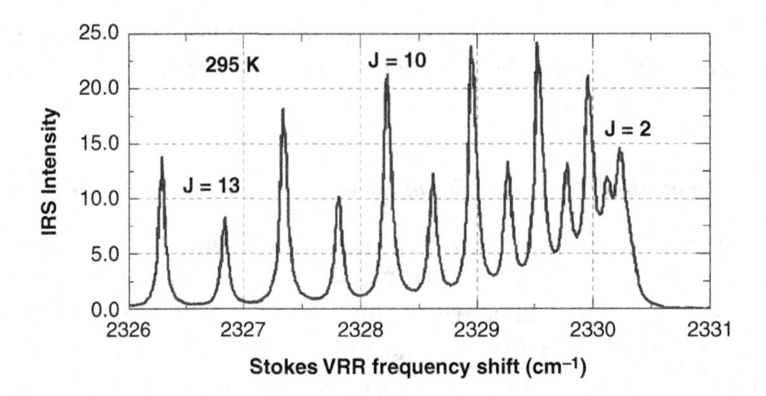

Fig. 4.5 An inverse Raman spectrum (IRS) of the N_2 Q-branch VRR lines at 295 K and 1 atm. The lines from $J = 2$ (right) to $J = 14$ (left) are clearly seen. Adapted with permission from [4.18]. © The Optical Society.

improvement over the incoherent (spontaneous scattering) technique [4.15]. Their inverse Raman spectrum (IRS) of N_2 at 1 atm. and 295 K in the middle panel of Fig. 2 of [4.18], is adapted here as Fig. 4.5. Since the center of the $J = 1$ line is only ~ 0.036 cm^{-1} from the band head at 2329.9 cm^{-1}, it is unresolved in Fig. 4.5. The line center positions of the IRS spectrum shown for $J = 2$ to $J = 14$ appear to agree with those presented in [4.15] by the incoherent method. We note that, as expected, the intensity of even-value J lines is a factor of 2 stronger than the neighboring odd-value J lines. The factors that determine the lineshape and width of these Q-branch VRR lines are complex; as mentioned in Section 2.3, in addition to Doppler broadening there are several effects due to collisions as density increases, including dephasing broadening, and Dicke narrowing, which result in a collapse of the inhomogeneous Doppler broadening. In addition, when the collision rate is comparable to the vibration-rotation coupling, the neighboring low J Q-branch VRR lines can merge and narrow as a result of inelastic collisions, termed collisional/motional narrowing, as discussed, for example, in [4.19]. As mentioned in Section 2.3, Rahn et al. were able to fit the resolved individual Q-branch VRR lines in nitrogen to the Galatry model [4.18]. However, under atmospheric conditions (at 1 atm. or less), most measured linewidths are not that different from the Doppler broadening of ~ 0.06 cm^{-1} at 295K. At any rate, only the relative strengths or intensities of the PRR and VRR are important for atmospheric lidar applications. On the other hand, the shape of the Cabannes line, along with its temperature and pressure dependence, is very important for lidar applications and will be further discussed in Section 4.4 and elucidated in Chapter 7.

4.4 Cabannes Scattering Spectra

The shape of the Cabannes scattering spectrum is the result of random (entropy fluctuations) and organized (propagating pressure perturbations) translational molecular motions. These lead respectively to Rayleigh and Brillouin scattering, as commonly referred to in the literature. Einstein, in 1910, suggested [4.20] that light scattering intensity is proportional to the square of permittivity fluctuations, $\delta\mathsf{e}(\vec{r},t)$. Since molecular permittivity is the sum of vacuum permittivity and the product of the molecular number density $\mathcal{N}(\vec{r},t)$ and electronic polarizability per molecule, α, it can be expressed as $\mathsf{e} \equiv \mathsf{e}_0 + \alpha\mathcal{N}$. The perturbation in permittivity, $\delta\mathsf{e}(\vec{r},t)$, like that of enthalpy (see (A.23.b)), may be expressed in terms of perturbations in entropy and in pressure as

$$\delta\mathsf{e}(\vec{r},t) = \alpha[\delta\mathcal{N}(\vec{r},t)] = \alpha\left[\left(\frac{\partial\mathcal{N}}{\partial S}\right)_p \delta S + \left(\frac{\partial\mathcal{N}}{\partial p}\right)_S \delta p\right]. \tag{4.25}$$

At very low gas density, when molecules barely make sufficient collisions to maintain thermal equilibrium, the two terms are considered to be (nearly) independent. In this regime (termed the Knudson regime), molecular density fluctuations mainly result from entropy fluctuations, and the molecules experience occasional and random collisions, which have minimal effect on the equilibrium (Maxwellian) speed distribution, and thus no organized motion (or sound wave) can transmit or propagate. As the gas density increases, the collision frequency increases appreciably. The ensemble of gas reaches an intermediate, kinetic regime, in which growing pressure perturbations add to entropy fluctuations. Finally, at still higher gas density, the gas is said to be in the hydrodynamic regime. Here, collisions occur so frequently that organized pressure perturbations, such as sound waves, initiated either spontaneously or driven by external forces, can propagate collectively. These give rise to the direction-dependent light scattering, termed Brillouin scattering.

The density perturbations in a dense gas are thought to consist of normal-mode components, each specified by a wave vector and angular frequency $(\vec{K},\ \Omega)$. When observing normal-mode fluctuations by spontaneous light scattering, the corresponding $\vec{K}$ wave-vector component is determined by the observation direction relative to the incident (exciting) laser beam direction. Alternatively, in a coherent light scattering experiment, the pump and probe laser beam directions, $\vec{k}_L$ and $\vec{k}_S$, select the normal mode wave vector through momentum conservation (i.e., $\vec{K} = \vec{k}_S \pm \vec{k}_L$). A schematic relating the wave vectors and associated frequencies is shown in Fig. 4.6.

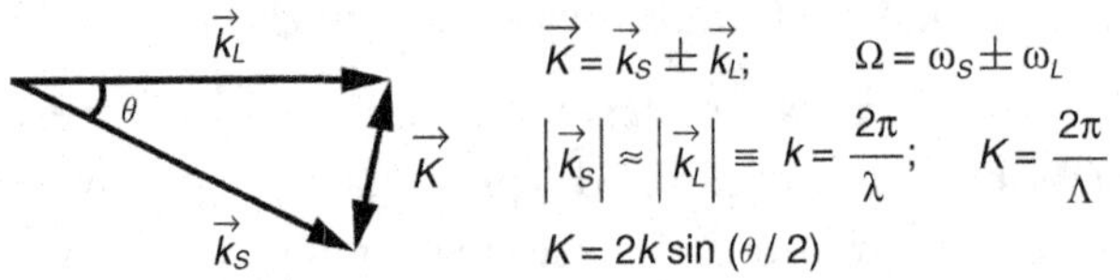

$\vec{k}_L$ θ $\vec{K}$ $\vec{k}_S$

$$\vec{K} = \vec{k}_S \pm \vec{k}_L; \qquad \Omega = \omega_S \pm \omega_L$$

$$\left|\vec{k}_S\right| \approx \left|\vec{k}_L\right| \equiv k = \frac{2\pi}{\lambda}; \qquad K = \frac{2\pi}{\Lambda}$$

$$K = 2k \sin(\theta/2)$$

Fig. 4.6 The scattering geometry showing the perturbation wave vector and frequency $(\vec{K}, \Omega)$ in the medium responsible for light scattering. These perturbation parameters are related to the medium's wave speed c_S as $\Omega = c_S K$; the perturbation frequency thus increases as the scattering angle θ increases, a property of Bragg scattering.

As is well known, for a real signal $f(t)$ and its Fourier transform $F(\omega)$, the Fourier transform of its autocorrelation function $g(t)$ is its power spectrum, $|F(\omega)|^2$ (see p. 242 of [4.21])). Mathematically, by writing $F(\omega)$ as the product of amplitude and phase (i.e., $F(\omega) = A(\omega)e^{i\phi(\omega)}$), one has

$$F(\omega) = \int_{-\infty}^{\infty} f(t)e^{i\omega t}dt\,; \quad g(t) = \int_{-\infty}^{\infty} f(t-\tau)f(\tau)d\tau \rightarrow$$

$$|F(\omega)|^2 = A^2(\omega) = \int_{-\infty}^{\infty} g(t)e^{i\omega t}dt.$$

Since the scattered light signal is proportional to the fluctuations of space-time permittivity perturbations $\delta\mathrm{e}(\vec{r}, t)$ and in turn to density perturbations $\delta\mathcal{N}(\vec{r}, t)$, the intensity of scattered light or its light scattering power spectrum $S(\vec{K}, \Omega)$ is then proportional to the double (time-space) Fourier transform of the time-displaced, density–density correlation function $G(\vec{r}, t)$ of number density perturbations $\delta\mathcal{N}(\vec{r}, t)$, $S(\vec{K}, \Omega) \propto \int_{-\infty}^{+\infty} dt \int dr^3 G(\vec{r}, t)e^{i(\vec{K}\bullet\vec{r}\, -\, \omega t)}$; see (1.5) of [4.20] and (5) of [4.22]. For a given $\vec{K}$ component, the power spectrum in frequency then maps out the strengths of perturbation normal modes as a function of frequency, with $\Omega = \omega_S \pm \omega_L$ complying with energy conservation. Here, ω_S and ω_L are, respectively, the scattering and inducing optical frequencies, and the plus (minus) sign corresponds to the gas losing (gaining) energy. Figuring out the wave vector–dependent power spectrum, $S(\vec{K}, \Omega)$, for Rayleigh and Brillouin scattering (termed Cabannes scattering in this book) of a molecular ensemble amounts to the determination of its density perturbations, $\delta\mathcal{N}(\vec{r}, t)$.

The theory of the light scattering spectrum $S(\vec{K}, \Omega)$, or the dynamics of its associated gas density perturbations, has been well known for the Knudson and hydrodynamic regimes. In the Knudson regime, with rare collisions, the spectrum can be related to the Doppler spectrum, or the Voigt spectrum with dephasing collision broadening as discussed in Section 2.3. In the hydrodynamic regime, the theory of the light scattering spectrum has been given, for example, by Herman and Gray [4.23]. The understanding of the kinetic regime turns out to be much more challenging. Indeed, it took the intensive efforts of many research groups from the beginning of the 1960s to the late 1970s to figure out the theoretical light scattering spectra for pure monoatomic and diatomic gases in order to agree with the experimental results. Yip and Nelson [4.22] were among the first to solve a single kinetic equation (also termed linearized Boltzmann equation) that complies with both the first and the second law of thermodynamics, and thus determined $S(\vec{K}, \Omega)$ for a monoatomic system in the kinetic regime. A model for molecular gases needs to address the effect of internal degrees of freedom, thus a single kinetic equation is obviously inadequate. Kinetic equations that accommodate these complications were then proposed, and, based on these, Tenti et al. [4.24] finally developed a kinetic model that produced agreeable Rayleigh–Brillouin (Cabannes) spectra from pure molecular gases in the kinetic regime. These efforts exemplify the exciting and sometimes complicated history of scientific research. A short and excellent review of this historical development may be found in a 2003 Ph.D. thesis [4.20].

Since the mean polarizability, $a = \left(\alpha_{\parallel} + 2\alpha_{\perp}\right)/3$, and trace scattering from the molecular ensemble is responsible for Cabannes scattering, the molecular system is considered isotropic as far as translational motion is concerned. In an isotropic medium, the scattering power spectrum depends only on the magnitude of the wave vector (i.e. $S(\vec{K}, \Omega) = S(K, \Omega)$). To avoid confusion, we call $\vec{K}$ and Ω, respectively, the perturbation wave vector and perturbation frequency (or frequency shift; not to be confused with solid-angle); see Fig. 4.6. At high gas density (hydrodynamic regime), the spectra of Rayleigh and Brillouin scattering are well separated; the location of the peak of the dominating Brillouin scattering spectrum, Ω, is related to the magnitude of the perturbation wave vector, K, via the speed of wave propagation (aka sound) c_S, as $\Omega = c_S K = 2c_S|k_L|\,\sin(\theta/2)$. Interestingly, in this regime, the Brillouin scattering intensity is proportional to square of the pressure, leading to linewidth narrowing proportional to $\sin^2(\theta/2)$. The angular dependence of the line center and the width of the Brillouin scattering spectrum have been experimentally demonstrated, for example, by Tang et al. [4.25], to be in agreement with theory [4.23] in the hydrodynamic regime.

4.4.1 Kinetic Equations for Density Perturbations of Monoatomic and Diatomic Gases

A medium-density gas (or kinetic regime) is of particular interest for backscatter lidar, as the lower atmosphere falls in this regime. As mentioned, here Rayleigh and Brillouin spectra cannot be separated, and the theory to explain the exact lineshape of the composite "Rayleigh-Brillouin spectrum" $S(K,\Omega)$ of Cabannes scattering has been an area of intensive research. Recalling that the Cabannes scattering spectrum, $S(\vec{K},\Omega)$, is proportional to the space-time Fourier transform of the correlation function $G(\vec{r},t)$ of number density perturbations, $\delta\mathcal{N}(\vec{r},t)$, in order to understand its structure for a molecular ensemble, we need to invoke kinetic equations of molecular gases and evaluate the associated density perturbations $\delta\mathcal{N}(\vec{r},t)$. We have thus provided a brief review of the kinetic theory and transport of ideal gases in Section A.4. Its application to Cabannes scattering spectral lineshape is given below.

The gas number density $\mathcal{N}(\vec{r},t)=\int f(\vec{\mathrm{v}},\vec{r},t)d^3\mathrm{v}$ in space-time coordinates may be expressed as the integral of the six-dimensional molecular distribution function $f(\vec{\mathrm{v}},\vec{r},t)$ over the three-dimensional velocity space. To determine the dynamics of $\delta\mathcal{N}(\vec{r},t)$, we then must solve the initial value problem of the appropriate kinetic equations for the distribution function with an appropriate two-body collision model that satisfies the first law while constrained by the second law of thermodynamics. For a monoatomic gas, there is only one collision integral leading to a single kinetic equation for the linearized deviation of the one-particle distribution function, $h(\vec{\mathrm{v}},\vec{r},t)$, from equilibrium defined from $f(\vec{\mathrm{v}},\vec{r},t)=\mathcal{N}_0\phi(\mathrm{v})\left[1+h(\vec{\mathrm{v}},\vec{r},t)\right]$, where $\mathcal{N}_0$ and $\phi(\mathrm{v})$ are, respectively, the average number density and the 3-D equilibrium Maxwellian speed distribution. Once this problem is solved, the density perturbation, $\delta\mathcal{N}(\vec{r},t)=\mathcal{N}_0\int\phi(\mathrm{v})h(\vec{\mathrm{v}},\vec{r},t)d^3\mathrm{v}$, may be evaluated and from that its correlation function of density-density perturbations, $G(\vec{r},t)$, after which the associated wave vector–dependent spectrum $S(K,\Omega)$ may be calculated. Though complicated, the power spectrum of the density perturbation for a monoatomic system can be expressed analytically; see [4.22] and Chapter 3 of [4.20]. In this manner, Yip and Nelkin [4.22] deduced the normalized Cabannes spectrum for a monoatomic gas, $C_0(x,y)$, as the function of two dimensionless variables, x and y:

$$x=\frac{\Omega}{K\mathrm{v}_0}\ \text{and}\ y=\frac{p}{K\mathrm{v}_0\mu},\quad \text{with}\ \mathrm{v}_0=\sqrt{\frac{2k_BT}{m}}\ \text{and}\ \int_{-\infty}^{\infty}C_0(x,y)dx=1,\quad (4.26.\mathrm{a})$$

where m, μ, and v_0, are, respectively, the molecular mass, the (dynamic) shear viscosity, and the speed at the peak of the Maxwellian distribution. Employing $K = 2\pi/\Lambda$, with Λ being the perturbation wavelength in the gas, along with (A.24.c), we can relate the mean free path $\overline{\ell}$ and collision frequency λ_c to viscosity and average speed of the Maxwellian distribution, $\overline{v}$, and in turn calculate the dimensionless y variable as

$$\rho = m\mathcal{N} = \frac{mp}{k_B T};\ \mu = D\rho = \left(\frac{1}{3}\overline{\ell}\overline{v}\right)\left(\frac{mp}{k_B T}\right) \rightarrow \frac{p}{\mu} = \frac{3}{2\overline{\ell}}\sqrt{\frac{\pi k_B T}{2m}} = \frac{3\pi}{8}\lambda_c.$$
$$\therefore\ x = \frac{\Omega}{Kv_0} = \frac{\nu\Lambda}{v_0};\ v_0 = \sqrt{\frac{2k_B T}{m}}\ \text{and}\ y = \frac{p}{Kv_0\mu} = \frac{1}{Kv_0}\frac{3\pi}{8}\lambda_c = \frac{3\sqrt{\pi}}{4}\frac{1}{K\overline{\ell}} = \frac{3}{8\sqrt{\pi}}\frac{\Lambda}{\overline{\ell}}. \tag{4.26.b}$$

In other words, the scattering spectrum is a function of x (perturbation angular frequency Ω, frequency ν, wavelength Λ) and y (or $3\pi/8$ times the collision frequency λ_c), both normalized to Kv_0, which is the angular frequency of sound waves in the gas. It is clear that the y parameter also equals $(3\sqrt{\pi}/4) \approx 1.33$ divided by $K\overline{\ell}$, as shown in Table 1 of [4.17] (where the two groups of y definitions were erroneously reversed); x/y is then the ratio of perturbation frequency, $\Omega/2\pi$, to (3/16) times the collision frequency, λ_c. Therefore, the y parameter may be used to classify the gas at a given temperature into three regimes [4.22; 4.24]: $y \ll 1$ (Knudson regime), $0.3 \leq y \leq 3$ (kinetic regime), and $y \gg 1$ (hydrodynamic regime). Note that the parameter y is also $3/(8\sqrt{\pi}) = (4.73)^{-1}$ times the ratio of perturbation wavelength Λ to the collision mean free path $\overline{\ell}$, suggesting that $y = 1$ corresponds to about five collisions in one perturbation wavelength. In this connection, since $K = 2k\sin(\theta/2)$, the perturbation wavelength Λ at $\theta = 1^\circ$ (nearly forward scattering) is about 115 times longer than Λ at $\theta = 180^\circ$ (back scattering). Thus, for the same gas density or same mean free path, the near forward scattering could be in the hydrodynamic regime, while it could be in the Knudson regime for backward scattering.

The main challenge of collision dynamics for a molecular system comes from the need to incorporate the internal degrees of freedom that affect collisions. At temperatures of interest, electronic transitions may be ignored, but molecular rotations and lower energy vibrations could be activated, leading to additional contributions to (the internal) specific heat, c_{int}. Since the vibrational frequencies of nitrogen and oxygen are, respectively, 2331 cm^{-1} and 1556 cm^{-1}, no vibrational state can be excited at atmospheric temperatures. To account for their two degrees of rotational freedom, their internal specific heat (per molecule) is $c_{\text{int}} = k_B$ (or the specific heat at constant volume, $c_v = 2.5\ k_B$), less than that of other complex molecules, such as CO_2. In addition to their effect on specific heat, the internal degrees of freedom make inelastic collisions possible, leading to finite bulk viscosity, μ_b. This also alters the relationship between thermal conductivity and molecular diffusion due to the change in c_v(see (A.24.c)); strictly speaking, for diatomic molecules c_v increases from $1.5k_B$ to $2.5k_B$ as temperature increases and the rotational degrees of freedom fully activate. Since both shear viscosity, μ, and thermal conductivity, K, are measurable quantities as a function of temperature, they may be independently input into an algorithm along with internal specific heat, c_{int}, and an educated guess of bulk to shear viscosity ratio, μ_b/μ, such as given in Table 4.1 of [4.20], in order to derive the distribution function $h(\vec{v}, \vec{r}, t)$.

From a theoretical viewpoint, the internal degrees of freedom of a molecule cause complications. There are many internal (mostly rotational) energy levels that can represent binary collision initial and final states. This means there are many possible collision integrals and cross sections, leading to many kinetic equations. A detailed discussion of these equations and their solutions is, of course, beyond the scope of this book. It suffices to say that it is advantageous to form an $N\times N$ matrix to represent the operator for binary collisions (with collision integrals as matrix elements) in order to solve the kinetic equations. The collision matrix is then transformed into Hilbert space, with eigenvectors constructed from eigenstates representing the atomic gas and polynomials representing the internal energy (see chapter 4 of [4.20]). The first six (dominating) eigenstates represent mass, momentum, translational temperature, internal temperature, translational heat flux, and internal heat flux. Tenti et al. [4.24] keep the first six eigenstates, thus the S6-moment model. When the seventh eigenstate of the collision operator is also kept in the analysis (S7 model), the numerical results show less agreement with the experimental spectra [4.24]. With all these complications, the Tenti S6 power spectra cannot be expressed in terms of analytic functions; thus, the Cabannes spectrum, $C_0(x, y)$, is the output of a numerical computer model.

Though it is still a function of the same two dimensionless variables depending on p, T, m, K, and μ, the functional form (or lineshape) of $C_0(x, y)$ is different for a molecular gas due to the need to incorporate the effects of the intrinsic internal degrees of freedom into the collision terms of the molecular kinetic equations. The Tenti S6 (6-moment) model [4.24] is now considered the best for predicting the spectral lineshape of Cabannes (or Rayleigh–Brillouin) scattering from molecular gases. Many of the theoretical derivations and solutions to the kinetic equations can be found in [4.20], and Vieitez et al. [4.26] have provided a brief account of these theoretical developments, along with experiments to demonstrate the validity of the Tenti S6 spectrum for molecular nitrogen and oxygen. To apply this model, we consider the atmosphere as a fictitious gas (of dry air, for example) and apply the

Tenti S6 model with effective particle mass of 28.964 kg-kmol^{-1} along with measured shear viscosity (see Table 4.5) and thermal conductivity as a function of temperature in order to determine the Cabannes spectrum. The bulk viscosity μ_b is, however, not easily measured at light scattering frequencies [4.26; 4.27]. Here, for the following calculation, we take the value of the ratio of bulk-to-shear viscosities for dry air, μ_b/μ, to be 0.76 as used in early work [4.28], though not in complete agreement with the value of 0.73 for nitrogen in Table 4.1 of [4.20], or value of 0.81 measured by recent light scattering experiments, in Table 1 of [4.27]. Once the Cabannes spectrum, $S(K,\Omega)$ or $C_0(x,y)$, is obtained for a given scattering geometry, $\vec{K}$, we deduce the spectral differential Cabannes backscattering cross section from the number of scattering particles per unit volume, $\sigma_{\Sigma/V}$, using the differential Cabannes (or polarized Rayleigh) backscattering cross section (per molecule), $d\sigma_\pi^{CS}/d\Omega \equiv \sigma_\pi^P$ in (4.19a). The $\sigma_{\Sigma/V} = \mathcal{N}\sigma_\pi^P$ can be expressed in x-domain or frequency ν-domain as

$$\begin{aligned} &\frac{d\sigma_\pi^{CS}}{d\Omega} \equiv \sigma_\pi^P = \frac{\omega^4}{16\pi^2 \mathrm{e}_0^2 c^4}\left(a^2\right) \rightarrow \frac{d^2\sigma_{\Sigma/V}}{dx\,d\Omega} = \mathcal{N}\frac{\omega_S^4}{c^4}\left(\frac{a}{4\pi\mathrm{e}_0}\right)^2 C_0(x,y), \text{ or} \\ &\frac{d^2\sigma_{\Sigma/V}}{d\nu_s\,d\Omega} = \mathcal{N}\frac{\omega_S^4}{c^4}\left(\frac{a}{4\pi\mathrm{e}_0}\right)^2 \mathscr{R}(\Delta\nu;T,P);\ \Delta\nu = \nu_S - \nu_L = \frac{\Omega}{2\pi}, \end{aligned} \tag{4.27}$$

where $\int_{-\infty}^{\infty} C_0(x,y)dx = \int_{-\infty}^{\infty} \mathscr{R}(\Delta\nu;T,P)d\nu = 1$, leading to $\mathscr{R}(\Delta\nu;T,P) = (2\pi/K\mathrm{v}_0)C_0(x,y)$, with $x = 2\pi\Delta\nu/K\mathrm{v}_0$ and $y = p/K\mathrm{v}_0\mu$. For backscattering, $K = 4\pi/\lambda$ (where λ is the excitation wavelength) and $\mathscr{R}(\Delta\nu;T,P) = (\lambda/2\mathrm{v}_0)C_0(x,y) = f_0^{-1}C_0(x,y)$, leading to the normalizing factor $f_o \equiv 2\mathrm{v}_o/\lambda = (2/\lambda)\sqrt{2k_BT/m}$, i.e., $\int_{-\infty}^{\infty} C_0(x,y)d\nu = f_o$. To compare with experiments, one typically expresses the normalized Cabannes spectral density, $\mathscr{R}(\nu;T,P)$ in the frequency domain, in convenient units such as cm^{-1} or GHz. The observed scattered power at given scattering and incident laser frequencies can then be determined by multiplying the above expression by the scattering volume.

4.4.2 Cabannes Scattering Spectra for a Standard Atmosphere

As pointed in the thesis of Pan [4.20], the difference between spontaneous and coherent Cabannes scattering lies in density fluctuations within the scattering gas. In spontaneous scattering, the fluctuations are generated randomly by thermal motion and then relax toward equilibrium through collisions between gas molecules. In coherent scattering, the scattering gas density perturbation is generated through the interaction of the medium with crossed laser beams, thus the amplitude of the perturbation modes varies sinusoidally in space and time. As a result, scattering perturbations separated much farther than the scattering wavelength are still correlated. The normalized coherent and spontaneous scattering power spectra are thus different, and Pan [4.20] studied both and compared them for pure gases (for a quick view of their differences, see Fig. 5.3 of [4.20]). For lidar applications, the spontaneous Cabannes scattering spectrum, $C_0(x, y)$ should be used.

Though the Tenti S6 model has been well verified by both spontaneous and coherent light scattering experiments [4.20] for pure gas at different temperatures and pressures, the verdict on the application of the Tenti S6 model to a gas mixture (such as air) is at this time the subject of ongoing research [4.26, 4.27]. Even for the major species (nitrogen–oxygen) air mixture, the collision cross section between different gases should be considered, leading to different relaxation processes and several collision frequencies to deal with. An obvious, apparently successful approach has been to consider the atmosphere as a fictitious gas and use the Tenti S6 model with an effective molar mass (28.964 kg-kmol^{-1}), as we have noted above in Section 4.4.1, and then experimentally determine effective transport coefficients of the atmosphere for calculation. Witschas et al. [4.27] have employed this approach and compared the S6 model successfully to the experimental light scattering spectra of dry and moist air. For this discussion, we use this approach and calculate $C_0(x, y)$ with the Tenti S6 model.

As an example of how the Cabannes spectrum of dry atmosphere changes as a function of temperature and pressure, we plot the normalized backscattering spectra at 532 nm, $\mathscr{R}(\Delta\nu; T, P)$, versus frequency shift in GHz for a standard atmosphere at selected (geometric) altitudes. These plots, as shown in Fig. 4.7(a), were adapted from a 2007 M.S. thesis [4.29]; they are the outputs of a computer program based on the Tenti S6 model used at Colorado State University through the 1990s. It is essentially the same as the program, also adapted from Tenti's work, presented in Appendix C of [4.20] for pure gases. The selected altitudes and the associated pressure, temperature, and shear viscosity are downloaded from Table 1 of the 1976 Standard Atmosphere on line, https://www.pdas.com/atmoscalculator.html, and are given in Table 4.5. Note that the important dimensionless y-parameter calculated from (4.26.a), the temperature-dependent shear viscosity and thermal conductivity, along with internal specific heat and bulk-to-shear viscosity ratio, c_{int}/k_B and μ_b/μ, taken to be 1 and 0.76 respectively, were provided as the inputs to the program. To test the effect resulting from the uncertainty in μ_b/μ, we have compared the atmospheric Cabannes spectra at 5 km with $\mu_b/\mu = 0.73,\ 0.76,$ and 0.81 and found their differences to be less than 0.02%.

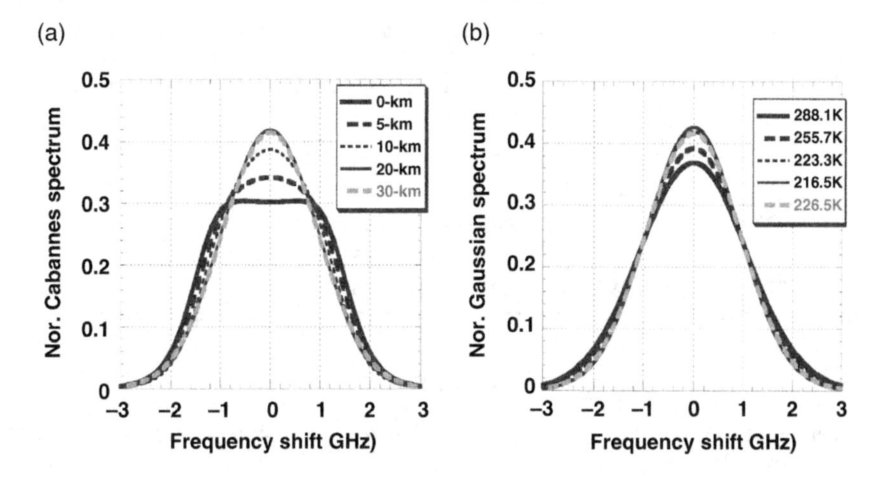

Fig. 4.7 (a) The normalized Cabannes scattering spectra for standard atmosphere at 0, 5, 10, 20, and 30 km (with temperature, pressure, and shear viscosity listed in Table 4.5, along with $\mu_b/\mu = 0.76$). (b) For comparison, the corresponding Gaussian (Doppler-broadened) spectra at the same temperatures are shown. Notice that the two (left and right) spectra are only very similar for conditions above 10 km.

The normalized backscattering Cabannes spectra for air molecules at selected altitudes, $\mathscr{R}(\nu;T,P) = f_o^{-1}C_0(x = \lambda\Delta\nu/2\mathrm{v}_0, y = \lambda p/2\pi\mathrm{v}_0\mu) \equiv f_o^{-1}R(\nu;T,P)$, are then calculated and shown in the left panel of Fig. 4.7 as a function of frequency shift, $\Delta\nu$. These may be compared to the corresponding Gaussian (Doppler-broadened) spectra for temperatures corresponding to those altitudes shown in the right panel of the same figure. At 532 nm, dry air at ground level ($y = 0.585$) and at 5 km ($y = 0.364$) are in the kinetic regime, and their Cabannes spectra are quite clearly broader than the corresponding Gaussian spectra. At 10 km altitude ($y = 0.214$), the Cabannes spectrum is only a bit broader. For standard dry air at and above 20 km, the Cabannes spectrum looks identical to the Gaussian spectrum. We point out that we use shear viscosities from the 1976 standard atmosphere for the calculation of the y variable in Table 4.5. Had we used the formula for the y variable given in (1b) of [4.30], we would obtain essentially the same y values as those presented in Table 4.5.

The detailed spectral lineshape of Cabannes scattering is very important for high-spectral-resolution atmospheric lidar applications for temperature and wind measurements in the troposphere, as well as for aerosol backscattering ratio and lidar ratio measurements. For this reason, a reliable theoretical spectral model, like the Tenti S6 model, that can represent realistic conditions of the atmosphere (free from aerosols) is indeed very useful.

Table 4.5 *The parameters used for calculated normalized Cabannes scattering spectra*

Altitude (km)	Temperature (K)	Pressure (Pa)	Viscosity (Pa-s)	Peak speed, v_0 (ms^{-1})	Dimensionless variable, y
0	288.1	101325	1.789×10^{-5}	406.7	0.590
5	255.7	54048	1.628×10^{-5}	383.1	0.367
10	223.3	26500	1.458×10^{-5}	358.0	0.215
20	216.5	5529	1.422×10^{-5}	352.6	0.0467
30	226.5	1197	1.475×10^{-5}	360.6	0.00953

References

4.1 Shen, Y. R. (1984). *The Principles of Nonlinear Optics.* Wiley-Interscience, ISBN: 0 471-88998-9.

4.2 Long, D. A. (2002). *The Raman Effect: A Unified Treatment of the Theory of Raman Scattering by Molecules.* John Wiley & Sons, Ltd., ISBN 0-471-49028-8 (Hardback); 0-470-84576-7 (Electronic).

4.3 Placzek, G. (1934). Rayleigh-Streuung und Raman-Effekt. In *Handbuch der Radiologie*, E. Marx, ed., **6**, 205–374, Academische Verlag: Leipzig.

4.4 Edmonds, A. R. (1957). *Angular Momentum in Quantum Mechanics*. Princeton University Press.

4.5 She, C.-Y. (2001). Spectral structure of laser light scattering revisited: Bandwidths of non-resonant scattering lidars. *Appl. Optics* **40**(27), 4875–4884.

4.6 Placzek, G., and E. Teller. (1933). Die Rotationsstruktur der Ramanbanden mehratomiger Moleküle. *Zeitschrift für Physik* **81**, 209–258. doi: https://doi.org/10.1007/BF01338366

4.7 Altmann, K., and G. Strey. (1972). Application of spherical tensors and Wigner 3-*j* symbols to the calculation of relative intensities of rotational lines in Raman bands of molecular gases. *J. Mol. Spectroscopy*, **44**(3), 571–577.

4.8 Wandinger, U. (2005). Raman lidar. Chapter 9 in *Lidar Range-Resolved Optical Remote Sensing of the Atmosphere*. C. Weitkamp, ed., Springer.

4.9 She, C.-Y., H. Chen, and D. A. Krueger. (2015). Optical processes for middle atmospheric Doppler lidars: Cabannes scattering and laser induced resonance fluorescence. *Jour. Opt. Soc. Am.* B **32**(9), 1575–1592, and Erratum, *ibid.*, p. 1954.

4.10 Young, A. T. (1982). “Rayleigh scattering.” In *Physics Today* (January), pp. 42–48.

4.11 Griffith, D. J. (1998). *Introduction to Electrodynamics*. 2nd ed. Prentice Hall.

4.12 King, L. V. (1923). On the anisotropic molecule in relation to the dispersion and scattering of light. *Proc. R. Soc. London*, **A104**(726), 333–357.

4.13 Tomasi, C., V. Vitali, B. Petrov, A. Lupi, and A. Cacciari. (2005). Improved algorithm for calculations of Rayleigh-scattering optical depth in standard atmospheres. *Appl. Optics* **44**(16), 3320–3341.

4.14 Butcher, R. J., D. V. Willetts, and W. J. Jones. (1971). On the use of a Fabry–Perot etalon for the determination of rotational constants of simple molecules – the pure rotational Raman spectra of oxygen and nitrogen. *Proc. R. Soc. London Ser. A* **324** (1557), 231–245.

4.15 Bendtsen, J., and F. Rasmussen. (2000). High-resolution incoherent Fourier transform Raman spectrum of the fundamental band of $^{14}N_2$. *J. Raman Spectroscopy* **31**(5), 433–438.

4.16 Loëte, M., and H. Berger. (1977). High resolution Raman spectroscopy of the fundamental vibrational band of $^{16}O_2$. *J. Mol. Spectrosc.* **68**(2), 317.

4.17 She, C. Y., G. C. Herring, H. Moosmüller, and S. A. Lee. (1985). Stimulated Rayleigh-Brillouin gain spectroscopy. *Phys. Rev. A*, **31**(6), 3733–3740.

4.18 Rahn, L. A., and R. E. Palmer. (1986). Studies of nitrogen self-broadening at high temperature with inverse Raman spectroscopy. *Jour. Opt. Soc. Amer. B*, **3**(9), 1164–1169.

4.19 Tam, R. C. H. and A. D. May. (1983). Motional narrowing of the rotational Raman band of compressed CO, N_2, and CO_2. *Can. J. Phys.* **61**, 1558–1566.

4.20 Pan, X.-G. (2003). "Coherent Rayleigh-Brillouin scattering," Ph.D. dissertation, Princeton University.

4.21 Papoulis, A. (1962). *The Fourier Integral and Its Applications*. McGraw-Hill.

4.22 Yip, S., and M. Nelkin. (1964). Application of a kinetic model to time-dependent density correlations in fluids. *Phys. Rev. A* **135**(5A), 1241.

4.23 Herman, R. M., and M. A. Gray. (1967). Theoretical prediction of stimulated thermal Rayleigh scattering in liquids. *Phys. Rev. Lett.* **19**(15), 825–827.

4.24 Tenti, G., C. Boley, and R. Desai. (1974). On the kinetic model description of Rayleigh -Brillouin scattering from molecular gases. *Can. J. Phys.* **52**(4), 285–290.

4.25 Tang, S. Y., C. Y. She, and S. A. Lee. (1987). Continuous-wave Rayleigh–Brillouin-gain spectroscopy of SF_6. *Optics Letters* **12**(11), 870–872.

4.26 Vieitez, M. O., E. J. van Duijn, W. Ubachs et al. (2010). Coherent and spontaneous Rayleigh–Brillouin scattering in atomic and molecular gases and gas mixtures. *Phys. Rev. A* **82**(4), 0438361–14.

4.27 Witschas, B., M. O. Vieitez, E.-J. van Duijn et al. (2010). Spontaneous Rayleigh–Brillouin scattering of ultraviolet light in nitrogen, dry air, and moist air. *Appl. Opt.* **49** (22), 4217–4227.

4.28 Krueger, D. A., L. M. Caldwell, R. J. Alvarez II, and C. Y. She. (1993). Self-consistent method for determining vertical profiles of aerosol and atmospheric properties using high-spectral-resolution Rayleigh–Mie lidar. *J. Atm. Oceanic Tech.* **10**(4), 533–545.

4.29 Yan, Z. A. (2007). "A study of iodine filtrated atmospheric temperature lidar," M. S. thesis, Ocean University of China (in Chinese).

4.30 Shimizu, H., S. A. Lee, and C. Y. She. (1983). High spectral resolution lidar system with atomic blocking filters for measuring atmospheric parameters. *Appl. Opt.* **22**(9), 1373–1381.

5

Introduction to Lidar Remote Sensing and the Lidar Equation

There are at least two motivations for investigating the Earth's atmosphere: scientific curiosity and practical necessity. It is of fundamental interest to understand the structure of nature and processes in it, and in some instances this leads to counterintuitive results, such as summer being colder than winter in the (upper) mesosphere and lower thermosphere (MLT). In order to sustain the inhabitability of Earth, it is of urgent necessity to evaluate the effects of human activity and their potential influences on the natural variability and fragile balance of her environment. These include the health effects of air pollution and the impacts of global warming caused by human emissions of greenhouse gases.

It is well known that science divides the atmosphere into layers by altitude according to its thermal structure and that this is strongly influenced by the effects of photoionization in the upper atmosphere, photodissociation of ozone molecules in stratosphere, and solar heating of water vapor in the troposphere and on the earth's surface. The annual mean temperature profile forms the S-shape curve shown in Fig. 5.1. The layer boundaries, marked by short horizontal lines, are denoted as "pause" with an identifying prefix referring to the layer below it. The typical height variation of atmospheric pressure also is shown in Fig. 5.1; the fact that it is nearly linear in a semilog plot suggests the hydrostatic nature of air density above the earth's surface. The altitude and thickness of the tropopause has a seasonal as well as a geographical dependence; its mean height in the mid-latitudes is about 11 km, less than 10 km in the polar regions, and as high as 16 km in the equatorial region.

Dynamically speaking, the transport of atmospheric constituents depends upon both molecular diffusion and eddy diffusion (turbulent mixing). At the turbopause, ~105 km in altitude, these two processes are equally effective, as demonstrated by rocket chemical release experiments: below 105 km, the released gas develops globular eddy or turbulent structures, and above, the released gas shows a smooth/laminar (and less dense) growth [5.1]. Locally, atmospheric pressure, p

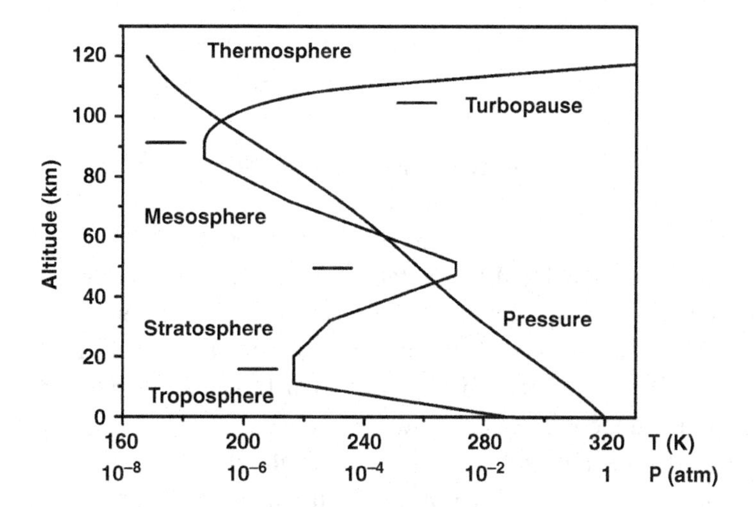

Fig. 5.1 The averaged atmospheric temperature and pressure at mid-latitudes. The solid S-shape curve gives temperatures, while the solid, almost-straight line denotes pressures. Adapted with permission from Fig. 20 of [5.1] © Wiley & Sons, and with permission from Fig. 1 of [5.2] © Taylor & Francis.

(in atm, 1 atm $= 1.013 \times 10^5$ Pa), temperature, T (in K), and air molecular (number) density, $\mathcal{N}$(in m^{-3}) or n (in kmol^{-1}), are known to be related by the ideal gas law. The vertical air distribution is essentially in hydrostatic equilibrium, under which the pressure gradient force due to density stratification balances the gravitational force. If local hydrostatic equilibrium is upset, convection occurs, and the atmosphere returns to static balance in a short time. Thus, with some spatial and temporal averaging, the atmosphere is thought to be in hydrostatic balance [5.2], leading to the following relations:

$$p = \mathcal{N} k_B T \text{ or, } \frac{R}{M} = \frac{k_B}{m} \rightarrow p = nRT = \rho\left(\frac{R}{M}\right) T \text{ and} \tag{5.1.a}$$

$$\frac{dp}{dz} = -\rho g = -mg\mathcal{N}(\text{z}) \rightarrow \frac{dp}{dz} = -\frac{mg}{k_B T} p; \; z_s = \frac{k_B T}{mg} \tag{5.1.b}$$

where $g = 9.5 - 9.8\,\text{m/s}^2$, $m = 4.8 \times 10^{-26}$kg, z, $k_B = 1.38 \times 10^{-23}\text{JK}^{-1}$ and $R = 8.31 \times 10^3\,\text{JK}^{-1}\text{kmol}^{-1}$ are, respectively, gravitational acceleration, mass of an air molecule, altitude, the Boltzmann constant, and the universal gas constant. The temperature-dependent quantity z_s is the scale height, about 8.5 km at 300 K. To satisfy (5.1.a) and (5.1.b), the fractional perturbations in the three atmospheric parameters, $p, T, \mathcal{N}$, in a height interval, Δz, are related as:

$$\frac{\delta p}{p} = \frac{\Delta z}{z_s}\left(\frac{\delta \mathcal{N}}{\mathcal{N}}\right); \quad \frac{\delta T}{T} = \frac{\delta p}{p} - \frac{\delta \mathcal{N}}{\mathcal{N}} \approx -\frac{\delta \mathcal{N}}{\mathcal{N}} \tag{5.1.c}$$

The relation (5.1.c) may also be applied to the measurement uncertainties. Since, if one parameter is measured with a vertical resolution Δz, the other two parameters may be determined by (5.1.a) and (5.1.b). The uncertainties of the two derived parameters are then related to the uncertainty of the measured parameter by (5.1.c). Since atmospheric pressure is the integration of air number density, the pressure uncertainty should be much smaller than the density uncertainty, in agreement with the fact that $\Delta z/z_s$ is typically much smaller than 1. The temperature and number density fractional uncertainties are, however, comparable.

To investigate the atmosphere with lidar (**li**ght **d**etection **a**nd **r**anging), we employ a pulsed transmitting laser at optical frequency ν_L, polarized in the plane $(\hat{e}_a, \hat{e}_b)$ perpendicular to its propagation direction, $\hat{k}_0$, specified by the polar angle, θ_0, and azimuthal angle, ϕ_0, with respect to a space-fixed coordinate system with unit vectors $(\hat{x}, \hat{y}, \hat{z})$, adapted from Fig. B.1 as shown in Fig. 5.2(a). Both transmitter and receiver are located on the ground at an altitude (elevation) H_0 above mean sea level (e.g., $H_0 = 1.57$ km in Fort Collins, CO), at the origin of the beam propagation vector. The scattering volume under consideration is located at polar

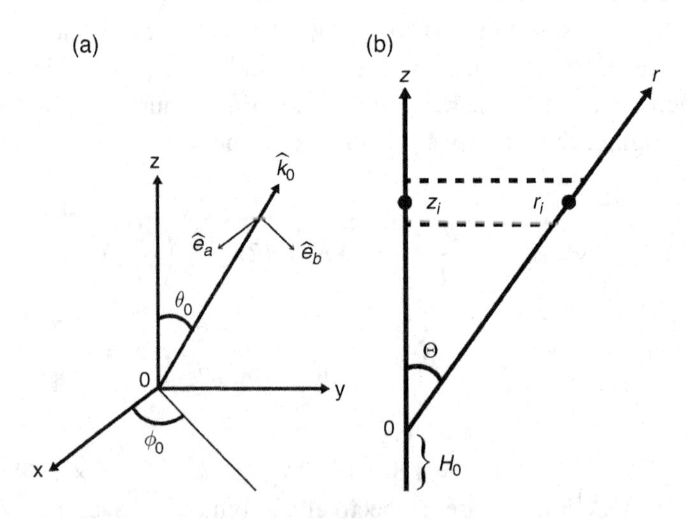

Fig. 5.2 (a) Lidar transmitting beam is pointing in the $\hat{k}_0$ direction from a ground-based lidar sitting at the origin of the coordinate system. (b) The relationship between slant range r and vertical height z is $z = r\cos\theta_0$; $\theta_0 = \Theta$. The lidar is located at the point 0 in both figures at elevation H_0 (not to scale). The altitude of the scatterers in the ith bin shown by both solid circles is Altitude $= z_i + H_0$, nominally also denoted by the same symbol, z.

coordinate (r, θ_0, ϕ_0), where we assume the atmosphere is horizontally homogeneous, resulting from the hydrostatic equilibrium condition. The atmospheric density is also a function of the vertical height z (altitude $= z + H_0$) with $\rho_a(r) = \rho_a(z)$, where the height is related to the beam's slant range r as $z = r \cos \theta_0$, or $r = z/\cos \Theta$ with $\theta_0 \equiv \Theta$ as depicted in Fig. 5.2(b).

The received photon counts scattered from a single laser pulse with energy E_L are recorded in counting bins, each with preset duration τ_B (or bin length $r_B = c\tau_B/2$, corresponding to vertical bin size, $z_B = c\tau_B \cos\Theta/2$), as discussed in Section 5.1. Before reaching the scattering volume at r, the incident number of photons, $E_L/h\nu_L$, is attenuated by a factor $A_\uparrow(\nu_L, r)$ by the atmosphere below r. Assuming the entire scattering volume is in the field of view of the receiving telescope, the backscattered photon counts at the scattering frequency, ν_S, from the scattering volume is the product of the attenuated incident photons, the number density of the scattering particles $\mathcal{N}$(r), differential scattering cross section of the scattering particles, $d\sigma(\nu_S)/d\Omega$, and the bin length r_B. The backscattered photons, attenuated by a factor of $A_\downarrow(\nu_S, r)$ through the intervening atmosphere, within the solid angle A_R/r^2 subtended by the receiver and further reduced by the detector quantum efficiency η, are collected by the receiving telescope and become the signal photon counts $N(\nu_S, r;\ r_B)$ – not to be confused with atmospheric molecular number density $\mathcal{N}(r)$. This leads to a rudimentary form of the lidar equation:

$$N(\nu_S, r; r_B) = \eta \frac{A_R}{r^2} A_\downarrow(\nu_S, r)[\beta(\nu_S, r) r_B] A_\uparrow(\nu_L, r) \frac{E_L}{h\nu_L} + B^{r_B};\ \beta(\nu_S, r) = \mathcal{N}(r) \frac{d\sigma(\nu_S)}{d\Omega} \tag{5.2}$$

where A_R is the area of the receiving telescope, $\nu_S \neq \nu_L$ for Raman scattering or laser-induced fluorescence, and B^{r_B} is the (sky) background counts within a counting bin, independent of incident laser energy. In (5.2), $\beta(\nu_S, r)$ is the backscattering coefficient for molecular scattering; it is given as the product of molecular density and differential cross section, as given above. The upward and downward attenuation coefficients, $A_\uparrow(\nu_L, r)$ and $A_\downarrow(\nu_S, r)$ are, respectively, the inverse exponential functions of optical thickness of the layer, $\tau_\uparrow(r_0, r)$ and $\tau_\downarrow(r_0, r)$. These, in turn, are the integrals of the extinction coefficients, $\alpha(\nu_L, r')$ and $\alpha(\nu_S, r')$ between initial (or reference) range r_0 (nominally at the position of the transmitter) and the range of the scatterors in question r as,

$$A_{\uparrow}(\nu_L, r) = \exp[-\tau_{\uparrow}(r_0, r)]; \; A_{\downarrow}(\nu_S, r) = \exp[-\tau_{\downarrow}(r_0, r)], \text{ with}$$

$$\tau_{\uparrow,\downarrow}(r_0, r) = \int_{r_0}^{r} \alpha(\nu_{L,S}, r')dr'. \tag{5.2.a}$$

In addition to molecular scattering (mainly due to nitrogen and oxygen), there is scattering from aerosols (including ice particles). Since molecules and aerosols have quite different scattering properties, we express both the backscattering coefficient and the extinction coefficient as the sum of aerosol (subscript *a*) and molecular (subscript *m*) contributions as $\beta(r) = \beta_a(r) + \beta_m(r)$ and $\alpha(r') = \alpha_a(r') + \alpha_m(r')$. Here, the aerosol extinction term $\alpha_a(r')$ that attenuates light is the result of absorption and scattering, which depend on particle size, shape, and, respectively, on the imaginary and real parts of the refractive index – assuming the particles are homogeneous. It is impossible to separate their contributions. Fortunately, for lidar applications and solving the lidar equation, we do not need to separate the aerosol absorption and scattering components. Since the absorption wavelength of oxygen molecules occurs at 193 nm and the strong absorption band of ozone peaks at 255 nm, absorption of light at wavelengths longer than 300 nm by atmospheric molecules is negligible. Thus, only the scattering contribution needs to be considered for the molecular extinction coefficient, $\alpha_m(r)$, which is related to the molecular backscattering coefficient as $\alpha_m(r) = 8\pi\beta_m(r)/3$ since air molecules are roughly spherical. The backward scattering coefficient, $\beta_m = \mathcal{N}(r)d\sigma_m/d\Omega$, is the product of molecular density and differential Rayleigh (Cabannes plus rotational Raman) scattering cross section of an air molecule.

This simple lidar equation (5.2) may be modified and extended to describe the various lidars to be discussed in the chapters to follow. We classify these lidars into two categories:

i. Broadband (simple) lidar types, including Rayleigh and Mie scattering lidars, polarization lidars, vibrational Raman scattering lidars, fluorescence lidars, and differential absorption lidars.
ii. Narrowband lidar types for profiling aerosol optical properties, temperature, and wind. With the exception of the vertical integration technique for atmospheric temperature measurement, lidars for sensing atmospheric parameters employ narrow bandwidth (less than 200 MHz) lasers. Here, the lidar signal is the result of either Cabannes and/or rotational Raman scattering from the troposphere, stratosphere, and lower mesosphere or of laser-induced resonance fluorescence from the upper mesosphere and lower thermosphere (MLT). The receivers in these lidars use narrowband filters whose bandwidth depends on whether and

how much rotational Raman scattering should be included, as well as whether observations take place in nighttime or daytime.

5.1 The Lidar Equation – An Overview

We now consider an atmospheric lidar employing a pulsed transmitting (probing) laser at frequency ν_L, which can scatter from air molecules or excite metal atoms such as Na in the mesopause region, along with a receiving telescope with area A_R. Both transmitter and receiver are located on the ground at $z = r = 0$ and point along the direction $\theta_0 \equiv \Theta$ off zenith; see Fig. 5.2. Here, thanks to the assumption of atmospheric horizontal homogeneity, we can, for ease of discussion, use both vertical height z above the ground and slant range distance r interchangeably, keeping in mind that they are related by $z = r \cos \Theta$ as shown in Fig. 5.2(b). Horizontal homogeneity, $\rho_a(r) = \rho_a(z)$, leads to identical number densities of the atmospheric or probing species (e.g., Na) in units of m^{-3} at range r (or height z) in km (i.e., $\mathcal{N}(r) = \mathcal{N}(z)$). The energy and width of the laser pulses are, respectively, E_L and τ_L. The receiving electronics record photocount signals digitized in timed bins, each with a fixed duration ($\tau_B = 1$ μs, for example) corresponding to a length $r_B = c\tau_B/2$ (0.15 km for $\tau_B = 1$ μs) that is much longer than the laser pulse width τ_L (10 ns, for example) and much shorter than the dwell or integration time for compiling a photocount profile, τ_D (1 minute, for example). The maximum probing range, r_{pb}, is predetermined (at 120 km or 300 km, for example) to cover the atmosphere of interest, with the maximum probing height, z_{pb}, related by $z_{pb} = r_{pb} \cos \Theta$. The probing range is divided into a number of digital bins (for our example with $r_B = 150$ m, it is 800 or 2,000 bins, respectively, for $r_{pb} = 120$ or 300 km). As with the relationship between r_{pb} and z_{pb}, the ith bin along r and z is shown in Fig. 5.2(b) as $z_i = r_i \cos \Theta$. This relationship extends from the ground to beyond the top of the metal layer, with the higher altitudes being included to provide bins with no signal, which we use as a baseline for background subtraction. In data analysis, we utilize the stored backscattered photocounts along the slant range in digital bins with increasing bin number and evaluate $\tau_\uparrow(r_0, r_i)$, $\tau_\downarrow(r_0, r_i)$, $A_\uparrow(\nu_L, r_i)$, and $A_\downarrow(\nu_S, r_i)$ progressively to retrieve atmospheric parameters of interest. We then convert the bin numbers to altitude and report both photocount profiles and analysis results as a function of altitude (i.e., $Altitude = z_i + H_0$). This form of presentation is also used for line-of-sight (LOS) wind measurements, along with the information of the pointing angle needed to assess the contribution of the LOS wind component to horizontal or vertical winds.

In addition to nitrogen and oxygen molecules, there are minor species in the atmosphere, such as aerosol and cloud particles, that are not clearly described by number density. We therefore no longer identify the volume backscattering coefficient, $\beta(\nu_S)$, as $\mathcal{N}(r)d\sigma(\nu_S)/d\Omega = \mathcal{N}(z)d\sigma(\nu_S)/d\Omega$ in (5.2) but rather use $\beta(\nu_S)$ more generally to represent the volume backscattering coefficient of all scattering particles in the atmosphere. Furthermore, there may be a need to probe more than one physical property for a given atmospheric measurement. This can often be performed with a single probing beam and by alternating among distinct transmitter or receiver settings to capture the different properties. Accordingly, the lidar return would consist of a number, μ, of profile channels, each defined by its distinct instrument settings, recording scattered photon counts versus range. In this case, the lidar equation has many channels as

$$N_i(\nu_{iS}, r; r_B) = \eta_i \frac{A_R}{r^2} A_\downarrow(\nu_{iS}, r)[\beta_i(\nu_{iS}, r) r_B] A_\uparrow(\nu_{iL}, r) \frac{E_{iL}}{h\nu_{iL}} + B_i^{r_B}; \quad i = 1, 2, \ldots \mu. \tag{5.3}$$

In (5.3), the required lidar profile channels/characteristics, $i = 1, 2, \ldots \mu$, may, for example, represent different laser frequencies, enabling a scan across a spectral feature of a target atom or molecule. This is indeed the case for several lidar types discussed in this book. In other lidar types, the probing frequencies between different channels are in fact the same, but the receiver settings could oscillate between different physical properties, such as polarization or filter bandwidth. We list below the varied applications of (5.3) to the different types of lidar we will discuss:

i. Broadband (simple) lidar types:
   a. Rayleigh and Mie scattering lidars: $\mu = 1$ with one frequency at which the atmosphere is transparent, and in which Mie scattering from aerosols and Rayleigh scattering from air molecules (majority species of N_2 and O_2) are recorded in a single profile channel.
   b. Polarization lidars: typically, $\mu = 2$, single laser frequency with mutually perpendicular receiving polarizations. Sometimes $\mu = 4$, with a single laser frequency and four receiving channels with specified polarization characteristics to extract the depolarization properties of the scattering medium (Mueller matrix) via the received scattered light (Stokes vector).
   c. Vibrational Raman scattering and fluorescence emission lidars: $\mu \geq 2$ with one probing laser frequency at which the atmosphere is transparent and multiple receiver channels at different frequencies using bandpass filters in order to determine the number density(ies), $\mathcal{N}_{i=1,2,\ldots\mu}(r)$, of the molecular (major and minor) species under study. All channels are recorded simultaneously.

d. Differential absorption lidars: typically, $\mu = 2$, with different laser frequencies specifically tuned to compare the difference between on-resonance and off-resonance scattering of an atmospheric molecular (minor) constituent of interest. In the case of ozone lidar (a common target of differential absorption measurements), due to its broad absorption line (~200 nm width), differences in aerosol-scattering cross sections at the two laser frequencies may contaminate the ozone absorption signal. In this case, one uses a third (lower) frequency ($\mu = 3$) to resolve the contamination.

ii. Lidar types for profiling aerosol optical properties, atmospheric temperature, and wind:

a. Profiling lidar ratio and aerosol optical properties: typically, $\mu = 2$, with a single laser frequency and different receiving filters to permit a comparison between total scatter (aerosols plus molecular) and molecular-only scattering.
b. Integration technique for temperature measurements with (broadband) Rayleigh and vibrational Raman scattering: $\mu = 2$, employing a single laser frequency, and there are two receiver channels with filters to discriminate Rayleigh scattering at higher altitudes (aerosol-free) from N_2 Raman scattering (aerosol signal filtered) at lower altitudes. Narrowband receiver filters are centered at different frequencies – one at the laser frequency and the second separated by nitrogen Raman resonance wavenumber of 2331 cm^{-1}.
c. Temperature profiling with rotational Raman and Cabannes scattering: $\mu = 3$, using a narrowband laser and receiver bandpass filters centered at different frequencies to reject aerosol scattering and receive two different sections of the rotational Raman and/or Cabannes spectrum.
d. Wind profiling with edge filter techniques: $\mu = 2$ or 3, with receiving filters shifted from the single-frequency laser line center so the laser falls on the filter edge in order to detect frequency shift from intensity change in the received scattered light. Fabry–Perot interferometers or molecular vapor filters are employed to determine frequency shifts in aerosol scattering signal in the lower troposphere and/or molecular scattering signal in the upper atmosphere.
e. Mesopause region temperature and wind profiling with laser-induced resonance fluorescence (LIF): by cycling the transmitting laser output between multiple predetermined frequencies, all spectrally within a Doppler-broadened atomic resonance transition. The received signals can be used to determine mesopause region temperature ($\mu = 2$) or both temperature and line-of-sight (LOS) wind ($\mu = 3$). Alternatively, one can scan ($\mu \geq 5$ usually) through the transition spectrum at preset steps to deduce the fluorescence

spectrum, thereby determining temperature and LOS wind with the associated Doppler broadening and shift.

As mentioned, the bin length, $r_B = c\tau_B/2$, and the probing length (or range), r_{pb}, along with the dwell time, τ_D (time required to complete the detection for a set of designated channels), are all preset appropriately; for example, they may be, respectively, 0.15 km, 250 km, and 1 min. We typically perform data analysis with photon count profiles integrated over a longer time interval, τ_{int}, (60 min for example) and in a coarser vertical bin length, Δr (or $\Delta z = 1$ km, for example), to reduce photon (shot) noise at the price of degraded temporal and vertical and/or range resolution [5.2, 5.3]. Using the integrated photocounts, physical properties of the atmosphere, such as density, temperature, and wind, are retrieved. Except for brief accounts in Sections 6.3, 7.2, and 7.3, the question of signal-to-noise [5.3, 5.4] or the analysis on the effect of one atmospheric parameter's uncertainty on the accuracy of another [5.5] will generally be referred to the literature and not be detailed in this book.

References

5.1 Campbell, I. M. (1977). *Energy and the Atmosphere: A Physical-Chemical Approach*, Wiley & Sons Ltd.
5.2 She, C.-Y. (1990). Remote measurement of atmospheric parameters: New applications of physics with lasers, *Contemp. Phys.* **31**(4), 247–260.
5.3 Krueger, D. A., C.-Y. She, and T. Yuan (2015). Retrieving mesopause temperature and line-of-sight wind from full-diurnal-cycle Na lidar observations, *Appl. Opt.* **54**(32), 9469–9489. doi: https://doi.org/10.1364/AO.54.009469.
5.4 Liu, Z.-Y., W. Hunt, M. Vaughan et al. (2006). Estimating random errors due to shot noise in backscatter lidar observations, *Appl. Opt.* **45**(18), 4437–4447. doi: https://doi.org/10.1364/AO.45.004437.
5.5 Zhang, Y., D. Liu, Z. Zheng et al. (2018). Effects of auxiliary atmospheric state parameters on the aerosol optical properties retrieval errors of high-spectral-resolution lidar, *Appl. Optics.* **57**(10), 2627–2637.

6

Common (Broadband) Lidar Types and Associated Applications

We first consider broadband atmospheric lidars. The linewidth of the probing laser in a common (broadband) lidar is typically in the range of 10–100 GHz, so that spectral features or properties, which typically have frequency characteristics narrower than the laser linewidth, cannot be distinguished. In this case, the many challenging issues of laser tuning and locking are not a concern. Though the instrument is relatively simple, broadband lidars are powerful tools capable of measuring many important properties of the atmosphere as well as of probing stationary (sometimes hidden) and moving structures of cultural and practical interest.

6.1 Rayleigh–Mie Scattering (Elastic Backscattering) Lidars

In a Rayleigh–Mie scattering lidar, we assume the laser beam is attenuated as it propagates through and scatters from aerosols and atmospheric molecules in its path (along with absorption at the laser wavelength only by selected minor species of interest). Aerosol and molecular scattering (denoted by subscripts a and m respectively) additively contribute to both the total backscattering coefficient, β, and the total extinction (including some possible absorption) coefficient, α. Due to the varying size, shape, and composition of aerosol particles, one cannot uniquely relate the backscattering coefficient, β_a, to the extinction coefficients, α_a, without detailed knowledge of the physical properties of the aerosols present – particularly particle size distribution and refract ive index. The lidar equation for such a Rayleigh–Mie lidar, (5.2), may be expressed as

$$N(\nu_L, r; r_B) = \eta \frac{A_R}{r^2} \beta(r) r_B \frac{E_L}{h\nu_L} \exp\left[-2 \int_{r_0}^{r} \alpha(r')dr'\right];$$

$$\beta(r) = \beta_a(r) + \beta_m(r) \text{ and } \alpha(r') = \alpha_a(r') + \alpha_m(r'), \tag{6.1}$$

where the upward and downward attenuations, $A_{\uparrow}(\nu_L, r) = A_{\downarrow}(\nu_S, r)$ in (5.2) are assumed to be the same. Unlike in the case of aerosol particles, atmospheric molecules may be reasonably treated as spherical particles. Thus, we may justifiably relate $\beta_m(r)$ and $\alpha_m(r)$ to molecular density, $\mathcal{N}(r)$, as $\alpha_m(r) = 8\pi\beta_m(r)/3 = (8\pi/3)\mathcal{N}(r)d\sigma_m/d\Omega$, where $d\sigma_m/d\Omega$ is the differential Rayleigh (Cabannes plus rotational Raman) scattering cross section of an air molecule. Although the air number density $\mathcal{N}(r')$ can be taken from a standard atmospheric profile (thus, it is known for this purpose), solving one equation, (6.1), cannot determine both atmospheric optical parameters, $\beta(r)$ and $\alpha(r)$.

To retrieve the optical scattering parameters including aerosol scattering from the elastic-backscattering lidar signal, Klett in 1981 [6.1] adapted the method used in the discussion of rain attenuation of radar signal and employed a simple relationship between β and α, $\beta(r) = B\alpha(r)^k$. In a later paper, Klett [6.2] discussed the dependences of $\beta_a(r)$ and $\alpha_a(r)$ on particle size distribution and gave an example of droplet size distributions from ice-cloud measurements that correspond to $B = 0.017$ and $k = 1.34$. In the same paper, he also introduced a linear dependence between β_a and α_a (i.e., $\beta_a = B_a(r)\alpha_a$), via a range-dependent proportionality function $B_a(r) = f[\alpha_a(r)]$. In terms of $B_a(r)$, sometimes referred to as the aerosol phase function, Klett then proceeds to solve the inversion problem, with the lidar signal resulting from both aerosol and molecular scattering. We will discuss this realistic case further below, as has been done in more recent literature [6.3], employing the lidar ratio, $L_a(r)$, which is the inverse of the aerosol phase function $B_a(r)$. For lidar retrieval, it is convenient to use the natural logarithm $S_{\ln}(\nu_L, r; r_B)$ of the range-adjusted (and power normalized) signal $S(\nu_L, r; r_B)$, defined in (6.2):

$$S(\nu_L, r; r_B) \equiv \frac{h\nu_L r^2}{\eta A_R E_L r_B} N(\nu_L, r; r_B) = \beta(r)\exp\left[-2\int_{r_0}^{r} \alpha(r')dr'\right];$$

$$S_{\ln}(\nu_L, r; r_B) = \ln[S(\nu_L, r; r_B)]. \tag{6.2}$$

As pointed out in [6.1], with B and k assumed to be constant (same type of aerosol, independent of r), the lidar equation may be expressed as a well-known differential equation in (6.2.a) with an analytic solution in (6.2.b):

$$\frac{dS_{\ln}}{dr} = \frac{1}{\beta}\frac{d\beta}{dr} - 2\alpha \xrightarrow{\beta = B\alpha^k;\ Const.\ B\ and\ k} \frac{1}{\beta}\frac{d\beta}{dr} = \frac{kB}{B\alpha^k}\alpha^{k-1}\frac{d\alpha}{dr}, \text{ and } \frac{dS_{\ln}}{dr} = \frac{k}{\alpha}\frac{d\alpha}{dr} - 2\alpha. \tag{6.2.a}$$

The last differential equation for $S_{\ln}(\nu_L, r; r_B)$ above is the Bernoulli equation; subjected to initial values at r_0 of $S_{\ln}(r_0)$ and $\alpha(r_0)$, its solution for $\alpha(r)$, and thus $\beta(r)$, is well known:

$$\alpha^{-1}(r) = \exp\left[-\int_{r_0}^{r} \frac{1}{k}\frac{dS}{dr'}dr'\right]\left[\alpha^{-1}(r_0) - 2\int_{r_0}^{r} \frac{1}{k}\exp\left(+\int_{r_0}^{r'} \frac{1}{k}\frac{dS_{\ln}(r'')}{dr''}dr''\right)dr'\right], \text{ or}$$

$$\alpha(r) = \frac{\exp[(S_{\ln}(r) - S_{\ln}(r_0))/k]}{\alpha(r_0)^{-1} - \frac{2}{k}\int_{r_0}^{r} \exp[(S_{\ln}(r') - S_{\ln}(r_0))/k]dr'}; \beta(r) = B\alpha(r)^k. \tag{6.2.b}$$

Indeed, one can show that (6.2.b) is a solution to the last equation in (6.2.a) by direct differentiation and substitution. Mathematically, the solution in (6.2.b) is equally valid whether the reference range is higher or lower than the range of interest (i.e., either $r \geq r_0$ or $r \leq r_0$). Given that the signal $S_{\ln}(r)$ decreases as the range increases, for $r \geq r_0$ both the numerator and the denominator of the second expression in (6.2.b) decrease. Then, at ranges much larger than r_0, the noise associated with the return signal – not shown in the expression (6.2.b) – will eventually render the retrieval unstable. For the case of $r \leq r_0$, both the numerator and the denominator of this expression increase, and the ratio of two larger (but finite number) signals will be increasingly stable against noise perturbations. This difference was first pointed out and numerically demonstrated by Klett [6.1]. In addition to noise-induced potential instability, inverting upward $(r \geq r_0)$ versus downward $(r \leq r_0)$ presents an additional stabilization issue resulting from the intrinsic stepwise (digitized) retrieval process. This inversion difference favors the downward inversion as well. Thus, a high-altitude (upper troposphere or above) reference range should be chosen, where the atmosphere is nearly aerosol-free.

The lidar inversion of $S_{\ln}(r)$ given in (6.2.b) can be applied to two situations: (1) in the stratosphere or above (i.e. 30 km$<r<r_0$), with $\beta_a = \alpha_a = 0$, and (2) in the lower troposphere, or $r \leq 3$ km, with low visibility (or aerosol dominance), that is, $\beta_m \ll \beta_a$ and $\alpha_m \ll \alpha_a$, along with the known relationship $\beta_a(r) = B\alpha_a(r)^k$ with B and k as constants. In both cases, there is only one type of scatterer, and the optical extinction coefficient is retrievable from the lidar signal given by (6.2.b), and from which the elastic-backscattering coefficient may be deduced via either $\beta_m(r) = (3/8\pi)\alpha_m(r)$ or $\beta_a(r) = B\alpha_a(r)^k$. In the sections that follow, we first perform the lidar inversion in Section 6.1.1 for the aerosol-free case, assuming the absence of aerosols in (6.2.b). We then expand the treatment of lidar inversion to

the case of both aerosol and molecular scattering by developing appropriate inversion formulae in Section 6.1.2, developing a realistic example for this case in Section 6.1.3. Here, we will first compare the errors (that result from stepwise retrieval) between upward and downward inversions.

6.1.1 Retrieval for the Aerosol-Free Region

For the aerosol-free region, $\beta_m(r)$ and in turn $S_{\ln}(r)$ in (6.2) may be related to $\alpha_m(r)$. Equation (6.2.b) then yields:

$$\alpha_m(r) = \frac{\exp[S_{\ln}(r) - S_{\ln}(r_0)]}{\alpha_m(r_0)^{-1} - 2\int_{r_0}^{r} \exp[S_{\ln}(r') - S_{\ln}(r_0)]dr'}, \text{ with}$$

$$S_{\ln}(r) = \ln\left\{\frac{3}{8\pi}\alpha_m(r)\exp\left[-2\int_{r_0}^{r}\alpha_m(r')dr'\right]\right\}. \tag{6.2.c}$$

With the extinction coefficient profile determined by lidar inversion (6.2.c), the result for the backscattering profile is simply $\beta_m(r) = 3\alpha_m(r)/(8\pi)$. We note that the factor of $3/(8\pi)$ assumes that both scattered polarizations are received by a photodetector and air molecules are spherical. Strictly speaking, air molecules are ellipsoids with relative anisotropy of 0.221, leading to $\alpha_m(r) = (4\pi/1.479)\beta_m(r)$ – see (4.9.g) in this book or (A8) of [6.4].

As an example, we take the standard air mass density from the 1976 Standard Atmosphere Table (www.digitaldutch.com/atmoscalc/table.htm) with molecular mass of 28.97 kg/kmol and Rayleigh backscattering cross section of 6.11×10^{-32} m^{2}sr^{-1} [6.4], and from this compute the profile of $\beta_m(r)$ and $\alpha_m(r)$ directly. We use this website based on geopotential altitude because it offers a calculator for data in any interval requested. We assume the data obtained are the same as for geometric altitudes, since their difference in the troposphere is negligible (e.g., at 4 km, the pressure is only 0.025% lower in geopotential altitude, as opposed to 22% at 84 km). Using the result, we construct the range-adjusted lidar signal $S(532\text{ nm}, z)$, with resolution of 0.05 km. We then compute $\alpha_m(r)$ from (6.2.c) for the range between 2.5 and 5.5 km. The results in Fig. 6.1(a) compare the three methods of extinction coefficient calculations: The solid curve shows the directly calculated (or input) $\alpha_m(r)$, the dashed curve retrieves α from below, and the dotted curve retrieves α from above by calculation of (6.2.c). We see that the three profiles overlap, suggesting negligible inversion errors.

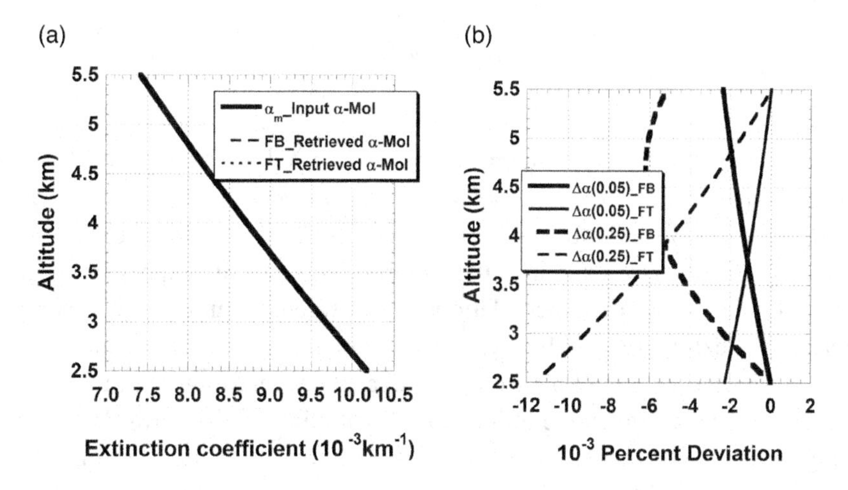

Fig. 6.1 Comparison between two deduced Rayleigh extinction coefficient (a) profiles (upward and downward) in comparison to the correct values, and (b) the deviations of the deduced profiles from the correct values.

We then repeat the signal construction and retrieval of $\alpha_m(r)$ with a larger step size of 0.25 km. The plots are virtually identical to those for 0.05 km and are thus not shown. However, if we plot the percent deviation between the results for different step sizes, the differences are apparent, as shown in Fig. 6.1(b). Notice that the scale on the x-axis in Fig. 6.1(b) is 10^{-3} percent, suggesting that the inversion error is of no practical importance. This being a case where the initial condition for retrieval is precisely known and the simulated signal contains no statistical noise, the deviations, however minute, must therefore come from the inversion process. In comparing the solid curves to the dashed curves, it is not surprising to see more accurate results from the finer retrieval step size. In addition, by comparing the thin curves to the thick curves we observe that the downward retrieval yields more accurate results, likely due to the propagation of the discrepancy in the initial guess through the first few steps. Thus, as with the noise perturbation, the extinction coefficient inversion also favors using downward retrieval. Here, we have discussed the inversion process without the influence of statistical noise in the data. Despite its importance, we do not discuss the details of statistical error propagation here. The procedures and measurement uncertainty calculation examples may be found in [6.5] and [6.6] as well as in cases in Chapter 7.

6.1.2 Retrieval for the Case with Both Aerosol and Molecular Scattering

The total backscattering coefficient at a given r may be written as the sum of aerosol and molecular scattering, each proportional to the associated extinction coefficient as,

$$\beta(r) = L_a^{-1}(r)\alpha_a(r) + L_m^{-1}\alpha_m(r)\ ,\ \text{or}\ \alpha(r) = \alpha_a(r) + \alpha_m(r) = L_a(r)\beta_a(r) + L_m(r)\beta_m(r). \tag{6.3}$$

Here, following the modern literature, we have used the inverse of the phase function, that is, the aerosol and molecular lidar ratios [6.3], $L_a(r) = B_a^{-1}(r)$ and $L_m \equiv B_m^{-1} = (8\pi)/3$, respectively. Before discussing the solution for the aerosol optical parameter, we examine the appropriateness of expressing $\beta_a(r) = L_a^{-1}(r)\alpha_a(r)$. It is true that the relationship between β_a and α_a, describing the scattering and extinction within a viewing volume, is more complicated than the simple linear relationship we have employed, as it depends on the complex refractive index and the particle size and shape distribution of the aerosols within the viewing volume. Further, these parameters can vary as a function of time and height. However, during the probing time and in a vertically resolved altitude bin, the aerosol characteristics within the volume do not change. This is what allows us to regard $\beta_a(r)$ and $\alpha_a(r)$ as directly proportional to one another, with a range-dependent proportionality factor. In this approach, the aerosol lidar ratio represents the inverse aerosol scattering phase function in the resolving volume within the measurement time. For this linear relationship to be meaningful, we must find an alternative method or measurement to determine the lidar ratio profile, independent of the Rayleigh–Mie lidar signal used for the retrieval. As discussed later in Section 7.1.1, the lidar ratio may indeed be measured by high spectral resolution lidar (HSRL) techniques. Once the range-dependent aerosol lidar ratio $L_a(r)$ is known, we can retrieve the total backscattering coefficient by the inversion outlined below.

As given in (6.2.a), the range gradient of the logarithm of the adjusted lidar signal can be written in terms of $\beta(r)$ and $\alpha(r)$; it can be further expressed as the first expression in (6.3.a) below. By introducing a new related signal, $\widetilde{S}(r)$, given in the second expression in (6.3.a), we arrive at an equation for the derivative of $\widetilde{S}\ (r)$ in the form of the Bernoulli equation. This leads to the solution for the quantity $L_a(r)\beta(r)$ and, in turn, the total backscattering coefficient, the aerosol backscattering coefficient, and the aerosol extinction coefficient as given respectively in (6.3.b).

$$\frac{dS_{\ln}(r)}{dr} = \frac{d\beta}{\beta dr} - 2\alpha = \left[\frac{1}{(L_a\beta)}\frac{d(L_a\beta)}{dr} - \frac{1}{L_a}\frac{dL_a}{dr}\right] - 2[L_a\beta - (L_a - L_m)\beta_m]$$

$$\widetilde{S}(r) \equiv S_{\ln}(r) + \ln L_a(r) - 2\int_{r_0}^{r}(L_a(r') - L_m)\beta_m(r')dr' \to \frac{d\widetilde{S}(r)}{dr} = \frac{1}{(L_a\beta)}\frac{d(L_a\beta)}{dr} - 2(L_a\beta) \tag{6.3.a}$$

$$L_a(r)\beta(r) \to \beta(r) = \frac{\exp\left(\widetilde{S}(r) - \widetilde{S}(r_0)\right)}{L_a(r)\left[L_a^{-1}(r_0)\beta^{-1}(r_0) - 2\int_{r_0}^{r}\exp\left(\widetilde{S}(r') - \widetilde{S}(r_0)\right)dr'\right]}$$

$$\beta_a(r) = \beta(r) - \mathcal{N}(r')\frac{d\sigma_m}{d\Omega};\ \alpha_a(r) = L_a(r)\beta_a(r) \tag{6.3.b}$$

To summarize the inversion process for Rayleigh–Mie lidar (scattering from both aerosols and air molecules) with the (aerosol) lidar ratio, $L_a(r)$, determined from an independent measurement – for example, by a HSRL (high-spectral resolution lidar) technique [6.3] – we first form the profile of the lidar signal $\widetilde{S}(r)$, which satisfies the Bernoulli equation given in (6.3.a). The inversion of this equation leads to the total backscattering coefficient, $\beta(r)$, from which we then deduce the aerosol backscattering coefficient, $\beta_a(r)$, and the aerosol extinction coefficient, $\alpha_a(r)$, all given in (6.3.b). Here, the molecular optical properties are assumed to be known explicitly or can be deduced from the standard air molecular density. With an additional properly designed lidar (or other) measurement, one can independently determine the lidar ratio profile, as will be discussed in Section 7.1.

If knowledge of the lidar ratio is absent, in order to employ the inversion formulae in (6.3.b), we must assume (with an educated guess) the lidar ratio profile, $L_a(r)$, and the initial condition for the total backscattering coefficient, $\beta(r_0)$. Typically, with some knowledge of the aerosol type, we can (at best) assume a constant value for the lidar ratio profile and assume $\beta(r_0)$ in relation to the known molecular counterpart, $\beta_m(r_0)$, depending on the height of the reference altitude. In this case, both the inversion details and our ignorance of the boundary conditions (BC) contribute to the uncertainties (deviations in this case) of the retrieved atmospheric optical properties.

6.1.3 An Example for the Case with Both Aerosol and Molecular Scattering

We consider an example of lidar retrieval of atmospheric optical properties. For this purpose, we simulate the signal from the output of an HSRL Rayleigh–Mie lidar, extracting vertical profiles of aerosol extinction coefficient, $\alpha_a(z)$, and (aerosol) lidar ratio, $L_a(z)$, at 532 nm. These data are adapted from the dataset that yielded Fig. 14 of Hair et al. [6.7] and were kindly provided by John Hair. The required air molecular number density, $\mathcal{N}(z)$, and observed aerosol backscattering coefficient, $\beta_a(z)$, are taken, respectively, from the 1976 standard atmosphere table and calculated from the ratio $\alpha_a(z)/L_a(r)$. From the information on $\mathcal{N}(z)$, $\beta_a(z)$, $\alpha_a(z)$, and $L_a(z)$, we simulate the lidar signals $S(r)$, $S_{\ln}(r)$, and $\widetilde{S}(r)$, from which we deduce the atmospheric optical properties $\beta(z)$, $\beta_a(z)$, and $\alpha_a(z)$ employing (6.3.b). We show the resulting profiles for $\beta(z)$ and $\alpha_a(z)$, respectively, in Fig. 6.2(a) and (b), of upward (in dashed black) and downward (in solid black) retrievals using 0.05 km steps between 2.5 km and 5.25 km. We include the input profile (observed, in gray) and uncertainty short bars (in black) for comparison.

Since in the simulation the retrieved total backscattering coefficient is the solution to the differential equation (6.3.a), the simulated profiles (upward dashed and downward solid) match the observation exactly at the respective end points, 2.5 km and 5.25 km, respectively. If a measured lidar ratio is not available, the retrieval of atmospheric optical properties with the Rayleigh–Mie signal alone, using (6.3.b), requires guesses of the lidar ratio profile, $L_a(r)$, and the initial value for the total backscattering coefficient, $\beta(r_0)$. Since we know the input parameters to the simulated lidar signal, we can make "best" educated guesses for $L_a(r) = 35$ (about its altitude mean value) and $\beta(r_0) = 2\beta_m(r_0)$ for downward, or $\beta(r_0) = 4\beta_m(r_0)$ for upward, retrievals. With these guesses, we obtain aerosol extinction profiles and compare them in Fig. 6.2(c). Again, we see that the retrieval downwards (from top, solid) fares better than that upwards (from bottom, dashed). In fact, the top-down aerosol extinction retrieval for the case with lidar ratio measurement yields results in agreement with the gray input data (within its error bars), as shown in Fig. 6.2(b). The results with educated guesses are shown in Fig. 6.2(c). These also perform well, particularly the retrieval from the top (solid). Notice that even for the "best possible" guesses, we were unable to match the initial values of $\alpha_a(r_0)$ for either upward (solid) or downward (dashed) retrievals. Nevertheless, the latter matches the correct result (gray) reasonably well at 5.25 km, suggesting that the initial condition from the top could correctly estimate the molecular backscatter coefficient, $\beta_m(r_0)$, should the retrieval start from 30 to 40 km where the atmosphere is aerosol-free.

The retrieved profiles in Fig. 6.2(c) do not match the deviations in the measurement, and this is due to the use of a constant lidar ratio. The percent deviations of the retrievals from the observed values are compared in Fig. 6.2(d). Overall, the deviations for the downward retrieval are on the order of 10% and are smaller than those from the upward retrieval. Of course, in most cases, we are unable to do such a good job with the Rayleigh–Mie lidar data alone. In fact, if, in (6.3.b), we use the molecular backscattering coefficient at the reference altitude as the initial guess for $\beta(r_0)$ but keep the lidar ratio at 35, or, alternatively, we change the value of the lidar ratio from 35 to 15 but keep the same boundary condition, the percentage deviation would be about 50% for the downward and about 100% for the upward retrieval (not shown). This example demonstrates that information on the lidar ratio should be independently provided if the retrieved atmospheric profiles are to be trusted.

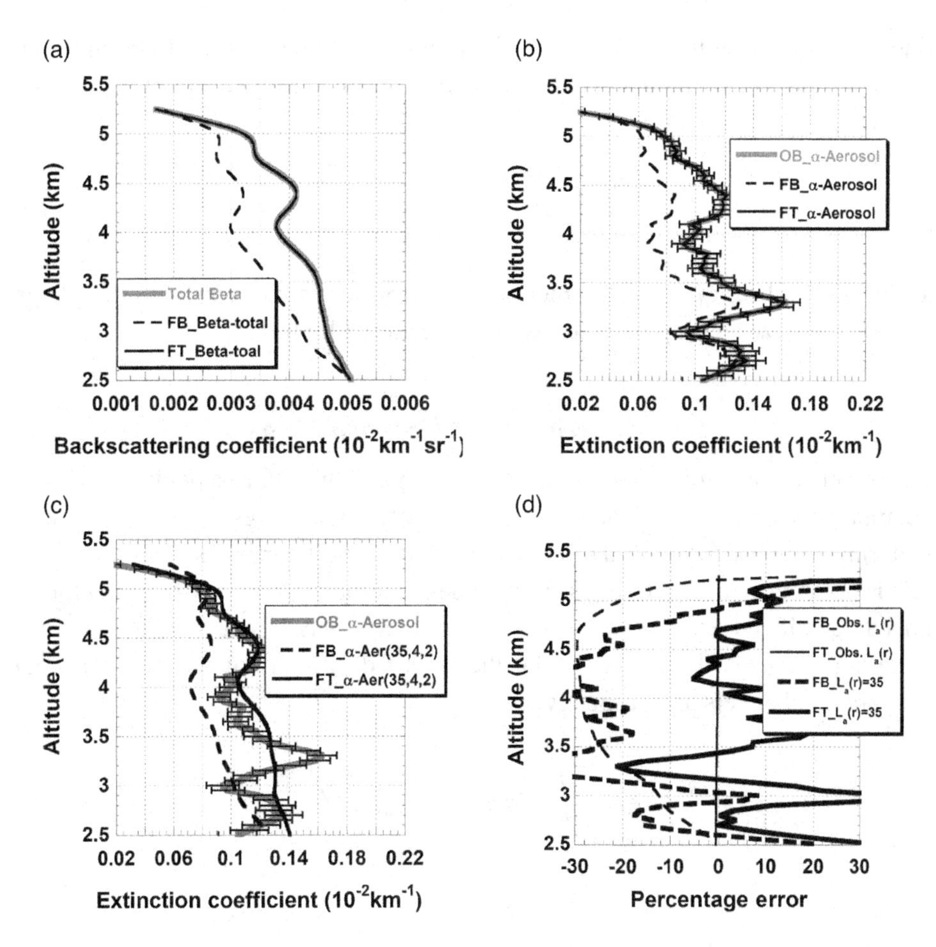

Fig. 6.2 Observed scattering coefficient (gray) compared to that retrieved from below (dashed) and above (solid) in (a) for $\beta(r)$ and (b) for $\alpha_a(r)$, with the information of observed lidar ratio profile and the correct initial value incorporated. (c) for $\alpha_a(r)$ with the lidar ratio set to 35, $\beta(r_0)$ to $2\beta_m(r_0)$ from above, and to $4\beta_m(r_0)$ from below. The percent deviations for the retrievals in (b) and (c) are compared in (d). The observed profiles and associated uncertainties are provided by John Hair and gratefully appreciated.

6.2 Polarization Lidars

Since an air molecule is a homonuclear diatomic molecule, there are two independent elements in its polarizability, the mean a and anisotropy γ, as dictated by symmetry (see Chapter 4). This means that the isotropic average of the dipole moment induced by an s-polarized electric field (perpendicular to the plane of incidence, or along $\hat{e}_a$ in Fig. B.1) has both s- (along $\hat{e}'_a$ or parallel) and p- (along $\hat{e}'_b$ or perpendicular) components in its scattering cross section. The ratio of the two backscattering coefficients (perpendicular to parallel) $\beta_\perp/\beta_\parallel$ has been termed the (linear) depolarization ratio, δ_ℓ, in the literature, for an ensemble of homonuclear diatomic molecules; it is dependent on the relative anisotropy R_A defined in (6.4); see (A10) of [6.4]:

$$\delta_\ell = \frac{\beta_\perp}{\beta_\parallel} = \frac{3R_A}{45 + 4R_A}; \quad R_A = \left(\frac{\gamma}{\alpha}\right)^2. \tag{6.4}$$

For air molecules, $R_A = 0.22$, which can be in principle measured by a Rayleigh lidar that analyzes both polarizations of the scattered light.

6.2.1 Conventional Polarization Lidars

Aerosol and cloud particles are bigger and less symmetric; thus, a polarization lidar that transmits a linearly polarized wave and separately receives the cross-polarized scattering signals can be used to characterize and study particle shapes and sizes. This is potentially useful to classify the types of atmospheric aerosols and clouds. Indeed, Schotland et al. [6.8] (in 1971) has described such a lidar with two scattering channels. Following (6.1), the received photons of the perpendicular and parallel channels are, respectively,

$$N_{\perp,\parallel}(\nu_L, r; r_B) = \eta \frac{A_R}{r^2} \beta_{\perp,\parallel}(r) r_B \frac{E_L}{h\nu_L} \exp\left[-2\int_{r_0}^{r} \alpha(r')dr'\right]; \quad \alpha(r) = \alpha_\parallel(r) + \alpha_\perp(r). \tag{6.5}$$

Since that time, this method has been used to identify desert dust in the free troposphere, shed light on ice nucleation [6.9], and to classify polar stratospheric clouds [6.10]. Thus, cross-polarization receiver channels are included in space-based lidars, such as CALIPSO, for the identification of cloud phases [6.11]. As another example, the Cloud, Aerosol Polarization, and Backscatter Lidar (CAPABL) was developed by Alvarez et al. [6.12] in 1998 for sustained observations to characterize and document cloud properties under varied polar atmospheric conditions. Polarization lidar for atmospheric remote sensing was reviewed by Sasson [6.13]. Determining the depolarization ratio from the observed scattering signals, $\delta_\ell(r) = N_\perp(r)/N_\parallel(r)$, employing the "legacy approach" described above, though straightforward, has several problems, as pointed out by Gimmestad et al. [6.14]. Though under most conditions of interest we can get by with single scattering from an aerosol particle and/or air molecule, we still need to worry about the fact that the ratio $\beta_\perp/\beta_\parallel$ is not the physical property of interest; that is, even for air molecules, $\beta_\perp/\beta_\parallel$ is related to R_A but is not the relative anisotropy, $R_A = (\gamma/a)^2$, of the atmosphere. The two are related via (6.4) only because we have a confirmed model for the atmospheric molecular system. Aerosol and cloud particles are much more complex. For one thing, they do not possess the symmetry of a homogeneous diatomic molecule. This means they generally have three independent elements for their polarizability. Thus, even if we assume the system in question is randomly oriented and we know a good model to describe the symmetry of the scattering particle, we need three measurements to determine two independent ratios. In this sense, the measurement of the "depolarization ratio" as previously defined ($\delta_\ell = \beta_\perp/\beta_\parallel$) is at best ambiguous. Furthermore, there is more than one constituent in the atmosphere, including air molecules, aerosols, clouds, and ice. Each constituent has two or three different polarizability tensor elements. In addition, the assumption that the observed particles are randomly oriented in the atmosphere may be questionable. This is especially true when one realizes certain ice crystals aerodynamically favor horizontal orientation [6.15] due to Earth's gravity. Thus, a more complete description for polarization of scattered light is needed. It has long been known that the most general way to describe light scattering employs the Stokes vector for the light beam and Mueller matrix for the scattering medium (see, for example, van de Hulst [6.16]). This approach started in the late 1970s and 1980s [6.17; 6.15]. Partially due to the interest in polar clouds and horizontally oriented ice crystals (HOIC), this approach was further theorized [6.18] and applied to atmospheric observations [6.19; 6.20]. This more general approach on polarization lidar is discussed in Section 6.2.2.

6.2.2 Stokes Vectors for Light Beams and Mueller Matrices for Atmospheric Media

We first discuss the use of Stokes vectors for describing a light beam and of the Mueller matrix for describing atmospheric media. Depending on the properties of the atmospheric medium (ensemble of particles), polarized laser light scattered from a single particle may become depolarized. We thus need a description that includes the depolarized scattered light along with the unpolarized natural light (background) in a way that is separable from the polarized contributions. In 1852, G. G. Stokes figured out that the state of a light beam may be represented by four measurable intensities (S_0, S_1, S_2, S_3) with S_0 being the total intensity (polarized plus unpolarized), while the components (S_1, S_2, S_3) combine to describe the polarized component, which requires three values to determine the magnitudes of the orthogonal components and their relative phases.

Putting the four intensities into a column matrix, the so-called Stokes vector provides the general description of a realistic scattered light beam. Using this representation, the medium (including any optical elements along the beam path through the medium that lie between the laser transmitter and polarization-sensing detector) is characterized by a 4×4 matrix, called a Mueller matrix, relating the input and output light beams. The symmetry of the Mueller matrix reflects the symmetry of the optical properties of the medium plus any optical elements [6.15, 6.16] through which the light beam propagates and scatters. Since the Stokes vector representing light fields lies on a plane perpendicular to the direction of light propagation, the Mueller matrix describing the effect of light scattering from an ensemble of particles will depend on the symmetry of the medium in relation to the incident and scattered light beams. However, in lidar studies, we are interested in backward scattering only and from an ensemble of identical particles; here, the Mueller matrix is termed the backscattering phase matrix (BSPM) [6.15]. If the particles are randomly oriented (such as air molecules and aerosols in hydrostatic equilibrium), the symmetry of the BSPM should be independent of the coordinate system and the direction of the light beams used to describe the scattering process. We thus typically use the geometric coordinate system with the z-axis pointing at the zenith. In contrast, particles with large azimuthal dimensions (HOIC being a case in point) are more likely oriented horizontally due to the Earth's gravity. Not only will the symmetry of the Mueller matrix that describes such a system be different from that for randomly oriented particles, but their matrix elements should generally depend on the beam's tilt angle, Θ, off zenith.

We start by considering the Stokes vector for a light field propagating along the z-axis, represented by a four-element column matrix $(S_0 S_1 S_2 S_3)^T$, where the superscript T stands for "transpose," turning a row matrix into a column matrix with the same elements. As described in (B.9.b), the four S_i elements for a right-handed, elliptically polarized light beam with principal axes $a \cos \chi$ and $a \sin \chi$, where χ is the ellipticity angle, oriented along axes rotated by an angle φ from the x-axis on the horizontal (x-y) plane, are given, respectively, as

$$S_0 = a^2;\ S_1 = pa^2 \cos(2\chi)\cos(2\varphi);\ S_2 = pa^2 \cos(2\chi)\sin(2\varphi);\ S_3 = pa^2 \sin(2\chi), \tag{6.6.a}$$

where $S_0 = a^2$ is the total intensity – the sum of the unpolarized portion $a^2(1-p)$ and the polarized portion a^2p – with $p = \sqrt{S_1^2 + S_2^2 + S_3^2}/S_0$ designating the degree of polarization (DOP) of the beam. By choosing specific values for χ, this Stokes vector can represent linear and circularly polarized waves, as explained in Appendix B. Without losing generality, this Stokes vector (6.6.a) can also be used to represent the electric field of a light beam propagating along $\hat{k}_0$, tilted at an angle Θ off zenith, that is, $(\theta_0 = \Theta,\ \phi_0 = \pi/2)$, illustrated in Fig. 6.3. In this case the electric field lies on the tilted plane $(x' - y')$ perpendicular to $\hat{k}_0$. We note that the rotated coordinate $\left(\hat{x}', \hat{y}', \hat{k}_0\right)$ is the original $(\hat{x}, \hat{y}, \hat{z})$ rotated with respect to the $\hat{x}$ axis by an angle Θ; the polarization axis $\hat{x} = \hat{x}'$ is often referred to as s-polarization, lying on the intersection between the $(x - y)$ plane and the new reference plane $(x' - y')$. In the language used in Appendix B, the $\left(\hat{x}', \hat{y}', \hat{k}_0\right)$ coordinate is that rotated from the original $(\hat{x}, \hat{y}, \hat{z})$ by Euler angles (0, Θ,0), as shown in Fig. 6.3. Projected onto the $(\hat{x}, \hat{y}, \hat{z})$ coordinates using (B.1.a), (B.1.c), and (B.1.d), we can relate the relevant unit vectors $\left(\hat{e}_a, \hat{e}_b, \hat{k}_0\right)$ and its backscattering counterpart $\left(\hat{e}_a', \hat{e}_b', \hat{k}\right)$. To be consistent with Fig. 6.3 by setting $\theta_o = \Theta$, $\phi_o = \pi/2$ and $\theta = \pi - \Theta$, $\phi = \pi/2 + \pi = 3\pi/2$, respectively, we have

$$\hat{k}_0 = (0, \sin\Theta, \cos\Theta); \hat{e}_a = (1, 0, 0) = \hat{x}'; \ \hat{e}_b = (0, \cos\Theta, -\sin\Theta) = \hat{y}';$$
$$\hat{k} = (0, \sin\Theta, -\cos\Theta) = -\hat{k}_0; \ \hat{e}_a' = (-1, 0, 0) = -\hat{x}' \ ; \ \hat{e}_b' = (0, \cos\Theta, -\sin\Theta) = \hat{y}'. \tag{6.6.b}$$

That $(\hat{e}_a, \ \hat{e}_b) = (\hat{x}', \ \hat{y}')$ and $(\hat{e}_a', \ \hat{e}_b') = (-\hat{x}', \ \hat{y}')$ in (6.6.b) suggests that the same Stokes vector given in (6.6.a) can also be used to represent the backscattered beam with the same elliptic parameter format, except that the sense of polarization (in the same coordinate) is reversed to left-handed.

We now turn to the 4×4 Mueller matrices for an atmospheric scattering system, comparing that for an ensemble of randomly oriented particles with that for HOICs. Unlike the system with random orientation, the ensemble of HOICs is not macroscopically isotropic, though it retains isotropic mirror symmetry with respect to the

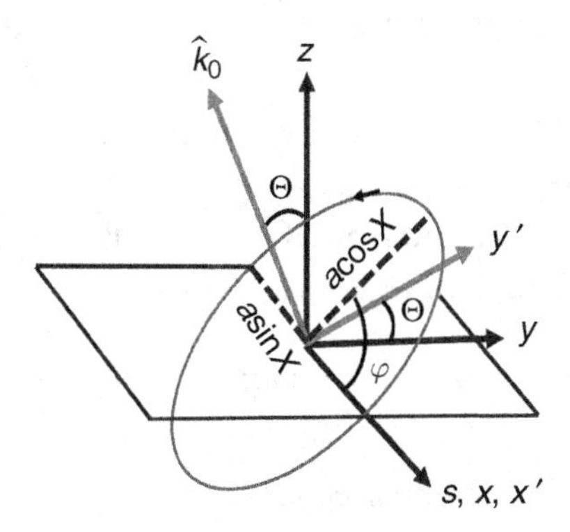

Fig. 6.3 The right-handed elliptically polarized field of a laser beam propagating along the direction $\hat{k}_0$ tilted from zenith and the y-axis by the same angle Θ, as shown geometrically on a plane tilted from the horizontal. The backscattering direction (not shown) points at $(\theta = \pi - \Theta, \ \ \phi = \phi_0 + \pi = 3\pi/2)$.

vertical or z-axis. The Mueller matrix (BSPM) for this system, which has been studied [6.15] and determined to have the form of F(6) given in (B.10), is given as $\vec{M}_o(\Theta)$ in (6.7.a). For the ensemble of randomly oriented particles, rotational symmetry with respect to the z-axis demands invariance in exchange between x- and y-axes, thus requiring $F_{12} = F_{34} = 0$ for zenith pointing, $\mathbf{M}_o(0^\circ)$. Other symmetry operations [6.20; 6.14] further simplify and reduce $\mathbf{M}_o(\Theta)$ for an ensemble of randomly oriented particles to $\vec{M}_r$ in (6.7.b), which is independent of light propagation direction, or Θ.

$$\mathbf{M}_o(\Theta) = \begin{bmatrix} F_{11} & F_{12} & 0 & 0 \\ F_{12} & F_{22} & 0 & 0 \\ 0 & 0 & F_{33} & F_{34} \\ 0 & 0 & -F_{34} & F_{44} \end{bmatrix}; \tag{6.7.a}$$

$$\mathbf{M}_r = \begin{bmatrix} f_{11} & 0 & 0 & 0 \\ 0 & f_{22} & 0 & 0 \\ 0 & 0 & -f_{22} & 0 \\ 0 & 0 & 0 & f_{11} - 2f_{22} \end{bmatrix}. \tag{6.7.b}$$

As alluded to in Appendix B, Lu and Chipman [B.6] have shown that a Mueller matrix can be decomposed into three different factors, the product of which is the final matrix: a diattenuator followed by a retarder followed by a depolarizer. Such a decomposition is beyond the scope of this book, but Hayman and Thayer [6.18] have worked out this decomposition of the BSPM for HOIC $\mathbf{M}_o(\Theta)$ and have given the result in their Eqs. (26), (27), and (28), from which they provide the following physical interpretations for the matrix elements: F_{11}, F_{12}, and F_{34} are, respectively, the consequences of backscattering (and depolarizing), diattenuation, and retardance in the medium. Thus, $\mathbf{M}_o(\Theta)$ can be diattenuating, retarding, and depolarizing. In contrast, the off-diagonal elements of the matrix $\mathbf{M}_r$ for randomly oriented particles, f_{12} and f_{34}, are zero; it is strictly depolarizing. On the same token, since $F_{12} = F_{34} = 0$ for $\mathbf{M}_o(0°)$, a zenith-pointing polarization lidar will miss the diattenuation and retardance attributes of HOIC, because their presence can only be detected by a lidar tilted off-zenith.

6.2.2(1) Stokes Vectors-Based Lidar Equation for Diattenuation Measurements

Armed with the knowledge of Stokes vectors and Mueller matrices $\mathbf{M}_o(\Theta)$ and/or $\mathbf{M}_r$, we can now proceed to the lidar equation for a full description of polarization lidar. To do so, we go to (5.3) and consider a lidar equation with four components ($\mu = 4$) by letting $E_{iL} = S_{Ti}E_L$ with $i = 0, 1, 2$, and 3, where E_L is the laser energy and S_{Ti} the components in the Stokes vector representing the polarized state of the transmitting laser light (the subscript T refers to transmitter; not to be confused with the superscript T for transpose). By further identifying the relevant quantities in matrix form in (6.8.a) and assuming the same detection efficiency for all four channels, $\eta_i = \eta$, the normalized received photon counts in four channels, $(h\nu_{iL}/\eta_i E_L)N_i(\nu_{iS}, r; r_B)$, give rise to the components of the receiving Stokes vector, S_{Ri}. The equation for the polarization lidar is then given in (6.8.b):

$$A_{\downarrow}(\nu_{iS}, r) \rightarrow \mathbf{T}_{atm}(\vec{k}_s, r);\ A_{\uparrow}(\nu_{iL}, r) \rightarrow \mathbf{T}_{atm}(\vec{k}_i, r);$$

$$\left[\mathcal{N}_i(r)\frac{d\sigma(\nu_{iS})}{d\Omega}\right] \rightarrow \mathbf{F}(\vec{k}_i, \vec{k}_s, r);\ E_{iL} = S_i E_L \rightarrow \vec{S}_T E_L; \tag{6.8.a}$$

$$\frac{h\nu_{iL}}{\eta_i E_L} N_i(\nu_{iS}, r; r_B) \text{ with } i = 0, 1, 2, 3 \rightarrow$$

$$\vec{S}_R(r) = \frac{A_R r_B}{r^2}\mathbf{T}_{atm}(\vec{k}_s, r)\mathbf{F}(\vec{k}_i, \vec{k}_s, r)\mathbf{T}_{atm}(\vec{k}_i, r)\vec{S}_T. \tag{6.8.b}$$

Equation (6.8.b) is consistent with the Stokes vector lidar equation (SVLE), Eq. (4) of [6.18], first introduced by Hayman and Thayer in 2012, except that $\vec{S}_T$ here represents the laser beam transmitted into the atmosphere, same as their $\mathbf{M}_{TX}\vec{S}_T$, and $\vec{S}_R(r)$ here is the Stokes vector representing light received before the polarization analysis, same as their $\mathbf{M}_{RX}^{-1}\vec{S}_{RX}(r)$.

As an illustrative example, we relate the attributes of HOIC to the measurable quantities of a polarization lidar. In order to focus on the essence of the problem, we assume unity atmospheric transmittances, $\mathbf{T}_{atm} = 1$, and consider a laser beam pointing at an angle Θ off-zenith, linearly polarized along an axis rotated φ degrees from the x'-axis on the rotated reference $(x' - y')$ plane (see Fig. 6.3). In this plane, the beam polarization is represented by a column matrix (or written as the transpose of the corresponding row matrix) $\vec{S}_T$ in (6.8.c), derived from (6.6.a) with $\chi = 0$ and $p = 1$; it impinges on a HOIC ensemble represented by a Mueller matrix of (6.7.a), $\mathbf{F}(\vec{k}_i, -\vec{k}_i, r) = \mathbf{M}_o(\Theta)$. The normalized receiving Stokes vector $\vec{S}_R(r)$ is derived from (6.8.b), yielding the result in (6.8.c):

$$\vec{S}_R(r) = \mathbf{M}_o(\Theta)\vec{S}_T;$$

$$\vec{S}_T = S_{T0}[1 \quad \cos(2\varphi) \quad \sin(2\varphi) \quad 0]^T \rightarrow$$

$$\vec{S}_R(r) = S_{R0}[F_{11} + F_{12}\cos(2\varphi) \quad F_{12} + F_{22}\cos(2\varphi) \quad F_{33}\sin(2\varphi) \quad -F_{34}\sin(2\varphi)]^T, \tag{6.8.c}$$

where both S_{T0} and S_{R0} are normalized constants, respectively proportional to laser energy and received signal. To understand the information contained in the received

Stokes vector, we need to interpret it with a set of three polarization-analyzed experiments done either simultaneously or sequentially within a short time before the atmospheric conditions change. We choose three linear polarizers as the analyzers: parallel to, 45° from, and perpendicular to the $\vec{S}_T$ polarization. These polarizers for backscattered light receiving channels are, respectively, represented by the following row matrices:

$$\vec{P}_{\parallel} = \begin{bmatrix} 1 & \cos(2\varphi) & -\sin(2\varphi) & 0 \end{bmatrix}; \; \vec{P}_{45} = \begin{bmatrix} 1 & \sin(2\varphi) & \cos(2\varphi) & 0 \end{bmatrix};$$
$$\vec{P}_{\perp} = \begin{bmatrix} 1 & -\cos(2\varphi) & \sin(2\varphi) & 0 \end{bmatrix}.$$

By projecting the received light onto the detection polarizers (i.e., performing dot products between $\vec{P}_{\parallel,45,\perp}$ and $\vec{S}_R$), we can obtain three normalized signal counts, $N_{\parallel}, N_{45}$ and $N_{\perp}$ along with the sum $N_{\parallel} + N_{\perp}$, as

$$\begin{aligned} N_{\parallel} &= F_{11} + 2F_{12}\cos(2\varphi) + F_{22}\cos^2(2\varphi) - F_{33}\sin^2(2\varphi), \\ N_{45} &= F_{11} + F_{12}(\cos(2\varphi) + \sin(2\varphi)) + (F_{22} + F_{33})\sin(2\varphi)\cos(2\varphi), \\ N_{\perp} &= F_{11} - F_{22}\cos^2(2\varphi) + F_{33}\sin^2(2\varphi), \text{ and} \\ N_{\parallel} + N_{\perp} &= 2[F_{11} + F_{12}\cos(2\varphi)]. \end{aligned} \quad (6.8.d)$$

This leads to the linear depolarization ratio δ_ℓ and diattenuation D_{45} as

$$\delta_\ell = \frac{N_\perp}{N_\parallel} = \frac{F_{11} - F_{22}\cos^2(2\varphi) + F_{33}\sin^2(2\varphi)}{F_{11} + 2F_{12}\cos(2\varphi) + F_{22}\cos^2(2\varphi) - F_{33}\sin^2(2\varphi)} \xrightarrow{\varphi=45^o} \delta_\ell = \frac{F_{11} + F_{33}}{F_{11} - F_{33}};$$
$$D_{45} = \frac{2N_{45} - (N_\parallel + N_\perp)}{N_\parallel + N_\perp} = \frac{\sin(2\varphi)F_{12} + (F_{22} + F_{33})\cos(2\varphi)}{F_{11} + F_{12}\cos(2\varphi)} \xrightarrow{\varphi=45^o} D_{45} = \frac{F_{12}}{F_{11}}. \quad (6.8.e)$$

Referring to Fig. 6.3, and considering we have a choice when selecting the angle φ – the orientation of the linearly polarized laser beam – we have chosen $\varphi = 45^o$ in order to simplify the expressions and obtain the elements for depolarization and diattenuation in (6.8.e), as indicated. We point out, for the case of backscattered light, $\hat{e}_a{}' = -\hat{e}_a$ and $\hat{e}_b{}' = \hat{e}_b$, a 45° rotation from $(\hat{e}_a{}', \hat{e}_b{}')$ amounts to a –45° rotation from $(\hat{e}_b, -\hat{e}_a)$, giving rise to $F_{33} < 0$. We also point out, since $F_{12} = 0$ for $\mathbf{M}_o(0°)$ and $\mathbf{M}_r$ in an atmosphere consisting of both randomly and horizontally oriented subsystems, these do not contribute to diattenuation; to detect the presence of HOIC, the beam direction for a polarization lidar must be off zenith. In addition, though $\mathbf{M}_o(\Theta)$ in (6.7.a) appears to be independent of Θ, its matrix elements do depend on Θ. The choice of the tilt angle would depend on physical limitations as well as signal-to-noise considerations. For more details, see for example, Figs. 4.2 and 4.3 of [6.21], which show the dependence of backscattering coefficient and diattenuation on beam tilt angle relative to the HOICs, a topic we shall not discuss further.

6.2.2(2) An Example for the Detection of HOIC

We now present an example of actual polarization lidar measurement in Greenland, employing a Depolarization and Backscatter Unattended Lidar (DABUL) [6.12], modified [6.19] and improved [6.22] for characterization of the full range of cloud conditions, including HOIC, dubbed as Cloud, Aerosol Polarization, and Backscatter Lidar (CAPABL). In operation, this lidar is pointed off-zenith with a tilt angle, Θ. The transmitter laser beam passes through a half-wave plate (HWP) before exiting into the atmosphere, to set the polarization of the transmitting laser as the reference polarization $\hat{e}_a$ for the instrument; the transmitting light is horizontally polarized when $\varphi = 0^{\circ}$, $\hat{e}_a = \hat{s}$, as can be seen in Fig. 6.3. To achieve the polarization analysis described above, the backward-tilted received Stokes vector $\vec{S}_R(r)$ is passed sequentially through a quarter-wave plate (QWP), a liquid crystal variable retarder (LCVR), and a polarizing beam splitter, with the polarization axes oriented, respectively, perpendicular to, rotated 45° from, and parallel to the reference polarization (i.e. the $\hat{e}'_a = -\hat{e}_a$ direction on the $(x' - y')$ plane). A detailed schematic of CAPABL given in Fig. 4 of [6.19] is reproduced here as Fig. 6.4. When the voltage-controlled phase shift of the LCVR, Γ_{WP}, is set to $(-\pi/2,\ 0,\ \pi/2)$, the combined analyzer performs the function of $\left(P_{\perp}, P_{45^{\circ}}, P_{\parallel}\right)$ analyses, respectively. Thus, by changing the voltage applied to LCVR, the three quantities in (6.8.d) are measured sequentially using single (low-altitude and high-altitude) detectors.

CAPABL carried out one of the first simultaneous observations of diattenuation D_{45} and linear depolarization δ_{ℓ} in Greenland on 18 February 2012 [6.19]. The results are illustrated in Fig. 8(a) and 8(c) of [6.19], which reports having "observed two diattenuation signatures (magenta circles I and II) that coincided with clouds at altitudes between 3000 and 4500 m from 0230 to 0530 UTC for circle 1 and between 1000 and 2500 m from 0500 to 0630 UTC for circle 2" (in Fig. 8(a) of [6.19]). "During this same period, signals at lower altitudes, between 1300 m to 2000 m from 0430 to 0510 UTC, and between 500 m to 1000 m from 0640 to 0720 UTC depict variable linear depolarization with no concomitant diattenuation" (Fig. 8(c) of [6.19]). The implication is that HOIC, while being present at higher altitudes (within the two magenta circles), were not present at lower altitudes. Of interest is, of course, a circumstance with both HOIC (denoted with subscript 'o') and randomly oriented particles (denoted with subscript 'r'). Its Mueller matrix, $\mathbf{F}(\vec{k}_i, -\vec{k}_i, r) = \mathbf{M}_t(\Theta)$, is the sum of the two contributions as given in (6.9.a), as is also the case for the total backscattering coefficient β_t, which

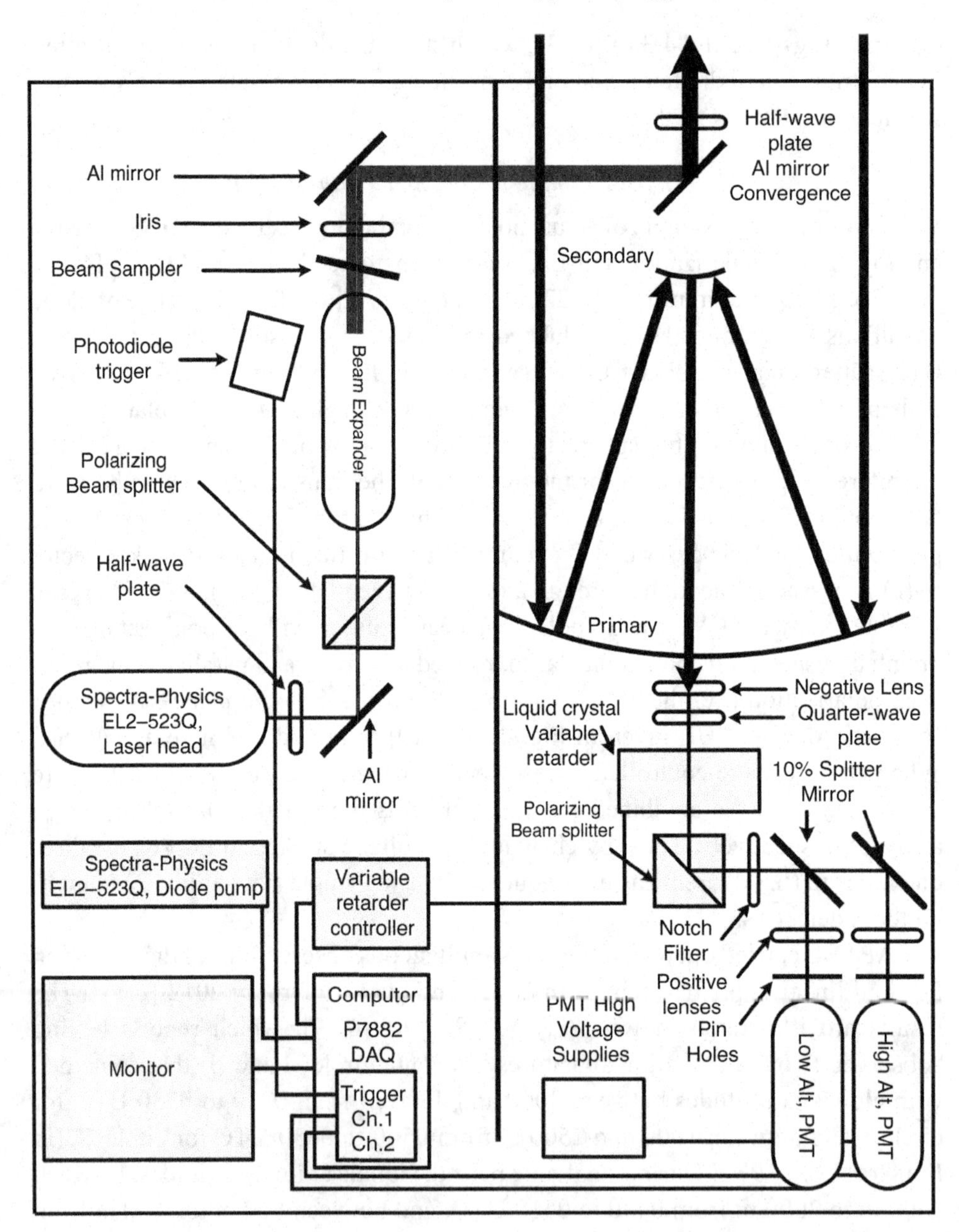

Fig. 6.4 Schematic layout of optics and main components of CAPABL. See text for details. Reproduced from Fig. 4 of [6.19] ©American Meteorological Society. Used with permission.

depends on the populations of the two groups, n_o and n_r, and their respective differential cross sections $d\sigma_o/d\Omega$ and $d\sigma_r/d\Omega$ (with their ratio A_o). Therefore, we have

$$\mathbf{M}_t(\Theta) = \mathbf{M}_o(\Theta) + \mathbf{M}_r;$$
$$\beta_t = f_{11}^t = F_{11} + f_{11} = n_o \frac{d\sigma_o}{d\Omega} + n_r \frac{d\sigma_r}{d\Omega}, \text{ with } \frac{d\sigma_o}{d\Omega} = A_o \frac{d\sigma_r}{d\Omega}. \quad (6.9.a)$$

Since f_{12} for the randomly oriented particles is zero, the diattenuation of the system for $\varphi = 45^o$ may be written as $D_{45}^t = f_{12}^t / f_{11}^t = F_{12}/f_{11}^t$, leading to $D_{45}^t = (F_{12}/F_{11})/(f_{11}^t/F_{11}) \rightarrow (F_{11}/f_{11}^t) = D_{45}^t/D_{45}^o$, where D_{45}^o is the diattenuation of the HOICs, F_{12}/F_{11}. We have

$$1 = \frac{F_{11}}{f_{11}^t} + \frac{f_{11}}{f_{11}^t} = \frac{D_{45^o}^t}{D_{45^o}^o} + \frac{n_r}{n_o A_o + n_r} \rightarrow \frac{n_r}{n_o} = A_o \left(1 - \frac{D_{45}^t}{D_{45}^o}\right) \frac{D_{45}^o}{D_{45}^t} \rightarrow \frac{n_o}{n_r} = \frac{D_{45}^t}{A_o (D_{45}^o - D_{45}^t)}, \quad (6.9.b)$$

which is same as Eq. (13) of [6.19], except expressed in terms of physical parameters. Note that the diattenuation of the system D_{45}^t is measured by the polarization lidar discussed in [6.18], while D_{45}^o is a property of the HOICs that may be obtained via laboratory experiment or modeling. For example, Fig. 4.3 of [6.21] presents modeling results for hexagonal crystal ice platelets with thickness of 120 μm and base radius 160 μm, as a function of the tilt angle. Although CAPABL uses an 11° tilt angle due to its physical limitations, the modeling shows that a larger tilt angle is preferred. With this information, CAPABL can be used to determine the fraction $n_o/(n_o + n_r)$ of horizontally oriented cloud particles. We note that, whereas two constituents, randomly and horizontally oriented, are considered in the above discussion, in practice the randomly oriented particles could comprise several components (e.g., air molecules, aerosols, and small ice particles).

6.3 Raman Lidar and DIAL for Monitoring Minor Species in the Atmosphere

There are two well-known lidar techniques for profiling minor species in the atmosphere. The first is the Raman technique. Vibrational motions in these molecules produce Raman scattering of the impinging laser beam. Whereas the molecular vibrational Raman scattering cross sections are small compared with Cabannes, their associated scattering frequencies, ν_{iS}, are well shifted from the incident laser frequency, ν_L, thus allowing their detection with minimal background noise contamination. This enabled operation of the first Raman lidars as far back as 1969.

As the molecular vibrational frequencies fall in the infrared, the probing frequency of a Raman lidar can be any visible or UV frequency for which the atmosphere is transparent. Since the Q-switched ruby laser was readily available at that time, and given that the scattering cross section is proportional to λ^{-4}, the second harmonic of ruby at 347 nm was used as the probing source in these early publications of Inaba and Kobayashi [6.23] and Melfi, Lawrence, and McCormick [6.24]. The scattering signal from individual species is received in separate channels, each with an appropriate narrowband filter to block the quasi-elastic scattered light.

The second technique is called DIAL – differential absorption (and scattering) lidar. To increase scattering signal, Schotland developed the DIAL method in 1966 [6.25]. In DIAL, the lidar signal results from a comparison between scattering from two separate frequencies: one on-resonance and a second off-resonance of a molecular absorption line, employing a tunable probing laser. Thus, DIAL requires two (or more) probing frequencies specific to a selected absorption line. As noted, we typically perform data analysis with photon count profiles in a longer time interval τ_{int} (60 min for example) and in a coarser vertical bin, Δr (associated range bin of $\Delta z = 1$ km, for example) in order to reduce photon (shot) noise at the price of a degraded temporal and vertical and/or range resolution. The equation for both Raman and DIAL lidar may be deduced from (5.3) by replacing r_B with Δr:

$$N_i(\nu_{iS}, r; \Delta r) = \eta_i \frac{A_R}{r^2} \beta_i(r) \Delta r \frac{E_L}{h\nu_L} \exp\left[-2 \int_{r_0}^{r} \alpha(r') dr' \right]; \quad \text{i} = 1, 2 \text{ or more};$$

$$\beta(r) = \beta_a(r) + \beta_m(r) + \mathcal{N}_i(r) \frac{d\sigma(\nu_{iS})}{d\Omega} \text{ and } \alpha(\text{r}') = \alpha_a(r') + \alpha_m(r') + \mathcal{N}_i(r) \sigma^A(\nu_{iS}). \tag{6.10.a}$$

Here, the subscripts a, m, and i refer respectively to aerosol, atmospheric molecules (N_2 and O_2), and minor species in the atmosphere. The symbol $\sigma^A(\nu_{iS})$ is the absorption cross section of the ith species of interest. The lowest range for a full (overlap) field of view r_0 should be lower than 200 m for tropospheric and boundary layer studies; this may be achieved with a monostatic lidar. That the first DIAL experiment was even possible is due to the fortuitous proximity of water vapor absorption lines at 6942.15 Å, 6942.37 Å, and 6943.8 Å to the ruby laser wavelength. The ruby laser wavelength, thermally tunable (via cooling water temperature) from 6943 Å to 6950 Å, can measure at and near the strongest absorption line centered at 6943.8 Å with an absorption width of 0.2 Å [6.25]. At a displacement of ~ 0.6 Å from the line center, the water vapor absorption of the laser light is negligible, while the Rayleigh-Mie scattering from water vapor, air molecules, and aerosols remains essentially the same as at the on-resonance frequency. Thus, the water vapor density is determined from the on-line/off-line difference measurement.

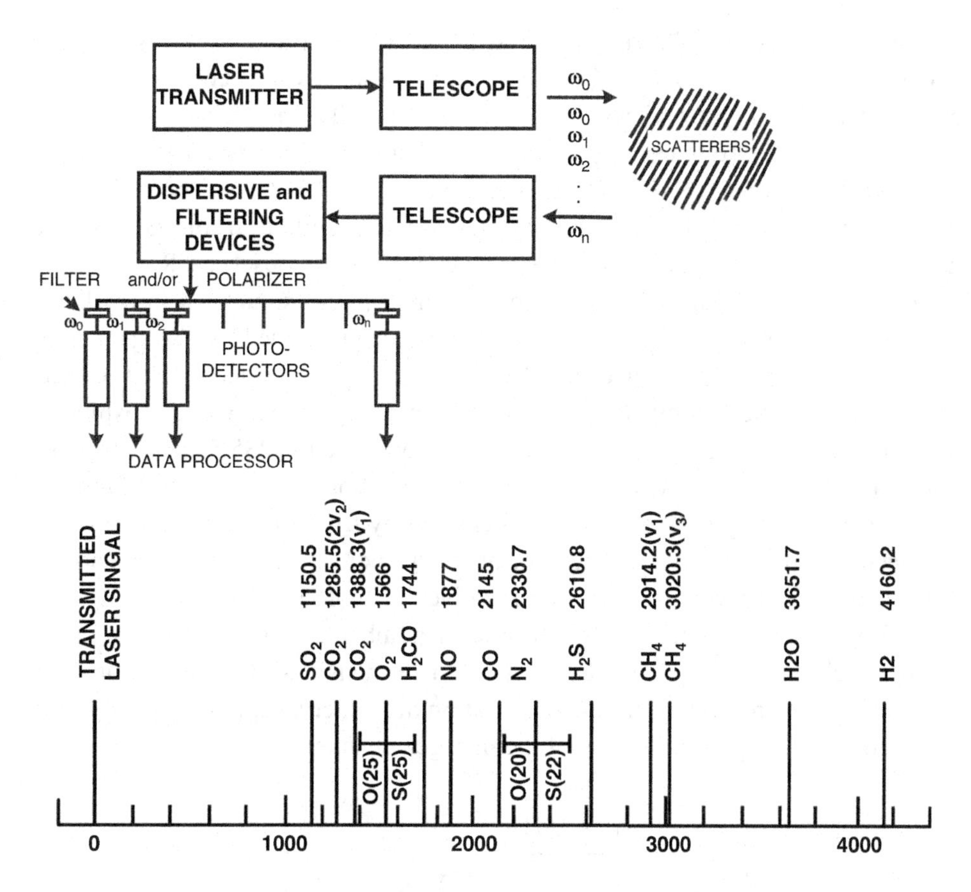

Fig. 6.5 Schematic of a Raman lidar is given in the upper panel as a conceptual system for the monitoring of chemicals in a polluted atmosphere. Reprinted with permission from Springer: Nature. Laser-Raman radar for chemical analysis of polluted air, Inaba, H and T. Kobayashi, © 1969. Given in the lower panel are the observed frequency shifts of vibrational Raman scattering (in units of cm^{-1}) of major atmospheric molecules (N_2 and O_2) along with minor species in a polluted atmosphere. See text for detailed discussion on observed CO_2 lines.

6.3.1 Raman Lidar Technique

Leonard [6.26] recognized the simplicity of the Raman scattering technique and its ability for "a range resolved measurement of atmospheric constituents, with respect to both species and number density, from a single remote location" and reported, in 1967, the vibrational Raman lidar observation of atmospheric nitrogen and oxygen with a pulsed nitrogen ultraviolet laser. At about the same time, Cooney (in 1968) [6.27] reported the atmospheric density profile of nitrogen by observing scattering of ruby laser light offset by vibrational-rotational Raman lines to a wavelength of 8285 Å. Using the second harmonic of a ruby laser at 3471 Å, Inaba and Kobayashi [6.23] and Melfi, Lawrence, and McCormick [6.24], respectively, reported profiling of pollutant chemicals and water vapor. Conceptually, the system and results of a Raman lidar are extremely simple. A schematic of a Raman lidar system is shown in the upper panel of Fig. 6.5, along with 13 receiver channels, each with a narrowband filter centered at a frequency shifted from the laser to measure a different constituent. The frequencies of the observed Raman lines with the species identifications given in Table 1 of [6.23] are shown in the lower panel of Fig. 6.5; the shift reported at 1285.5 cm^{-1} identified as CO_2 (ν_2) must have been a typo, given that the CO_2 (ν_2) is Raman inactive – a correct identification being the overtone of the CO_2 bending mode (ν_2). As explained in Section A.3.3, the frequencies of the pair reported here, 1285.5 cm^{-1} ($2\nu_2$) and 1388.3 cm^{-1} (ν_1), each shifted from the respective overtone ($2\nu_2$) 667×2 = 1334 cm^{-1} and fundamental (ν_1) 1340 cm^{-1} by ~ 48 cm^{-1} away from each other due to Fermi resonance interaction, are in agreement with more recent measurements of 1285.4 cm^{-1} ($2\nu_2$) and 1388.2 cm^{-1} (ν_1) as given in Table 1 of [A.17].

Each receiving channel yields a profile of number density for the associated minor species of interest. To quantify the number density of the ith chemical for a specified range resolution Δr and observation time interval τ_{int}, we normalize the received signal in question to that of the nitrogen channel as follows:

$$\frac{N_i(\nu_{iS}, r)}{N_{N_2}(\nu_{N_2}, r)} = \frac{\eta_i \mathcal{N}_i(r)\left(d\sigma^{Ram}(\nu_{iS})/d\Omega\right)}{\eta_{N_2}\mathcal{N}_{N_2}(r)\left(d\sigma^{Ram}(\nu_{N_2})/d\Omega\right)} \rightarrow$$

$$\frac{\mathcal{N}_i(r)}{\mathcal{N}_{N_2}(r)} = \left[\frac{\eta_{N_2}\left(d\sigma^{Ram}(\nu_{N_2})/d\Omega\right)}{\eta_i\left(d\sigma^{Ram}(\nu_i)/d\Omega\right)}\right]\frac{N_i(\nu_{iS}, r)}{N_{N_2}(\nu_{N_2}, r)}. \tag{6.10.b}$$

Here, we assume the differential Raman cross sections of both nitrogen $(d\sigma^{Ram}(\nu_{N_2})/d\Omega)$ and the species of interest $(d\sigma^{Ram}(\nu_{iS})/d\Omega)$ are known, and that the profile of nitrogen number density is known or independently measured. Unlike the species density, the atmospheric nitrogen density, though dependent on temperature, does not change much during the course of the experiment. The efficiencies of the receiving channels (including the filter transmission), η_{N_2} and η_i, were measured in the laboratory. In addition to this information, to arrive at (6.10.b) from (6.10.a), we also assume that the integrated round-trip extinction coefficient is zero so that the round-trip transmission in (6.10.a) is basically unity. With the help of the information tabulated in Table 6.1(a), we can see the validity of this assumption.

For example, using the Rayleigh extinction cross section at 532 nm, $\sigma^R = (4\pi/1.479)\cdot 6.11\times 10^{-32} = 5.19\times 10^{-31}\mathrm{m}^2$ (see Table 6.1(a)), and the standard air number density of $1.53\times 10^{25}\mathrm{m}^{-3}$ at 5 km altitude, we calculate $\alpha_m(r') = 7.94\times 10^{-6}\ \mathrm{m}^{-1}$. This leads to a round-trip transmission between the ground and 5 km of $\exp[-0.0794] = 0.924$. We note that the vibrational Raman scattering cross sections of N_2 and CO_2 are about three orders of magnitude smaller than their Rayleigh scattering cross sections, the round-trip transmission may be set to unity, especially for minor species, whose number densities are typically less than 0.1% of the atmosphere. With the narrowband filter rejecting Rayleigh–Mie scattering and passing only the Raman signal of interest, $N_i(\nu_{iS}, r)$ and $N_{N_2}(\nu_{N_2}, r)$, the validity of (6.10b) is thus justified.

Shown in Table 6.1(b) are the vibrational frequencies of minor atmospheric species of interest, CO_2, H_2O, and O_3. Because of inversion symmetry, only the symmetric mode of CO_2 is Raman active, see Section A.3.3. There exists no inversion symmetry in H_2O and O_3, so all three vibrational modes are both Raman and infrared active. In theory, Raman lidar can be used to probe the Raman active modes of minor molecular species, including some aerosols. In practice, due to its minute scattering cross section, the traditional Raman lidar technique is most popular for air pollution studies in the lower troposphere. However, for precision measurements, the temperature (and pressure) dependence of the vibrational line-shape of molecular species must be accounted for. Such considerations can be complicated, and the interested reader may consult [6.28; 6.29], where the temperature effect on water vapor mixing ratio and aerosol scattering ratio is considered.

Table 6.1(a) *Rayleigh, vibrational Raman, and absorption cross sections of Air, CO_2 and O_3*

Gas	Cab. (m^2/sr)	Ray. (m^2/sr)	Gas	Vib. Ram. (m^2/sr)*	Gas	Abs. (m^2)
N_2	6.02×10^{-32}	6.13×10^{-32}	N_2 (2331 cm^{-1})	4.81×10^{-35}	CO_2 (667 cm^{-1})	5×10^{-22}
O_2	4.92×10^{-32}	5.19×10^{-32}	O_2 (1556 cm^{-1})	6.25×10^{-35}	O_3 (255 nm)	1×10^{-21}
Air	5.96×10^{-32}	6.11×10^{-32}	CO_2 (1388 cm^{-1})	6.73×10^{-35}	O_3 (300 nm)	4×10^{-23}

*Values for N_2 are taken from Table 2 of [6.4]; those for O_2 and CO_2 are 1.3 and 1.4 times larger, see Table II of [6.30].

Table 6.1(b) *Vibrational modes and frequencies of minor atmospheric species, CO_2, H_2O, and O_3*

Gas	Symmetric stretch ν_1 (cm^{-1})	bending ν_2 (cm^{-1})	asymmetric stretch ν_3 (cm^{-1})
CO_2	1388	667	2349
H_2O	3657.1	1594.7	3755.9
O_3	1103.2	701.42	1042.2

In theory, a broadband laser can also induce fluorescence (LIF) from atmospheric gases as well as aerosols. Although the LIF cross section is much larger than the Raman cross section, due to quenching induced by collisions with air molecules, it is not useful for lower atmospheric studies. Nonetheless, a fluorescence lidar system is similar in its simplicity to a Raman lidar, and it is useful for surveys of plants, rivers, and oceanic contaminants. Thus, we do not discuss LIF in this book for lower atmospheric lidar. However, we present an in-depth description of its use in Chapter 7 for the exploration of temperature and winds in the mesopause region, where the air is dilute and incapable of quenching fluorescence of resident metal atoms. Because of the minute Raman scattering cross section, most recent studies and profiling of CO_2, H_2O, and O_3 have been carried out by DIAL. In this case, the round-trip transmission factor for the on-resonance wavelength can no longer be set to unity, due to much larger absorption cross section. For example, the absorption cross section of the important 15 μm (667 cm^{-1}) CO_2 line is about 5×10^{-22} m^2 with 0.05% number density (7.65×10^{21} m^{-3}). As a result, we have $\alpha_{CO_2}(5\text{ km}) = 3.8\text{ m}^{-1}$. In the case of ozone DIAL, one typically uses its dissociating electronic transition $X\,^1A_1 \rightarrow {}^1B_2$, which has maximum absorption cross section at 255 nm of about 1×10^{-21} m^2 (from Fig. 3 of [6.31]); the maximum number density that occurs at about 30 km is roughly $3.74\times10^{17}\text{m}^{-3}$, or $\alpha_{O_3}(30\text{ km}) = 7.4\times10^{-3}\text{ m}^{-1}$. The advantage and complexity of DIAL is explored below.

6.3.2 Differential Absorption Lidar (DIAL) Technique

DIAL techniques have been used extensively to investigate important minor atmospheric species, CO_2, H_2O, and O_3. Depending on the availability of lasers capable of tuning on and off a resonance absorption line of the species of interest, a DIAL can be built to interact with a selected molecular vibration in the infrared range or with an electronic transition in the ultraviolet range [6.32]. In mid-infrared frequencies, DIAL has been used to investigate molecules such as CH_4, CO_2, CO, N_2O, and others, while in ultraviolet DIAL can probe molecules like O_3, NO_2, and SO_2 [6.33]. The lidar equation that leads to the determination of species number density is the same, derivable from (6.10.a), with the correction and error bar details depending on the specific molecule, atmospheric conditions, and the lidar system. We repeat (6.10.a) with on- and off-resonance frequencies and separate out the species of interest from possible interfering gases as

$$N_i(\nu_i, r; \Delta r) = \eta_i \frac{A_R}{r^2} \beta_i(r) \Delta r \frac{E_L}{h\nu_L} \exp\left[-2\int_{r_0}^{r} \alpha(r', \nu_L) dr'\right]; i = on \text{ or } off, \text{ where} \tag{6.10.c}$$

$$\beta(r) = \beta_a(r) + \beta_m(r) + \mathcal{N}(r)\frac{d\sigma(\nu_i)}{d\Omega} + \mathcal{N}_{IG}(r)\frac{d\sigma_{IG}(\nu_i)}{d\Omega};$$

$$\alpha(r', \nu_L) = \alpha_a(r') + \alpha_m(r') + \mathcal{N}(r')\sigma^A(\nu_i) + \mathcal{N}_{IG}(r')\sigma^A_{IG}(\nu_i).$$

Here, the quantities $\alpha, \beta, \sigma, \mathcal{N}$ without a subscript refer to the species under consideration, while with subscript a, m, and IG refer to, respectively, aerosol, air molecules (N_2 and O_2), and interfering gas, which is a gas with finite absorption cross section at the on- and/or off-resonance frequencies of the species in question. The quantity $\alpha_m(r) = \mathcal{N}_m(r)\sigma^R_m(\nu_i)$ is the extinction coefficient (or the product of number density and Rayleigh cross section) of air. As the detection efficiency, backscattering coefficient, and laser energy can be different between the two frequencies, (6.10.c) becomes

$$\frac{d}{dr}\left\{\ln\left[\frac{N_{on}(r; r_B)\eta_{off}\beta_{off}(r)E_{off}}{N_{off}(r; r_B)\eta_{on}\beta_{on}(r)E_{on}}\right]\right\} = -2\mathcal{N}(r)\Delta\sigma^A - 2\Delta\alpha_m - 2\Delta\alpha_a - 2\Delta\alpha_{IG}, \tag{6.10.d}$$

where $\Delta\sigma^A = \sigma^A(\nu_{on}) - \sigma^A(\nu_{off})$ is the difference between absorption cross sections of the species of interest; $\Delta\alpha_m$, $\Delta\alpha_a$ and $\Delta\alpha_{IG}$ are similarly defined for air molecules, aerosols, and interfering gas. Taking the derivative in steps of Δr, the vertical range bin, and noting that $\eta_{off}, \eta_{on}, E_{off}$, and E_{on} are independent of range, we can derive the number density of interest, $\mathcal{N}(r)$ as

$$\mathcal{N}(r + \Delta r/2) = \frac{1}{2\Delta\sigma^A\Delta r}\ln\left[\frac{N_{off}(r + \Delta r)N_{on}(r)}{N_{on}(r + \Delta r)N_{off}(r)}\right] - B - D - E - F, \text{ where}$$

$$B = \frac{1}{2\Delta\sigma^A\Delta r}\ln\left[\frac{\beta_{off}(r + \Delta r)\beta_{on}(r)}{\beta_{on}(r + \Delta r)\beta_{off}(r)}\right];\ D = \frac{\Delta\alpha_m}{\Delta\sigma^A};\ E = \frac{\Delta\alpha_a}{\Delta\sigma^A};\ F = \frac{\mathcal{N}_{IG}\Delta\sigma^A_{IG}}{\Delta\sigma^A}. \tag{6.10.e}$$

The terms B, D, E, and F are the corrections resulting from the differences between on- and off-resonance frequencies in backscattering coefficient, atmospheric Rayleigh extinction, aerosol extinction, and extinction by interfering gas (such as SO_2 when measuring O_3); ideally, they should all be zero. A comprehensive discussion of these correction terms, which are in fact not negligible, can be found in a book chapter by Gimmestad [6.32]. Briefly, the D, E, and F terms are independent of the number density of species of interest, $\mathcal{N}(r)$, and in principle may be estimated from supplemental information individually. For example, with the knowledge of atmospheric density $\mathcal{N}_m(r)$ and using the λ^{-4} dependence in Rayleigh cross section, the term D can be determined exactly. Since aerosol extinction varies considerably, especially in the boundary layer, the E term has the largest uncertainty. Thus, the aerosol inversion algorithm plays a significant role in DIAL retrieval [6.33], particularly in the case of ozone DIAL, when the difference between on- and off-frequency is of necessity large. Unlike these terms, the backscattering correction term B is not independently additive; it also depends on $\mathcal{N}(r)$. The aerosol correction, combining B and E terms, is negligible in the free troposphere, but it can be large where a large aerosol gradient exists, such as in the atmospheric boundary layer. The need to better assess the aerosol correction is the main reason one or even two additional (i.e., three or four total) off-resonance wavelengths are used in DIAL to provide more information.

The stratosphere contains about 90% of all atmospheric ozone. The ozone layer protects us from the harmful UVB radiation; its density peaks at about 30 km with concentration as high as 15 ppm. In the troposphere near Earth's surface, the natural ozone concentration is about 10 ppb. This is a good thing, as, according to the US Environmental Protection Agency, exposure to ozone levels greater than 70 ppb for 8 hours or longer is unhealthy. The relevant electronic transition bands in O_3 are all dissociative and the molecule splits into $O + O_2$ after absorbing a photon. The strongest absorption is in the Hartley band ($X\,^1A_1 \rightarrow {}^1B_2$, width of about 100 nm), with absorption cross section of ~1×10^{-21} m^2 peaking at 255 nm [6.31]. Transitioning to the longer-wavelength Huggins band, the cross section falls off by six orders of magnitude at ~360 nm. The broad and deep Hartley band provides both challenges and opportunities for DIAL, leading to the following guidelines for wavelength optimization, following [6.32]:

a) The on wavelength, λ_{on}, is selected to reach the maximum required range
b) $\Delta\lambda = \lambda_{off} - \lambda_{on}$ should be chosen to have large $\Delta\sigma^A$ for the required measurement resolution
c) $\Delta\sigma^A/\Delta\lambda$ should be maximized, and
d) λ_{on} and λ_{off} should be chosen to minimize $\Delta\sigma^A_{IG}$.

For example, on/off wavelengths of 308/353 nm have been used for stratospheric ozone DIAL [6.34], while 266/299 nm were selected for tropospheric ozone DIAL [6.35]. In the former, a XeCl excimer laser at 307.9 nm was used for λ_{on} and ~30% of its energy was used to pump a high-pressure H_2 cell to generate 353 nm by stimulated Raman scattering for λ_{off}. The absorption cross section at 308 nm is about 5×10^{-24} m^2, low enough for the laser pulse not to be depleted before reaching 15 km and high enough for signal to reach 50 km. In the latter case, the 4th harmonic of an Nd:YAG laser at 266 nm is used for λ_{on} and H_2-cell Raman-shifted pulses from the same 4th harmonic at 299 nm for λ_{off}. The absorption cross section at 266 nm, about 1.0×10^{-21} m^2, 2000 times larger than that at 308 nm, is strong enough to probe ozone in the troposphere with 1000x lower concentration. Furthermore, the sum-frequency of the Nd:YAG's fundamental and 2nd harmonics can be mixed to generate 355 nm laser light. That, combined with stimulated Raman generation from the 4th harmonic at 266 nm through high-pressure D_2 and HD cells to generate 289 nm and 295 nm pulses, respectively, can create a source for multiple-wavelength DIAL for simultaneous tropospheric ozone and aerosol observation and profiling [6.35]. Since its first implementation [6.25], and due mainly to the dynamical and chemical importance of water, as well as its involvement in cloud formation in the atmosphere, water vapor DIAL has been of great and continued interest. DIAL has been an important tool for the investigation of water vapor. However, the complexity in the water infrared absorption spectrum and its temperature dependence is a considerable challenge, and it is beyond the scope of this book. We refer interested readers to a comprehensive book chapter by Bösenberg [6.36] for further investigation.

6.3.3 Measurement Uncertainty Comparison between Raman and DIAL Methods

To wrap up Section 6.3, we discuss the measurement uncertainty due to photon noise and perform simple calculations comparing statistical errors between CO_2 Raman and O_3 DIAL retrieval. We first calculate the uncertainty of CO_2 Raman retrieval. We set the subscript i to CO_2 and ignore the frequency dependence in (6.10.b). The number density of CO_2 is related to the number density of N_2, and thus, the ratio of the observed Raman counts of the former, $N_{CO_2}(r)$, to the latter $N_{N_2}(r)$. Assuming the observed photon counts are the only source of uncertainty, we can deduce the observed uncertainty $\Delta N_{CO_2}(r)$ for a specified resolution Δr and τ_{int}, as:

$$\Delta\mathcal{N}_{CO_2}(r) = \mathcal{N}_{N_2}(r)\left[\frac{\eta_{N_2}\left(d\sigma_{N_2}^{Ram}/d\Omega\right)}{\eta_{CO_2}\left(d\sigma_{CO_2}^{Ram}/d\Omega\right)}\right]\Delta\left[\frac{N_{CO_2}(r)}{N_{N_2}(r)}\right] \rightarrow$$

$$\Delta\mathcal{N}_{CO_2}(r) = \mathcal{N}_{N_2}(r)\left[\frac{\eta_{N_2}\left(d\sigma_{N_2}^{Ram}/d\Omega\right)}{\eta_{CO_2}\left(d\sigma_{CO_2}^{Ram}/d\Omega\right)}\right]\frac{N_{CO_2}(r)}{N_{N_2}(r)}\left(\frac{\sqrt{N_{CO_2}(r)+B^{\Delta r}}}{N_{CO_2}(r)}+\frac{\sqrt{N_{N_2}(r)+B^{\Delta r}}}{N_{N_2}(r)}\right)$$

$$\text{or } \frac{\Delta\mathcal{N}_{CO_2}(r)}{\mathcal{N}_{CO_2}(r)} = \left(\frac{\sqrt{N_{CO_2}(r)+B^{\Delta r}}}{N_{CO_2}(r)}+\frac{\sqrt{N_{N_2}(r)+B^{\Delta r}}}{N_{N_2}(r)}\right), \tag{6.11.a}$$

where $B^{\Delta r}$ is the associated background count. To arrive at the middle expression of (6.11.a) from the upper, we assume that the uncertainty in photon counts is due only to the shot-noise of the photodetector. As mentioned, due to the weakness of Raman scattering, the round-trip attenuation of Raman signal may be ignored. Recognizing that the product of (N_{CO_2}/N_{N_2}) and the square bracket that precedes it leads to $(\mathcal{N}_{CO_2}/\mathcal{N}_{N_2})$, we can write the final expression for fractional error in retrieved $\mathcal{N}_{CO_2}$, as shown in the final line of (6.11.a).

As a numerical example, we first calculate the CO_2 Raman (1388 cm^{-1}) backscattering coefficient $\beta_{CO_2}(r) = 5.15 \times 10^{-13} m^{-1}$ as the product of the CO_2 number density – 0.5% air number density at 5 km (7.65×10^{21} m^{-3}) – and the Raman differential cross section (6.73×10^{-35} m^2/sr). From this we calculate the photocounts received in the Raman channel, using a 1-m-diameter telescope, 50% efficient photodetector, and scattering from a 1-W laser at 266 nm as

$$N_{CO_2}(r;\Delta r) = \eta_{CO_2}\frac{A_R}{r^2}\beta_{CO_2}(r)\Delta r\frac{E_L}{h\nu_L} = \eta_{CO_2}\frac{A_R}{r^2}\lambda_L\beta_{CO_2}(r)\Delta r\frac{E_L}{hc} \tag{6.11.b}$$

For a range resolution of 0.1 km and τ_{int} = 1.0 hr, we estimate $N_{CO_2}(5\text{ km};\ 0.1\text{ km}) = 3,900$. The scaled photocounts in the N_2 Raman channel should be $N_{N_2}(5\text{ km}; 0.1\text{ km}) = 4,400,000$, in view of $N_2(5\text{ km}) = 0.79 \times 1.53 \times 10^{25}\ m^{-3}$, $\left(d\sigma_{N_2}^{Ram}/d\Omega\right) = 4.81 \times 10^{-35} m^2/sr$ and $\beta_{N_2}(r) = 5.81 \times 10^{-10} m^{-1}$. Using this information and assuming $B^{\Delta r} = 10$, we deduce that the fractional observed CO_2 number density from the formula in the bottom of (6.11.a) is 0.016 (1.6%).

To work out an example in DIAL retrieval uncertainty, we start with (6.10.e) by assuming the correction factors to be nil (i.e., $B = D = E = F = 0$); the standard procedure for retrieval uncertainty yields the following expression for that of the number density of interest:

$$\Delta\mathcal{N}(r+\Delta r/2) \approx \frac{1}{\Delta\sigma^A\Delta r}\frac{\sqrt{N_{off}(r+\Delta r)+B^{\Delta r}}+\sqrt{N_{on}(r)+B^{\Delta r}}}{N_{off}(r+\Delta r)N_{on}(r)}. \tag{6.11.c}$$

As a numerical example, we consider O_3 at 5 km with the same resolution as before: 0.1 km and τ_{int} =1.0 hr. In this case, assuming the atmosphere at 5 km is aerosol-free, Rayleigh backscattering from air molecules weighted by attenuation due to O_3 absorption is mainly responsible for the received signal. To obtain the necessary information, we use the standard atmospheric number density at 5 km, $\mathcal{N}_{Air} = 1.53 \times 10^{25} m^{-3}$, and differential Rayleigh cross section, $d\sigma_{Air}^{Ray}/d\Omega = 6.11 \times 10^{-32} m^2/sr$, to derive $\beta_{Air}^{Ray}(5\text{ km}) = 9.35 \times 10^{-7} m^{-1}$, a factor 1.82×10^6 larger than $\beta_{CO_2}^{Ram}(5\text{ km}) = 5.15 \times 10^{-13} m^{-1}$, corresponding to $1.82 \times 10^6 \cdot 3900 = 7.098 \times 10^9$ signal counts. Since $\mathcal{N}_{O_3} = 1.53 \times 10^{17} m^{-3}$(10 ppb at 5 km) and the on- and off-absorption cross-sections are $\sigma^A(266\text{ nm}){\sim}10^{-21} m^2$ and $\sigma^A(299\text{ nm}) \sim 4 \times 10^{-23} m^2$, the 10 km round-trip transmissions are calculated to be 0.217 and 0.941, giving rise to signal counts of $N_{on} = 1.54 \times 10^9$ and $N_{off} = 6.68 \times 10^9$, respectively. Using $B^{\Delta r} = 1000$, for example, and calculating $\Delta\sigma^A\Delta r = 9.6 \times 10^{-20} m^3$, we deduce from (6.11.c) the observed uncertainty of O_3 number density, $\Delta\mathcal{N}_{O_3}(5\text{ km}) = 3.25 \times 10^9 m^{-3}$, giving rise to a fractional error of $(\Delta\mathcal{N}_{O_3}/\mathcal{N}_{O_3}) = 2.14 \times 10^{-8}$. This much smaller fractional uncertainty demonstrates a clear superiority of DIAL over the Raman technique from the point of view of photon noise. However, unlike in the Raman technique, DIAL suffers many additional causes of uncertainty, leaving us with the challenging problem to figure out the right correction factors and associated uncertainties as well as that of interfering gas. Various techniques and algorithms for the treatment of correction factors and measurement error for DIAL are found in the references of two book chapters in [6.32; 6.36].

6.4 Other Important Lidar Types Not Discussed in This Book

There are other types of important lidars that are not covered in this book. Here, we provide a narrative discussion on airborne/spaceborne lidar, lidar for cloud, particulate and air pollution monitoring, as well as lidars illuminating cultural and modern living.

6.4.1 Airborne and Spaceborne Lidars

An airborne lidar, see [6.7] for example, is often deployed for specific campaigns or for reliability tests of a planned spaceborne lidar. Since spaceborne lidars can cover the atmosphere on a global scale, these will no doubt play an increasingly important role in scientific research. The introductory principle of lidar probing of the atmosphere presented here is the same for both ground-based and spaceborne applications. The price to pay for the advantage of broad coverage over a large area quickly lies in the associated loss in its ability to investigate localized phenomena that require extended observations from fixed locations, such as atmospheric gravity waves and tidal waves (which ideally require, respectively, temporal resolution in seconds and daily 24-hour-cycle local time coverage). Nevertheless, this price applies only to relatively narrow objectives, which may be seen as minute compared with the broad science objectives that can be investigated by global coverage. An intrinsic difference between an airborne/spaceborne lidar and a ground-based lidar is that an airborne/spaceborne generally probes the space below the transmitter (Nadir-viewing) as it transits over the Earth below, whereas a ground-based lidar sits at a single location probing the atmosphere above.

As an example, we compare the signal returns of a ground-based lidar and two spaceborne lidars, one located at the Space Shuttle altitude (250 km) and the other at a satellite altitude (800 km). These have the same Power-Aperture Product, looking at the same aerosol distribution in the troposphere and stratosphere with a background 1976 US standard atmosphere between the ground and 100 km. The common lidar system used consists of a frequency-doubled Nd:YAG laser at 532 nm with 500 mJ, 10 Hz pulses, and a 1-m-diameter telescope (P-A product = 3.9 Wm^2). The detection (including collection, interference filter, and PMT) efficiency is assumed to be 2.5%. The aerosol distribution is modeled from the SAGE II extinction measurements [6.37] as a function of altitude at 550 nm, which represents a September 1994 mid-latitude stratosphere for the use of the LITE experiment. The lidar signal in Fig. 6.6 is the photon counts collected in a 10-min interval at 0.1 km vertical resolution. A differential Rayleigh backscattering cross section (including PRR) of 6.11×10^{-32} m^2/sr is employed, and an aerosol lidar ratio of 20 (L.R. = 20) is assumed for the calculation of the total Rayleigh–Mie backscattering coefficient. The solid lines are received photon count signals (a+m), from right to left, for (1) ground-based, (2) Nadir-view from shuttle-height (250 km), and (3) satellite height (800 km). The associated dashed lines are the corresponding returns assuming clean air (in the absence of an aerosol layer). Although the lidar return in the satellite nadir geometries is much weaker (due to the long distance between the scattering region and the transmitter), they actually see the aerosol structure more clearly, because the range of their aerosol distribution is nearly constant. We should not forget that the weak space-based signal is more susceptible to interference from background noise (a fact not accounted for here). Therefore, a narrow bandpass, notch-type filter should be employed to reject background noise, especially for daytime observations.

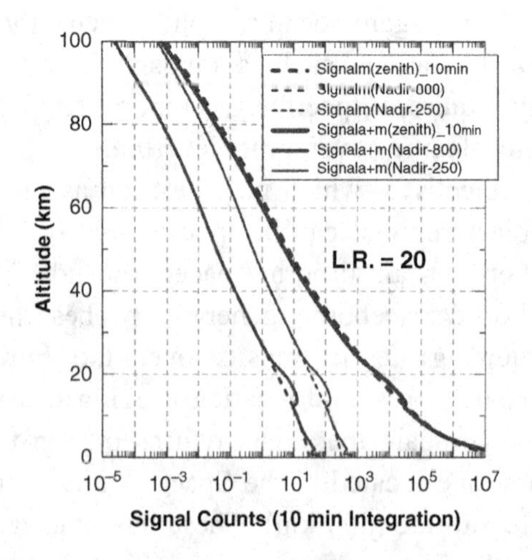

Fig. 6.6 Photon count profiles of ground-based and spaceborne (Nadir from 250/800 km) lidars

In addition to basic lidar system design, a meaningful spaceborne deployment requires special attention to system robustness and longevity, minimal power consumption, and weight. We refer to the expert book chapter by McCormic [6.38] on past developments and future prospects of spaceborne lidar. Here, the author describes the very successful first spaceborne lidar experiment, termed LITE (Lidar In-space Technology Experiment), which was launched onboard the Space Shuttle Discovery flight STS-64 on September 9, 1994. In addition to atmospheric studies, the use of spaceborne lidar for investigation of marine systems, for example, has been evaluated [6.39].

6.4.2 Particulate Matter and Air Pollution Monitoring

Simple lidars are used ubiquitously to investigate tropospheric and stratospheric aerosols and clouds and to monitor air-polluting chemicals. With the exception of the detection of horizontally oriented ice crystals (HOIC), employing (the still-evolving) polarization lidar via the Stokes vector/ Mueller matrix approach described in this chapter, and the profiling of lidar-ratio and aerosol optical properties described in Chapter 7, we have chosen to exclude the important studies of particulates, clouds, and ice as being outside the realm of the book. Of fundamental importance in aerosol particulate research is the distribution in types, shapes, and sizes, which can be very complex. In some situations when the particulate matter is monodispersed, we can assume it consists of a collection of spherical particles of the same type (refractive index). In this case, the particulate matter could be characterized by a monomodal lognormal size distribution, which is completely defined by the mean radius and width of the distribution, r_{med} and σ, along with the particle density $\mathcal{N}$. These three parameters may be experimentally determined by a three-color backscattering lidar, as described, for example, by von Cossart et al. for the study of noctilucent clouds [6.40]. They used the measured backscattering coefficients of the fundamental, 2nd harmonic, and 3rd harmonic wavelengths, at 1064 nm, 532 nm, and 355 nm, of a Nd:YAG laser to form two color ratios (CRs), β_{1064}/β_{532} and β_{355}/β_{532}. By comparing the relationship between the two CRs deduced from Mie calculations for a set of curves with constant r_{med} and σ with those from the measured points, they determined the value of r_{med} and σ of the noctilucent clouds in question. Then, assuming a lognormal particle size distribution, they determined the particle density from the lidar backscattering coefficient.

The presence of aerosols and clouds in the atmosphere affects both $\beta_a(r)$ and $\alpha_a(r')$ in the lidar equation (6.1), in which $\beta_a(r)$ is the single aerosol backscattering coefficient and $\alpha_a(r')$ is the aerosol extinction coefficient, for determining the round-trip attenuation of the excitation and return light. As the concentrations of aerosols and cloud particles increase, the effects of multiple scattering events must be considered, as they impact lidar signal. Multiple scattering increases attenuation, and its effect is present in the detected lidar signal for the single-scattering lidar equation; its importance relative to the single backscattering coefficient reflects upon the associated lidar ratio. The amount of multiple backscattering detected in the receiver can be minimized by keeping the receiving field-of-view as narrow as possible. Multiple scattering also increases the (apparent) single scattering linear depolarization ratio, δ_ℓ. In figure 3.2 of a book chapter discussing lidar and multiple scattering [6.41], one sees δ_ℓ increasing from zero to 60% within 2° (35 mrad) from the exact backward direction, suggesting that setting the telescope's full field of view to 0.18 mrad is sufficient to reject the unwanted multiple backscattering signal. A more extensive coverage of aerosol and cloud monitoring in the troposphere by elastic lidars can be found in a book chapter by Takeuchi [6.42]. In addition to polluting aerosols, harmful chemicals in the atmosphere, resulting from human and industrial activities, such as nitrogen oxides, SO_2, Cl_2, Hg, and aromatic hydrocarbons, may be monitored by lidars. Like naturally occurring trace gases, these chemicals can also be detected by the vibrational Raman scattering and differential absorption (DIAL) lidar techniques discussed in Section 6.3. For more extensive coverage of their detection with lidar and possible pollution abatement strategies, a book chapter written by Calpini and Simionov may be consulted [6.43].

6.4.3 Lidars for Cultural Heritage and Modern Living

There has been considerable excitement and public interest in vastly different lidar applications; these include the investigation of historic buildings [6.44] and lost civilizations [6.45] as well as for 3-D mapping and autonomous vehicles, just to name a few. The excitement and amazingly brilliant stories behind these developments and innovations "that are changing the world" have been captured in a narrative monograph written by a popular science writer, Todd Neff [6.46].

References

6.1 Klett, J. D. (1981). Stable analytical inversion solution for processing lidar returns. *Appl. Opt.* **20**(2), 211–220.

6.2 Klett, J. D. (1985). Lidar inversion with variable backscatter/extinction ratios. *Appl. Opt.* **24**(11), 1638–1643.

6.3 Ansmann, A. and D. Müller. (2005). Lidar and atmospheric aerosol particles. Chapter 4 in *Lidar Range-Resolved Optical Remote Sensing of the Atmosphere*, C. Weitkamp, ed., Springer Science+Business Media, Inc., vol. 102.
6.4 She, C.-Y. (2001). Spectral structure of laser light scattering revisited: Bandwidths of nonresonant scattering lidar. *Appl. Opt.* **40**(27), 4875–4884.
6.5 Krueger, D. A., C.-Y. She, and T. Yuan. (2015). Retrieving mesopause temperature and line-of-sight wind from full-diurnal-cycle Na lidar observations. *Appl. Opt.* **54** (32), 9469–9489.
6.6 Liu, Z.-Y. et al. (2006). Estimating random errors due to shot noise in backscatter lidar observations. *Appl. Opt.* **45**(18), 4437–4447.
6.7 Hair, J. W. et al. (2008). Airborne high spectral resolution lidar for profiling aerosol optical properties. *Appl. Opt.* **47**(36), 6734–6753.
6.8 Schotland, R. M., K. Sassen, and R. Stone. (1971). Observations by lidar of linear depolarization ratios for hydrometeors. *J. Appl. Meteorol.* **10**(5), 1011–1017.
6.9 Murayama, T. (2001). Ground-based network observation of Asian dust events of April 1988 in east Asia. *J. Geophys. Res.* **106**(D16), 18345–18359.
6.10 Poole, L. R., G. S. Kent, M. P. McCormick et al. (1990). Dual-polarization airborne lidar for observations of polar stratospheric cloud evolution. *Geophys. Res. Lett.* **17** (4), 389–392.
6.11 Hu, Y., D. Winker, P. Yang et al. (2001). Identification of cloud phase from PICASSO-CENA lidar depolarization: A multiple scattering sensitivity study. *J. Quant. Spectrosc. Radiat. Transfer* **70**(4–6), 569–579.
6.12 Alvarez, R. J., W. L. Eberhard, J. M. Intrieri et al. (1998). A depolarization and backscatter lidar for unattended operation in varied meteorological conditions. *Proc. 10th Symp. on Meteorological Observations and Instrumentation*. Phoenix, AZ, *Amer. Meteor. Soc.*, 140–144.
6.13 Sassen, K. (2005). Polarization in Lidar, in *Lidar: Range-Resolved Optical Remote Sensing of the Atmosphere*, C. Weitkamp, ed. (Springer), pp. 19–42.
6.14 Gimmestad, G. G. (2008). Reexamination of depolarization in lidar measurements. *Appl. Opt.* **47**(21), 3795–3802.
6.15 Kaul, B. V., I. V. Samokhvalov, and S. N. Volkov. (2004). Investigating particle orientation in cirrus clouds by measuring backscattering phase matrices with lidar. *Appl. Opt.* **43**(36), 6620–6628.
6.16 van de Hulst, H. (1981). *Light Scattering by Small Particles*. Wiley.
6.17 Houston, J. D. and A. I. Carswell. (1978). Four-component polarization measurement of lidar atmospheric scattering. *Appl. Opt.* **17**(4), 614–620.
6.18 Hayman, M., and J. P. Thayer. (2012). General description of polarization in lidar using Stokes vectors and polar decomposition of Mueller matrices. *J. Opt. Soc. Amer.*, **29A**(4), 400–409.
6.19 Neely, R. R., M. Hayman, R. Stillwell et al. (2013). Polarization lidar at Summit, Greenland, for the detection of cloud phase and particle orientation. *J. Atmos. Oceanic Technol.* **30**(8), 1635–1655.
6.20 Hayman, M., S. Spuler, B. Morley et al. (2012). Polarization lidar operation for measuring backscatter phase matrices of oriented scatterers. *Opt. Express* **20**(28), 29553–29567.
6.21 Hayman, M. (2011). Optical theory for the advancement of polarization lidar. Ph.D. thesis, University of Colorado.
6.22 Stillwell, R. A., R. R. Neely III, J. P. Thayer et al. (2018). Improved cloud-phase determination of low-level liquid and mixed-phase clouds by enhanced polarimetric lidar. *Atmos. Meas. Tech.*, **11**(2), 835–859. doi: https://doi.org/10.5194/amt-11-835-2018.

6.23 Inaba, H. and T. Kobayashi. (1969). Laser-Raman radar for chemical analysis of polluted air. *Nature* **224**(5215), 170–172. doi: https://doi.org/10.1038/224170a0.
6.24 Melfi, S. H., J. D. Lawrence, and M. P. McCormick (1969). Observation of Raman scattering by water vapor in the atmosphere. *Appl. Phys. Lett.* **15**(9), 295–297.
6.25 Schotland, R. M. (1969). Some observations of the vertical profile of water vapor by means of a laser optical radar. *Proc. 4th Symposium on Remote Sensing of Environment*, 273–283, PRIM.
6.26 Leonard, D. A. (1967). Observation of Raman scattering from the atmosphere using a pulsed nitrogen ultraviolet laser. *Nature* **216**(5111), 142–143.
6.27 Cooney, J. A. (1968). Measurements on the Raman component of laser atmospheric backscatter. *Appl. Phys. Lett.* **12**(2), 40–42. doi: https://doi.org/10.1063/1.1651884.
6.28 Whiteman, D. N. (2003). Examination of the traditional Raman lidar technique. I. Evaluating the temperature-dependent lidar equations. *Appl. Opt.* **42**(15), 2571–2592.
6.29 Whiteman, D. N. (2003). Examination of the traditional Raman lidar technique. II. Evaluating the ratios for water vapor and aerosols. *Appl. Opt.* **42**(15), 2593–2608.
6.30 Fenner, W. R., H. A. Hyatt, J. M. Kellam and S. P. S. Porto (1973), Raman cross section of some simple gases, Jour. Opt. Soc. Am. **63**, 73–77.
6.31 Gorshelev, V., A. Serdyuchenko, M. Weber et al. (2014). High spectral resolution ozone absorption cross-sections – Part 1: Measurements, data analysis and comparison with previous measurements around 293 K. *Atmos. Meas. Tech.*, **7**(2), 609–624. doi: https://doi.org/10.5194/amt-7-609-2014.
6.32 Gimmestad, G. G. (2005). Differential-absorption lidar for ozone and industrial emissions. In Chapter 7 in *Lidar: Range-Resolved Optical Remote Sensing of the atmosphere*, C. Weitkamp, ed., Springer-Verlag.
6.33 S. Godin, A. I. Carswell, D. P. Donovan et al. (1999). Ozone differential absorption lidar algorithm inter-comparison. *Appl. Opt.* **38**(30), 6225–6236.
6.34 McDermid, I. S., S. M. Godin, and L. O. Lindqvist. (1990). Ground-based laser DIAL system for long-term measurements of stratospheric ozone. *Appl. Opt.* **29**(25), 3603–3612.
6.35 Zhao, Y. Z., R. M. Hardesty, and M. J. Post. (1992). Multibeam transmitter for signal dynamic range reduction in incoherent lidar systems. *Appl. Opt.* **31**(36), 7623–7632.
6.36 Bösenberg, J. (2005), Differential-absorption lidar for water vapor and temperature profiling. Chapter 8 in *Lidar: Range-Resolved Optical Remote Sensing of the Atmosphere*, C. Weitkamp, ed., Springer-Verlag.
6.37 Thomason, L. and M. Osborn. (1992). Lidar conversion parameters derived from SAGE-II extinction measurements. *Geophysical Research Letters*, **19**(16), 1655–1658.
6.38 McCormic, M. P. (2005). Airborne and spaceborne Lidar. Chapter 13 in *Lidar: Range-Resolved Optical Remote Sensing of the Atmosphere*, C. Weitkamp, ed., Springer-Verlag.
6.39 Hostetler, C. A., M. J. Behrenfeld, Y.-X. Hu et al. (2018). Spaceborne lidar in the study of marine systems. *Annual Review Marine Science*, **10**(1): 121–147, https://doi.org/10.1146/annurev-marine-121916-063335.
6.40 von Cossart, G., J. Fielder, and U. von Zahn. (1999). Size distributions of NLC particles as determined from 3-color observations of NLC by ground-based lidar. *Geophys. Res. Lett.*, **26**(11), 1513–1516.
6.41 Bissonnette, L. R. (2005). Lidar and multiple scattering. Chapter 3 in *Lidar: Range-Resolved Optical Remote Sensing of the Atmosphere*, C. Weitkamp, ed., Springer, pp. 43–100.

6.42 Takeuchi, N. (2005). Elastic lidar measurement of the troposphere (2005). In *Laser Remote Sensing*, Fujii, T. and T. Fukuchi, eds., CRC Press, Taylor & Francis Group, pp. 123–168.
6.43 Calpini, B. and V. Simionov. (2005). Trace gas species detection in the lower atmosphere by lidar: From remote sensing of atmospheric pollutants to possible air pollution abatement strategies. In *Laser Remote Sensing*, Fujii, T. and T. Fukuchi, eds., CRC Press, Taylor & Francis Group, pp. 123–168.
6.44 Raimondi, V., G. Cecchi, L. Pantani, and R. Chiari. (1998). Fluorescence lidar monitoring of historic buildings. *Appl. Opt.* **37**(6), 1089–1098.
6.45 Clynes, T. (2018). "Exclusive: Laser Scans Reveal Maya 'Megalopolis' below Guatemalan Jungle," *National Geographic*, February 1 issue.
6.46 Neff, T. (2018). *The Laser That's Changing the World: The Amazing Stories behind Lidar, from 3D Mapping to Self-Driving Cars*. Prometheus Books.

7
Lidars for Profiling Aerosol Optical Properties, Atmospheric Temperature and Wind

It is well known that standard dry air contains about 79% nitrogen and 21% oxygen by volume with less than 1% argon and other trace gases such as carbon dioxide and ozone. Air also contains a varied amount of water vapor: around 1% at sea level and 0.4% over the entire atmosphere. Molecular absorption of radiation by CO_2 and H_2O lies in the near-infrared longer than 2.5 μm, while dissociative transitions of ozone peak in the ultraviolet region at 255 nm, becoming negligible above ~350 nm; see Section 6.3.2. Therefore, infrared and visible light between 1.55 μm (wavelength for optical fiber communication) and 355 nm (tripled Nd:YAG laser) propagates through the atmosphere generally without suffering absorption. The extinction of light is then mainly the result of Rayleigh scattering from nitrogen and oxygen. The strength, spectral width, and frequency shift of this Rayleigh scattering can be utilized to measure air density, temperature (random air motion), and wind (collective air motion). The presence of other heavier species, including aerosols, cloud particles, ice crystals, and contrails, causes additional (Mie) scattering, which may overwhelm molecular scattering at lower atmospheric altitudes and absorb species-specific wavelengths. Fortuitously, light scattered by these heavier objects – collectively termed aerosol scattering for simplicity – is essentially elastic (falling within a bandwidth of ~10 MHz). This is in contrast with molecular scattering, which falls within a bandwidth of ~ 3 GHz due to Cabannes scattering. (For reference, see the following: Fig. 4.7 showing the thermally broadened atmospheric Cabannes spectrum; Fig. 4.4 showing shifted peaks between ~ 4 cm^{-1} and 200 cm^{-1} due to pure rotational Raman scattering; and Table 4.3 showing N_2 and O_2 vibrational Raman scattering around 2330 and 1556 cm^{-1}.) Thus, aerosol and molecular scattering can be separated using a high-spectral resolution lidar (HSRL), and in so doing, the profiles of aerosol optical properties on the one hand, and atmospheric parameters (density, temperature, and winds) on the other, can be determined. By separating the contributions of aerosol and molecular scattering profiles, one can characterize the lidar ratio and determine optical properties (including visibility) of the atmosphere, and independent analysis of aerosol-free molecular scattering allows determination of profiles of atmospheric parameters (density, temperature, and wind). In what follows, we discuss the principles and lidar measurement methods for the determination of aerosol/optical properties and atmospheric parameters.

Before we treat lidar retrieval of these atmospheric and aerosol parameters, we first divide the transmitters into broadband versus narrowband by pulsed laser linewidth and roughly regard those between 1 and 100 GHz (1 cm^{-1} = 30 GHz) as broadband, and those around 200 MHz or less as narrowband. In terms of laser technology, the former is typically a pulsed laser without injection seeding, while the latter is that with injection seeding by a single-mode CW laser. Measurements of pulsed laser linewidths are rare in the published literature, but examples for the narrowband pulsed laser bandwidth can be found for Nd:YAG at 1064 nm in Fig. 5 of [7.1], which showed FWHM ~ 40 MHz, implying ~ 80 MHz bandwidth for its second harmonic at 532 nm. The linewidth of an injection-seeded pulsed dye laser at 589 nm of ~ 110 MHz can be found in Fig. 2 of [7.2], and in Fig. 7.27(a) below. Since the FWHM of the backward Cabannes scattering spectrum of air (at tropospheric temperature of ~ 300 K) at 532 nm is around 2.6 GHz, the pulsed laser linewidth should be narrower if atmospheric parameters are to be retrieved with Cabannes scattering. This is similarly true for measurement of Na atoms (at mesospheric temperatures of ~ 200 K) at 589 nm, which has FWHM around 2.2 GHz. Furthermore, to measure temperature and wind to within ~ 1 K and 1 m/s accuracy, the long-term stability of the center of the pulsed laser line, and thus the variation of the CW seed laser frequency, must be maintained to be within ~ 1 MHz. For this reason, the seed laser of a narrowband lidar has to be locked to an atomic or molecular absorption line.

We shall not discuss various laser locking techniques here, except in providing references to some of these techniques in the literature. The Doppler-free absorption spectroscopic technique has been used for some time to lock the laser to the Na transition at 589.15 nm [7.3] and the Ba transition at 553.7 nm [7.4]. Because of the importance of diode-laser-pumped Nd:YAG lasers in a variety of applications requiring high spectral purity, these lasers have been frequency stabilized by locking with a phase modulation technique to the Doppler-free absorption [7.5] or Doppler-broadened peak [7.6] of the $^{127}I_2$ molecular transition to within 1 kHz, an accuracy much better than necessary for lidar applications. To eliminate long-term frequency drift in lidar applications, the second harmonic of the Nd:YAG seed is stabilized to the steep half-power point of an iodine absorption line [7.7], leading to an RMS variation of 0.84 MHz in a 66-hour test. With the phase modulation technique [7.6] applied to an airborne HSRL, the long-term frequency stability was found to be within 0.1 MHz during flights [7.1]. To facilitate the monitoring of

Doppler-free fluorescence, the CW Doppler-free features in Na were determined to an accuracy of 0.1 MHz [7.8], and the associated frequency jitter in the Na laser-induced fluorescence (LIF) lidar operation [7.2] was ~ 1 MHz. With the seed frequency locked and pulsed laser lineshape function measured in sufficient detail, the effect of the pulsed laser lineshape on the theoretical light scattering spectrum may be accounted for in lidar data analysis.

For atmospheric temperature measurements from the upper stratosphere (~ 30 km) through the mesosphere (~ 80 km), the method of Rayleigh integration with a broadband lidar may be utilized. Below 30 km (or below a higher altitude during periods of volcanic eruptions), either pure rotational Raman (PRR) or Cabannes scattering (CS) must be used in order to reject the influences of aerosols. Although in theory a broadband laser with bandwidth less than 10 GHz may be adequate for PRR lidar, narrowband lasers are commonly used for both PRR and CS lidars. Above 80 km, due to the weakness of Rayleigh scattering, the laser-induced fluorescence (LIF) technique requires that a narrowband lidar be used. For atmospheric wind measurements, since the Doppler shift from an air molecule with a beam-aligned wind component of 1 m/s is about 1 MHz in visible wavelengths, Cabannes scattering and LIF (also termed resonance fluorescence) techniques are the only choices for, respectively, below and above ~ 80 km; both require a narrowband lidar. We shall discuss the observation of aerosol and atmospheric parameters below ~ 80 km first, in Sections 7.1 – 7.4, and then of temperature and winds in the mesopause region between ~ 80 km and 110 km in Section 7.5.

It was Fiocco et al. [7.9] in 1971 who first used a high-spectral-resolution scanning Fabry–Perot interferometer to obtain the backward scattering spectrum from aerosol and molecular scattering of a CW argon-ion laser at 488 nm, resulting in a central peak with pedestals on both sides. They suggested that by analysis of such a spectrum, one can determine atmospheric temperature and the aerosol-to-molecular ratio in the troposphere. This spectrum is similar to the Mie–Cabannes spectrum shown in Fig. 7.1, with a normalized Gaussian central peak (×0.02) having a RMS width of ~ 10 MHz to emulate aerosol (Mie) scattering and a normalized pedestal backscattering (Cabannes) shoulder to emulate standard air at sea level (thick solid). Along with the Mie–Cabannes spectrum, we show a Gaussian spectrum from air at the same temperature (288.15 K, FWHM width of 2.55 GHz) with the thick-dashed line for comparison (see Fig. 4.7 for more information). By separating the lidar-observed Mie spectral component from the Cabannes (or Doppler-broadened) component with an ideal 2 GHz-wide notch filter (solid thin), for example, in principle one can clearly determine atmospheric temperature T as well as the aerosol-to-molecular ratio $\mathrm{R}_a = \beta_a/\beta_m$. Practical methods for this separation follow.

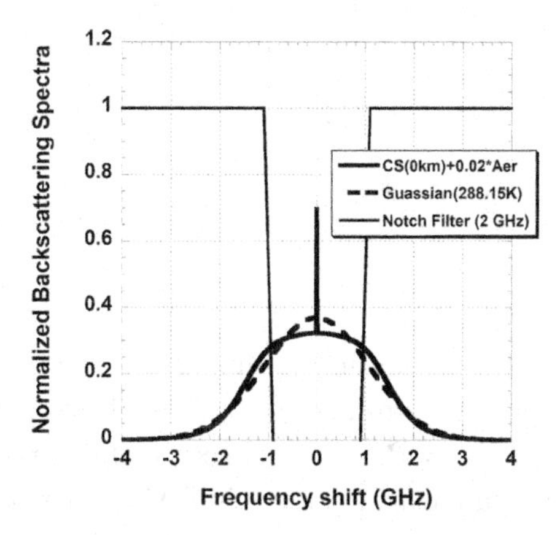

Fig. 7.1 A backscattering spectrum of aerosol (with $\mathrm{R}_a = 1$) plus normalized molecular Cabannes scattering at sea level (thick solid line). Also shown are a molecular Gaussian spectrum at 288.15 K (thick dashed line) and a 2 GHz notch filter (thin line).

A decade later, Schwiesow and Lading in 1981 [7.10] proposed measuring the atmospheric temperature profile by Rayleigh scattering using two stabilized Michelson interferometers. And two years later, Shimizu et al. [7.11] proposed the use of high-spectral-resolution lidar (HSRL) to do the same by using an atomic vapor filter (AVF) for better blocking of aerosol scattering. Both papers predicted tropospheric temperature measurement capability to an accuracy of ±1 K. To achieve these proposed temperature measurements in practice, as explained below, turned out to be challenging. Nonetheless, it was first demonstrated in 1992 [7.12] with barium filters at 553.7 nm and in 2001 realized with iodine filters at 532 nm [7.13]. Notwithstanding, the proposed HSRL techniques for blocking and separating aerosol scattering have proven powerful for measuring the less-demanding profiles of backscatter ratio and aerosol optical properties, provided that the associated atmospheric density profile is given. This is particularly true with the use of an iodine filter, as will be discussed in Section 7.1. The simultaneous profiling of aerosol lidar ratio and atmospheric temperature with HSRL via Rayleigh–Mie scattering using vapor filters as the frequency discriminator, along with its future prospects, will be discussed in Section 7.3.

With this aim in mind, we first consider lidar systems deriving signal from aerosol, Rayleigh, and vibrational Raman scattering from the atmosphere. In this case, we typically deal with two or three receiving channels ($\mu = 2$ or 3) and return to (5.3), expressing photon count profiles with a given resolution as a function of altitude z (instead of its equivalent range r), with upward and downward attenuations given by $A_\uparrow(\nu_{iL}, z)$ and $A_\downarrow(\nu_{iS}, z)$, and explicitly written-out backscattering and extinction coefficients: $\beta_i(\nu_{iS}, z)$, $\alpha(\nu_{iS}, z')$, and $\alpha(\nu_{iL}, z')$, as in (7.1):

$$N_i(\nu_{iS}, z; r_B) = \eta_i \frac{A_R}{z^2} \exp\left[-\int_{z_0}^{z} \alpha(\nu_{iS}, z')dz'\right] [\beta_i(\nu_{iS}, z) r_B] \exp\left[-\int_{z_0}^{z} \alpha(\nu_{iL}, z')dz'\right] \frac{E_{iL}}{h\nu_{iL}}; \; i = 1, 2, 3$$

$$\text{with } \beta(\nu_{iS}, z) = \beta_a(\nu_{iS}, z) + \mathcal{N}_m(z) \frac{d\sigma^{mol}(\nu_{iS})}{d\Omega}; \; \alpha(\nu_{iS,iL}, z) = \alpha_a(\nu_{iS,iL}, z) + \mathcal{N}(z)\sigma^R(\nu_{iS,iL}), \tag{7.1}$$

where $N_i(\nu_{iS}, r; r_B)$ is the background subtracted photon counts for the ith receiving channel. For the determination of aerosol optical profiles (backscatter ratio, $\mathbb{R} = (\beta_a + \beta_m)/\beta_m = \mathbb{R}_a + 1$ and aerosol lidar ratio, $L_a = \alpha_a/\beta_a$), we consider two receiving channels of a Rayleigh–Mie lidar, one being a total (or combined) light scattering channel, and the other being either a vibrational nitrogen Raman channel (termed Raman lidar) or a Rayleigh/Cabannes scattering channel (termed HSRL lidar). Since the vibrational Raman shift for molecular nitrogen is 2331 cm^{-1} (see Table 4.2), a broadband lidar is adequate to separate molecular scattering from the total (combined aerosol and molecular) scattering in a Raman lidar [7.14]. In this case, the molecular density, $\mathcal{N}_m(z)$, responsible for molecular backscattering (with differential cross section $d\sigma^{mol}(\nu_{iS})/d\Omega$), is less than the atmospheric density $\mathcal{N}(z)$, which is responsible for molecular Rayleigh scattering with cross section $\sigma^R(\nu_{iS,iL})$ in the extinction coefficient. To do so with an HSRL lidar [7.1; 7.15], a narrowband lidar in conjunction with a spectral filter to sort out aerosol from temperature-dependent molecular scattering must be employed. For these measurements, at least two receiving channels, one for total (or combined) scattering and one for molecular scattering, must also be used. For atmospheric temperature measurements (see Section 7.2), owing to the need to analyze the width (shape) of the molecular scattering spectrum, at least three receiving channels (one for total scattering and two for molecular scattering) must be used.

7.1 Profiling Lidar-Ratio and Aerosol Optical Properties

To determine (total) backscatter ratio, $\mathbb{R} = (\beta_a + \beta_m)/\beta_m$, and aerosol lidar ratio, $L_a = \alpha_a/\beta_a$ (and/or their alternatives, aerosol backscatter ratio, $\mathbb{R}_a \equiv \beta_a/\beta_m = \mathbb{R} - 1$, and aerosol phase function, $L_a^{-1} = \beta_a/\alpha_a$, or aerosol extinction coefficient, $\alpha_a = L_a\beta_a = L_a\mathbb{R}_a\beta_m$), we need two separate receiving channels: the total and molecular scattering channels from (7.1), denoted by $i = \tau$ and m. We first express the range-power-adjusted or normalized signals, $\widetilde{N}_\tau(z)$ and $\widetilde{N}_m(z)$, for the two channels as

$$\widetilde{N}_\tau(\nu_L, z) = \frac{h\nu_L z^2}{A_R E_L z_B} N_\tau(\nu_L, z; r_B) = \eta_\tau[\beta_a(\nu_L, z) + \beta_m(\nu_L, z)]\exp\left[-2\int_{z_0}^{z} \alpha(\nu_L, z')dz'\right], \text{ and} \tag{7.2.a}$$

$$\widetilde{N}_m(\nu_S, z) = \frac{h\nu_L z^2}{A_R E_L z_B} N_m(\nu_S, z; r_B) = \eta_m[\beta_m(\nu_S, z)]\exp\left[-\int_{z_0}^{z} \{\alpha(\nu_L, z') + \alpha(\nu_S, z')\}dz'\right], \tag{7.2.b}$$

where $\beta_m(\nu_L, z) = \mathcal{N}(z)\frac{d\sigma^R(\nu_L)}{d\Omega}$, $\beta_m(\nu_S, z) = \mathcal{N}_m(z)\frac{d\sigma^{mol}(\nu_S)}{d\Omega}$ and $\alpha(\nu_L, z) \equiv \alpha_a(\nu_L, z) + \mathcal{N}(z)\sigma^R(\nu_L)$ and $\alpha(\nu_S, z) \equiv \alpha_a(\nu_S, z) + \mathcal{N}_m(z)\sigma^{mol}(\nu_S)$. Here we note that $A_\uparrow(\nu_L, z) = A_\downarrow(\nu_L, z)$, $A_\uparrow(\nu_L, z) \neq A_\downarrow(\nu_S, z)$ and η_τ and η_m are the collection efficiencies for the total and molecular channels, respectively. The collection efficiencies are generally not identical; thus, they must be provided/measured independently. Assuming we are given or have obtained from a companion measurement the atmospheric molecular density, $\mathcal{N}(z)$, along with the density of scattering molecules, $\mathcal{N}_m(z) < \mathcal{N}(z)$, and differential cross section, $d\sigma^{mol}/d\Omega = d\sigma^{Ram}/d\Omega$, for Raman lidar, and $\mathcal{N}_m(z) = \mathcal{N}(z)$ and $d\sigma^{mol}/d\Omega = d\sigma^R/d\Omega$ for HSRL, we can derive the backscattering ratio and the aerosol extinction coefficient, respectively, as (7.3.a) and (7.3.b):

$$\mathbf{R}(\nu_L, z) = \frac{\beta_a(\nu_L, z) + \beta_m(\nu_L, z)}{\beta_m(\nu_L, z)} = \frac{(\eta_m/\eta_\tau)\widetilde{N}_\tau(\nu_L, r)}{\widetilde{N}_m(\nu_S, r)} \frac{\beta_m(\nu_S, z)}{\beta_m(\nu_L, z)}; \tag{7.3.a}$$

$$\alpha_a(\nu_L, z) = \left[1 + \frac{\alpha_a(\nu_S, z)}{\alpha_a(\nu_L, z)}\right]^{-1} \left\{\frac{d}{dz}\ln\left[\frac{\eta_m \beta_m(\nu_S, z)}{\widetilde{N}_m(\nu_S, z)}\right] - \alpha_m(\nu_S, z) - \alpha_m(\nu_L, z)\right\}. \tag{7.3.b}$$

7.1.1 Profiling Aerosol Optical Properties with Raman Lidar

If a broadband nitrogen Raman lidar is used to retrieve the aerosol lidar ratio, $L_a(\nu_L, z) = \alpha_a(\nu_L, z)/\beta_a(\nu_L, z)$, at the transmitting laser frequency ν_L (or wavelength λ_L), we need to replace $\mathcal{N}_m(z)$ by $\mathcal{N}_{N_2}(z) = 0.79\mathcal{N}(z)$ and assume the wavelength dependence of the aerosol extinction coefficient. This latter assumption converts the molecular backscattering and extinction coefficients measured at the scattered frequency ν_S (or wavelength λ_S) explicitly to the same quantities at the transmitting frequency ν_L, as employed in (7.4.a) and (7.4.b); (7.4.b) is in agreement with Eq. (3) of [7.16]. The aerosol lidar ratio $L_a(\nu_L, z)$ is then given in (7.4.c):

$$\frac{\alpha_m(\nu_S,z)}{\alpha_m(\nu_L,z)} = \left(\frac{\lambda_L}{\lambda_S}\right)^4; \ \frac{\beta_m(\nu_S,z)}{\beta_m(\nu_L,z)} = \frac{\mathcal{N}_{N_2}(z)}{\mathcal{N}(z)}\left(\frac{\lambda_L}{\lambda_S}\right)^4 \rightarrow$$
$$\mathbf{R}(\nu_L,z) = \frac{0.79(\eta_m/\eta_\tau)\widetilde{N}_\tau(\nu_L,r)}{\widetilde{N}_m(\nu_S,r)}\left(\frac{\lambda_L}{\lambda_S}\right)^4, \tag{7.4.a}$$

$$\frac{\alpha_a(\nu_S,z)}{\alpha_a(\nu_L,z)} = \left(\frac{\lambda_L}{\lambda_S}\right)^k \rightarrow \alpha_a(\nu_L,z) = \left[1+\left(\frac{\lambda_L}{\lambda_S}\right)^k\right]^{-1} \times$$
$$\left\{\frac{d}{dz}\ln\left[\frac{\eta_m\beta_m(\nu_S,z)}{\widetilde{N}_m(\nu_S,z)}\right] - \alpha_m(\nu_S,z) - \alpha_m(\nu_L,z)\right\}, \text{ and} \tag{7.4.b}$$

$$L_a(\nu_L,z) = \frac{\alpha_a(\nu_L,z)}{\beta_a(\nu_L,z)}; \frac{[\beta_a(\nu_L,z)+\beta_m(\nu_L,z)]}{[\beta_a(\nu_L,z_R)+\beta_m(\nu_L,z_R)]} = \frac{\widetilde{N}_\tau(\nu_L,z)\exp\left[-2\int_{z_0}^{z_R}\alpha(\nu_L,z')dz'\right]}{\widetilde{N}_\tau(\nu_L,z_R)\exp\left[-2\int_{z_0}^{z}\alpha(\nu_L,z')dz'\right]} \tag{7.4.c}$$

where $\beta_a(\nu_L,z)$ is the aerosol backscattering coefficient; its determination involving comparing the signal of total scattering $N_\tau(\nu_L,z)$ with that at a reference height $z_R \gg z, N_\tau(\nu_L,z_R)$ as given in (7.4.c) with $\beta_a(\nu_L,z_R)$ negligible compared to $\beta_m(\nu_L,z_R)$. Notice that in (7.4.a) we have assumed the strength of aerosol scattering is proportional to λ^{-k}, where the exponent k depends on the type of aerosol. This exponent is well known to be equal to 4 for particles much smaller than the scattering wavelength; it may be taken to be ~ 1 for particles, like water droplets with size comparable to the wavelength, and justifiably taken to be zero for particles like ice, with dimensions much larger than the wavelength [7.16]. Also to be noted in (7.4.a), it is Rayleigh scattering by all air molecules and vibrational Raman scattering by nitrogen molecules that are, respectively, responsible for the extinction and backscattering of the molecular return signal. For further study on the retrieval of atmospheric optical properties of a variety of aerosols by Raman lidar, we refer the reader to its original source [7.16] and a good review [7.17]. As we will explain, though one pays a price for needing a much more demanding narrowband lidar system, complications in wavelength scaling of the retrieved signal can be totally avoided if the HSRL technique is employed.

7.1.2 Profiling Aerosol Optical Properties with HSRL

In 1983, a group at the University of Wisconsin [7.18], realizing that the transmission bandwidth of a high-spectral-resolution scanning interferometer [7.9] is much narrower than the width of the molecular spectrum, thus blocking much of the spectrum, used a fixed Fabry–Perot étalon with its transmission locked at the laser frequency and efficiently performed HSRL observations. They used both the transmission through the étalon and the reflection from the étalon to separate the aerosol backscattering β_a (transmission channel) from the molecular backscattering β_m (reflection channel). In this arrangement, the entire Mie–Cabannes spectrum is detected all the time, and the efficiency of the backscattering ratio measurement is thus very much increased. However, the spectral separation by a fixed Fabry–Perot étalon is not complete; the fraction of aerosol scattering and molecular scattering in each of the aerosol and molecular channels must be carefully calibrated; see Eqs. (3a) and (3b) in [7.18]. This calibration is carried out by comparing the observed molecular lidar return to the lidar return computed from the theoretical Rayleigh/Cabannes scattering spectrum at a given or assumed atmospheric temperature (and pressure). Since their pioneering work, the Wisconsin group has developed a fixed Fabry–Perot étalon HSRL and performed valuable observations of backscattering ratio as well as depolarization ratio under various atmospheric conditions; see [7.15]. Motivated by the initial work at Colorado State University with barium AVF [7.19] (methodology to be discussed below in Section 7.3.1), they switched to the use of an iodine absorption filter, which has better performance than the Fabry–Perot étalon that it replaces [7.20]. The clean measurement of optical depth through cloud layers, shown in Fig. 5 of their 1994 seminal paper, has demonstrated the superiority of the atomic/molecular filter techniques for HSRL.

The advantages of the iodine vapor filter (IVF) over the barium AVF are twofold: (1) the IVF can produce much higher vapor pressure than barium at lower vapor temperature and (2) the IVF has many absorption lines that match the tunable second harmonic of a pulsed Nd:YAG laser at 532 nm. This clearly avoids the need to operate vapor temperatures as high as 750–800 K as in the case of barium, wherein temperature instability and gradients produce considerable fluctuations in filter transmission. For this reason, we restrict ourselves in the remaining portion of this section to discussion of the iodine filter and examples of its use in HSRL observations. Notwithstanding, the barium filter has a transmission spectrum approaching an ideal notch filter; we shall discuss its use in Section 7.2 in conjunction with temperature measurements.

The first comprehensive atlas of iodine absorption was reported by Gerstenkorn and Luc [7.21], and the I_2 spectrum was extensively studied in the region of the frequency-doubled Nd:YAG wavelength by the Princeton group [7.22], with a useful simulation model. In Fig. 7.2, we exhibit a section of its relevant absorption spectrum taken from a 2017 publication [7.23] with strong absorption lines 1104–1112 seen clearly. The strong 1109 line was used in the initial HSRL

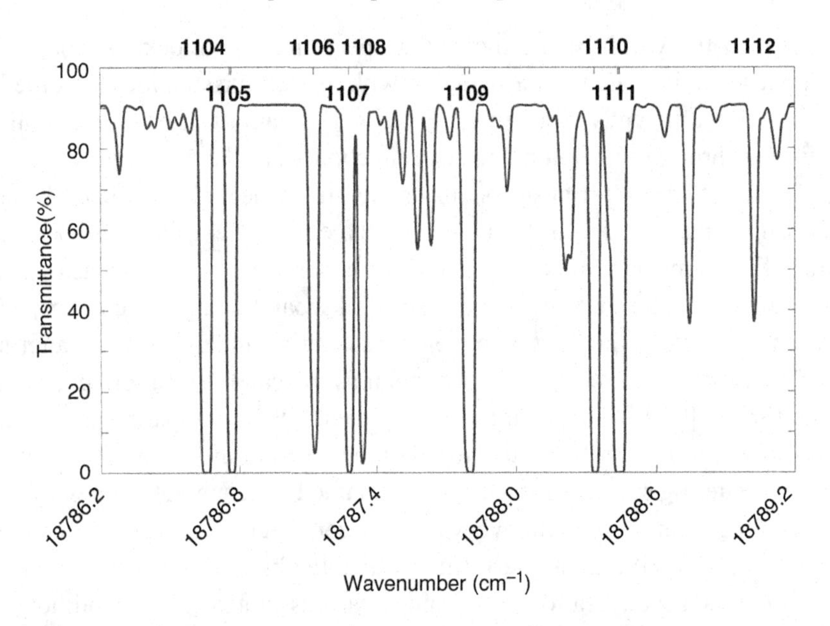

Fig. 7.2 The transmittance profile of a 15-cm-long iodine cell at 38°C and 0.7 torr near the frequency-doubled Nd:YAG wavelength at 532.27 nm (18,787.3 cm^{-1}). Reproduced with permission from Fig. 2 of [7.23] © the Optical Society.

observation [7.20] for observing various optical properties of interest, and its Fig. 5 is reproduced in Fig. 7.2 here; for details, see the original paper [7.20].

Another example, from 2008, is an airborne tropospheric HSRL measurement [7.1]. For this airborne lidar, the laser frequency is set by locking the second harmonic of a CW Nd:YAG injection seed laser near iodine absorption line 1111 at 532.242 nm [7.1] by a phase modulation technique [7.6]. This laser seeds at the desired frequency with spectral purity greater than 5000:1, even in the high-vibrational environment of the aircraft. The lidar sends out both the linearly polarized fundamental and second harmonic of the Nd:YAG laser, and it detects both parallel and perpendicular polarizations, thus measuring depolarization ratio at both 532 and 1064 nm, along with aerosol optical properties at 532 nm. As shown in the schematic of the detection system in Fig. 8 of [7.1], similar to that shown in Fig. 7.10, the 532 nm light passes through a solid Fabry–Perot étalon (FWHM ~ 60 pm) along with an interference filter to reject O- and S-branches of PRR, leaving Mie and Cabannes scattering to be detected. As the authors state: "Significant and critical engineering designs were implemented to provide a set of internally-calibrated measurements that could be performed during flight in relatively short times (<5 min total)." These provide necessary real-time calibration of filter transmission and relative efficiencies of the 532 nm channels. In a period of two years, the absolute transmission of the iodine filter varied by less than 0.3%.

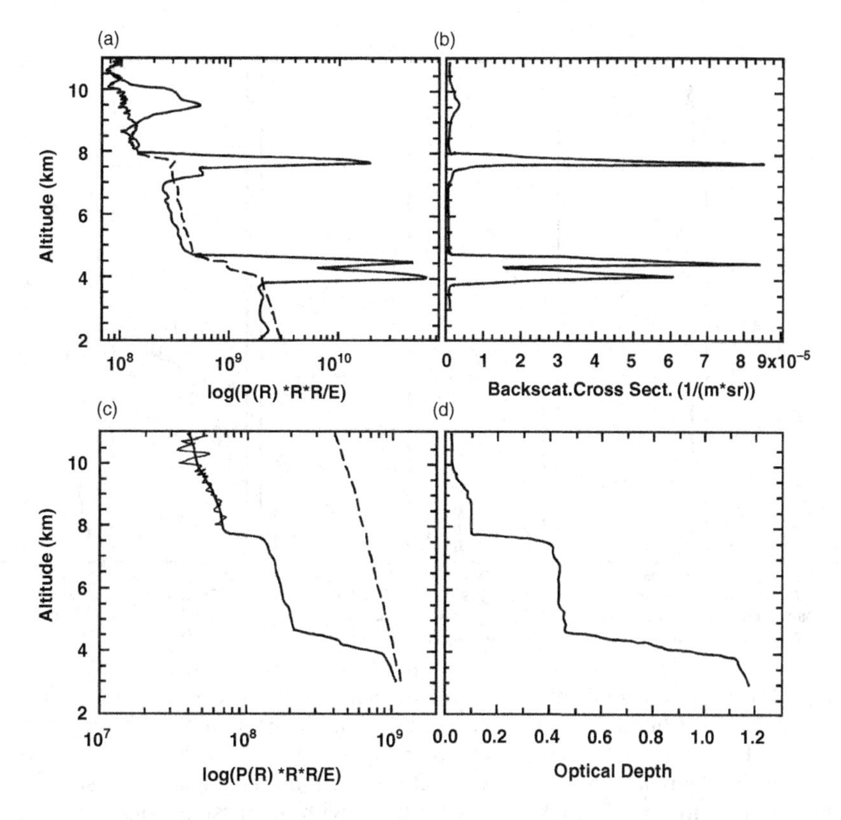

Fig. 7.3 HSRL Measurements obtained on July 21, 1993. (a) Signals detected with PMT1 (solid curve) and PMT2 (dashed curve). (b) Aerosol backscatter cross section. (c) A constrained nonlinear regression fit (solid curve) to the inverted molecular signal (wiggly curve at left) together with the calculated clear air molecular return (dashed curve at right). The quantity $P(R) \times R^2/E$ is the range-adjusted counts in (7.2). (d) Optical depth through the cloud layers. Reproduced with permission from Fig. 5 of [7.20]. © the Optical Society.

To retrieve aerosol optical properties, we set $\lambda_S = \lambda_L$ in (7.4) and account for the filtering by an atomic or molecular vapor filter in addition to the detection efficiency for the molecular channel, (7.3.b). Since the Cabannes spectrum depends on atmospheric pressure and temperature, proper retrieval (explained in the next section) requires their values; hence, atmospheric density, $\mathcal{N}(z)$, at each altitude must be known and may be obtained from radiosonde data or an assimilation model [7.1]. Fig. 7.4, reprinted from Fig. 14 of [7.1], provides an example. It shows the resulting aerosol depolarization ratio at 532 nm and 1064 nm (panels c and d), and wavelength dependence of aerosol backscattered signal (panel f) as defined in (13) of [7.1], as well as β_a, α_a, and L_a, respectively, at 532 nm in panels (a), (b), and (e).

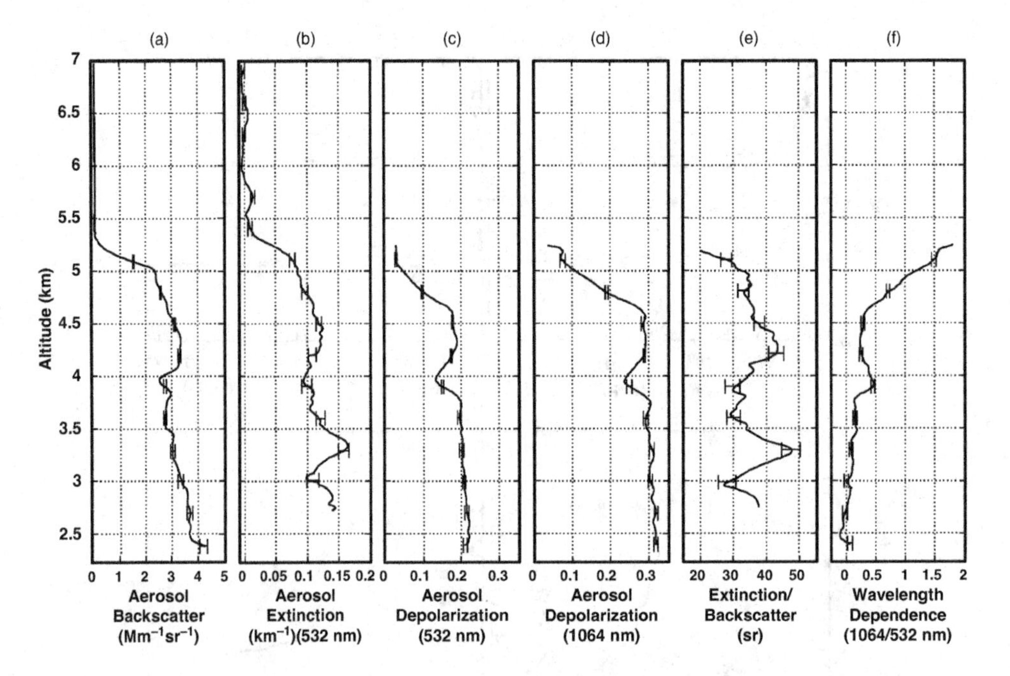

Fig. 7.4 HSRL measurement profiles plotted versus altitude above mean sea level at 9:15 UT on March 7, 2006. The minimum altitude plotted, 2.27 km, is set at ground level for this location. Error bars on all plots represent shot noise 1-σ standard deviations for 300 m vertical and 6 km horizontal spatial resolution. $Mm^{-1} = 10^{-6}\ m^{-1} = 0.001\ km^{-1}$. Reproduced with permission from Fig. 14 of [7.1]. © the Optical Society.

The stability of the system has allowed the lidar to perform over 650 operational flight hours with minimal maintenance. This type of HSRL data from nine major field experiments, according to [7.1], "have been archived and are being used in various scientific studies, including investigations of aerosol properties and aerosol-cloud interactions, satellite validation, assessment of chemical transport models, and investigation of new lidar retrieval techniques." At the time of this writing, interest in cloud phase studies involving horizontally oriented ice crystals (HOICs) has increased, leading to new developments in Stokes vector-based polarization lidars; see Section 6.2 and [6.19; 6.23]. In this connection, HRSL scanning at multiple angles within 20° of zenith while detecting both parallel and perpendicular polarizations is also highly relevant. Preliminary observations of high backscatter cross section and low depolarization ratio at zenith have indicated the presence of oriented crystals in clouds [7.24]. Future investigations are likely to include methods to quantify relative HOIC and aerosol concentrations, along with comparisons with the Stokes vector-based polarization technique and including incorporating multiple scattering corrections to retrieve the extinction coefficient from both methods [7.25].

In reporting of total or aerosol to molecular scattering ratio, $\mathbb{R}$ or $\mathbb{R}_a$ with HSRL, the question of how to include pure rotational Raman (PRR) scattering in the calculations has been somewhat ambiguous in the literature. Though this is a small effect, we briefly discuss it here for theoretical understanding. For clarity, we include both polarized (parallel) and depolarized (perpendicular) components (i.e., for aerosols, $\beta_a = \beta_a^{\|} + \beta_a^{\perp} = \beta_a^{\|}(1 + \delta_a)$), with the aerosol linear polarization ratio $\delta_a = \beta_a^{\perp}/\beta_a^{\|}$ [7.1], and for molecules, $\beta_m = \beta_C + \beta_R^{\|} + \beta_R^{\perp}$, with the subscript C for Cabannes and R for PRR scattering, respectively. We recall the PRR consists of the unshifted Q-branch (~25%) and shifted O- and S-branches (~ 75%), see [4.5] and Fig. 4.1; the former cannot be separated from aerosol scattering and the latter backscattering coefficient is denoted as $\beta_{RR} = 0.75\beta_R$. Using the relative anisotropy $R_A = (\gamma/a)^2 = 0.22$ for dry air, see Table 1 of [4.5], (4.9.f) allows us to relate the backscattering coefficient of PRR to Cabannes scattering: $\beta_R^{\|} = 0.22 \times 4\beta_C/45$ and $\beta_R = 0.22 \times 7\beta_C/45$, leading to $\beta_m = (1 + 0.22 \times 7/45)\beta_C = 1.0342\beta_C$. Also note, the linear polarization ratio for air molecules is $\delta_\ell = (0.22 \times 3/45)/1.0342 = 0.0142$. With this background, we compare the difference in the measurement of aerosol to molecular ratio, $\mathbb{R}_a \equiv \beta_a/\beta_m$, between two different experimental configurations here. Neither employs a narrowband filter. The first, denoted as $\mathbb{R}_{am}$, does not discriminate polarization in its receiver, and the second, denoted as $\mathbb{R}_{am}^{\|}$, employs a polarization analyzer to detect only parallel polarization of scattered light. We note that experimentally the Q-branch of PRR is by necessity removed from molecular scattering and included in the aerosol scattering. We derive the relations for $\mathbb{R}_{am}$ and $\mathbb{R}_{am}^{\|}$ in terms of $\mathbb{R}_a$ in (7.5.a) and (7.5.b), respectively:

$$\text{Case A}: \mathbb{R}_{am} = \frac{\beta_a + (0.22 \times 0.25 \times 7/45)\beta_C}{(1 + 0.22 \times 0.75 \times 7/45)\beta_C} = \frac{\beta_a}{1.0257\beta_C} + \frac{0.385}{46.155}$$
$$= 1.0171\mathbb{R}_a + 0.00834\,; \tag{7.5.a}$$

$$\text{Case B}: \mathbb{R}_{am}^{\|} = \frac{\beta_a^{\|} + (0.22 \times 0.25 \times 4/45)\beta_C}{(1 + 0.22 \times 0.75 \times 4/45)\beta_C} = \frac{\beta_a/(1 + \delta_a)}{1.0147\beta_C} + \frac{0.22}{45.66}$$
$$= \frac{1.019}{(1 + \delta_a)}\mathbb{R}_a + 0.00482\,. \tag{7.5.b}$$

Because of the inclusion of the Q-branch PRR, the measured ratio is about 1% smaller than the correct value, $\mathbb{R}_a \equiv \beta_a/\beta_m$, by definition, noting $\beta_a/\beta_C = 1.0342\,\mathbb{R}_a$; the difference between the two measured values is only about 0.1%, likely smaller than measurement uncertainty (~ 1% perhaps).

7.2 Temperature Measurements with Rayleigh and Rotational Raman Scattering

We now turn to the measurement of the most important atmospheric parameter: temperature. The most popular lidar atmospheric temperature measurement technique is commonly known as the integration technique by the inversion of Rayleigh backscattering signal. This technique was pioneered by Elterman [7.26] using a searchlight in 1954, and later by Hauchecorne and Chanin using a dye laser in 1980 [7.27] and a Nd:YAG laser in 1981 [7.28]. Since the integration technique does not depend on the spectral properties of the scattered light, a narrowband laser is not required, making it a less challenging lidar. Nowadays, the integration technique combines with the rotational Raman technique [7.29; 7.30] to filter out aerosol scattering in the lower atmosphere, which requires a narrower linewidth (< 10 GHz), and as such uses a narrowband laser in practice. In this section, we first describe the integration technique, then the rotational Raman technique, followed by a realistic example employing both techniques, reporting a temperature profile from near-ground to 85 km [7.31].

7.2.1 Rayleigh Inversion and the Integration Technique

In an aerosol-free atmosphere, the relation between pressure, p, temperature, T, and density, ρ, follows the ideal gas law, and buoyancy is balanced by Earth's gravity in hydrostatic equilibrium. Mathematically, these relations have been given in (5.1.a) and (5.1.b), repeated in (7.6.a):

$$p(z) = \rho(z)\left(\frac{R}{M}\right)T(z); \quad \frac{dp(z)}{dz} = -\rho(z)g(z); \quad \frac{dp(z)}{p(z)} = -\frac{Mg(z)}{RT(z)}dz = d\left(\ln p(z)\right). \tag{7.6.a}$$

Assuming horizontal uniformity, we follow [7.27] and divide the atmosphere into n layers with thickness Δz each. Applying the second and third equations of (7.6.a) to the ith layer of the atmosphere, we can integrate between the lower edge, $z_i - \Delta z/2$, and the upper edge, $z_i + \Delta z/2$, and obtain the pressure of the lower edge of this layer $p(z_i - \Delta z/2)$ and the temperature within this layer, $T(z_i)$, given in (7.6.b):

$$\begin{aligned} p(z_i - \Delta z/2) &= \rho(z_i)g(z_i)\Delta z + p(z_i + \Delta z/2); \\ T(z_i) &= \frac{Mg(z_i)\Delta z}{R\ \ln[\ p(z_i - \Delta z)/p(z_i + \Delta z/2)]}\ . \end{aligned} \tag{7.6.b}$$

Since gravity acceleration $g(z_i)$ is known (or treated as a constant) – ~9.6 ms^{-2} for the middle atmosphere – if both the atmospheric density $\rho(z_i)$ and the pressure of the upper edge of this layer $p(z_i + \Delta z/2)$ are known, we can determine $p(z_i - \Delta z/2)$ and $T(z_i)$ from (7.6.b). Invoking the fact that in an aerosol-free environment, Rayleigh lidar signal is proportional to atmospheric density, we can determine the proportionality constant by the atmospheric density at the upper edge of the top layer either by calculating from pressure $p_b(z_n + \Delta z/2)$ and temperature $T_b(z_n + \Delta z/2)$, by using a standard air model, or by an educated guess. With the proportionality constant and the atmospheric density profile $\rho(z_n)$ determined, the boundary value of $p_b(z_n + \Delta z/2)$ enables us to determine $T(z_n)$ and $p(z_n - \Delta z/2)$.

Since the lower edge pressure of the upper layer is the upper edge pressure for the layer below it (i.e., $p(z_{n-1} + \Delta z/2)$), and the density at the layer below it, $\rho(z_{n-1})$, is now known, the process can be iterated to obtain the profile of atmospheric temperature, $T(z_i)$, and pressure, $p(z_i)$, for all $i = 1, 2, \ldots n$. Since (7.6.b) determines the pressure at either the upper or lower edge of a layer, we define the pressure of a layer as the mean value of its two edges, $p(z_i) \equiv 0.5 [p(z_i - \Delta z/2) + p(z_i + \Delta z/2)]$. In this manner, the temperature and pressure profiles between the top and bottom altitudes, z_n and z_1, is retrieved by the integration technique, using data from the received Rayleigh lidar signal. We typically set z_n at an altitude where the signal/noise (SNR) of received photon counts (at a specified resolution) is around 5 to 10, conservatively, and z_1 at an altitude above which no aerosols are expected to contribute to the signal.

7.2.1(1) Temperature Uncertainty in the Integration Technique

We now discuss the temperature measurement error, again following [7.27]. To simplify the math, we define a quantity X_i at altitude z_i and deduce the temperature and associated relative uncertainty as

$$X_i = \frac{\rho(z_i)g(z_i)\Delta z}{p(z_i + \Delta z/2)} \rightarrow (1 + X_i) = \frac{p(z_i + \Delta z/2) + \rho(z_i)g(z_i)\Delta z}{p(z_i + \Delta z/2)} = \frac{p(z_i - \Delta z/2)}{p(z_i + \Delta z/2)}, \tag{7.7.a}$$

$$\therefore T(z_i) = \frac{Mg(z_i)\Delta z}{R \ln(1 + X_i)} = -\frac{Mg(z_i)\Delta z}{R}\ln(1 + X_i); \tag{7.7.b}$$

$$\frac{\delta T(z_i)}{T(z_i)} = \xi_i \frac{\delta X_i}{X_i}, \text{ with } \xi_i = \frac{X_i}{(1 + X_i)\ln(1 + X_i)};$$

$$\frac{(\delta X_i)^2}{X_i^2} = \left|\frac{\delta\rho(z_j)}{\rho(z_j)}\right|^2 + \frac{[\delta p(z_i + \Delta z/2)]^2}{[p(z_i + \Delta z/2)]^2}, \tag{7.7.c}$$

$$\text{where } \frac{[\delta p(z_i+\Delta z/2)]^2}{[p(z_i+\Delta z/2)]^2} = \frac{\sum_{j=i+1}^{n} |\delta\rho(z_j)|^2 \left(g(z_j)\Delta z\right)^2 + |\delta p_b(z_n+\Delta z/2)|^2}{[p(z_i+\Delta z/2)]^2}. \quad (7.7.d)$$

The parameter ξ_i is a multiplicative factor relating the fractional uncertainty between temperature T_i and the parameter X_i; for a standard atmosphere between geometric heights 0 and 86 km (see below), ξ_i varies between 0.841 and 0.884 (i.e., approximating it as unity is a bit of an overestimation). Since, in the above expressions, $\delta\rho(z_j)$ and $\delta p(z_j)$ are the uncertainties due to photon noise, which are much smaller than the incremental values $\Delta\rho(z_j)$ and $\Delta p(z_j)$, for N_m signal counts received at a given range resolution above B background counts, $\delta\rho/\rho = \delta p/p \sim \sqrt{N_m+B}/N_m$, which reduces to $1/\sqrt{N_m}$ when the background is negligible. Therefore, both contributions to $[\delta p(z_i+\Delta z/2)]^2$ in (7.7.d) become negligibly small compared to $[p(z_i+\Delta z/2)]^2$. In this case, if we choose to take $\xi_i \approx 1$, we have $\delta T/T \approx \delta p/p$, and the temperature uncertainty (in simple calculation) is approximately equal to temperature divided by the photon signal-to-noise ratio (SNR), that is, $\delta T \approx T/SNR = T\sqrt{N_m+B}/N_m$. To see how the temperature and its associated uncertainty may be derived from the integration method, we consider the following analysis based on the 1976 standard atmosphere data.

7.2.1(2) An Analysis Example of the Integration Technique

For this example, we download the temperature, pressure, and mass density of the 1976 standard atmosphere between 0 and 86 km at a 2 km interval ($z_1 = 0$ km, $z_n = 86$ km, and $\Delta z = 2$ km) from the website (www.pdas.com/atmosTable1SI .html), a public domain aeronautical software. With those data, we conceptually produce three sets of synthesized atmospheric density profiles as synthetic normalized Rayleigh lidar returns. These are (a) identical to, (b) 90% of, and (c) 150% of the downloaded density profile. We then calculate the "correct" boundary value pressure at $z_n + \Delta z/2$ (87 km), that is, $p_c(87\text{ km})$, by first computing $p(85\text{ km})$ from the average of $p(84\text{ km})$ and $p(86\text{ km})$, and then deducing $p_c(87\text{ km})$ applying the knowledge of $p(85\text{ km})$, $g(86\text{ km})$ and $\rho(86\text{ km})$ from the first equation in (7.6.b). We now proceed to retrieve the best educated guesses of the temperature profile from these synthetic profiles.

From profile (a), we set the initial pressure value $p_b(z_n+\Delta z/2) = p_c(87\text{ km})$ and follow the procedures to retrieve our "best educated guess" temperature profile, which is plotted in Fig. 7.5(a) in solid black. For profile (b), since the synthesized atmospheric densities are 90% of those of (a), we use the best previous initial value and set $p_b(z_n+\Delta z/2) = 0.9p_c(87\text{ km})$ to initialize the temperature retrieval, and

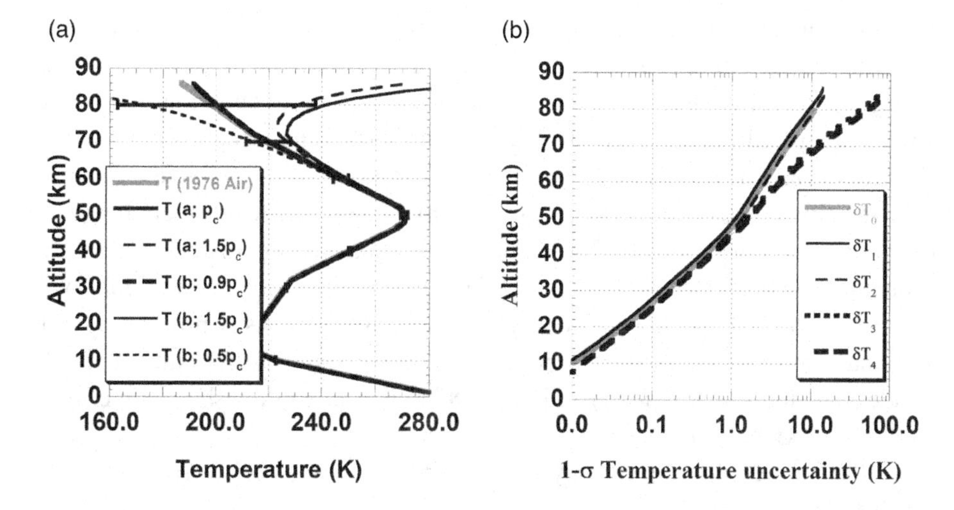

Fig. 7.5 (a) Comparison between the 1976 standard air temperature and retrieved profiles with selected uncertainties δT_i using different normalizations (see text) between air density and Rayleigh scattering photon-count profiles along with the top layer atmospheric pressure assumptions. Also shown are error-bars corresponding to the simply calculated δT_0. Note that the black solid and black dashed (buried below solid) curves are nearly identical, while the thin solid and short dashed curves diverge oppositely at high altitudes (explained in text). (b) Five different temperature uncertainties were calculated from (7.7.e) for δT_0 (simple calculation) with the first term only (and setting $\xi_i = 1$), followed by four calculations keeping the correct expression of $\xi_i(z_i)$, for δT_1, δT_2, δT_3 and δT_4, as shown in the figure legends and defined in the text, resulting in temperature uncertainty profiles in gray, thin solid, thin dashed, short dashed, and dashed lines. They show that pressure variations (2nd term) and additional error in the initial pressure guess have negligible effect on the uncertainty of the temperature retrieval.

we plot the result as black dashes in Fig. 7.5(a). The resulting temperature profiles (a) and (b) are indistinguishable, as the black dashes have disappeared into the black solid line. We also computed profile (c), with the synthesized atmospheric densities 150% of those of (a), that is, $p_b(z_n + \Delta z/2) = 1.5p_c(87\text{ km})$ (not plotted) and obtained precisely the same temperature profile. Compared to the downloaded temperature profile shown in gray in the same figure, we find only a slight difference near the top of the profile (where the gray line is cooler by ~ 5 K), a difference that disappears at lower altitudes. The three synthetic densities, differing only by a proportionality constant (0.9, 1.0, and 1.5), can yield the same retrieved temperatures points to the success of the integration method. This results from the fact that the retrieved temperature at z_i depends on the ratio between (and not on the values of) the pressures of the lower edge to the upper edge of the layer, $p(z_i - \Delta z)/p(z_i + \Delta z/2)$, as in (7.6.b).

Unfortunately, in practical lidar inversion we do not have the benefit of the atmospheric pressure at the upper edge of the top altitude, z_n, $p_b(z_n + \Delta z/2)$. Setting this initial pressure by guessing can lead to a huge deviation in retrieved temperature profile at the top altitudes. Nevertheless, the integration with such a deviation converges to real atmospheric temperatures with acceptable errors once the integration procession reaches an altitude more than one scale height below the top altitude in order to resolve the demand for thermal and hydrostatic equilibrium of the atmosphere. The result is clearly depicted in Fig. 7.5(a), in addition to the correct T(1976 Air) and those retrieved from the synthetic density profiles, by three additional profiles: thin dashed curve with case (a) and $1.5p_c$(87 km) at the top altitude, thin solid with case (b) and $1.5p_c$(87 km), and short dashes with case (b) and $0.5p_c$(87 km).

The 1-σ temperature uncertainty is given in (7.7.c) and (7.7.d), which may be combined as:

$$\delta T(z_i) = \xi_i T(z_i) \times$$

$$\left\{ \frac{\sqrt{N_m(z_i)+B}}{N_m(z_i)} + \frac{\sum_{j=i+1}^{n} \left(\rho(z_j) g(z_j) \Delta z\right)^2 \frac{\sqrt{N_m(z_i)+B}}{N_m(z_i)} + |\delta p_b(z_n + \Delta z/2)|^2}{[p(z_i + \Delta z/2)]^2} \right\};$$

$$\delta T_0(z_i) = \frac{T(z_i)\sqrt{N_m(z_i)+B}}{N_m(z_i)}; \ \delta T_1(z_i) = \xi_i(z_i)\frac{T(z_i)\sqrt{N_m(z_i)+B}}{N_m(z_i)};$$

$$\delta T_{2,3,4}(z_i) = \delta T(z_i; \xi_i(z_i)). \tag{7.7.e}$$

where $\delta T_{2,3,4}(z_i)$ is given in (7.7.e) with the altitude dependent $\xi_i(z_i)$ given in (7.7.c), along with an added term for initial pressure (IP) uncertainty in $\delta T_4(z_i)$. The quantity $N_m(z_i)$ is the signal in photon counts for a given PA product and resolution, plus the associated background counts, B.

The signal count profile for Cabannes scattering at 532 nm produced by our hypothetical reference lidar with $P_L A_R = 0.5$ Wm^2 and 1-hr/2-km resolution, N_m(532 nm, z; 2 km) $= 5.746 \times 10^{-15} \mathcal{N}(z)(\#\text{m}^{-3})/z^2(\text{km})^2$, may be calculated from (7.8) with the substitution of $d\sigma^C/d\Omega = 5.96 \times 10^{-32}(\text{m}^2\text{sr}^{-1})$, $\tau_{\text{int}} = 1$ hr, with $\Delta z = 2$ km, $B = 0$, and $\eta_m \exp[\ldots] = 0.01$. Equation (7.8) below is derived from (7.1) by adding the background count term B and substituting $\eta_i = \eta_m$ and $E_L = P_L \tau_{\text{int}}$ as:

$$N_m(\nu_L, z; r_B) = \frac{\eta_m P_L A_R \tau_{\text{int}} \Delta z}{h\nu_L z^2} \left[\mathcal{N}(z) \frac{d\sigma^C(\nu_L)}{d\Omega} \right] \times$$
$$\exp\left[-2 \int_{r_0}^{r} \alpha(\nu_L, z') dz' \right] + B; \; h\nu_L = \frac{1.9864 \times 10^{-16}}{\lambda(\text{nm})} . \tag{7.8}$$

The background count in a dark night by a 35-cm diameter telescope is assigned to be 3600 for the assumed resolution, an empirical result that is scaled from the observed background counts for 150 m/1 min resolution at 589 nm with a 40% efficient PMT (see Fig. 7(a) in [7.2]). The five different temperature uncertainties given in (7.7.e) and shown in Fig. 7.5(b) are δT_0 defined by setting $\xi_i = 1$ and ignoring the second term in the equation, and the remaining four estimates by employing the height dependent $\xi_i(z_i)$: δT_1 ignoring the second term; δT_2, δT_3, and δT_4 keeping both terms with background count B and initial pressure guess uncertainty $\delta p_b(z_n + \Delta z/2)$ explicitly imposed. For the curves in Fig. 7.5(b) the values for δT_2, δT_3, and δT_4 are $B = 0$, $B = 3600$, and $\delta p_b(z_n + \Delta z/2) = 0$, and $B = 3600$, respectively. The resulting temperature uncertainties, on the same order as those in Fig. 7.5(b), and with $\delta p_b(z_n + \Delta z/2) = 0.5 p_c(87\,\text{km})$, are, respectively, 0.147 K, 0.127 K, 0.165 K, 0.171 K, and 0.173 K at 30 km, and 8.37 K, 4.29 K, 5.26 K, 11.0 K, and 12.8 K at 70 km. Notice that δT_0 is an overestimate compared to δT_1 (due mainly to setting of $\xi_i = 1$ instead of ≈ 0.86), while δT_2, δT_3 and δT_4 are not that different, a reflection of the fact that the signal counts, $N_m(30\,\text{km}) = 4.44 \times 10^6$ and $N_m(70\,\text{km}) = 1957$, are at 30 km much larger and at 70 km are slightly smaller than the background count $B = 3600$.

To summarize, we note from Fig. 7.5(a) that ignorance in initial guesses in the retrieval leads to a large deviation in the retrieved temperature profile for the first one-to-two scale heights downward from the top altitude (see the comparison between gray solid and thin dashed curves as well as that between the thin and thin short-dashed curves). A comparison between five temperature uncertainties shows that the two without background counts (thin and dashed) are comparable, and this is similarly the case for the two with background counts (thick short- and long-dashed). The simpler expression $\delta T_0(z_i)$ (with $\xi_i = 1$) lies between the two sets and approaches δT_3 and δT_4 when background counts are included, suggesting this simpler expression may be a good approximation, and in fact it is commonly used in practical lidar temperature retrieval with the integration method. For a quick reference, we note that $\delta T_0(z_i)$ is $10^{-4} \times T(z)$ for a measurement that receives $N_0 = 10^8$ photocounts. For later reference, the estimated error of our hypothetical lidar with 1-hr/2-km resolution, that is, $(PA_R)\tau_{\text{int}}(\Delta z) = 1.0\,\text{W m}^2\,\text{hr km}$, is $\delta T_3 = 0.17$ K at 30 km and 11 K at 70 km.

7.2.2 Rotational Raman Technique for Temperature Measurements

Since the pure rotational Raman (PRR) scattering lines of nitrogen and oxygen are sufficiently separated from the quasi-elastic Cabannes–Mie scattering and their intensity distribution is temperature dependent, one can measure atmospheric temperature by probing the intensity distribution of these lines with passband filters, as proposed by Arishinov et al. in 1983 [7.32]. However, as noted by Cooney in 1984 [7.33], the existence of much stronger aerosol scattering in the troposphere can strongly interfere with the PRR signal and render it useless. When aerosol scattering is effectively filtered out, an PRR temperature lidar probes the intensity ratio of two specified spectral bands of the anti-Stokes PRR spectrum. In this way, it records the temperature dependence; this was done in 1993 by Neldejevoc et al. [7.29] and in 2000 by Behrendt et al. [7.30, 7.31]. Our discussion follows the last two references with the equivalent idealized filter transmission functions for simplicity.

We start by considering the S-branch or Stokes ($\nu_S < \nu_L$) lines with $\Delta J = +2$, and the O-branch, or anti-Stokes lines with ($\nu_S > \nu_L$) with $\Delta J = -2$, of the pure rotational Raman (PRR) scattering spectrum near the standard air conditions of 275 K and 0.75 atm. Since the lineshape of each rotation component is not resolved, its pressure dependence may be ignored. Like in Fig. 4.4, we plot these spectra for temperature 270 K and 280 K in Fig. 7.6. By way of review, we recall that using the formulae in Section 4.3.3, each spectral line strength given in Fig. 7.6 may be calculated, and that the sum of all spectral line strengths for either N_2 or O_2 is normalized to unity. The share of their Q-branches (not shown) is somewhat greater than its classical limit of 25%. For example, the sum of all lines in the Q-branch at 270 K is calculated to be 0.2605 for O_2 and 0.2556 for N_2; thus, correspondingly the O+S share is a bit smaller than 75%. We also recall that the atmosphere consists of 78.1% nitrogen and 20.9% oxygen. The square of mean and anisotropy Rayleigh scattering parameters for nitrogen and oxygen are given in Table 4.2. Weighted by the species concentration, the information in Table 4.2 permits calculation of the ratio of PRR to CS cross-sections as $\sigma(PRR)/\sigma(CS) = (7/45) \times (0.781 \times 0.52 + 0.209 \times 1.26)/(0.781 \times 3.17 + 0.209 \times 2.66) = 0.03435$, leading approximately to $\sigma(RR) \simeq 0.75 \times \sigma(PRR) = 0.02576\ \sigma(CS)$, which is consistent with the relation $\beta_{RR} = 0.75\beta_R$, with $\beta_R = 0.22 \times 7\beta_C/45 = 0.0342\beta_C$ and given in 7.1.2.

To facilitate the atmospheric temperature measurement with the rotational Raman technique, the (PRR) scattered signal is divided into two (molecular) channels, each equipped with a passband filter placed in different spectral regions of the anti-Stokes PRR spectrum. The ratio of the signal passing through the filter in the higher-frequency region (in Channel 1, termed the H.T. filter), $N_1(z)$, to that through the filter in the lower-frequency region (in Channel 2, termed the

L.T. filter), $N_2(z)$, $N_1(z)/N_2(z)$, is a monotonically increasing function of temperature. It can be determined theoretically by using the filter functions in conjunction with the temperature-dependent PRR spectrum. To see this dependence, we derive the power-altitude-adjusted, spectral-filtered molecular signals for the two channels from (7.2.b). These signals may be expressed as

$$\widetilde{N}_{mi}(\nu_S, z) = \frac{h\nu_L z^2}{A_R E_L z_B} N_i(\nu_S, z; r_B) = \eta_i f_i(T, \Delta\nu_{iPB}) \left[\mathcal{N}_m(z) \frac{d\sigma(PRR)}{d\Omega} \right] \times \exp\left[-2 \int_{r_0}^{r} \alpha(\nu_L, z')dz' \right]; \; i = 1, 2, \tag{7.9.a}$$

$$\text{with } f_i(T, \Delta\nu_i) = \int_{-\infty}^{\infty} \mathscr{R}^{PRR}(\nu; T) F_i(\nu) d\nu = \tau_i \mathscr{R}^{aRR}_{Air}(\Delta\nu_i; T) \text{ with } i = 1, 2, \ldots, \text{ where} \tag{7.9.b}$$

$$\mathscr{R}^{aRR}_{Air}(\Delta\nu_i; T) = 0.781 \times \frac{0.52}{1.78} \times \left[\mathscr{R}^{aRR}_{N_2-Even}(\Delta\nu_i; T) + \mathscr{R}^{aRR}_{N_2-Odd}(\Delta\nu_i; T) \right] + 0.209 \times \frac{1.26}{1.78} \times \mathscr{R}^{aRR}_{O_2-Odd}(\Delta\nu_i; T).$$

Here, we express the filter function $F_i(\nu)$ as the product of its peak transmission and an ideal passband filter with pass-bandwidth $\Delta\nu_{iBP}$, $\tau_i F_i^{ideal}(\Delta\nu_{iBP})$; by doing so, the attenuation function $\mathscr{R}^{aRR}_{Air}(\Delta\nu_i; T)$ may be expressed in terms of the anti-Stokes portion of individual species' normalized PRR spectra, $\mathscr{R}^{PRR}_{N_2-Even}(\Delta\nu_i; T) + \mathscr{R}^{PRR}_{N_2-Odd}(\Delta\nu_i; T)$ for nitrogen and $\mathscr{R}^{PRR}_{O_2-Odd}(\Delta\nu_i; T)$ for oxygen. We construct the equation (7.9.b) based on knowledge of nitrogen and oxygen species concentrations, the difference in the PRR cross sections between these major species as given in Table 4.2, and constrained by the fact that only odd-J oxygen rotational states are Raman active. Writing out the filter function more explicitly, with τ_{HT} and τ_{LT} as the peak transmissions of the respective filters, $f_1(T, \Delta\nu_1) = f_{H.T.}(T, \Delta\nu_{HPB}) = \tau_{HT} \mathscr{R}^{aRR}_{Air}(\Delta\nu_{HT}; T)$ and $f_2(T, \Delta\nu_2) = f_{L.T.}(T, \Delta\nu_{LPB}) = \tau_{LT} \mathscr{R}^{aRR}_{Air}(\Delta\nu_{LT}; T)$, we obtain the photocount ratio as

$$\frac{\widetilde{N}_1(z)}{\widetilde{N}_2(z)} = \frac{N_1(z)}{N_2(z)} = \frac{\eta_1 f_{H.T.}(T, \Delta\nu_{HPB})}{\eta_2 f_{L.T.}(T, \Delta\nu_{LPB})} = \frac{(\eta_1 \tau_{HT}) \mathscr{R}^{aRR}_{HT}}{(\eta_2 \tau_{LT}) \mathscr{R}^{aRR}_{LT}}, \tag{7.9.c}$$

where η_1 and η_2 are the efficiencies of the receiving channels, and the symbols of the attenuation factor of these idealized filters are simplified as $\mathscr{R}^{aRR}_{LT} \equiv \mathscr{R}^{aRR}_{Air}(\Delta\nu_{LT}; T)$ and $\mathscr{R}^{aRR}_{LT} \equiv \mathscr{R}^{aRR}_{Air}(\Delta\nu_{LT}; T)$, which are dependent on atmospheric temperature and the pass-band frequency widths. For example, if we

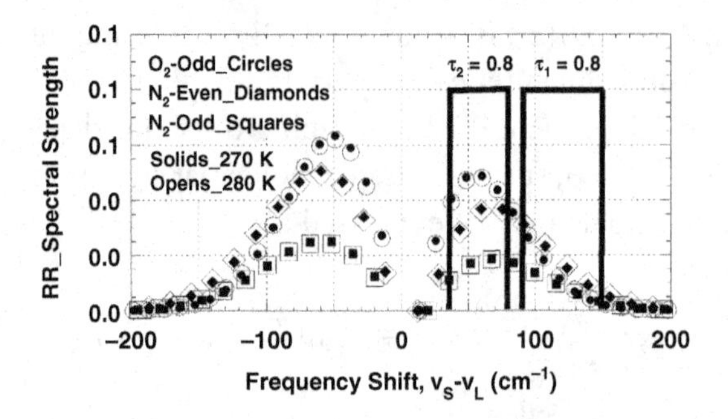

Fig. 7.6 The S- and O-branches of pure rotational Raman scattering spectra at 270 K and 280 K with $\nu_S < \nu_L$ and $\nu_S > \nu_L$, respectively (Q-branch at $\nu_S = \nu_L$ not shown). Also shown are two idealized bandpass filters with $\tau = 0.8$: L.T. from 35 to 80 cm^{-1} and H.T. from 90 to 150 cm^{-1}.

choose $\Delta\nu_{HT}$ to be between 90 and 150 cm^{-1} and $\Delta\nu_{LT}$ to be between 35 and 80 cm^{-1}, only one oxygen line ($J = 13$) and one nitrogen line ($J = 9$) fall between the two filters as can be seen in Fig. 7.6; this choice is then one that retains substantial scattering signal. Again, the sum of all line strengths including the Q-branch (not shown) for each molecular species is unity. Numerical calculation shows the attenuation at 275 K to be $\mathscr{R}^{aRR}_{Air}(\Delta\nu_{HT}; 275\ K) = 0.0359$ and $\mathscr{R}^{aRR}_{Air}$ $(\Delta\nu_{LT}; 275\ K) = 0.0605$.

7.2.3 Temperature Uncertainty for the Ratio Measurement

To assess the figure of merit for temperature measurement based on measuring the ratio of two spectrally integrated intensities, we need to estimate the resulting temperature uncertainty. Unlike the error propagation in the integration method, for a measurement based on observing an intensity ratio this process is quite straightforward. We first define the signal ratio R_T and calculate its relative uncertainty in terms of the temperature and pressure dependence of the atmosphere, as reflected in the attenuation factors intrinsic to the experimental technique. From this, we can assess the uncertainty (error) of the resulting temperature as a function of signal received. From (7.9.c), assuming for this brief discussion that η_1 and η_2 are constants, we express R_T and its variation as

$$R_T = \frac{N_1}{N_2} = \frac{\eta_1 f_1(T,P,z)}{\eta_2 f_2(T,P,z)} \to \frac{\delta R_T}{R_T} = \frac{\delta N_1}{N_1} - \frac{\delta N_2}{N_2} \to \delta\ \ln(R_T) = \delta\ \ln\left(\frac{N_1}{N_2}\right), \text{with} \tag{7.10.a}$$

$$\delta \ln(R_T) = \delta \ln\left(\frac{N_1}{N_2}\right) = \frac{\partial}{\partial T}\ln\left(\frac{f_1}{f_2}\right)\delta T + \frac{\partial}{\partial p}\ln\left(\frac{f_1}{f_2}\right)\delta p = S_{T12}\delta T + S_{P12}\delta P,$$
$$\therefore\ \delta T_{Air} = \frac{1}{S_{T12}}\delta \ln\left(\frac{N_1}{N_2}\right) = \frac{1}{S_{T12}}\sqrt{\frac{1}{N_1}+\frac{1}{N_2}};$$
$$S_{T12} = \frac{\partial}{\partial T}\ln\left(\frac{f_1}{f_2}\right) = \left(\frac{\delta f_1}{f_1}\right)_{\Delta T=1K} - \left(\frac{\delta f_2}{f_2}\right)_{\Delta T=1K}. \tag{7.10.b}$$

Here, the temperature measurement sensitivity S_{T12} is the difference in temperature sensitivity, defined as $S_{Ti} = (\delta f_i/f_i)$ per 1 K, for channel 1 (i=1) versus that for channel 2 (i=2). The pressure sensitivity is defined similarly as the difference in channel sensitivities, $S_{Pi} = (\delta f_i/f_i)$ per 1 Pa pressure change for the two channels. The pressure channel sensitivity is generally very small and, for rotational Raman scattering technique, it is in fact zero, as PRR depends only on atmospheric temperature. The variation of the observed signal ratio is the result of photon noise, which follows Poisson statistics and for which the two channels are independent; it can be expressed in terms of N_1 and N_2 as given above in (7.10.b). If we assume the received photocounts before physically dividing into two channels (with $\eta_1 + \eta_2 = 1$) is N_0, we may express the photocounts in terms of N_0, channel efficiencies, η_1 and η_2, and attenuation factors, f_1 and f_2. The photon variation and associated temperature error then become

$$S_{T12} = \frac{\partial}{\partial T}\ln\left(\frac{f_1}{f_2}\right);\ \delta \ln\left(\frac{N_1}{N_2}\right) = \sqrt{\frac{1}{N_1}+\frac{1}{N_2}} = \frac{\xi}{\sqrt{N_0}};\ \xi = \sqrt{\frac{1}{\eta_1 f_1}+\frac{1}{\eta_2 f_2}};$$
$$\delta T_{Air} = \frac{\xi}{S_{T12}\sqrt{N_0}}.$$
$$\text{For RRL, } S_{T12} = \frac{\partial}{\partial T}\ln\left(\frac{\mathscr{R}^{aRR}_{HT}}{\mathscr{R}^{aRR}_{LT}}\right);$$
$$\xi = \sqrt{\frac{1}{(\eta_1\tau_{HT})\mathscr{R}^{aRR}_{HT}} + \frac{1}{(\eta_2\tau_{LT})\mathscr{R}^{aRR}_{LT}}};\ \delta T_{Air} = \frac{\xi}{S_{T12}\sqrt{N_0^{PRR}}}. \tag{7.10.c}$$

Here, the parameter ξ is an important property of the lidar receiving system; we term it the single-photon uncertainty of the lidar measurement. In this sense, the intrinsic figure of merit of the lidar atmospheric temperature measurement technique is the ratio of single-photon uncertainty ξ to the measurement sensitivity, S_{T12}. We wish to make the former smaller and the latter bigger. In order to make a temperature measurement with small error bars, we also need to have a lidar with large PA product to increase the value of N_0. For the convenience of comparison with the Cabannes scattering (CS) process (to be considered), we use N_0 for CS and $N_0^{PRR} = 0.0342 N_0$ for PRR.

For the example depicted in Fig. 7.16, we have $\mathscr{R}_{Air}^{aRR}(\Delta\nu_{HT}, 275\text{ K}) = 0.0359$ and $\mathscr{R}_{Air}^{aRR}(\Delta\nu_{LT}, 275\text{ K}) = 0.0605$, as well as $\eta_1 = \eta_2 = 0.5$ and $\tau_{HT} = \tau_{LT} = 0.8$. With these values, we calculate the single-photon uncertainty to be $\xi = 10.5$. The one-channel temperature sensitivities for a 10 K difference (between 270 K and 280 K) were calculated for H.T. and L.T. to be $ST_{HT} = +0.00343$ and $ST_{LT} = -0.00182$, respectively, leading to temperature measurement sensitivity $ST_{12} = 0.52\%\text{ K}^{-1}$. If we have a modest lidar that could produce CS photocounts of $N_0 = 100,000,000$, the associated PRR-received photocount would be $N_0^{PRR} = 0.0342 N_0 = 3,420,000$, giving rise to $\delta T_{Air} = 1.1$ K.

7.2.4 A Combined Temperature Lidar for the Troposphere, Stratosphere, and Mesosphere

Since a typical separation between rotational Raman lines is more than ~ 2 cm^{-1} (or 60 GHz), they can be resolved with a laser with 10–20 GHz bandwidth. Because a multilayer dielectric narrowband filter does not reject elastic scattering from aerosols as well as an atomic vapor filter, one uses a transmitting laser with much narrower bandwidth. For example, the first rotational Raman lidar employed a laser with bandwidth of ~ 0.2 pm [7.29]. Ultra-narrowband injection-seeded pulsed lasers with < 0.1 GHz (< 0.003 cm^{-1}) are now commonly used for RRS lidars [30; 31], which allow the absolute laser wavelength to be specified to better than the 0.01-nm accuracy necessary to define the transmission functions of narrowband filters.

To give a realistic example of lidar atmospheric temperature profile measurements, we describe briefly a Rayleigh-rotational Raman lidar ($P_L A_R = 15.84$ Wm^2) developed by Behrendt et al. [7.31] at the Radio Atmospheric Science Center (RASC) in Kyoto, Japan. This lidar combines the downward Rayleigh integration method with simultaneous measurements of rotational Raman scattering to cover the troposphere, stratosphere, and mesosphere from about 2 km to 80 km in altitude. The schematic for this combined lidar is shown in Fig. 7.7. It employs eight custom-designed narrowband filter-beam splitters (BSs) with characteristics given in Table 2 of [7.31] and presented here in Table 7.1. These divide the return signal into five separate optical detection channels to achieve the multiple-purpose measurement. Channels 1 and 3 are elastic scattering channels detecting Rayleigh–Mie scattering for low and high altitudes, respectively; they are combined to assess aerosol parameters. The Rayleigh signal above 30 km (aerosol-free) in Channel 3 is

Table 7.1 *Optical properties of the filters used in the RASC lidar (adapted with permission from Table 2 of [7.31]. © the Optical Society.)*

Wavelength (nm)	Parameter	BS1	BS2	BS3	BS4a + BS4b Combined	BS5	IFa + IFb Combined
	AOI (deg)	45	45	4.8	5.0	7.2	3.5 & 0
	CWL (nm)			532.34	531.14	528.76	660.65
	FWHM (nm)			0.80	0.65	1.10	1.2
660		$\tau \approx 0.9$	$\rho > 0.95$				$\tau = 0.48$
532.25		$\tau \approx 0.9$	$\tau \approx 0.85$	$\tau = 0.82$	$\tau < 10^{-6}$	$\tau < 10^{-6}$	$\tau < 10^{-8}$
		$\rho \approx 0.05$		$\rho = 0.11$			
531.1		$\tau \approx 0.9$	$\tau \approx 0.85$	$\rho > 0.95$	$\tau \approx 0.72$	$\tau < 2 \cdot 10^{-4}$	
528.5		$\tau \approx 0.9$	$\tau \approx 0.85$	$\rho > 0.96$	$\rho > 0.96$	$\tau \approx 0.87$	

[a] τ, transmission; ρ, reflectivity. Transmission values $\tau < 10^{-3}$ are estimates from the manufacturers.

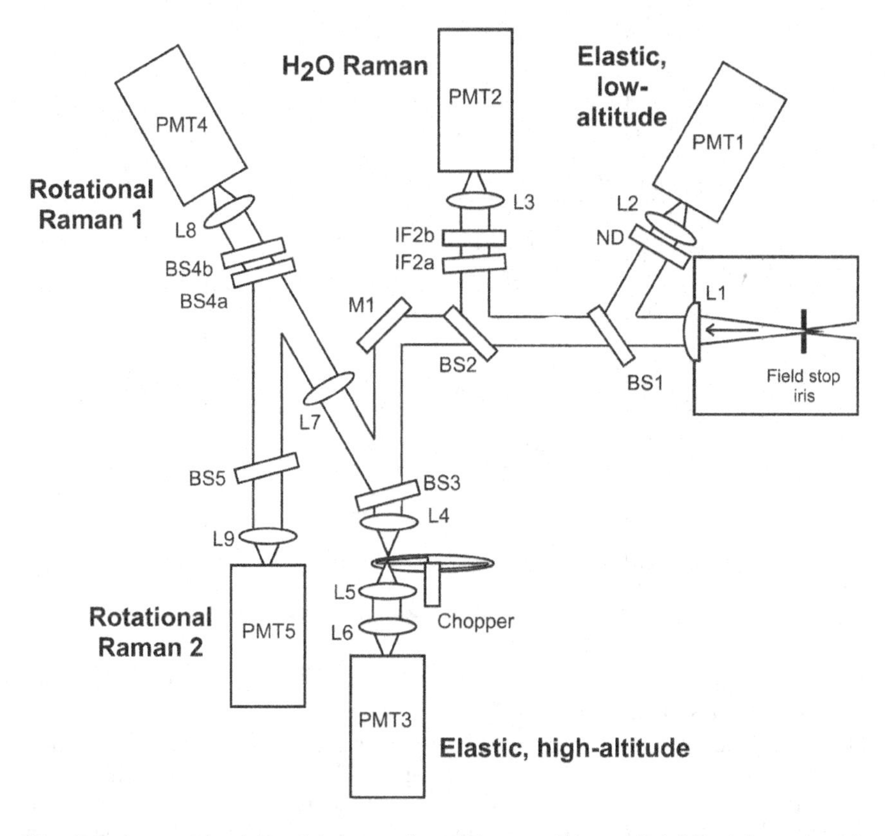

Fig. 7.7 A combined Rayleigh-rotational Raman lidar at RASC. Adapted with permission from Fig. 2 of [7.31]. © the Optical Society.

used for temperature measurements through the upper stratosphere and mesosphere with the integration method.

About 95% of rotational Raman signal is reflected from BS3 to Channels 4 and 5, where they are further filtered by BS5 and BS4a+BS4b for H.T. and L.T., passing the respective anti-Stokes PRR signal for the measurement of temperatures between 2 km and ~ 40 km. Channel 2 is for H_2O Raman density measurements (not discussed here).

An example of atmospheric temperature profile measurement with this lidar is shown in Fig. 7.8. Notice the upper end of rotational Raman and lower end of Rayleigh-integration measurements, between 30 and 40 km altitude, blend into one other nicely. They are in general agreement with local radiosonde measurements below 30 km and model predictions above 30 km. The central achievement of this lidar lies in its use of eight different custom-made narrowband filters, to form

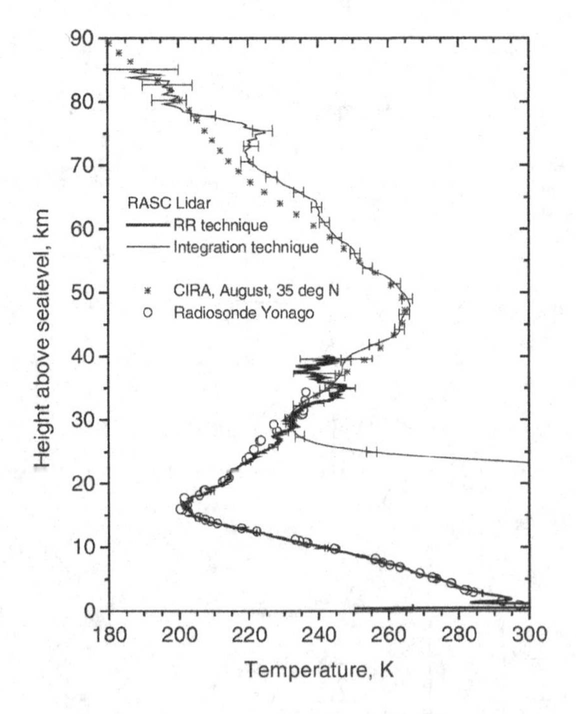

Fig. 7.8 Temperature measured with the RASC combined Rayleigh and Raman lidar on 9 and 10 August 2002, 23.15–00.27 Japan standard time (JST). Rotational Raman temperature values were derived with both analog detection (not shown) and photon-counting (solid curves) channels. Lidar data with a height resolution of 72 m were used up to 15-km height; the data between 15 and 20 km, 20 and 30 km, and above 30 km were smoothed with sliding average lengths of 360, 1080, and 2952 m, respectively. Reproduced with permission from Fig. 6 of [7.31]. © the Optical Society.

a network in a "polychromator," making it possible to measure atmospheric temperatures and aerosol optical properties from 2 km to 85 km, along with the important tropospheric H_2O density measurements in a single receiver beam train. Rotational Raman measurements with these filters were initially implemented at the GKSS Research Center in Germany [7.30] and with improved filters at the Radio Atmospheric Science Center (RASC) in Japan. They are compared in Fig. 7.9(a).

This excellent paper covers many detailed technical and theoretical considerations that are necessary for prudent lidar deployment. Readers who are interested in lidar deployment are advised to study [7.31] carefully. In the following list, we present selected salient features of this lidar with discussions, including a test utilizing realistic parameters in paragraph (5).

(1) To make real dielectric filters look more like the ideal ones with sharp cutoffs, two complementary filters are used. See filters BS4a+BS4b as an example (Fig. 7.9(a)). The optical properties of each filter in the RASC lidar polychromator are given in Table 7.1.

(2) The beauty of a custom-built interference filter (CIF) lies in the choice of its center wavelength, tailored to the application at hand. For example, the RASC lidar shifted the center wavelength of the H.T. filter, BS5, from near 529.5 nm for the GKSS lidar to 528.5 nm because the emphasis of the GKSS system was to detect condensation of polar stratospheric clouds (with temperature ~190 K), while the interest of RASC lidar was temperature profiling from the troposphere to the stratosphere. The filter choice for RASC was meant to minimize measurement uncertainties at temperatures near 240 K [7.31]. Using the center wavelength and FWHM given in Table 7.1, BS5 ($\tau_1 = 0.96 \times 0.87 = 0.835$, 528.5 nm, 0.80 nm) and BS4a+BS4b ($\tau_2 = 0.72$, 531.1 nm, 0.65 nm), the idealized representation of these filters is shown in Fig. 7.9(b) as two top-hat boxes. Comparing the optimized filter choices in Fig. 7.9(b) to the filter choices to increase photon signal as shown in Fig. 7.6, we see the former clearly places the L.T. above and the H.T. below the spectral peak in wavelength. This changes the temperature sensitivities, increasing ST_1 (H.T.) and decreasing ST_2 (L.T.) as temperature increases, thereby increasing the combined temperature sensitivity, ST_{12}, considerably. To compare these two filters sets quantitatively, we need to calculate both the temperature sensitivity ST_{12} and single-photon uncertainty, ξ, of both lidars. (We note that ST_{12} is the same as $Q^{-1}\partial Q/\partial T$ in Eq. (2) of [7.31], where channel 2 is the H.T. channel.) The computations were performed for the filters shown in Fig. 7.6, with $\mathscr{R}^{aRR}_{Air}(\Delta\nu_{HT}) = 0.0359$ and $\mathscr{R}^{aRR}_{Air}(\Delta\nu_{LT}) = 0.0605$, along with $\eta_1 = \eta_2 = 0.5$ and $\tau_1 = \tau_2 = 0.8$ (or $\tau^{eff}_{HT} = \tau^{eff}_{LT} = 0.4$), resulting in $\xi = 10.5$ and $ST_{12} = 0.52\%$ K^{-1} ($ST_1 = +0.00343$, $ST_2 = -0.00182$) as stated earlier. Using the same procedures for the filter choices shown in Fig. 7.9(b), with $\mathscr{R}^{aRR}_{Air}(\Delta\nu_{HT}) = 0.0149$ and

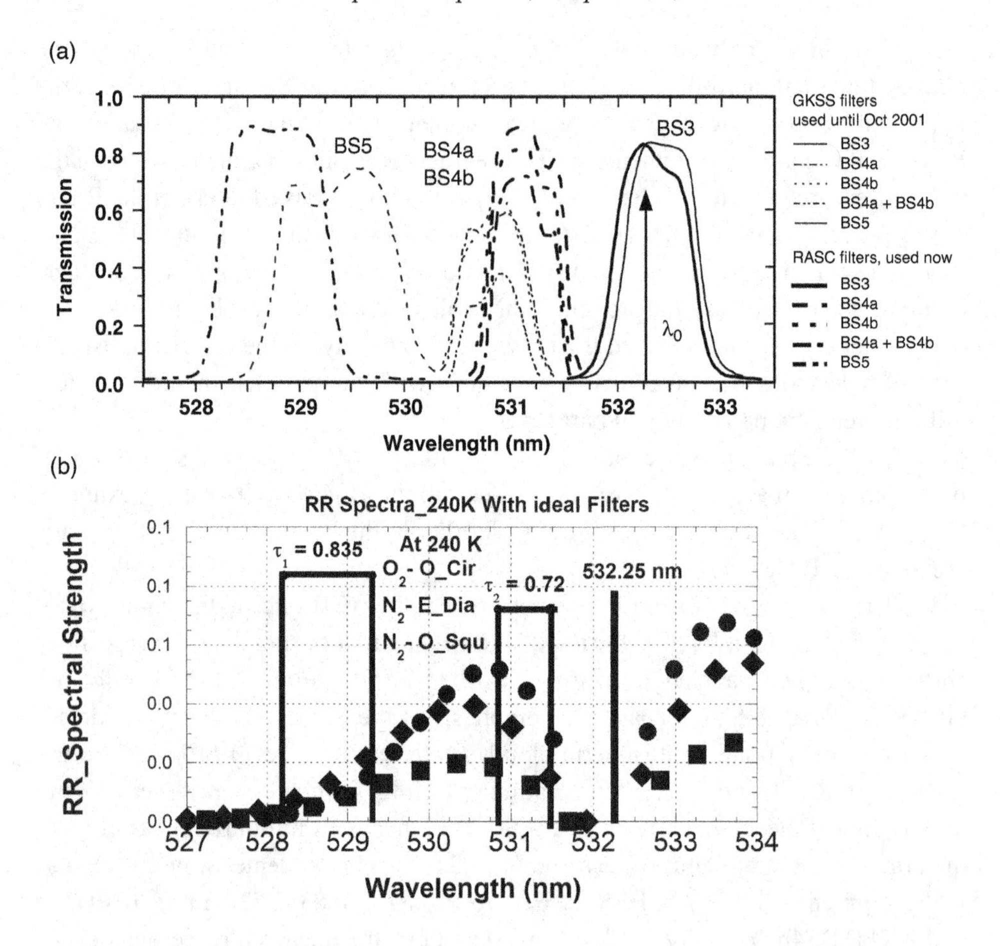

Fig. 7.9 (a) Transmission versus wavelength for two sets of bandpass filters BS3, BS4a-BS4b, and BS5 (in GKSS lidar and RASC lidar, respectively) used to extract the quasi-elastic signal and rotational Raman signals. Reproduced with permission from Fig. 3 of [7.31]. © the Optical Society. (b) Rotational Raman spectra along with idealized H.T. and L.T. to mimic BS5 and BS4a+BS4b, respectively. The laser wavelength λ_0 is assumed to be 532.25 nm as marked. Discussions follow in the text.

$\mathscr{R}_{Air}^{aRR}(\Delta\nu_{LT}) = 0.0578$, along with $\eta_1 = 0.96$, $\eta_2 = 1$ and $\tau_{HT} = 0.87$, $\tau_{LT} = 0.72$ (or $\tau_{HT}^{eff} = 0.835$, $\tau_{LT}^{eff} = 0.72$), results in $\xi = 10.2$ and $ST_{12} = 0.93\%\ \mathrm{K}^{-1}$ ($ST_1 = +0.00653$, $ST_2 = -0.00277$). For a lidar that produces $N_0 =$ 100,000,000 CS photocounts, the associated PRR received photocount would be $N_0^{PRR} = 0.0342 N_0 = 3,420,000$, it gives rise to $\delta T_{Air} = 1.1$ K for the CIFs of Fig. 7.6 and to $\delta T_{Air} = 0.59$ K for Fig. 7.9(b), where to cover a wide range of temperatures, the mean spectral characteristic at 200 K and 280 K is used in the calculation. Clearly, the optimization process of the CIFs in Fig. 7.9(b) increases measurement sensitivity greatly at little cost to the single-photon uncertainty, leading to nearly a twofold reduction of δT_{Air}. This improvement is in part due to an additional advantage of using the custom-built interference filters, which is that the condition $\eta_1 + \eta_2 = 1$ is no longer a limiting factor because the frequency ranges of the H.T. and L.T. filters are fully separated. In fact, for the case considered here, $\eta_1 + \eta_2 = 1.96$, helping to reduce the single-photon uncertainty.

(3) Because of saturation in photomultiplier tubes caused by high signal from lower altitudes, the combined lidar uses both analog and photon counting detection techniques in each of the low- and high-altitude channels for detecting elastic scattering, fostering true counting rates in lower altitudes for both the RRS lidar and aerosol parameter measurements. The RRS lidar was able to measure temperatures up to 7 km altitude with uncertainty under 1 K at nighttime, with time resolution of 5 min and height resolution of 72 m, this in the presence of a cloud layer with a backscatter ratio of 47.

(4) The stability problems in channel efficiency and filter functions encountered in early CS lidar, as experienced in [7.13] – to be elaborated later in Section 7.3.3 – and improved in later work [7.1], exists in principle in RRS lidars as well. These appear to be minimal in the deployed RRS lidars [7.30; 7.31], which have installed precision tilting mounts to tune the filter passbands, resulting in excellent mechanical stability in the polychromator. However, some calibration procedures may still be required and should be investigated on a case-by-case basis. We quote the following statements from published work: "The theoretical calculation of the method led to an analytic calibration function which, once adjusted with a radiosonde, can provide the temperature on successive days of measurement" in [7.29]; "For the temperature determination we calibrated the instrument with data from a local radiosonde" in [7.30]; and "We found the ratio of rotational Raman signal efficiencies experimentally by comparing the theoretical and experimental calibration functions for temperature measurements" in [7.31]. Thus, it appears to be prudent to test the stability of each lidar during operations.

(5) We now provide a reality check on our formulation of signal counts, (7.8), and error estimations, (7.7.e) and (7.10.c), by comparing our estimation with the measured errors. We choose to compare the measurement uncertainties at 70 km, as determined by the integration technique, and at 20 km by the RRS technique.

(a) We note from (7.8), the estimated signal photocount at 70 km for our reference lidar with $\tau_{int}\Delta z = 2$ hr km and $P_L A_R = 0.5$ W m^2, $A_R \approx 0.1$ m^2, that is, $P_L A_R \tau_{int} \Delta z = 3.6 \times 10^6$ Wm^{3}s and $\eta_m \exp[\ldots] = 0.01$, based on standard air with $\mathcal{N}(70\text{ km}) = 1.72 \times 10^{21}$ m^{-3}, calculated from (7.8), is $N_m = 1957$. Using the background count of $B = 3600$, we calculate a photon variation of 0.038. Using (7.7.e) with $T(70\text{ km}) = 219.6$ K from Fig. 7.5(a), the temperature error is

$\delta T_0 \sim 8.4$ K. We compare the RASC lidar ($P_L A_R = 15.84$ Wm^2 and $A_R = 0.53$ m^2) measurement shown in Fig. 7.8. If we assume that the integration time and vertical resolution are the same as the companion RRS lidar, that is, $\tau_{\rm int} = 1$ hr and $\Delta z = 2952$ m, we have $P_L A_R \tau_{\rm int} \Delta z = 1.68 \times 10^8 \mathrm{Wm^3s}$, giving rise to $N_m = 91,327$ and $B = 19,080$, from which we calculate photon variation to be 0.0036, leading to $\delta T_0 \sim 0.0036 \times 8.4/0.038 = 0.76$ K. The 1-σ error bar at 70 km, read from Fig. 7.8, is about 1.8 K, which is 2.4 times larger. In view of the uncertainties in the estimation, this difference has provided some confidence in our analysis.

(b) From Fig. 7 of [7.31], we deduce from the $\Delta T = 1$ K, $\Delta z = 360$ m curve an integration time of ~ 0.8 hr at $z = 20$ km. For this check (with $P_L A_R = 15.84$ Wm^2 and $A_R = 0.53$ m^2), we thus consider PRR scattering and insert $P_L A_R \tau_{\rm int} \Delta z = 1.64 \times 10^7 \mathrm{Wm^3s}$, $h\nu_L = 3.743 \times 10^{-22}$J, $\eta_m \exp[\ldots] = 0.01$, $d\sigma(PRR)/d\Omega = 0.0342 \times 5.96 \times 10^{-32} \mathrm{m^2/sr}$, and $\mathcal{N}(20\,\mathrm{km}) = 1.85 \times 10^{24}$ m^{-3} into (7.8) and obtain $N_0^{PRR} = 4.15 \times 10^9$. At 20 km, the contribution from background (scaled to be $B = 810$) may be ignored. Since the single-photon uncertainty and % sensitivity is $\xi = 10.2$ and 0.93% K^{-1}, respectively, the estimated temperature uncertainty is $\delta T_{air} = 1020/(0.93 \times \sqrt{4.15 \times 10^9}) = 0.017$ K. Because of the complexity of the system designed to perform multiple tasks (see Fig. 7.7), the receiver throughput is much reduced by the time the received scattered light gets to the end of the receiver chain. Still, it is difficult to explain the two orders of magnitude difference between the estimated and the observed temperature errors.

7.3 HSRL Profiling of Aerosol Optical Properties and Atmospheric State Parameters

In 1983, researchers at the University of Wisconsin first reported HSRL measurements of aerosol optical properties employing a fixed Fabry–Perot étalon (FPE) [7.18]. That same year, the idea of using an atomic vapor filter (AVF) to block aerosol scattering was proposed [7.11] by researchers at Colorado State University (CSU). They realized that the transmittance spectrum of an AVF is much more like an ideal notch filter than the bandpass function of a Fabry–Perot étalon. Citing better blocking of aerosol scattering with AVFs, the CSU group also proposed atmospheric temperature measurement by comparing the backscattering signal passing through two AVFs with different blocking widths, thus sampling different portions of the Cabannes spectrum. In Fig. 7.10 we show a simplified generic schematic that could be used for several experiments (as deployed by the CSU group between 1983 and 2001) to retrieve aerosol and atmospheric state parameter profiles (temperature, pressure, and wind) from lidar signals. Here, a linearly x-polarized single-frequency pulsed laser with bandwidth $\lesssim 200$ MHz with center frequency locked to an atomic or molecular transition is transmitted into the troposphere. Shown in the middle of the figure is a sketch of the spectrum of scattered light, which includes both Mie scattering from aerosols and Rayleigh scattering from atmospheric molecules. The central peak is the sum of Mie scattering and the Q-branch of pure rotational Raman (PRR) scattering, riding on the top of the Cabannes spectrum pedestal with FWHM ~3 GHz. Further out in the wings are the O- and S-branches of PRR with spectral lines between 4 and 100 cm^{-1} (1 cm^{-1} = 30 GHz), see Fig. 4.4 for details.

The main lidar receiver layout (without the FPE and polarizing beam splitter, PBS, in the beam path) shows the undifferentiated detection of both parallel (*x*-) and perpendicular (*y*-) polarizations in three separate channels. For different experiments, either a 100% reflecting mirror (RM) or a non-polarizing beam splitter (NPBS) is inserted into positions marked $\underline{3}$ and $\underline{2}$. We consider four different experiments. When the PBS is not inserted, we employ a NPBS for simplicity; if a dielectric beam splitter is employed instead, the difference between the *s*-polarized and *p*-polarized transmittance and reflectance must be accounted for in the data analysis [7.34]. With a RM in position $\underline{3}$ and inserting nothing in position $\underline{2}$, we have experiment A with one detection channel and detector (DET 2), which yields total and molecular scattering signals, respectively, by tuning the laser off- and on-resonance of the filter absorption line. By inserting a NPBS (say 30%T, 70% R) into position $\underline{2}$ followed by another RM in position $\underline{1}$, we have experiment B with two detection channels (DET 1 and DET 2). This experiment yields two separate total and molecular scattering signals, respectively, by tuning the laser off- and on-resonance of the filter absorption line. Using lidar returns with the transmitting frequency alternating between on and off (a shift of ~10 GHz) of the filter center position (at an atomic transition), experiments A and B allow, respectively, aerosol parameter [7.19] and both aerosol and atmospheric parameter measurements [7.12; 7.35]. With a NPBS (say 10%T, 90%R) in $\underline{3}$, we can then perform experiment C by inserting nothing in position $\underline{2}$. In this case, the laser transmitting frequency is fixed to the filter center position, so that DET 3 receives the total and DET 2 the molecular scattering signal. By inserting a NPBS (say 30%T, 70%R) into position $\underline{2}$ and another RM in position $\underline{1}$, we have experiment D. This version also fixes the laser on the absorption transition, and it receives one total scattering and two molecular scattering signals.

As to be explained below, experiments A and B are appropriate when atomic filters like barium AVF are used, while C and D are more applicable when molecular filters, such as iodine (IVF), are employed. (It is difficult to tune sufficiently off-resonance with the IVF without running into another absorption line.) If a Fabry–Perot étalon (FPE) filter (~ 60 pm in width) is inserted between the

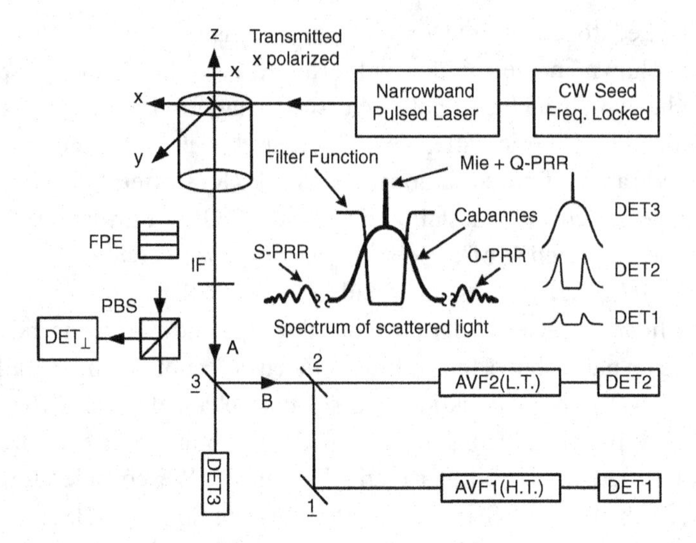

Fig. 7.10 A simplified generic schematic could be used to present several experiments (from the CSU group between 1983 and 2001) for aerosol and atmospheric parameter measurements using an AVF (for experiments A and B) or an IVF (for experiments C and D) as the frequency discriminator. For details, see text.

telescope and the interference filter (IF), then the same experiments would proceed with the O- and S-branches of PRR scattering excluded. If a PBS and detector ($DET_{\perp}$) are inserted from the left between the IF and position 3, with or without the FPE, parallel- and perpendicular-polarized scattered light are then separated [7.1]. Then, the experiment can deduce the linear polarization ratio of total scattering (in the troposphere due mostly to aerosols).

7.3.1 Barium Vapor Filter for Profiling Tropospheric Optical Properties: Experiment A

The singlet $(\mathrm{x}s)^2 - (\mathrm{x}s\mathrm{x}p)$ transitions in alkali Earth metals behave as two-level systems, and at a strong dipole-allowed transition an alkali in vapor phase can be used to construct a notch-like filter. The $(6s)^2\ {}^1S_0 - (6s6p)\ {}^1P_1^o$ transition in barium at 553.7 nm is a good example. In order to obtain high enough vapor density, the barium needs to be heated to a range between 740 K and 820 K. To this end, the vapor is contained in a stainless-steel tube (38 cm long) built into a heat pipe oven design and including a buffer gas. In this temperature range (below the melting point of barium), it is not technically a heat pipe [7.34]. However, by first heating the tube to ~ 1000 K and rotating it to distribute liquid barium throughout a mesh lining the inner circumference of the tube, the vapor density is reasonably uniform at

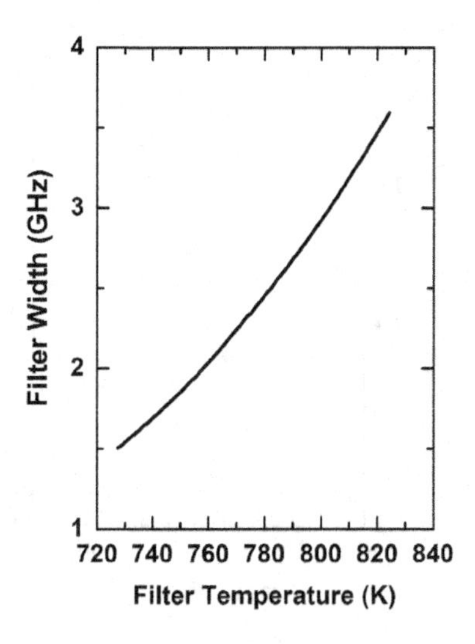

Fig. 7.11 Best fit to 350 measured filter widths as a function of vapor temperature over a period of two years. Reproduced with permission from Fig. 9 of [7.34]. © Raul J. Alvarez II.

operating temperature within the heated region (7.6 cm long) and drops off sharply outside of the heated area. The barium vapor filter is characterized by its measured FWHM. Fig. 7.11 shows 350 filter widths as a function of oven temperature [7.34], ranging from 1.7 to 3.7 GHz. These were measured over a period of two years with different thermocouple contacts to the filter tube; despite the associated systematic temperature uncertainties, a typical uncertainty in filter width was about 16 MHz in five consecutive runs at the same temperature setting [7.34]. Two barium filter transmission functions, with FWHM of 2.0 and 2.2 GHz, are shown in Fig. 7.12.

Despite the need for high-temperature operation with its attendant bandwidth fluctuations, this filter presents distinct advantages over other AVFs, particularly I_2. The lack of additional nearby absorption lines allows the laser frequency to be switched quickly between the absorption center ("locked" or "on-frequency") and an offset more than 10 GHz away ("off-frequency"). This allows the comparison between molecular and total (aerosol and molecular) scattering without worrying about the difference in detection efficiency between the on- and off-frequency channels. In experiment A, both polarizations of the scattered light propagate through the same optical path via AVF2 to Det2. We can derive the lidar equations, respectively, from (7.2.a) by replacing η_τ with η_2 for the total scattering channel 2off-frequency, and from (7.2.b) by replacing η_m with $\eta_2 f_{m2}(T, P, z)$ for the molecular scattering channel 2on-frequency:

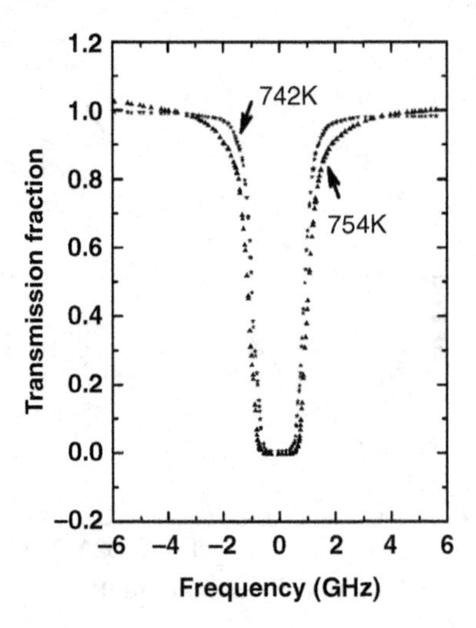

Fig. 7.12 The transmission functions of a barium filter at two vapor temperature settings (stars for 742 K and triangles for 754 K) with FWHM about 2 GHz. Reproduced from Fig. 2 of [7.19]. © American Meteorological Society. Used with permission.

$$\widetilde{N}_{2off}(\nu_L, z) = \frac{h\nu_L z^2}{A_R E_L z_B} N_{2off}(\nu_L, z; r_B) = \eta_2[\beta_a(\nu_L, z) + \beta_C(\nu_L, z) + \beta_R(\nu_L, z)] \times \exp\left[-2\int_{r_0}^{r} \alpha(\nu_L, z')dz'\right]; \tag{7.11.a}$$

$$\widetilde{N}_{2on}(\nu_L, z) = \frac{h\nu_L z^2}{A_R E_L z_B} N_{2on}(\nu_L, z; r_B) = \eta_2 f_2(T, P, z)[\beta_C(\nu_L, z) + \beta_R(\nu_L, z)] \times \exp\left[-2\int_{r_0}^{r} \alpha(\nu_L, z')dz'\right]. \tag{7.11.b}$$

Here, though only AVF2 is used for experiment A, we note that $f_i(T, P, z)$ is the attenuation factor for all AVFi ($i = 1$ or 2) when the laser is at the on-frequency. It is either $f_{Ci}(T, P, z)$ or $f_{Ci}(T, P, z) + \gamma_R$ with $\gamma_R \equiv \beta_{RR}(\nu_L, z)/\beta_C(\nu_L, z) = 0.0257$, depending on whether or not the O- and S-branches of PRR scattering are blocked by an ultranarrow filter: an FPE [7.1] or a daystar filter (see Fig. 3 of [7.13]). Also, we denote $\beta_R(\nu_L, z)$ and $\beta_{RR}(\nu_L, z)$, respectively, as backscattering coefficients of all and the sum of O- and S-branches of PRR spectra, and $\beta_C(\nu_L, z)$ as the backscattering coefficient of the Cabannes spectrum; they are related by $\beta_R(\nu_L,z)/\beta_C(\nu_L,z) = 0.22 \times 7/45 = 0.0342$ and $\gamma_R = 0.22 \times 0.75 \times 7/45 = 0.0257$ for dry atmosphere. The filter attenuation factor for Cabannes backscattering, $f_{Ci}(T,P,z)$, as given in (4.27), is the frequency integration of the product of the filter function $F_i(\nu)$ and the normalized Cabannes spectral density, $\mathscr{R}(\nu;T,P)$, where ν (instead of $\Delta\nu$) is now the frequency relative to laser lock position:

$$f_{Ci}(T,P,z) = \int_{-\infty}^{\infty} \mathscr{R}(\nu;T,P)F_i(\nu)d\nu; \quad \int_{-\infty}^{\infty} \mathscr{R}(\nu;T,P)d\nu = 1\,, \qquad (7.11.c)$$

where $F_i(\nu)$ should be replaced by $F_i(\nu)D(\nu)$ where $D(\nu)$ is the filter function of the ultranarrow FPE or Daystar filter, if either is used. We follow [7.13] and use the normalized Cabannes function $\mathscr{R}(\nu;T,P)$ instead of the un-normalized function $R(\nu;T,P)$ in earlier publications [7.12; 7.36; 7.37]; they are related by $\mathscr{R}(\nu;T,P) = f_o^{-1}R(\nu;T,P)$ with $f_o = (2/\lambda)\sqrt{2k_BT/m}$ as given after (4.27) for backscattering with $C_0(x,y)$ replaced by $R(\nu;T,P)$.

The effectiveness of measuring backscatter ratio profiles with a barium filter has been clearly demonstrated by Alvarez et al. in 1990 [7.19], where in Fig. 3(a), they show a raw lidar profile with a cloud layer near 5 km detected through the barium cell. When the laser is tuned from off-resonance to on-resonance, the cloud layer instantly disappears from an otherwise identical profile, seen in Fig. 3(b) of [7.19]. Using either of the filters, whose transmissions are shown in their Fig. 7.12, the backscatter ratio profiles can be obtained from the ratio of (7.11.a) to (7.11.b), independent of detection efficiency (because the same DET 2 is used for the total and molecular signals) as

$$\mathbf{R}(\nu_L,z) = \frac{\beta_a(\nu_L,z)+\beta_C(\nu_L,z)+\beta_R(\nu_L,z)}{\beta_C(\nu_L,z)+\beta_R(\nu_L,z)} = \frac{f_2(T,P,z)\widetilde{N}_{2\,off}(\nu_L,z)}{\widetilde{N}_{2\,on}(\nu_L,z)}\,, \qquad (7.11.d)$$

where $f_2(T,P,z)$ is $f_{C2}(T,P,z)$ or $f_{C2}(T,P,z)+\gamma_R$ with $\gamma_R \equiv \beta_{RR}(\nu_L,z)/\beta_C(\nu_L,z)$. The normalized Cabannes spectrum $\mathscr{R}(\nu;T,P)$ required for (7.11.c) is calculated with (T,P) either measured simultaneously as in Experiment B, or from a standard atmosphere table.

One interesting set of profiles from [7.19] is reproduced as Fig. 7.13. It shows a high backscatter ratio value of about 1.7 throughout the profile, likely due to high water vapor content in the air, which swells the aerosols. Ten minutes later the cloud layer disappears completely, but a high backscatter ratio and the bump near 1 km persist, as shown in the 22:59 LST profile. The subsequent profiles show similar aerosol structure but with a much lower backscatter ratio of 1.2 beyond 1 km, indicating restored normal Colorado dry air conditions. Either filter shown in

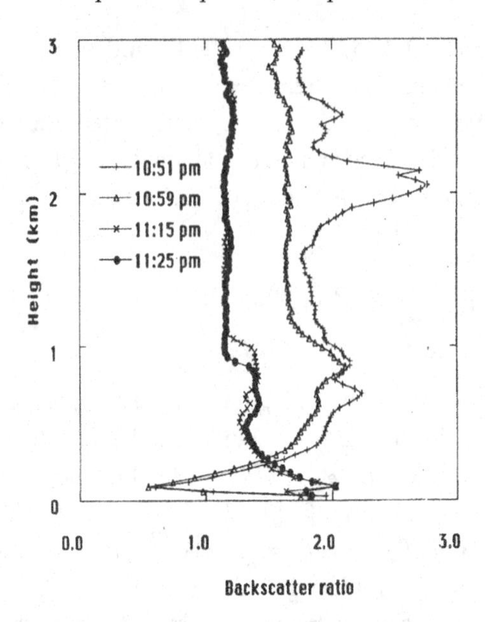

Fig. 7.13 Backscattering ratio for runs between 22:51 and 23:25 LST October 1989 show considerable dynamic variations in atmospheric content (see text). Reproduced from Fig. 6 of [7.19]. © American Meteorological Society. Used with permission.

Fig. 7.12 can be placed in channel 2. As long as $F_i(\nu)$ and the profile of atmospheric (T, P) are provided for $\mathscr{R}(\nu; T, P)$, the appropriate $f_2(T, P, z)$ may be calculated with (7.11.c), and the backscatter ratio profiles can be determined from (7.11.d).

7.3.2 Use of Barium Vapor Filters for Profiling Tropospheric Temperature: Experiment B

For experiment B, two vapor filters are used: the high-temperature filter, AVF 1 and the low-temperature filter, AVF 2, respectively with FWHM of 3 GHz and 2 GHz [7.19], similar to those in Fig. 7.12. By tuning the laser on- and off-resonance of the 553.7 nm barium transition, the two physical channels become four electronic channels with return signals given by (7.11.a) and (7.11.b) with subscript 2 replaced by i, with values of 1 and 2 for each set of off- and on-resonance signals detected, respectively, by Det1 and Det2. Since the atmospheric extinction coefficient and the backward total scattering cross section are the same for both channels, the ratio of off-resonance signals is equal to the ratio of their detection efficiencies. For a set of on- and off-resonance (only 10 GHz apart) signals on the same channel, the ratio of the on- and off-resonance backscattering signals, termed C_i following [7.12] (not to be confused with $C_0(x, y)$), will be independent of the detection efficiency as given in (7.12.a), since they are through the same filter. The ratio C_1/C_2 is then the ratio of the filtered molecular scattering to the total scattering of Det1 to that of Det2, leading to the equation for the determination of atmospheric T and P profiles, as that given in (7.12.b):

$$C_i(z) = \frac{\widetilde{N}_{ion}(z)}{\widetilde{N}_{ioff}(z)} = \frac{N_{ion}(z)}{N_{ioff}(z)} = \frac{\beta_C(\nu_L, z)[f_{Ci}(T, P, z) + \gamma_R]}{\beta_a(\nu_L, z) + \beta_m(\nu_L, z)} \rightarrow \tag{7.12.a}$$

$$\frac{C_1(z)}{C_2(z)} = \frac{N_{1on}(z)/N_{2on}(z)}{N_{1off}(z)/N_{2off}(z)} = \frac{f_{C1}(T, P, z) + \gamma_R}{f_{C2}(T, P, z) + \gamma_R} \equiv \frac{f_1(T, P, z)}{f_2(T, P, z)}; \; \frac{N_{1off}(z)}{N_{2off}(z)} = \frac{\eta_1}{\eta_2}, \tag{7.12.b}$$

where $f_i(T, P, z)$, with $i = 1$ or 2, is the attenuation factor of AVFi when the laser is at the on-resonance frequency, and $\gamma_R = 0$ if a FPE is inserted.

To facilitate data processing and numerical calculation of $f_i(T, P, z)$, we expand the normalized Cabannes spectrum $\mathscr{R}(\nu; T, P)$ in a Taylor series around a standard point with $T_o = 275$ K and $P_o = 0.75$ atm $=$ 76 kPa up to the second order:

$$\begin{aligned}\mathscr{R}(\nu; T, P) &= \mathscr{R}_{\mathrm{std}}(\nu) + (T - T_o)\mathscr{R}_T(\nu) + (P - P_o)\mathscr{R}_P(\nu) \\ &+ 0.5(T - T_o)^2\mathscr{R}_{TT}(\nu) + (T - T_o)(P - P_o)\mathscr{R}_{TP}(\nu) + 0.5(P - P_o)^2\mathscr{R}_{PP}(\nu),\end{aligned} \tag{7.12.c}$$

where $\mathscr{R}_\mu(\nu)$ with subscripts μ as *std, T, P, TT, TP,* and *PP,* are, respectively, the function and the first- and second-order partial derivatives of $\mathscr{R}(\nu; T, P)$ with respect to T and/or P all evaluated at the standard point. Eq. (7.12.c) is valid for typical near-ground conditions (see Table 4.5 for example). We can then evaluate the attenuation function $f_{Ci}(T, P, z)$ in (7.11.b) in terms of the expansion of the corresponding integrals for each AVF as

$$\begin{aligned}f_{Ci}(T, P, z) &= f_{stdi} + (T - T_o)f_{Ti} + (P - P_o)f_{Pi} + 0.5(T - T_o)^2 f_{TTi} \\ &+ (T - T_o)(P - P_o)f_{TPi} + 0.5(P - P_o)^2 f_{PPi},\end{aligned} \tag{7.12.d}$$

with $f_{\mu i} \equiv \int \mathscr{R}_\mu(\nu)F_i(\nu)d\nu$; $F_i(\nu)$ includes the effect of the FPE, if inserted; $i = 1$ or 2.

Note that, expressing temperature and pressure in the units of K and Pa, f_{std} is dimensionless, and the units for f_T, f_P, f_{TT}, f_{TP} and f_{PP} are, respectively, K^{-1}, Pa^{-1}, K^{-2}, $\mathrm{K}^{-1}\,\mathrm{Pa}^{-1}$ and Pa^{-2}. A comprehensive theoretical study [7.35] has determined that terminating the Taylor series in its second order is more than accurate enough for the retrieval intended here. With a known pressure at a chosen reference height, (7.12.b) with (7.12.c) substituted becomes a quadratic equation in temperature, which can then be solved for the temperature at height z. The atmospheric pressure at the neighboring height (range bin) may then be calculated by assuming hydrostatic equilibrium and the ideal gas law. Successive iterations yield vertical atmospheric temperature and pressure, and thus density $\mathcal{N}(z)$ profiles. For tropospheric atmospheric conditions, a pair of barium filters with widths of 2 and 3 GHz is a reasonable choice. The parameters for a pair of filters of widths 2.99 GHz and 2.04 GHz are listed in Table 2 of [7.35]: the attenuation factors $f_{std1} = 0.1951$ and $f_{std2} = 0.4644$, one-channel sensitivities 0.59% K^{-1} and 0.41% K^{-1}, lead to a temperature measurement sensitivity of $S_{T12} \approx 0.18\%\ K^{-1}$.

By expressing $\alpha(\nu_L, z')$ as $\alpha_a(\nu_L, z') + \mathcal{N}(z)\sigma^R(\nu_L)$ in (7.11.b), we can solve for aerosol extinction coefficient and aerosol lidar ratio as

$$\alpha_a(\nu_L, z) = \frac{1}{2}\frac{d}{dz}\ln\left\{\frac{\eta_i f_i(T,P,z)\beta_C(\nu_L,z)}{\widetilde{N}_{ion}(\nu_L,z)}\right\} - \mathcal{N}(z)\sigma^R(\nu_L)\text{ , and} \tag{7.12.e}$$

$$L_a(\nu_L, z) = \frac{\alpha_a(\nu_L,z)}{\alpha_m(\nu_L,z)} = \frac{1}{2}\left\{[\mathcal{N}(z)\sigma^R]^{-1}\frac{d}{dz}\ln\left[\frac{\eta_i f_i(T,P,z)\beta_c(\nu_S,z)}{\widetilde{N}_{ion}(\nu_S,z)}\right] - 2\right\}, \tag{7.12.f}$$

where the constant η_i may be ignored as in (3) of [7.12], and $\sigma^R(\nu_L)$ is the Rayleigh scattering cross section. Since the atmospheric state parameters, T, P, $\mathcal{N}(z)$, at each altitude are now determined, both $f_i(T,P,z)$ and $\beta_C(\nu_L,z) = \mathcal{N}(z)\left(d\sigma^C(\nu_L)/d\Omega\right)$ can be calculated, leading to the profile of the aerosol extinction and aerosol lidar ratio, respectively, by computing (7.12.e) and (7.12.f). Though these may be determined from either channel 1 or 2, use of low-temperature AVF (channel 2) should result in smaller photon noise, and thus smaller error bars.

In a demonstration experiment performed in February 4–5, 1991 [7.12], three temperature and aerosol extinction coefficient profiles along with a nearby balloonsonde launched at the time marked are shown in Fig. 7.14. The lidar temperature profiles appear to be somewhat lower than the balloonsonde profile, but they are in general agreement. The reference height was chosen at 1.5 km, and the pressure from the balloonsonde at this level was used for the lidar inversion. The steep rise below 1 km is an artifact that is due to photon counter-saturation in low-altitude bins. To minimize this problem at closer range, the energy per laser pulse is reduced to 150 μJ. Each lidar profile was deduced from signal integrated over 20 min, and vertically smoothed to an effective resolution of 375 m. If rotational Raman scattering is ignored (i.e., setting $\gamma_R = 0$), the same lidar signal would lead to a temperature determination 45 K too high [7.35] with this set of AVFs. Were the FPE included in Fig. 7.10, the PRR would have been experimentally removed.

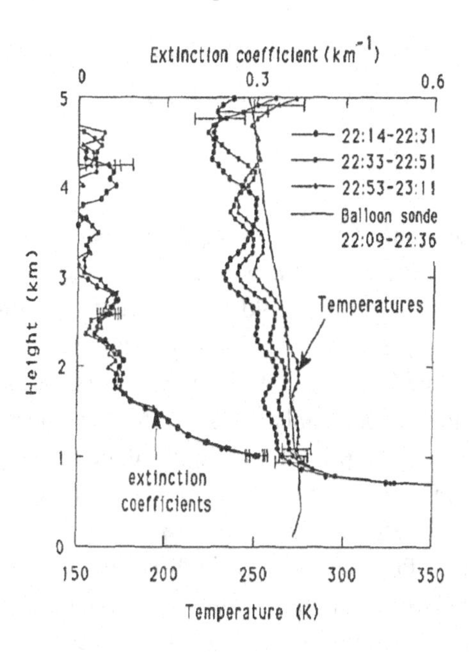

Fig. 7.14 Temperature and aerosol extinction coefficient profiles taken on February 4–5, 1991, along with a nearby balloonsonde. Reproduced with permission from Fig. 2 of [7.12]. © the Optical Society.

We now expand the discussion of the uncertainties in atmospheric temperature measurement as expressed in (7.10.a) – (7.10.c) by including the possible variations in η_1 and η_2, that is, $\delta\ \ln(\eta_1/\eta_2)\neq 0$, along with those resulting from our inability to maintain a stable vapor filter (temperature), that is, $\delta\ln(f_1/f_2)\neq 0$. Therefore, not only is the variation of R_T a function of temperature and pressure because of atmospheric scattering, it also depends on quadrature sum of the independent variations in efficiency η_1/η_2 and attenuation factor f_1/f_2 in addition to the photon statistics, $\xi/\sqrt{N_0}$. With this generalization (7.10.a) – (7.10.c) becomes:

$$\delta\ \ln(R_T)=\delta\ \ln\left(\frac{N_1}{N_2}\right)\rightarrow\delta\ \ln\left(\frac{N_1}{N_2}\right)+\delta\ \ln\left(\frac{\eta_1}{\eta_2}\right)+\delta\ \ln\left(\frac{f_1}{f_2}\right)=S_{T12}\delta T+S_{P12}\delta p \tag{7.13.a}$$

$$\delta\ \ln(R_T)=S_{T12}\delta T+S_{P12}\delta p:\ S_{T12}=\frac{\partial}{\partial T}\ln\left(\frac{f_1}{f_2}\right)=\left(\frac{\delta f_1}{f_1}-\frac{\delta f_2}{f_2}\right)_{\Delta T=1K},$$

$$S_{P12}=\left(\frac{\delta f_1}{f_1}-\frac{\delta f_2}{f_2}\right)_{\Delta P=1Pa} \tag{7.13.b}$$

$$\therefore \delta T_{Air} = \frac{1}{S_{T12}}\sqrt{\frac{\xi^2}{N_0} + \left(\frac{\delta\eta_1}{\eta_1}\right)^2 + \left(\frac{\delta\eta_2}{\eta_2}\right)^2 + \left(\frac{\delta f_1}{f_1}\right)^2 + \left(\frac{\delta f_2}{f_2}\right)^2}; \xi = \sqrt{\frac{1}{\eta_1 f_1} + \frac{1}{\eta_2 f_2}} \tag{7.13.c}$$

where the temperature sensitivity and pressure sensitivity, S_{T12} and S_{P12}, are defined as in (7.13.b). Since Cabannes scattering is pressure dependent in the atmospheric range considered, it should be accounted for throughout that range, while taking the effect of pressure variation at one altitude to be negligible. For the vapor filter pair considered, a relatively large atmospheric pressure change (~7.7 kPa) yields a temperature error of only ~ 1 K [7.35], thus the pressure term in (7.13. a) may be ignored; this leads to the temperature uncertainty given in (7.13.c). As compared with the temperature uncertainty, δT_{Air}, in (7.10.c), the effects of two more variations (channel efficiencies $\delta \ln(\eta_1/\eta_2)$ and filter stabilities $\delta \ln(f_1/f_2)$) are included in (7.13.c). For experiments with barium filters, since the efficiency ratio is measured in real time with the off-resonance count ratio as given in (7.12.b), $\eta_1/\eta_2 = N_{1off}(z)/N_{2off}(z)$, we can set $\delta \ln(\eta_1/\eta_2)$ to zero in (7.13.c) for the calculation of 1-σ temperature error, δT_{Air}.

For the observations presented in Fig. 7.14, the one-channel temperature sensitivities for the 2.04 GHz and 2.99 GHz filters are respectively measured to be 0.0041 and 0.0059 (see Table 2 of [7.35]); the sensitivity of the measurement is 0.18% K^{-1}. The instability in the barium filter is reflected in the fluctuations in the FWHM of the filter function $\Delta\nu_f$, which in turn give rise to fluctuations in the attenuation function $f_i(275\text{ K}, 76\text{ kPa})$. From ~ 700 measured barium filter functions, $F(\nu)$, we have obtained a relationship for $f_i(275\text{ K}, 76\text{ kPa})$ as a function $\Delta\nu_f$, from which we can calculate $\delta f_i/f_i$ using (19) of [7.35], giving rise to $\delta \ln(f_1/f_2)$ of ~1.5% for the filter pair. This may be added in quadrature to the photon fluctuations in signal ratio $\delta \ln(N_1/N_2)$, 0.3% and 2.2%, respectively, at 1 and 5 km. Using $S_{T12} \approx 0.2\%\ K^{-1}$, the temperature uncertainties were assessed at one standard deviation to be 8 and 14 K at 1 km and 5 km, respectively [7.12; 7.36].

7.3.3 Use of Iodine Vapor Filter for Optical and State Parameters: Experiments C and D

Mainly because of the high fluctuations in barium filter transmission, $\delta \ln(f_1/f_2)$, along with the convenience of matching the frequency-doubled Nd:YAG laser wavelength at 532 nm owing to the existence of many lines within the laser tuning range of 3 cm^{-1} (90 GHz) – see Fig. 7.2 – we turn to iodine filters for aerosol and atmospheric parameter measurements [7.13; 7.40]. However, the density of I_2

absorption lines makes it impossible to tune the laser ~10 GHz off-resonance without bumping into a neighboring line. This prevents us from using the same channel for sequentially receiving both total and molecular scattering and using the ratio of off-resonance (total) scattering of two channels for a real-time measurement of the ratio of channel efficiencies. Thus, the variation of count ratio $R_T = N_1/N_2$ depends on both the variation of (η_1/η_2) and (f_1/f_2) as described in (7.13.a). Though the vapor temperature to produce enough iodine vapor is much lower (<350 K), filter stability cannot be assumed, and transmission errors still exist.

Since, for the system depicted in Fig. 7.10, there exists generally a freedom to choose the values of the efficiencies η_1 and η_2, subject to the condition $\eta_1 + \eta_2 = 1$, we would do so in such a way as to minimize the single-photon uncertainty ξ. This process gives rise to the optimum efficiencies η_1^{opt} and η_2^{opt}, and single-photon uncertainty ξ_{opt} [7.37] as:

$$\eta_1^{opt} = \frac{\sqrt{f_2}}{\sqrt{f_2}+\sqrt{f_1}};\ \eta_2^{opt} = \frac{\sqrt{f_1}}{\sqrt{f_2}+\sqrt{f_1}} \rightarrow \xi_{opt} = \frac{1}{\sqrt{f_2}} + \frac{1}{\sqrt{f_1}}. \tag{7.14.a}$$

Furthermore, if the off-resonance transmission is not unity, for which iodine is a case in point due to the presence of continuum absorption (to be discussed in more detail later in this section), then it is more convenient to write the filter function as $F_i(\nu) = \tau_i F_i^n(\nu)$, where $F_i^n(\nu)$ is a filter with unity off-resonance transmission, giving rise to

$$f_i(T,P,z) = \int_{-\infty}^{\infty} \mathscr{R}(\nu;T,P)F_i(\nu)d\nu = \tau_i \int_{-\infty}^{\infty} \mathscr{R}(\nu;T,P)F_i^n(\nu)d\nu = \tau_i f_i^n(T,P,z)\,. \tag{7.14.b}$$

The three photocount signals (all on-resonance) may now be written with $i = 1$ and 2 for molecular channels as in (7.15.a) and channel 3 for the total scattering in (7.15.b) as

$$\widetilde{N}_i(\nu_L,z) = \frac{h\nu_L z^2}{A_R E_L z_B} N_i(\nu_L,z;r_B) = \eta_i f_i(T,P,z)\beta_C(\nu_L,z)\exp\left[-2\int_{r_0}^{r}\alpha(\nu_L,z')dz'\right],\ \text{and} \tag{7.15.a}$$

$$\widetilde{N}_3(\nu_L,z) = \frac{h\nu_L z^2}{A_R E_L z_B} N_3(\nu_L,z;r_B) = \eta_3[\beta_a(\nu_L,z)+\beta_C(\nu_L,z)]\exp\left[-2\int_{r_0}^{r}\alpha(\nu_L,z')dz'\right]. \tag{7.15.b}$$

With these signals, similar to (7.11.d) and (7.12.f), we can derive aerosol optical properties, the backscattering ratio $\mathbf{R}(\nu_L, z)$, and aerosol lidar ratio $L_a(\nu_L, z)$, respectively, as (7.15.c) and (7.15.d):

$$\mathbf{R}(\nu_L, z) = \frac{\beta_a(\nu_L, z) + \beta_C(\nu_L, z)}{\beta_C(\nu_L, z)} = \frac{f_i(T, P, z)\widetilde{N}_3(\nu_L, z)}{\widetilde{N}_i(\nu_L, z)}; \; i = 1 \text{ or } 2 \; ; \qquad (7.15.c)$$

$$L_a(\nu_L, z) = \frac{\alpha_a(\nu_L, z)}{\alpha_m(\nu_L, z)} = \frac{1}{2}\left\{ \left[\mathcal{N}(z)\sigma^C\right]^{-1} \frac{d}{dz} \ln\left[\frac{\eta_i f_i(T, P, z)\beta_C(\nu_S, z)}{\widetilde{N}_i(\nu_S, z)}\right] - 2 \right\}. \qquad (7.15.d)$$

The work of Hair et al. [7.13] in 2001 still appears to be the only tropospheric atmospheric temperature measurements based on Cabannes scattering at 532 nm using iodine filters. This measurement covers the entire troposphere, despite the presence of aerosols. Though the resulting profiles compare well with balloonsonde measurements in relative temperatures, it showed a consistent bias (different for each night) for temperature measurements. One such example is shown in Fig. 7.15(a). After a constant shift in temperature to compensate for the bias, the lidar profile is seen to yield excellent agreement with the balloonsonde in Fig. 7.15(b). A somewhat complicated but necessary calibration (termed normalization) procedure was carried out each night, the results of which allow estimates of various systematic uncertainties in channel efficiency. The authors identify error sources and determine those due to efficiency normalization $\delta \ln(\eta_1/\eta_2)$, laser frequency drift, filter transmission $\delta \ln(f_1/f_2)$, laser spectral purity, and reference pressure, leading to percent errors of, respectively, about 0.7, 0.15, 0.95, 1.0, and 0.12, as shown in Table 3 of [7.13]. Treating these as independent errors, the standard deviation of the total offset is calculated to be 5.8 K for $S_{T12} = 0.42\%\ \mathrm{K}^{-1}$. Indeed, because of the difficulty in the measurement of the relative efficiencies between total scattering and molecular-scattering channels, there exists a possible offset of a few percent each night. By comparing to the nearby balloonsonde measurements for the 11 nights with data, this bias, as summarized in Table 2 of [7.13], was found to range from –10 K to +12 K. This huge swing in bias between individual nights is hardly statistical, rather it is likely due to unreliable normalization measurements (perhaps due to mechanical mirror flippers) introducing the variation from one night to the next. Thus, for practical measurements with this lidar, a known reference temperature at a set altitude is required for each night in order to determine absolute measurement profiles. More careful airborne aerosol optical property measurements, without the use of mirror flippers in normalization measurements, yield an average variation in

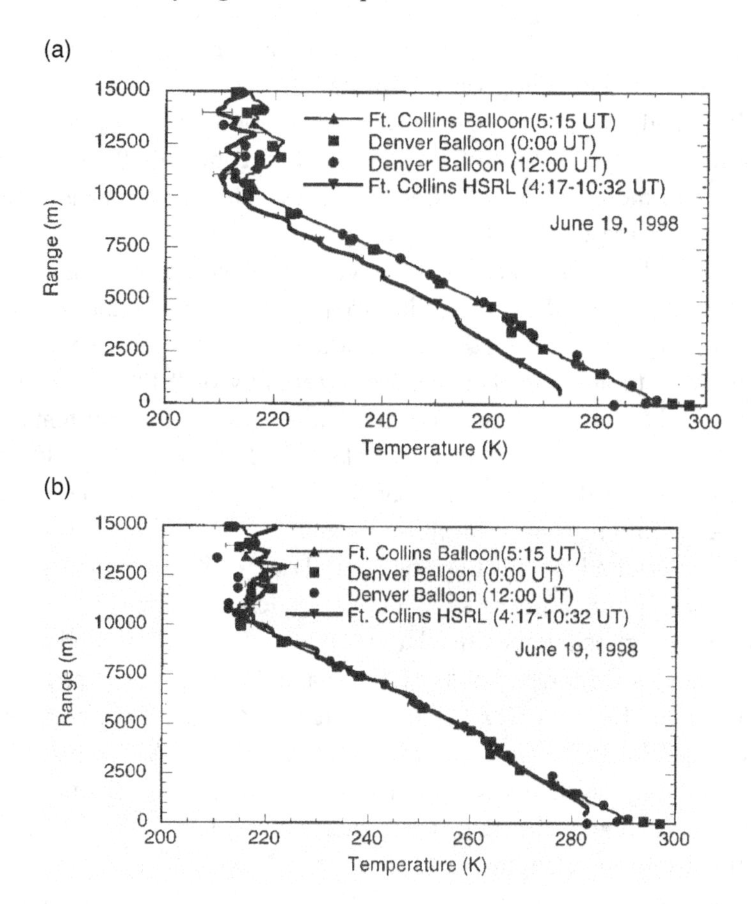

Fig. 7.15 HSRL 5-hr, 300-m averaged temperature profile for 19 June 1998 plotted along with the Fort Collins and Denver balloonsondes. The error bars give the effect of photon-counting statistical variations at selected heights for the HSRL data. (a) The lidar temperature profile with its bias relative to the balloon sounding. (b) With the adjustment of a constant shift bias correction, excellent agreement is shown. Reproduced with permission from Fig. 9 of [7.13]. © the Optical Society.

normalization of less than 0.3% [7.1]. Another remedy is that used for the barium filter system [7.36], using a single channel and tuning the laser frequency to on- and off-resonance, measuring molecular and total scattering from the same channels. Although this avoids the need for the normalization measurement, it is unfortunately impossible with iodine vapor filters.

We briefly compare the errors (measurement uncertainties) resulting from filter transmission, detection efficiency normalization, and photon noise at 1 km between the two systems (using barium and iodine filters, respectively). As mentioned, for the temperature measurements at 553.7 nm shown in Fig. 7.9, these errors [7.12] were respectively 1.5%, 0%, and 0.3% (with laser power 3 mW, 20 min, and 375 m resolution). The corresponding values for the 532 nm system shown in Fig. 7.10 from Tables 3 and 4 of [7.13] were 0.95%, 0.7%, and 0.1% (with 6 W, 1 hr, and 300 m resolution). The diameter of the receiving telescope is 0.203 m in both cases. The existence of the errors in detection efficiency calibration for the iodine filter system (0.7%) increases the filter transmission errors from 0.95% to 1.18% as compared to 1.5% for the barium filter case. The corresponding temperature uncertainty is 2.8 K for the iodine filter ($S_{T12} = 0.42\%\ \mathrm{K}^{-1}$) and 7.5 K for the barium filter ($S_{T12} = 0.2\%\ \mathrm{K}^{-1}$). More importantly and certainly unexpected is the fact that in order to achieve photon noise uncertainty of 0.1% (not counting for the loss of a third channel and Daystar filter) with the iodine filter system compared to 0.3% for the barium filter system, a much larger $PA\tau_{\mathrm{int}}$ value (by ~ 4800 times) is required. The principal culprit here (for the loss of signal) is the existence of continuum absorption resulting from the bound–free transitions ${}^1\Pi_{1u}\leftarrow X({}^1\Sigma_{o^+g})$ that overlap the bound–bound transitions $B({}^3\Pi_{o^+u})\leftarrow X({}^1\Sigma_{o^+g})$ [7.38] in iodine vapor [7.39; 7.40], which increase as the vapor density increases. For example, the off-resonance transmittance (excluding the passive loss of windows) for the low-temperature and high-temperature iodine vapor filters are, respectively, 0.5 and 0.1, as shown in Fig. 6 of [7.13] or in Fig. 7.18 here. This compares with a transmittance of 1.0 for the barium filter shown in Fig. 7.12. This is because, unlike in a molecular system, continuum absorption in atomic transitions occurs at frequencies much higher than the strong discrete dipole-allowed line transitions found in IR-visible wavelengths.

Another paper in the literature that uses Cabannes scattering for atmospheric temperature measurements is that of Hua et al. [7.41] in 2004. Their lidar transmits eye-safe single-frequency light at 355 nm. The lidar backscattering signal is detected through three FPEs, each with a different frequency shift and FWHM. These are denoted as Rayleigh-1 (frequency shift = 1 GHz, FWHM = 0.3 GHz), Rayleigh-2 (3 GHz, 0.6 GHz), and Mie (0 GHz, 0.2 GHz) as shown in Fig. 7.16, and detected, respectively, by photomultipliers PMT1, PMT2, and PMT3. A fourth channel, PMT4, detects total scattering and is shown in their filter box in Fig. 2 of [7.41]. The aerosol ratio, $\mathrm{R}_a \equiv \beta_a/\beta_m$, can be determined from the ratio of PMT3/PMT4, incorporating the appropriate efficiency calibrations. Although the FPEs for Rayleigh-1 and Rayleigh-2, respectively, receive low-frequency and high-frequency portions of Cabannes scattering, unlike for the AVF, they do not reject aerosol scattering completely. After aerosol correction with the measured aerosol ratio, the ratio of PMT1/PMT2 can then be used in combination with $\mathscr{R}(\nu; T, P)$ to determine atmospheric temperatures, as described earlier. One benefit of this arrangement: there is nearly complete design flexibility for the FWHM of FPEs.

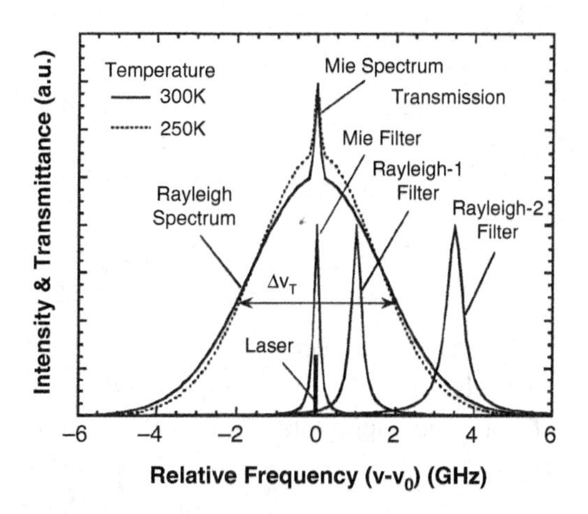

Fig. 7.16 Spectral diagram of the Mie and Cabannes scattering and the transmittance of the FPEs in a lidar system at 355 nm. Reproduced with permission from Fig. 1 of [7.41]. © the Optical Society.

Of course, to achieve the required sensitivity, the temperature of these FPEs has to be stabilized to within 0.001°C.

7.3.4 Search for an Optimum Vapor Filter for Atmospheric Temperature Measurements

To take the advantage of the strength of Cabannes-scattering atmospheric temperature measurements, about 40 (γ_R^{-1}) times better than the O+S branches of PRR, the use of AVFs or FPEs appears to be challenging in different ways. This is perhaps why, other than the works from the early 2000s outlined here, they have not been much pursued. After mentioning the different problems in barium and iodine filters, we now consider a strategy for the selection of much better vapor filters and associated absorption lines. We return to the expression of intrinsic temperature measurement uncertainty δT_{air} in (7.10.c), which showed its dependence on the interplay between signal level, single-photon uncertainty and measurement sensitivity. The intrinsic temperature uncertainties for Cabannes lidar and rotational Raman lidar are summarized respectively in (7.16.a) and (7.16.b):

For CS lidar:

$$\delta T_{Air} = \frac{\xi_{opt}}{S_{T12}\sqrt{N_0}};\ S_{T12} = \frac{\partial}{\partial T}\ln\left(\frac{f_1}{f_2}\right);\ \xi_{opt} = \frac{1}{\sqrt{f_2}} + \frac{1}{\sqrt{f_1}},\ \text{with}\quad f_i = \tau_i f_i^n. \tag{7.16.a}$$

For PRR lidar:

$$\delta T_{Air} = \frac{\xi}{S_{T12}\sqrt{N_0^{PRR}}}; \; S_{T12} = \frac{\partial}{\partial T}\ln\left(\frac{\mathscr{R}_{HT}^{aRR}}{\mathscr{R}_{LT}^{aRR}}\right);$$

$$\xi = \sqrt{\frac{1}{(\eta_1\tau_{HT})\mathscr{R}_{HT}^{aRR}} + \frac{1}{(\eta_2\tau_{LT})\mathscr{R}_{LT}^{aRR}}}. \tag{7.16.b}$$

Here $f_i(T,P,z)$ and $f_i^n(T,P,z)$ are, respectively, the attenuation factors for the filters $F_i(\nu)$ and $F_i^n(\nu)$; they are related as $f_i(T,P,z) = \tau_i f_i^n(T,P,z)$ with the off-resonance transmittance $\tau_i = 1$ for atomic vapor filters and $\tau_i < 1$ for molecular filters due to continuum absorption.

Since the temperature uncertainty depends on both the photon uncertainty, $\delta\ln(N_1/N_2)$ – which depends not only on PA product of the lidar but also on the single-photon uncertainty – and the temperature sensitivity of the measurement S_{T12}, we hope to keep the former small (less photon noise) and the latter large (more sensitive to air temperature change). That this represents a conflicting demand can be seen in Fig. 7.17. Here the filtered (fraction of) the Cabannes signal transmitted through an ideal blocking filter (solid) along with one-channel temperature sensitivity ($S_{Ti} = (\delta f_i/f_i)_{\Delta T=1\,K}$, dashed) under standard atmospheric conditions (T = 275 K and P = 0.75 atm) at λ = 532 nm are plotted as a function of the bandwidth of such a filter $\Delta\nu_B$.

We see that the transmitted Cabannes-scattered light decreases gradually to nearly zero as the bandwidth of the filter approaches 6 GHz, while the one-channel sensitivity S_{Ti} increases accordingly and rises most steeply between $\Delta\nu_B$ = 2.0 and 5.0 GHz, representing the region of preferred operation. For the pair of ideal filters with widths of 2 GHz and 3 GHz, at λ = 553.7 nm the measurement sensitivity is ~ 0.2% K^{-1} (assuming barium filters) – see Table 2 of [7.35] – while for filters (ignoring the wavelength effect for the time being) with the same widths at λ = 532 nm the measurement sensitivity is about 0.35% K^{-1}, read from Fig. 7.17. Although it is desirable to increase the width of the high-temperature filter from 3 GHz to 4 GHz, to do so in barium requires vapor temperature too high to be practical.

For iodine vapor, the required temperature of the high-temperature filter is not a problem, but its continuum absorption excessively attenuates the scattered light. Fortunately, there are many absorption lines in iodine vapor, and it is possible to increase the width by combining two neighboring absorption lines and maintain adequate off-line transmission with moderate vapor temperature. Hair et al. [7.13] have combined iodine lines 1107 and 1108 with a cell of length 15.42 cm employing two temperature controllers: one for the body, and the second for the finger of

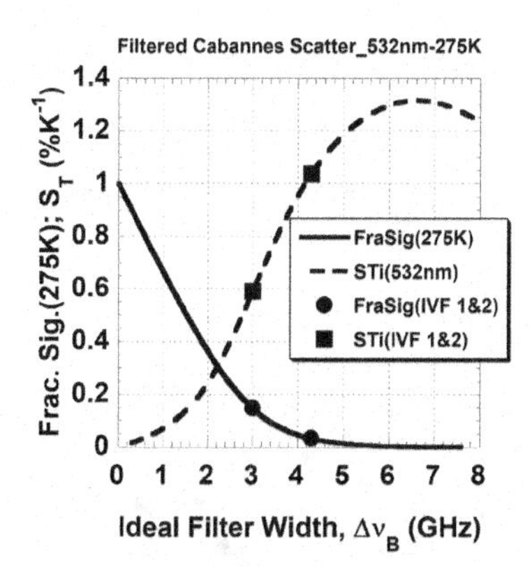

Fig. 7.17 One-channel sensitivity and the fraction of the Cabannes signal (at 275 K and 0.75 atm) at 532 nm transmitted through an ideal square filter as a function of the filter full width. Adapted with permission from Fig. 5 of [7.13]. © the Optical Society.

the cell, to set, respectively, the operation temperature and vapor pressure. The cell/finger temperatures used were (82.19°C/72.03 °C) and (56.18°C/47.74°C) for the high-temperature (H.T.) filter (cell 1) and low-temperature (L.T.) filter (cell 2), respectively. Their corresponding FWHM are 4.3 and 3.0 GHz; see Fig. 7.18 above. At these two bandwidths, the transmissions through two ideal filters are about 0.15 and 0.03 (Fig. 7.17, filled circles), respectively, and the one-channel sensitivities are 1.04% K^{-1} and 0.59% K^{-1}, leading to the measurement sensitivity of 0.45% K^{-1}, comparable to the 0.42% K^{-1} for iodine filters. Though the one-channel sensitivity calculated for ideal filters may be used approximately for the corresponding iodine filters, to use the calculated fractional signal from ideal filters for the real AVFs, their off-resonance transmittance (due to continuum absorption) must be accounted for. Thus, for the iodine filters in question, the attenuation factors of the associated ideal filter are $f_1 = 0.1 \times 0.033 = 0.0033$ for the H.T. channel and $f_2 = 0.5 \times 0.148 = 0.074$ for the L.T. channel. The optimum single-photon uncertainty is then $\xi_{opt} = 1/\sqrt{f_2} + 1/\sqrt{f_1} = 1/\sqrt{0.074} + 1/\sqrt{0.0033} = 21.08$. Since S_{T12} depends on the location of the laser frequency relative to the center frequency of the filter, one can choose from more than 20 Doppler-free spectroscopy lines to lock the laser wavelength. In so doing, it is possible to maximize the sensitivity, $S_{T12} = S_{T1} - S_{T2}$, leading to a value between 0.05% K^{-1} and 0.42% K^{-1}, as shown in Fig. 7.18. This results in a maximum sensitivity very near that of two ideal filters with the same widths, 4.3 and 3.0 GHz.

In spite of the innovative usage of two neighboring iodine lines to increase the filter's FWHM (and as such, its temperature sensitivity), the system still suffers

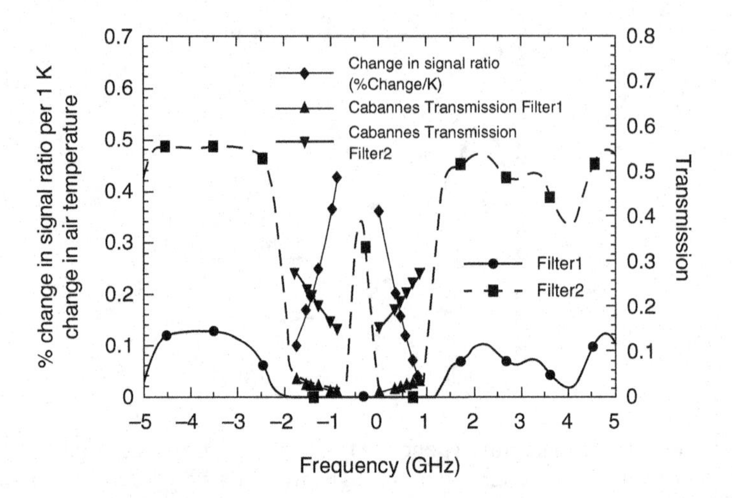

Fig. 7.18 Iodine filter (1 and 2) with combined 1107 and 1108 lines, transmission near 532 nm. The S_{T12} (solid diamonds) and the fractional transmissions of the Cabannes signal through the filters at selected frequencies. Reproduced from Fig. 6 of [7.13] with permission. © the Optical Society.

from low off-resonance transmittance, suggesting the need to search for better filters, employing atomic vapors where the absorption continuum is not a problem. The alkali Earth metals are attractive because their singlet $(\mathrm{x}s)^2 - (\mathrm{x}s\mathrm{x}p)$ transition forms a two-level system. Unfortunately, they all need high vapor temperature to produce sufficient vapor density. Alkali metals can produce sufficient vapor at reasonable vapor temperature, but their transitions are in the form of a doublet with more complex transition structures. Since for most alkali metals the ground state splitting is relatively small and the difference between doublet transitions much smaller than the width of Cabannes scattering of air, they could be good candidates for AVFs. For example, the ground-state splitting is $\Delta v = 1.772$ GHz for Na and $\Delta v = 0.4618$ GHz for K; their respective D-doublet transitions are 16956.172 cm^{-1} ($\mathrm{NaD_1}$) and 16973.368 cm^{-1} ($\mathrm{NaD_2}$), and 12985.186 cm^{-1} ($\mathrm{KD_1}$) and 13042.896 cm^{-1} ($\mathrm{KD_2}$). What makes these vapors particularly attractive is that reliable laser technologies are available. Resulting from the ongoing interest in Na laser guide stars, there are commercially available high-power single-frequency CW tunable lasers at 589 nm. Nd:YAG-based, laser-diode-seeded, all-solid-state Q-switched systems employing sum-frequency generation have enabled practically maintenance-free operation of Na LIF lidar [7.42]. In addition, a diode-pumped, Q-switched, single-longitudinal mode alexandrite ring laser has been successfully developed and lidar operation at 770 nm demonstrated [7.43]. Therefore, it is timely to consider the use of Na and K AVFs for lidar aerosol and state-parameter measurements in the troposphere based on Mie–Cabannes scattering with atomic vapor filters. Obviously, one has a choice of using D_1 or D_2 lines of sodium (589 nm) or of potassium (770 nm). The analysis and performance of these possibilities are expected to be similar. Below, for simplicity, we discuss the performance of one such lidar operated at the D_1 transition of potassium at 770 nm. Before we do so, we digress for a discussion of the effect of metal mass limiting development of single-absorption-line AVFs.

7.3.4(1) Vapor Mass: A Fundamental Limit on Vapor Filter Performance

Briefly, we consider the transmission function of a typical cylindrical vapor cell with length L and Pyrex windows for passing the light beam, along with a finger as the reservoir partially filled with the metal of interest. The saturated vapor density $n(\#\mathrm{m}^{-3})$ is controlled by the cell temperature, and the vapor pressure regulated by controlling the temperatures of the cell and finger respectively [7.44]. The transmittance of the AVF, $\mathscr{T}(\nu)$, is an exponential function of the product $(n\sigma^A(\nu)L)$, as given in (7.17.a), where $\sigma^A(\nu)$ is the absorption cross section of a resonance transition at frequency ν_0 between ground state |1> and excited state |2> with degeneracies, respectively, g_1 and g_2. It is proportional to the Einstein A-coefficient of the transition, A_0, and a Gaussian function $G(\nu-\nu_0)$, representing Doppler broadening of the absorption line. The Doppler-broadened line-shape function, $G(\nu-\nu_0)$, and its variance, σ_ν^A, which is proportional to the square root of ratio of temperature to molecular weight of the vapor atom ($\sqrt{T/M}$), is shown in (7.17.b), where k_B is the Boltzmann constant:

$$\mathscr{T}(\nu) = \exp\left[-n\sigma^A(\nu)L\right] = \exp\left\{-C\exp\left[-(\ln 2)\left(\frac{\nu-\nu_0}{\Delta\nu_{HWHM}}\right)^2\right]\right\}, \tag{7.17.a}$$

$$\sigma^A(\nu) = \frac{\lambda^2}{8\pi}\frac{g_2}{g_1}A_0 G(\nu-\nu_0),\ \text{ and } G(\nu-\nu_0) = \frac{1}{\Delta\nu_{HWHM}}\sqrt{\frac{\ln 2}{\pi}} \times \exp\left[-(\ln 2)\left(\frac{\nu-\nu_0}{\Delta\nu_{HWHM}}\right)^2\right],$$

$$\text{where } C = \frac{\lambda^2 nL}{8\pi}\frac{g_2}{g_1}A_0;\ \Delta\nu_{HWHM} = \frac{1}{\lambda(\mathrm{nm})}\sqrt{\frac{11455.5T\ (\mathrm{K})}{M\ (\mathrm{amu})}}. \tag{7.17.b}$$

Assuming one can adjust the cell length and vapor temperature and pressure, thus vapor density, then the filter transmission depends on only two parameters: C, the absorption/attenuation on-resonance, and $\Delta\nu_{HWHM}$, the half width at half-maximum of the vapor Doppler broadening. To understand the filter function and its limitations, we plot the transmission function in normalized frequency (in units of

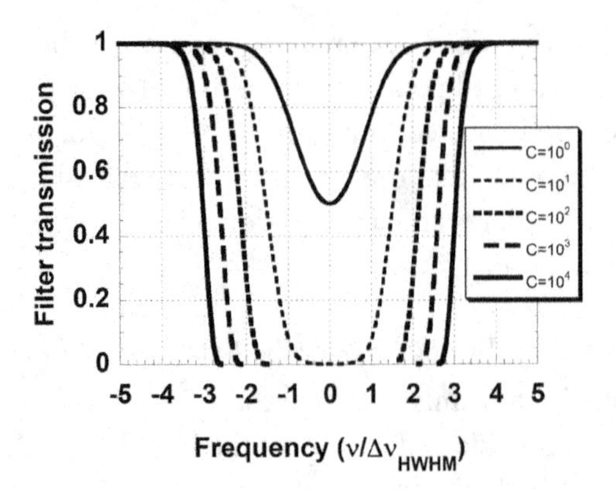

Fig. 7.19 Generic atomic vapor filter transmission as a function of normalized frequency at different values of attenuation C. In units of $\Delta\nu_{HWHM}$, the HWHM of the filter $\Delta\nu_{HWAVF}$ is about 1.55, 2.15, 2.65, and 3.05 respectively for $C = 10, 100, 1000, 10{,}000$. Reproduced with permission from Fig. 3 of [7.46]. © the Optical Society.

$\Delta\nu_{HWHM}$ for the vapor) in Fig. 7.19 for various values of C. Notice that the $\Delta\nu_{HWHM}$ of the backward light scattering spectrum is twice the width of that given in (7.17.b) for atmospheric mean mass of 28.97 u and that for the $\Delta\nu_{HWAVF}$ of the vapor filter to be the same as the $\Delta\nu_{HWHM}$ of the atmospheric backscattered spectrum, the temperature of the vapor cell must be $T_{cell} = \left(M_{vapor}(\text{amu})/28.97\right) T_{Air}$. Ironically, typical molecular and atomic vapors used for lidar applications all have masses considerably higher than ideal: iodine (253.8 u) and barium (137.3 u) implemented, and lead (207.2 u) attempted [7.45], all require the cell temperature to be 8.8x, 4.7x, and 7.2x higher, respectively, than the atmospheric temperature. Fortunately, the mass of potassium, 39.1 u, is more manageable, and the required temperature is less than twice the atmospheric temperature. Sodium (23 u) in this respect is better than potassium, but its large ground state splitting (1.772 GHz) more than negates this advantage. Since the FWHM of the AVF, $\Delta\nu_{HWAVF}$, also depends on the C value, in view of Fig. 7.19, a high value of C, say $C = 100$, should be enough for potassium, as it yields $\Delta\nu_{HWAVF} \sim 2.15\ \Delta\nu_{HWHM}$. Further increasing of its C value is not effective. As can be seen in Fig. 7.19 and has been pointed out [7.46], the increase of the filter's $\Delta\nu_{HWAVF}$ by increasing C from 100 to 10,000 is only about 50%. This discussion clearly demonstrates why it is not possible to get a bandwidth substantially broader than 3 GHz with barium, and though we can get around this fundamental limitation with iodine due to the existence of neighboring lines, continuum absorption negates this advantage.

7.3.4(2) Potassium Vapor as a Viable Filter Choice for Lidar

Since there is no Hanle effect in alkali D_1 transitions (see Section 3.3.2), the absorption cross-section, $\sigma^A(\nu)$, is simply 4π times the associated differential scattering cross section given in (3.22.a). Using the parameters for the D_1 transitions of a potassium atom, given in Tables (A.1) and (A.2), and assuming each transition lineshape is Doppler-broadened at a specified temperature, we plot in Fig. 7.20 the absorption cross-section spectrum of the K D_1 transition in a 50 K vapor cell (thin solid gray line, right scale). Its higher peak occurs at –0.18 GHz. Also shown in the figure are two normalized Cabannes-scattering spectra of standard air near ground (at 275 K and 0.75 atm), centered at zero (medium black solid line, left axis), and at –0.18 GHz (medium black dashed line, left axis). Fig. 7.20 also includes three potassium vapor filter functions, from cells constructed with 10 cm-long Pyrex tubes and including a reservoir tip for storing the metal. The temperatures of the reservoir and cell are controlled separately and for these plots are (T_{res}/T_{cell}): (300 K/305 K, thick short-dashed line), (350 K/355 K, thick dashed line), and (400 K/405 K, thick line) for LTB, LTA, and HT filters, respectively. To calculate a filter transmission function, we first obtain the vapor pressure, p_K, for T_{res} in solid and liquid phases using Eq. 22 of [7.44], given here by (7.18), then determine the vapor density, $n = \mathcal{N}_K = p_K/k_B T_{cell}$, with the ideal gas law and calculate the transmission function $\mathscr{T}(\nu) = \exp[-n\sigma^A(\nu)L]$ for each (T_{res}/T_{cell}) using (7.17.a).

$$\log_{10}(p_K/\mathrm{Torr}) = 69.53 - 10486/T_{res} + 1.8658 \times 10^8/T_{res}^3 + 0.0027286 T_{res} - 8.5732 \ln(T_{res}) \tag{7.18}$$

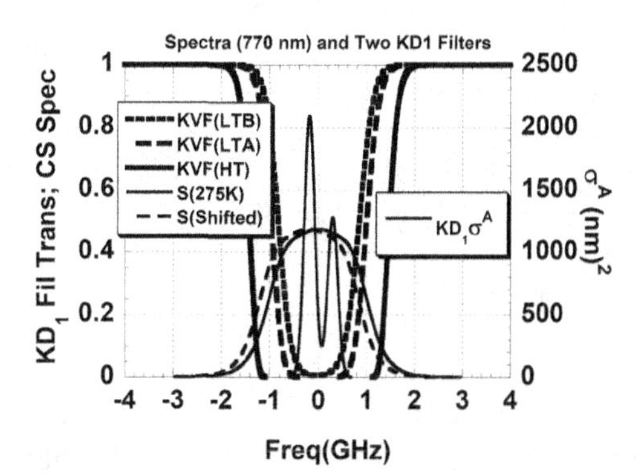

Fig. 7.20 A normalized KD_1 spectrum of a 50 K vapor cell (gray thin solid, right axis), transmission of 10-cm KVFs at the D_1 transition (770 nm) of two LT (A in thick dashed and B in short thick dashed lines, left axis) and one HT (thick solid line, left axis) filter cells, along with both centered and shifted Cabannes spectra of atmosphere near ground (medium black solid and dashed lines, left axis). Reproduced with permission from Fig. 4 of [7.46] © the Optical Society.

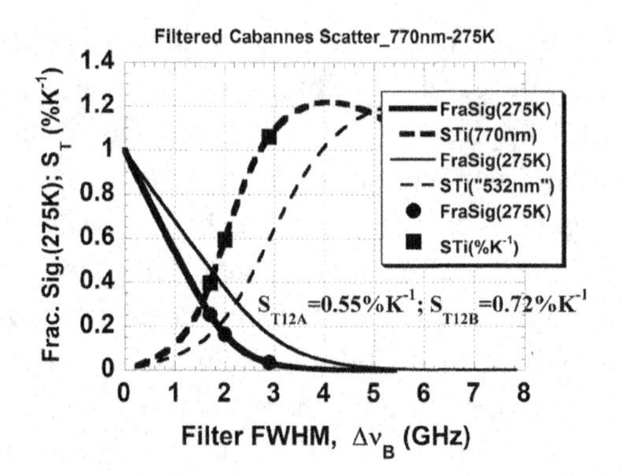

Fig. 7.21 Fractional filtered Cabannes scattering spectra of air at 275 K and 0.75 atm (thick solid) and one-filter temperature sensitivity S_{Ti} (thick dashed) at 770 nm. The circles and squares indicate respective values for three (1, 2A, 2B) selected KVFs. The thin solid and dashed lines are the same except when scaled to 532 nm. See text for details. Reproduced with permission from Fig. 5 of [7.46]. © the Optical Society.

At the center (0 GHz) of the absorption well, the filters block 21.4, 43.8, and 763 dB, respectively, for LTB, LTA, and HT, while they block 20.4, 41.9, and 701 dB at –0.18 GHz.

It is instructive to compare the use of a potassium cell (Fig. 7.21) against that of an iodine cell for atmospheric temperature measurements at 770 nm and 532 nm, respectively. To see the impact of wavelength difference, we revisit the nature of the normalized Cabannes spectrum, $\mathscr{R}(\nu; T, P)$, which is related to a model-dependent theoretical function, $C_0(x, y)$, of the two dimensionless parameters x and y given in (4.24):

$$\mathscr{R}(\nu; T, P) = \frac{2\pi}{K\mathrm{v}_0} C_0(x, y), \text{ with } x = \frac{2\pi\nu}{K\mathrm{v}_0} \text{ and } y = \frac{P}{K\mathrm{v}_0\,\mu}, \quad \mathrm{v}_0 = \sqrt{\frac{2k_B T}{m}} \text{ and}$$
$$\int_{-\infty}^{\infty} C_0(x, y)dx = 1,$$

where m, K, μ, and v_0, are respectively, the molecular mass, the magnitude of the perturbation wave-vector, (dynamic) shear viscosity, and the speed at the peak of the Maxwellian speed distribution. For backscattering, $K = 4\pi/\lambda$ (λ being the laser wavelength), we have $\mathscr{R}(\nu; T, P) = (\lambda/2\mathrm{v}_0)C_0(x, y)$, with $x = \nu\lambda/(2\mathrm{v}_0)$ and $y = \lambda P/(4\pi\mathrm{v}_0\mu)$. The theoretical spectrum, $C_0(x, y)$, is a bell-shaped function normalized in the x-space (sometimes called dimensionless frequency). The extent of the contribution due to collective scattering increases with the y value, which is proportional to the product of P and λ. The x-axis is proportional to the product of

ν and λ. Thus, to compare the effect of Cabannes scattering at two different frequencies between Fig. 7.17 and Fig. 7.21 for the same value of x and ν_0 (or atmospheric temperature), we consider a scaled frequency $\nu_{sca} = (770/532)\nu = 1.45\nu$. By replotting the two curves in Fig. 7.21 against the scaled frequency, we transform the thick curves (for 770 nm) into the corresponding thin curves (for 532 nm). Now, we can see that an ideal filter of say 2 GHz width for scattering at 770 nm is equivalent to a filter of width 2.9 GHz at 532 nm, because under otherwise identical atmospheric conditions, the Cabannes spectrum at 770 nm in frequency space is much narrower. Indeed, the thin curves (532 nm) in Fig. 7.21 look very much like the curves in Fig. 7.17. The very small difference between the two sets of curves may be accounted for by the difference in the y value, with the higher y value at 770 nm reflecting more collective scattering contributions. This bit of physical understanding hopefully provides more confidence in the plots shown in Fig. 7.20 and Fig. 7.21.

Shown in Fig. 7.21 are the fractional transmissions of Cabannes scattering from air (with the laser locked at 0 GHz) at standard atmosphere conditions of 275 K and 0.75 atm (solid lines) through the vapor filter, and the one-filter temperature sensitivity S_{Ti} at 770 nm (thick dashed) plotted as a function of the width of the ideal filter. The FWHM widths of the LTB, LTA, and HT filters as measured from Fig. 7.20 are, respectively, 1.7, 2.0, and 2.9 GHz. Their calculated fractional (or filtered) signal transmitted and one-filter sensitivity are, respectively, 29.02%, 18.6%, and 3.74% (filled circles), and 0.295% K^{-1}, 0.466% K^{-1}, and 1.012% K^{-1} (filled squares). When plotting these values in Fig. 7.21, we can see they are very similar to the ideal filter curves with width $\Delta\nu_B$ set to the vapor filter FWHM, especially for the HT filter. The temperature sensitivities for the two pairs of filters are $ST_{12A} = 0.55\%\ K^{-1}$ and $ST_{12B} = 0.72\%\ K^{-1}$, and their corresponding optimum single-photon uncertainties calculated from (7.16.a) are $\xi^{12A}_{opt} = 7.49$ and $\xi^{12B}_{opt} = 7.03$. If we lock the laser at the KD_1 transition peak at –0.18 GHz, the corresponding one-filter sensitivities would be somewhat smaller, with 0.256% K^{-1}, 0.387% K^{-1}, and 0.883% K^{-1}, giving rise to $ST_{12A} = 0.50\%\ K^{-1}$ and $ST_{12B} = 0.63\%\ K^{-1}$, and $\xi^{12A}_{opt} = 6.80$ and $\xi^{12B}_{opt} = 6.40$.

Encouraged by the higher sensitivity of $ST_{12B} = 0.72\%\ K^{-1}$, we would urge a new potassium filter–based temperature lidar at 770 nm, using the KVF-1 and KVF-2B filters. The temperature uncertainty can be easily estimated by (7.16.a); for example, with a signal level of $N_0 = 100,000,000$, it is $\delta T_{Air} = 0.097$ K. For data processing of this proposed lidar with three detection channels, like in Fig. 7.10, we lock the laser at 770 nm (i.e., $\nu_L = \nu_0$) and first retrieve the atmospheric temperature profile. We express the measured temperature ratio $R_T(z)$ as the count ratio between channels 1 and 2 in terms of efficiency ratio η_1/η_2 and attenuation ratio f_1/f_2, very much like that in (7.12.b) for the barium filter–based lidar as

$$R_T(z) = \frac{N_1(z)}{N_2(z)} = \frac{\eta_1^{opt} f_1[T(z), P(z)]}{\eta_2^{opt} f_2[T(z), P(z)]}, \text{ with } \eta_1^{opt} = \frac{\sqrt{f_{2std}}}{\sqrt{f_{2std}} + \sqrt{f_{1std}}};$$
$$\eta_2^{opt} = \frac{\sqrt{f_{1std}}}{\sqrt{f_{2std}} + \sqrt{f_{1std}}}. \tag{7.19.a}$$

Here, we assume an optimum division of the scattered photons according to the attenuation factors of KVFs at the standard atmospheric state with 275 K and 0.75 kPa. To facilitate continued data analysis, we follow (7.12.d) and evaluate the attenuation function $f_i(T, P, z)$ needed in (7.19.a) in terms of six integral coefficients of the quadratic equation in T and P for each AVF, resulting in:

$$f_i(T, P, z) = f_{stdi} + (T - T_o)f_{Ti} + (P - P_o)f_{Pi} + 0.5(T - T_o)^2 f_{TTi}$$
$$+ (T - T_o)(P - P_o)f_{TPi} + 0.5(P - P_o)^2 f_{PPi},$$
$$\text{with } f_{\mu i} \equiv \int \mathscr{R}_\mu(\nu) F_i(\nu) d\nu . \tag{7.19.b}$$

Here, $F_i(\nu)$ includes the effect of the FPE, if inserted; $i = 1$ or 2. Note that, expressing temperature and pressure in units of K and Pa, f_{std} is dimensionless, and the units for f_T, f_P, f_{TT}, f_{TP} and f_{PP} are, respectively, K^{-1}, Pa^{-1}, K^{-2}, $K^{-1}\,Pa^{-1}$, and Pa^{-2}. For the proposed potassium filter temperature lidar, utilizing KVF-1 and KVF-2B, the values of $f_{\mu i}$ were determined from measured filter functions $F_i(\nu)$ and theoretical normalized Cabannes spectra, $\mathscr{R}(\nu; T, P)$, along with their associated derivatives. The results are tabulated in Table 7.2.

As discussed in a comprehensive theoretical study [7.35], terminating the Taylor series in its second order as in (7.19.a) is more than accurate enough for data retrieval for realistic temperature and pressure ranges ($245\ K \le T \le 305\ K$ and $50\ kPa \le P \le 100\ kPa$). For expected T, P values beyond these, more T, P ranges centered at different standard points may be needed for the Taylor expansion. Once the $f_i(T, P, z)$ are determined from (7.19.b) with the help of values in Table 7.2, the procedure for retrieving an observed temperature profile may proceed. Starting with a known pressure at a chosen reference height from the top of the altitude range, (7.19.b) is a quadratic equation of temperature, which can be solved for the temperature at that height. The atmospheric pressure at the neighboring height (range bin Δz) may be calculated by assuming hydrostatic equilibrium and the ideal gas law. Successive iterations yield vertical atmospheric temperature and pressure and thus, in turn, density profiles. Once the temperature and pressure profiles are

Table 7.2 *The coefficients of (10b) for the L.T. and H.T. KVFs*

$f_{Ci}(T,P,z)$	f_{stdi}	f_{Ti} (T^{-1})	f_{Pi} (Pa^{-1})	f_{TTi} (K^{-2})	f_{TPi} $(K^{-1}Pa^{-1})$	f_{PPi} (Pa^{-2})
KVF (LT-2B)	2.90×10^2	8.58×10^{-1}	1.53×10^{-1}	-3.14×10^{-3}	1.10×10^{-3}	-1.04×10^{-5}
KVF (HT-1)	3.74×10^1	3.80×10^{-1}	-1.30×10^{-1}	1.50×10^{-3}	4.72×10^{-4}	7.12×10^{-4}

obtained along with $f_i(T, P, z)$, the signal from the total scattering channel (channel 3 with aerosol plus molecular) can be used to determine the profiles of aerosol extinction coefficient and lidar ratio using (7.12.e) and (7.12.f), respectively.

7.3.5 Comparison between Four Atmospheric Temperature Lidars

We are now able to compare four atmospheric lidar temperature measurement methods. For the three that are based on Cabannes scattering, we first list their respective single-photon uncertainties, ξ_{opt}, and sensitivities, S_{T12}, and then calculate the respective temperature uncertainties, δT_{air}, using (7.16.a) based on $N_0 = 10^8$ photon counts arriving at point B in Fig. 7.10. For the RRS lidar, since as explained the condition $\eta_1 + \eta_2 = 1$ is no longer a requirement, we list ξ calculated from (7.16.b) instead. The photon counts arriving at point B would be those by PRR instead (i.e., $N_0^{PRR} = 0.0342N_0 = 3,420,000$). The single-photon uncertainties and temperature sensitivities for each lidar have been calculated in Section 7.3.2 for the barium filter lidar, CS (553.7 nm)/Ba, in Section 7.3.4 for the iodine filter lidar, CS (532 nm)/I_2, in Section 7.2.4 for rotational Raman lidar with custom-made interference filters, PRR (532 nm)/CIF, and in Section 7.3.4(2) for the proposed potassium filter lidar, CS(770 nm)/K. These have been summarized in Table 3 of a recent paper [7.46] and reproduced as Table 7.3 here, along with calculation of temperature uncertainties at the lidar wavelength $\delta T_{Air}(K)$ and those scaled to 532 nm, $N_0^{PRR} = 0.0342N_0 = 3,420,000$.

Compared to the barium and iodine filter systems, the proposed potassium filter (M = 39.1 amu) lidar performs better for a variety of reasons. For example, iodine (M = 253.8 amu) and barium (M = 173.3 amu) suffer the problem of very narrow Doppler width, thus making it difficult to achieve the blocking bandwidth required for the H.T. filter. Coupled with the need for unreasonably high temperature to produce sufficient Ba vapor, the temperature sensitivity ended up being about four times lower than that for the potassium filter. With the creative use of two neighboring absorption lines, the iodine system has achieved sensitivity only a factor of 1.7 less, but its single-photon uncertainty is three times larger than that for the potassium filter due to the presence of continuum absorption. Even though the signal strength of anti-Stokes PRR scattering is 78 times weaker than that of CS, suggesting a potential temperature error 8.8 times higher, the resulting uncertainty

Table 7.3 *Figure of merit comparison between 4 scattering temperature lidars. Reproduced with permission from Table 3 of [7.46]. © the Optical Society.*

Lidar/Filter	CS(554 nm)/Ba	CS(532 nm)/I_2	PRR(532 nm)/CIF	CS(770 nm)/K†
f_1 / f_2	0.1951/0.4644	0.0033/0.0740	0.0124/0.0416	0.0374/0.2902
ξ_{opt} or ξ	3.73	21.1	10.2	7.03
$\Delta\nu_{B1}/\Delta\nu_{B2}$(GHz)	2.99 / 2.04	4.3 / 3.0	848/ 689	2.9 / 1.7
S_{T12} (%K^{-1})	0.18	0.42	0.93	0.72
N_0	100,000,000	100,000,000	3,420,000	100,000,000
$\xi/\sqrt{N_0}$ (%)	0.037	0.211	0.551	0.070
δT_{Air}(K)††	0.206	0.502	0.592	0.097
δT_{Air}(K)$_{_sca}$†††	0.219	0.502	0.592	0.169

$^{\dagger}\ \nu_L = \nu_0$; $^{\dagger\dagger}\delta T_{Air}(\mathrm{K}) = \xi/\left(S_{T12}\sqrt{N_0}\right)$; $^{\dagger\dagger\dagger}\left(\lambda(\mathrm{nm})/532\right)^{1.5}\delta T_{Air}(\mathrm{K})$

of the δT_{Air} of the PRR lidar system with custom-made interference filters (CIFs) manages to be comparable to that of the CS/iodine system. This is mainly because in addition to the fact that $\eta_1 + \eta_2$ can exceed unity by a considerable amount, the beauty of being able to tailor the center wavelength and bandwidth of the H.T. and L.T. filters for the application at hand makes it possible to optimize the temperature sensitivity.

As an example, the center wavelength of the H.T. filter, BS5, has moved from near 529.5 nm for the GKSS lidar to 528.5 nm for the RASC lidar because the emphasis of the former was the condensation of polar stratospheric clouds (with temperature ~190 K), while the interest of RASC lidar was temperature profiling in the troposphere and stratosphere. Thus, the authors have selected filters designed to minimize temperature uncertainties at ~240 K [7.31]. In fact, this optimization has made the one-channel sensitivity negative for the L.T. filter and positive for the H.T. filter (see last column in Table 2 of [7.46]), resulting in a temperature sensitivity of 0.93% K^{-1}, 1.3 times higher than that for the proposed potassium filter temperature lidar. Of course, the strength of Cabannes scattering over anti-Stokes PRR still presents an advantage when utilized properly as can be seen in Table 7.3, for systems with the same PA product; the temperature error of the proposed potassium filter lidar (scaled to 532 nm) is a factor of 3.5 smaller than that for the PRR lidar.

We next briefly discuss an estimation of the system parameters required to deliver the 100,000,000 photons by Cabannes scattering assumed for Table 7.3 above, for temperature measurements. For this, we return to (7.8) and derive from it the received photon counts of the molecular channels $N_m(\nu_S, z; r_B)$ in terms of the $P_L A_R$ product and the atmospheric density $\mathcal{N}(z)$ as

$$N_m(\nu_L, z; r_B) = \frac{\eta_m P_L A_R \tau_{\text{int}} \Delta z}{h\nu_L z^2} \left[\mathcal{N}(z) \frac{d\sigma^C(\nu_L)}{d\Omega} \right] \exp\left[-2 \int_{z_0}^{z} \alpha(\nu_L, z') dz' \right],$$

$$\text{with } P_L = \frac{E_L}{\tau_{\text{int}}},\ h\nu_L = \frac{1.9864 \times 10^{-16}\text{J}}{\lambda(\text{nm})}, \tag{7.20}$$

where P_L, τ_{int}, Δz, $\mathcal{N}(z)$, and $d\sigma^C(\nu_L)/d\Omega$ are, respectively, laser power, integration (measurement) time, measurement vertical resolution, atmospheric density, and differential Cabannes scattering cross section. We use a hypothetical lidar system employing a 5 W laser at 532 nm and a 14-in telescope, giving us $P_L A_R = 0.5\ \text{Wm}^2$, as reference, along with $d\sigma^C(\nu_L)/d\Omega = 5.96 \times 10^{-32}\text{m}^2/\text{sr}$ at 532 nm. The total collection efficiency (optical and electronic including PMT quantum efficiency), including clear air attenuation, is $\eta_m\exp[\ldots] = 0.01$, and the chosen vertical resolutions are 10% of the height z (i.e., $\Delta z = 0.1$ km, 0.5 km, and 1 km) for $z = 1$ km, 5 km, and 10 km. We then calculate the required measurement time, τ_{int}, to acquire 10^8 photons from the standard atmosphere at 1 km, 5 km, and 10 km with $\mathcal{N}(z)$ of $2.31{\times}10^{25}$, $1.53{\times}10^{25}$, and $5.42{\times}10^{24}\ \text{m}^{-3}$. Using $h\nu_L = 1.9864 \times 10^{-16}\text{J}/\lambda$ (with λ in nm), we determine from (7.20) the required measurement time for the tabulated temperature error for 532 nm at 1 km, 5 km, and 10 km to be about 0.90, 6.83, and 38.5 min, respectively. For these same integration times and vertical resolutions required to produce $N_m = N_0 = 10^8$, the temperature error for the potassium filter (center) system scaled to 532 nm is estimated to be 0.17 K, while that for iodine filter system is 0.5 K; see Table 7.3. The calculated $(PA_R)\tau_{\text{int}}(\Delta z)$ product for the hypothetical system at 532 nm at $z = 1$ km turns out to be 0.045 Wm^2-min-km; this may be compared to the same for the experimental barium system (3 mW, $A_R = 0.0324\ \text{m}^2$, 20 min, and 375 m resolution) and iodine (6 W, $A_R = 0.0324\ \text{m}^2$, 1 hr, and 300 m resolution) system of $7.3{\times}10^{-4}$ and 3.5 Wm^2-min-km, respectively. At 1 km, the measured temperature error for the barium filter system at 553.7 nm [7.19] is 8 K (8.5 K when scaled to 532 nm), which becomes 1.08 K when scaled to the $(PA_R)\tau_{\text{int}}(\Delta z)$ of the hypothetical reference system. Its temperature uncertainty is 6.4x larger than the 0.17 K for the potassium filter (center) system in Table 7.3. The measured error of the experimental iodine filter system [7.13] at 1 km is 0.25 K, which becomes 0.7 K when scaled to the shared $(PA_R)\tau_{\text{int}}(\Delta z)$, 1.4x larger than the 0.5 K for the iodine filter system in Table 7.3. Owing to the possible differences in the total collection efficiency, $\eta_m\exp[\ldots]$, and the real single-photon uncertainty ξ_{opt} of the experimental systems compared to those given in Table 7.3, it is gratifying that with educated assumptions, the results of both measurements at 532 nm and at 553.7 nm appear to be consistent (both larger by factors under 10) with the hypothetical system. Despite what PA_R value is needed to deliver 10^8 received Cabannes photons, it will not affect the relative figures of merit between the four lidars as outlined in Table 7.3.

Since the estimated measurement times for the modest hypothetical lidar system appear to be reasonable, the proposed potassium filter system in Table 7.3 could lead to a realistic lidar if implemented. It may be also beneficial to consider other pairs of (H.T./L.T.) filters. Similar analysis can be made for a system at 589 nm using sodium filters, which will enjoy a more favorable vapor mass limitation and wavelength scaling for Cabannes scattering from 532 nm. However, its wider ground-state hyperfine splitting, $\Delta\nu = 1.772$ GHz, compared to $\Delta\nu = 0.4618$ GHz for K, limits the available bandwidth of the L.T. filter, thus reducing the filtered signal and sensitivity.

7.4 Wind Profiling with Cabannes–Mie Scattering

When a monochromatic laser at frequency ν impinges upon aerosols or an air parcel moving with a mean velocity along the laser beam, V (line-of-sight wind), the frequency of the backscattered light, ν_s, is Doppler shifted by an amount $\Delta\nu \equiv \nu_D = \nu_s - \nu$ that depends on the line-of-sight (LOS) wind and laser wavelength λ, as given in (3.1.a) and repeated as (7.21):

$$\nu_s = \nu - 2V/\lambda; \quad \Delta\nu \equiv \nu_D = -2V/\lambda\,. \tag{7.21}$$

Note that the LOS wind, V, is the average radial velocity of all scatterers in the viewing volume during a measurement with integration time τ_{int} and range (vertical) resolution $\Delta r(\Delta z)$. Strictly speaking, though averaged to the value V, the LOS velocities of individual scatterers and their Doppler shifts ν_D are different; they fall within velocity and frequency distributions of aerosols and molecules. For aerosols moving with the wind, the distribution is taken to be the normalized laser line-shape function, $L(\nu)$ (assumed to be Gaussian). This, as can be seen in Fig. 7.1, sits atop the normalized Cabannes backscattering spectrum, $\mathscr{R}_i(\nu)$, for air molecules – also moving with the wind but with additional random thermal motion. By monitoring the mean Doppler shift, $\overline{\nu_D}$, the LOS wind, V, can be measured. Based on this principle, various lidar Doppler wind measurement schemes have been devised. Different techniques for measuring air flow for specific laboratory or similar applications were reviewed in the book chapter by Christian Werner [7.47]. Our emphasis is on pulsed Doppler wind lidar, which is capable of measuring atmospheric wind velocity in different atmospheric layers. These wind lidars are classified according to their detection schemes as coherent and incoherent lidar, which employ, respectively, heterodyne and direct detection techniques. Lidar systems that utilize coherent detection methods – typically at longer wavelengths to reduce the effect of atmospheric turbulence – are a demonstrated effective tool for wind measurements in the lower troposphere, typically at the planetary boundary layer where considerable aerosol scattering is present. Those deployed include, for example, ground-based systems with CO_2 laser transmitters at 10.6 μm [7.48] and airborne [7.49] employing a Ho:Tm:YLF laser near 2.05 μm [7.50]. The direct detection method has been employed at visible (532 nm) and UV (355 nm) wavelengths to benefit from higher Cabannes-scattering cross section to probe the upper troposphere, stratosphere, and mesosphere, where aerosols are absent. These varied techniques and their comparisons have been discussed extensively in the literature up to 2005, as well as reviewed in two book chapters [7.47, 7.51]. Briefly, when scattered signal power $P_s(t) = h\nu N_s(t)/\tau_{\text{int}}$, with signal photocounts $N_s(t)$ in an integration time τ_{int} (comprising a number of laser pulses), is received by a photodetector, a photocurrent $i(t)$ is produced. These are related as

$$i(t) = \frac{\eta\, e\, g\, N_s(t)}{\tau_{\text{int}}} \rightarrow \bar{i} \equiv \langle i(t) \rangle = \frac{\eta\, e\, g \langle N_s(t) \rangle}{\tau_{\text{int}}} = \frac{\eta\, e\, g\, \overline{N_s}}{\tau_{\text{int}}}. \tag{7.22.a}$$

$$\therefore \overline{(\Delta i)^2} = \frac{(\eta\, e\, g)^2 \left\langle \left(N_s(t) - \overline{N_s}\right)^2 \right\rangle}{(\tau_{\text{int}})^2} = \left(\frac{\eta\, e\, g}{\tau_{\text{int}}}\right)^2 \overline{N_s} \rightarrow SNR \equiv \frac{\bar{i}}{\sqrt{\overline{(\Delta i)^2}}} = \frac{\overline{N_s}}{\sqrt{\overline{N_s}}}. \tag{7.22.b}$$

In the presence of additional noise counts, N_n, $SNR = \dfrac{\overline{N_s}}{\sqrt{\overline{N_s} + \overline{N_n}}}$, (7.22.c)

where e, η, and g are, respectively, the electron charge, photodetector quantum efficiency, and gain (for a photomultiplier, for example). The angle bracket stands for average over a measurement time, which is also designated by an over bar. As is well known, the incoming photocounts, $N_s(t)$, obey Poisson statistics, and thus have standard deviation $\overline{|\Delta N_s|} = \sqrt{\overline{N_s}}$ and variance $\overline{(\Delta N_s)^2} \leq \left(N_s(t) - \overline{N_s}\right)^2 \geq \overline{N_s}$, as employed in (7.22.b). Notice that we define signal-to-noise ratio, SNR, as the ratio of signal counts to the standard deviation of the signal counts, though others in the literature have defined SNR as the square of this ratio. In our definition, SNR is the inverse of fractional uncertainty of the signal. The presence of additive (sky) background noise counts enhances fluctuations but does not increase signal, resulting in a decrease in SNR as given in (7.22.c).

Realizing that the optical signal detected comes in the form of a series of random pulses following Poisson statistics, the simplest way to detect it is to count them, that is, photon counting with a fast photodetector, such as a photomultiplier at visible or UV wavelengths; this is called *direct detection*. These incoming pulses (including optical background) are amplified and sorted by amplitude, discriminating with low and high thresholds, in order to accept pulses within a specified peak voltage range, deeming pulses too small or large to have been initiated by other (noise) sources. Each accepted pulse is then turned into a standard square pulse with fixed amplitude and width, say –1 V and 10 ns. Detection then amounts to counting photons (standardized electrical pulses), with all other noise except the photon count fluctuations eliminated. This is the process of quantum (shot) noise limited direct detection.

Coherent detection lidars employ longer wavelengths (10.6 or 2 μm, for example) because those wavelengths are less impacted by atmospheric turbulence, which degrades spatial coherence across the detector. At these wavelengths, while the photon number is higher for the same laser energy, the available detectors are also much slower than at shorter wavelengths. Thus, photon counting is not practical and analog detection is used. Heterodyne mixing of the signal with a local oscillator then diminishes the influence of other noise sources and actually achieves quantum (shot noise)-limited detection. Such a lidar system uses injection seeding to precisely control the frequency of the pulsed laser. The received pulsed scattered light is then routed to the detector together with a sample of the CW seed laser, called the local oscillator (LO), which has been frequency upshifted from the seeded pulsed laser. The detector receives an oscillatory signal at their difference frequency, which is in the range of radio frequencies (RF). The strength of the LO greatly enhances this RF heterodyne signal, thus its shot noise then dominates fluctuations, rendering contributions from other noise sources negligible. The detection of the desired scattered light signal approaches its quantum (shot noise) limit. The photocurrent is the sum of the scattered optical signal, the RF-shifted LO, and an RF beat resulting from the optical heterodyning:

$$i(t) = i_s(t) + i_{LO}(t) + i_{RF}(t)\cos\left[2\pi(\nu_s - \nu)t + \phi\right], \text{ with } i_{RF}(t) = 2\sqrt{i_s(t)i_{LO}(t)}. \tag{7.23.a}$$

where ϕ is the relative phase between the signal and LO optical electric fields. Following Section 6.1 of Kingston [7.52], we can calculate $< i_{RF}^2(t) >$ and $\overline{(\Delta i)^2}$ as

$$\left\langle i_{RF}^2(t)\right\rangle = \overline{i_{RF}^2} = 4\overline{i_s}\,\overline{i_{LO}}\left(\frac{1}{2}\right) = 2\overline{i_s}\,\overline{i_{LO}};$$

$$\overline{(\Delta i)^2} = \left\langle\left(i(t) - \overline{i}\right)^2\right\rangle \approx \left\langle\left(i_{LO}(t) - \overline{i_{LO}}\right)^2\right\rangle = \overline{i_{LO}}, \tag{7.23.b}$$

Similar to (7.22), the various components of the current may be related to respective photocounts in the integration time, τ_{int}, and the heterodyne detection signal-to-noise ratio, termed carrier-to-noise ratio (*CNR*), determined as

$$CNR = \frac{\overline{i_{RF}}}{\sqrt{\overline{(\Delta i)^2}}} = \frac{\sqrt{2\overline{i_s}\,\overline{i_{LO}}}}{\sqrt{\overline{i_{LO}}}} = \sqrt{\frac{2\overline{N_s}\,\overline{N_{LO}}}{\overline{N_{LO}}}} = \frac{\sqrt{2}\,\overline{N_s}}{\sqrt{\overline{N_s}}} \xrightarrow{\text{Adding } N_n} CNR = \frac{\sqrt{2}\,\overline{N_s}}{\sqrt{\overline{N_s}+\overline{N_n}}}. \tag{7.23.c}$$

For a more detailed discussion comparing shot-noise limited coherent and incoherent detection, see Sections 4.4 and 6.1 of [7.52]. The main difference lies in whether the shot-noise limit includes or not, respectively, the factor $\sqrt{2}$ in (7.23.c) and (7.22.c), implicating that in the shot-noise limited case, the noise equivalent power for coherent detection is half that for direct detection; see p. 126 [7.52].

Depending on the scattering process and the measurement technique for the determination of mean Doppler shift $\overline{\nu_D}$, there exist intrinsic root-mean-square (*RMS*) fluctuations, statistically standard deviation $(\delta\nu_D)_{rms}$, for the measured mean in the frequency domain. One can achieve this intrinsic level of uncertainty without much effort (i.e., it can be attained with unity *SNR* or *CNR*). Thus, with better signal-to-noise, the accuracy of the mean Doppler shift or LOS wind can be improved. The uncertainty of the measured LOS wind may be expressed as for direct and coherent systems:

$$\begin{aligned} \left(\delta\overline{V}\right)_{dir} &= \frac{\lambda}{2}\frac{(\delta\nu_D)_{rms}}{SNR} = \frac{\lambda}{2}(\delta\nu_D)_{rms}\frac{\sqrt{\overline{N_s}+\overline{N_n}}}{\overline{N_s}}, \quad \text{and} \\ \left(\delta\overline{V}\right)_{Coh} &= \frac{\lambda}{2}\frac{(\delta\nu_D)_{rms}}{CNR} = \frac{\lambda}{2}(\delta\nu_D)_{rms}\frac{\sqrt{\overline{N_s}+\overline{N_n}}}{\sqrt{2}\,\overline{N_s}}, \end{aligned} \tag{7.24}$$

where $\overline{N_s}$ and $\overline{N_n}$ are, respectively, the total detected signal photocounts and total noise counts during the experiment. For direct detection, the noise counts mainly result from sky background (i.e., $\overline{N_n} = \overline{B}$). In the case of coherent detection, since the signal is the result of beating between the signal field and the LO field, its effectiveness demands having temporal and spatial coherence between both optical fields. However, in practice, the random nature of the scatterers (moving or stationary, aerosol or solid target) in the viewing volume will produce speckle patterns [7.53] on the detector, which degrades the coherence and is an important contribution to $\overline{N_n}$. In addition, the speckle-induced phase fluctuations or speckle saturation in the signal will cause additional uncertainty.

Including these effects, the variance of mean LOS velocity may be written as in Eqn. 7.162 of [7.51] or Eqn. 4 of [7.54] and is repeated here as (7.25):

$$\mathrm{var}(\overline{V}) = \left(\frac{\lambda}{2}\right)^2 (\delta v)^2_{rms} \left[\frac{\kappa}{\overline{N}_s} + \frac{\kappa \overline{N}_n}{\overline{N}_s^2} + \frac{1}{2M_e}\right], \tag{7.25}$$

where κ is a constant that depends on the design and efficacy of the individual receiver. The last term in (7.25) is the result of speckle-induced phase fluctuations on the signal [7.53], or of speckle saturation [7.54]; it is inversely proportional to M_e, described in [7.54] as "the total effective diversity of the measurement, including independent pulses, independent coherence times within a range gate, independent range gates, independent polarizations, independent spatial samples (detectors), etc." The topic of speckle-modulation and the definition and analysis of M_e, as well as the associated effects on coherent wind lidar design is involved and subtle, and is beyond the scope of this book. It suffices to point out that the dominant $\overline{N}_n$ in coherent detection also depends on M_e, suggesting that the second and third terms of (7.25) represent a trade-off in determining the optimum value of M_e for a coherent detection experiment. As Henderson et al. show in Fig. 7.35 and described on p. 603 of [7.51]: "For any total detected photon count, $\overline{N}_s$, the optimal performance is achieved when the number of independent modes, or diversity, M_e, closely matches that count, so that only a few photons per diversity mode are detected." In addition, since coherent detection demands spatial coherence, and a light beam propagating in the atmosphere is no longer diffraction-limited due to atmospheric turbulence that causes random fluctuations in refractive index, the domain of spatial coherence is reduced. This makes a telescope with radius larger than the wavelength-dependent Fried radius ($r_0 \propto \lambda^{6/5}$) less effective for coherent detection. We will revisit this topic briefly in Chapter 8 when we discuss Adaptive Optics (AO) for large astronomical telescopes.

Our emphasis is on wind measurement based on Cabannes scattering with techniques that filter out aerosol scattering. For this case, because the spectral width of Cabannes scattering is much broader than that of aerosol scattering, direct detection is preferred. Here, the receiving telescope is effectively a photon bucket and the questions of spatial coherence and diversity in detection are not of fundamental concern.

Before we turn to this main topic, we shall briefly consider examples of aerosol scattering and atmospheric scattering, hoping to shed light on the complimentary nature of direct and coherent detection techniques. For this purpose, we consider the single-photon velocity uncertainty, defined as the LOS wind uncertainty retrieved by a single signal photon without noise by substituting $\overline{N}_s = 1$ and $\overline{N}_n = 0$ into (7.24), yielding the single-photon velocity uncertainties,

$\xi_{Coh}^V = \lambda(\delta\nu_D)_{rms}/(2\sqrt{2})$ and $\xi_{Dir}^V = \lambda(\delta\nu_D)_{rms}/2$. Thus, in the shot-noise limit, for the same laser wavelength λ and intrinsic RMS spectral width $(\delta\nu_D)_{rms}$, there is only a $\sqrt{2}$ difference in the single-photon LOS wind uncertainty (i.e., $\xi_{Coh}^V = \xi_{Dir}^V/\sqrt{2}$). The main determining factor for choosing between coherent versus direct detection lies in the nature of the scatterers, which determines the intrinsic spectral width, thus the single-photon LOS-wind uncertainty. For aerosol scattering, $(\delta\nu_D)_{rms}$ is determined by laser-shape function $L(\nu)$ – for a Gaussian laser beam, it is the RMS width of its power spectral density $(\delta\nu_D)_{rms} = (\Delta\nu)_{rms}$ – which is related to the RMS pulse width $(\Delta t)_{rms}$ by $(\Delta t)_{rms}(\Delta\nu)_{rms} = 1/(4\pi)$, or equivalently by $(\Delta t)_{FWHM}(\Delta\nu)_{FWHM} = 0.44$ [7.55]. For example, for a 10 Hz 2 μm laser with $(\Delta t)_{FWHM} = 180$ ns, the single photon LOS-wind uncertainty is $\xi_{Coh}^V = 1.73$ m/s, and it is $\xi_{Coh}^V = 11.3$ m/s for a 20 Hz laser at 532 nm with $(\Delta\nu)_{FWHM} = 60$ MHz. The pulsed narrowband scattered light from aerosols is likely strong, yielding bursts of highly overlapped photons that make photon counting difficult, if not impossible. Thus, the coherent detection process with designed optimal diversity and working with RF modulation can be used to advantage. In the case of Cabannes atmospheric scattering, the small Cabannes-scattering cross section ensures no scattering photon pileup within the scattered return from a given laser pulse. The *RMS* width of the broad Cabannes spectrum, assuming Doppler-broadened for simplicity (see Section 2.2) is $\sigma_\nu^{CS} = (2/\lambda)\sqrt{k_B T/m} = (2/\lambda)\sqrt{8314T/M}$ GHz, where λ, T, and M are expressed in nm, K and amu, respectively. For atmosphere ($M = 28.97$ amu) at 275 K and $\lambda = 532$ nm, $(\delta\nu_D)_{rms}^{532nm} = 1.056$ GHz. Compared to the width of aerosol spectrum of 60 MHz, the single photon velocity uncertainty of Cabannes scattering is ~ 25 times larger (i.e., $\xi_{Dir}^V = 532 \times 1.056/2 = 281$ m/s), requiring 625 times more photons to obtain the same accuracy in retrieved LOS velocity. In direct detection, one can use a large telescope despite atmospheric turbulence and accumulate signal from a great number of laser pulses by integration, while in the coherent detection scheme one would have to worry that the increase in diversity, M_e, could cause $\overline{N_n}$ in the second term in (7.25) to increase. Furthermore, to accommodate the entire broad Cabannes spectrum, the required RF frequency for the coherent detection would be too high to be practical. The basic conclusion is that the coherent detection scheme is more suitable for aerosol scattering, while direct detection works better for atmospheric scattering.

7.4.1 An Overview of Direct Detection Methods

We are interested in line-of-sight (LOS) wind measurements based on atmospheric Cabannes scattering, thus we concentrate on direct detection methods. We first briefly compare the classical use of a Fabry–Perot interferometer (FPI) with an iodine vapor filter (IVF) as the frequency discriminator. A method employing dual FPIs for Cabannes lidar frequency analysis was first demonstrated by Chanin et al. [7.56]. Korb et al. later performed analyses on wind measurements with a single [7.57] and a dual [7.58, 7.59] FPI, denominating them as single-edge and (symmetric) double-edge FPS, respectively. Since the free spectral range (FSR) of a FPI may be chosen freely, it is a common practice to use a FPI with a smaller FSR at longer laser wavelengths, for example, 1064 nm for wind measurement where aerosol scattering dominates [7.60], and a FPI with larger FSR at shorter laser wavelengths, for example, 355 nm, where molecular scattering dominates [7.61]. Utilizing the iodine filter technique pioneered at Colorado State University (for temperature measurements), Liu et al. proposed the use of IVF for frequency analysis of backscattered light from both aerosols and air molecules in the troposphere for wind measurements with a laboratory demonstration in 1997 [7.62]. In the same year, this wind measurement technique was independently demonstrated in the tropical stratosphere, where aerosol influence is negligible [7.63]. Unlike the FPI, the location of an absorption line of an IVF cannot be chosen freely, although its linewidth may be adjusted by changing the cell length and/or vapor temperature and pressure. Fortunately, since the wavelength of a modern monolithic doubled Nd:YAG laser at 532 nm can be tuned over a 60 GHz range, any of several strong absorption lines (see Fig. 7.2) may be implemented around 532 nm for wind measurements. The frequency of the laser for the work cited was tuned to the midpoint of an absorption edge of the IVF, giving rise to a single-edge method, termed se-IVF here. To employ the double-edge method with an IVF, termed de-IVF, the laser frequency must then be shifted alternately to the midpoint of the other absorption edge of the same absorption line. The se-IVF for tropospheric wind measurements was implemented in a van in the early 2000s as a mobile ground-based incoherent Doppler wind lidar (MIDWiL) by the group at the Ocean University of China [7.64], leading to research and applications with lidar wind measurements that continue at the time of this writing [7.65, 7.66]. The se-IVF was also implemented in 2010 [7.67] at the ALOMAR RMR lidar facility [7.68], upgrading the facility's atmospheric wind measurement capability at 532 nm with a measurement range from 20 km to beyond 80 km.

To gain a general understanding of direct detection Cabannes-scattering lidar wind techniques, we summarize the analysis of She et al., 2007 [7.37] and compare the performance of atmospheric wind measurements with a Cabannes–Mie lidar using filters with nearly the same maximum transmission of ~ 80%, fielded at different wavelengths. We configure these frequency analyzers into four receiving scenarios for analysis and comparison:

(a) a double-edge Fabry–Perot interferometer (FPI) at 1064 nm (IR-FPI)
(b) a double-edge Fabry–Perot interferometer at 355 nm (UV-FPI)
(c) a single-edge iodine vapor filter (se-IVF)
(d) a double-edge iodine vapor filter (de-IVF) both at 532 nm.

These are shown schematically in Fig. 7.22 along with the appropriate Cabannes–Mie spectrum.

Briefly, we consider direct detection LOS measurements with a received signal of N photon counts for a specified temporal and vertical resolution and altitude split into two channels with fractions η_1 and η_2 ($\eta_1 + \eta_2 = 1$). For simplicity, we assume unity quantum efficiency for the photodetectors, so the detected photons in the two channels, N_1 and N_2, written in terms of filter transmittance, $f_i(\nu_D, z)$, and aerosol backscattering ratio, $R_b = \beta_a/\beta_c$, may be expressed as:

$$\begin{aligned} &N_i = N\eta_i f_i(\nu_D, z),\ i = 1,\ 2, \quad \text{with} \\ &f_i(\nu_D, z) = R_b f_{ai}(\nu_D, z) + f_{ci}(\nu_D, z),\ \text{with}\ R_b = \beta_a/\beta_c\,, \end{aligned} \tag{7.26.a}$$

where β_a and β_c are aerosol (Mie) and Cabannes (molecular) backscattering coefficients. Equation (7.26.a) is intended to replicate Eq. (1a) of [7.37], except for the notation change in detection efficiency (from ξ to η). The $f_{ai}(\nu_D, z)$ and $f_{ci}(\nu_D, z)$ are the transmittance for aerosol and molecular scattering through the filter function, $F_i(\nu)$, given respectively as

$$f_{ai}(\nu_D, z) = \int G_i(\nu - \nu_D) F_i(\nu) d\nu;\ f_{ci}(\nu_D, z) = \int \mathscr{R}_i(\nu - \nu_D, T, P) F_i(\nu) d\nu. \tag{7.26.b}$$

Irrespective of our choice of IR-FPI or UV-FPI, both channels use the same filter with peak transmission symmetrically shifted to higher or lower frequencies with $\eta_1 = \eta_2 = 0.5$, and the filter functions, $F_i(\nu)$, are the transmission functions of the Fabry–Perot étalons used, analytically expressed as the Airy function [7.69]:

$$\begin{aligned} F_{FPI}(\nu - \nu_i, \Delta\nu_{FWHM}) &= T_{\max}\left[1 + \left(\frac{2F}{\pi}\right)^2 \sin^2\left(\pi\frac{\nu - \nu_i}{\Delta\nu_{FSR}}\right)\right]^{-1}; \\ \Delta\nu_{FWHM} &= \frac{2\Delta\nu_{FSR}}{\pi}\sin^{-1}\left(\frac{\pi}{2F}\right), \end{aligned} \tag{7.26.c}$$

where ν_i ($i = 1, 2$) are the frequencies at the peak transmissions, up- and down-shifted, respectively, from the laser frequency for FPI1 and FPI2. The symbols $T_{\max}$, $\Delta\nu_{FSR}$, $\Delta\nu_{FWHM}$, and F are, respectively, maximum

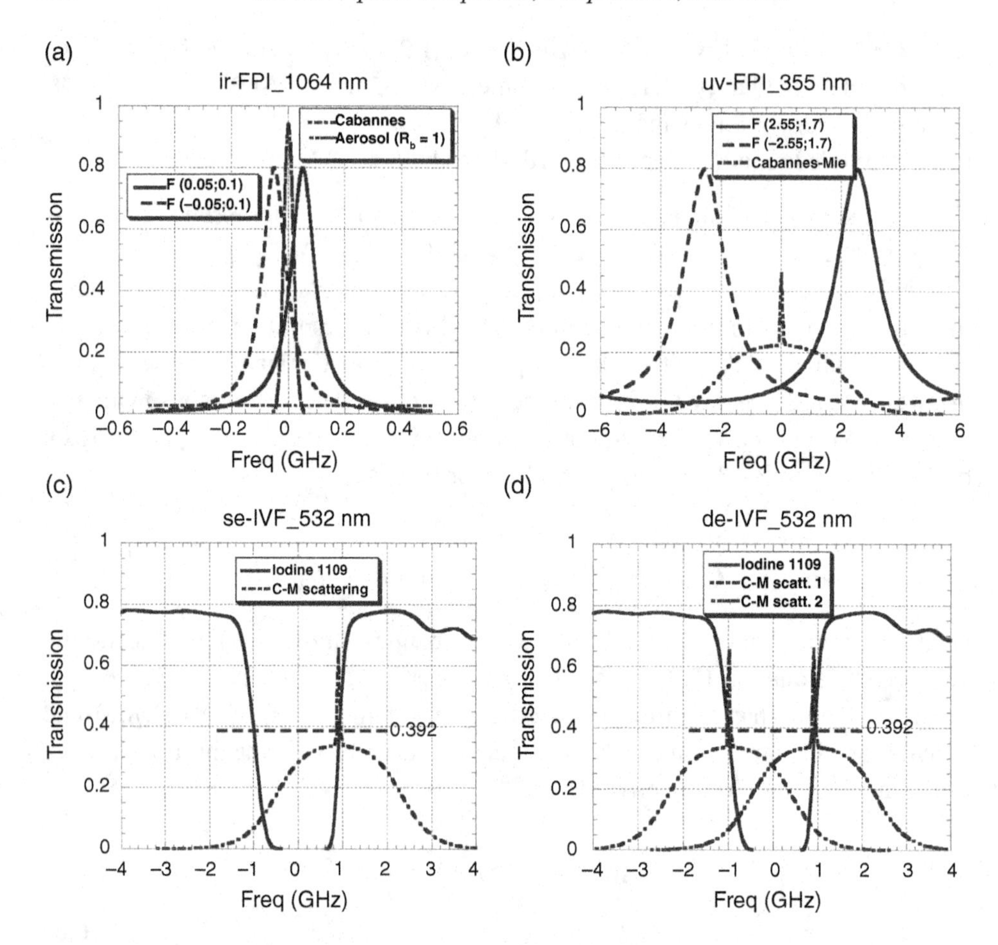

Fig. 7.22 Transmission functions of Fabry–Perot interferometers (FPI) and iodine vapor filters (IVF) together with the appropriate Cabannes–Mie scattering spectrum for (a) IR-FPI at 1064 nm, (b) UV-FPI at 355 nm, (c) se-IVF at 532 nm, and (d) de-IVF at 532 nm. Reproduced with permission from Fig. 1 of [7.37]. © the Optical Society.

transmittance, free-spectral-range, spectral resolution, and finesse of the respective FPI. Note that $F = \Delta\nu_{FSR}/\Delta\nu_{FWHM}$ for a high-finesse interferometer, such as the IR-FPI.

In the case where an iodine filter is the frequency analyzer, there is only one channel (say channel 1) with the iodine filter, and the other channel (say channel 2) is a reference channel without filter (but otherwise identical to channel 1). The transmission function of the IVF, $F_i(\nu) = F_{IVF}(\nu - \nu_1)$, is the measured transmission function of a 10-cm-long iodine filter with cell and finger temperatures set at 55 C and 50 C, respectively. We direct most of the received photons to channel 1 for the wind measurement, setting, for example, $\eta_1 = 0.9$ and $\eta_2 = 0.1$. For the se-IVF case, the laser frequency is locked at the center of one filter edge as shown in Fig. 7.22(c), and the count ratio for LOS wind retrieval is $R_W = N_1/N_2$, where the reference channel signal, N_2, as explained later, offers no wind measurement sensitivity, because $f_{a2}(\nu_D, z) = f_{c2}(\nu_D, z) = 1$. On the other hand, for the de-IVF case both N_1 and N_2 are derived from the IVF channel, with the difference being that they correspond to the respective integrations of odd and even pulses, for which the laser is tuned to the right (e.g., odd) and left (e.g., even) edge of the IVF; see Fig. 7.22(d). In order to account for the fact that half the laser pulses produce signal for each edge, we set $\eta_1 = \eta_2 = 0.5$ for the de-IVF. Note that the total number of counts N is shared between the two physically separated channels for IR-FPI or UV-FPI, while they are directed to the same physical channel for the de-IVF filter but are shared between odd and even laser shots. For se-IVF, though most signal is directed into channel 1, the reference channel (channel 2) offers no wind measurement sensitivity. By using the same number of signal counts, N, for all wind measurement methods, we compare the performance of the four methods on the same footing. The fact that 10% of the photons are used for the se-IVF reference channel, the figure of merit comparison, as presented in [7.37], directs 100,000 Cabannes-scattering photons to the IR-FPI, UV-FPI, and se-IVF, and 90,000 photons to de-IVF, as indicated in the text: "Figure 3(b) shows the SNR resulting from 100,000 $(R_b + 1)$ photons and 90,000 $(R_b + 1)$ photons for de-IVF, as a function of aerosol mixing ratio, $0<R_b<10$" [7.37]. This minor difference should not alter the qualitative difference exhibited between the four scenarios.

With this understanding, we present a simple description of lidar LOS wind measurements based on the detection of signal count ratios, from which we determine the mean LOS wind $\overline{V}$ and its uncertainty $\delta\overline{V}$ (for simplicity, we hide the over-bars). From this, we present and compare simulated results of the four scenarios discussed in [7.37], along with that of a proposed potassium vapor filter wind lidar [7.46].

7.4.2 Signal Ratio and the Figure of Merit of Direct Detection Wind Measurement

Like in temperature measurements in Section 7.2.3, the count ratio between the two channels R_W is used for the LOS wind retrieval. Similar to (7.10.a) and (7.10.b), we can derive the measurement sensitivity per unit (Doppler) frequency shift, S_{ν_D}, or per unit LOS wind $S_V = -2S_{\nu_D}/\lambda$, and LOS wind uncertainty as follows:

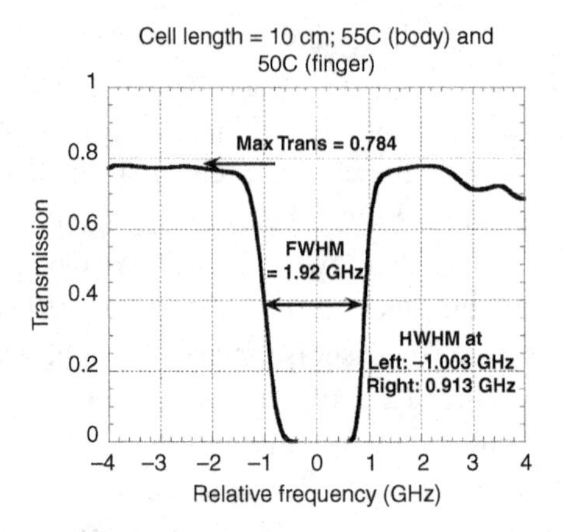

Fig. 7.23 Transmission function of the 1109 iodine absorption line with a 10 cm-long vapor cell. Reproduced with permission from Fig. 2(a) of [7.37]. © Optical Society.

$$R_W = \frac{N_1}{N_2} = \frac{\eta_1 f_1(\nu_D, z)}{\eta_2 f_2(\nu_D, z)} \rightarrow \frac{\delta R_W}{R_W} = \frac{\delta N_1}{N_1} - \frac{\delta N_2}{N_2} = \left[\frac{\partial f_1}{f_1 \partial \nu_D} - \frac{\partial f_2}{f_2 \partial \nu_D}\right]\delta\nu_D, \text{ and} \tag{7.27.a}$$

$$S_{\nu_D} \equiv \left[\frac{\partial f_1}{f_1 \partial \nu_D} - \frac{\partial f_2}{f_2 \partial \nu_D}\right] \rightarrow \delta\nu_D = \frac{1}{S_{\nu_D}}\sqrt{\frac{(\delta N_1)^2}{N_1^2} + \frac{(\delta N_2)^2}{N_2^2}} = \frac{1}{S_{\nu_D}}\sqrt{\frac{1}{N_1} + \frac{1}{N_2}}. \tag{7.27.b}$$

$$\therefore \delta V = \frac{\lambda}{2S_{\nu_D}}\sqrt{\frac{1}{N_1} + \frac{1}{N_2}} = \frac{\lambda}{2S_{\nu_D}}\frac{\xi}{\sqrt{N}} = \frac{\xi}{S_V\sqrt{N}} = \frac{\xi_{Div}^V}{\sqrt{N}}, \text{ with } \xi = \sqrt{\frac{1}{\eta_1 f_1} + \frac{1}{\eta_2 f_2}},$$
$$\text{when optimized, } \eta_1^{opt} = \frac{\sqrt{f_2}}{\sqrt{f_2} + \sqrt{f_1}} \text{ and } \eta_2^{opt} = \frac{\sqrt{f_1}}{\sqrt{f_2} + \sqrt{f_1}} \rightarrow \xi_{opt} = \frac{1}{\sqrt{f_2}} + \frac{1}{\sqrt{f_1}}. \tag{7.27.c}$$

Like (7.10.c), the quantity ξ is the single-photon uncertainty, while ξ_{Div}^V, the single-photon velocity uncertainty for direct detection LOS wind lidar, defined earlier as the uncertainty that can be attained with unity SNR (i.e., $\xi_{Div}^V = \xi/S_V$). Notice from (7.27.b) and (7.27.c), both the sensitivity, S_{ν_D}, and the single-photon velocity uncertainty, ξ_{Div}^V, are independent of the total photon counts, N. For the se-IVF wind retrieval, channel 2 is a reference channel with no filter, and the transmittance of aerosol and molecular scattering, $f_{a2}(\nu_D, z) = f_{c2}(\nu_D, z) = 1$, is independent of Doppler shift ν_D. This leads to the second term in S_{ν_D} being precisely zero.

7.4.3 Figure of Merit Comparison between the Performance of FPI and IVF

With these preliminaries discussed, we now summarize the results of the analysis and comparison in Fig. 7.24 based on [7.37], comparing the measurement sensitivity, S_V, (panel a) and LOS wind uncertainty, δV, (panel b) between the four measurement scenarios shown in Fig. 7.22. The parameters for these four analysis methods are taken from Table 1 of [7.37] and repeated here in Table 7.4.

As is typical, we evaluate the measurement sensitivity per unit LOS wind, at zero wind (i.e., $v_D = 0$). We summarize the comparison between the four scenarios by mostly quoting [7.37]: "For pure molecular scattering (stratospheric and lower mesospheric applications, i.e., $R_b = 0$, the sensitivities are 0, 0.0064, 0.0019, and 0.0038, in $(\mathrm{m/s})^{-1}$, respectively for IR-FPI, UV-FPI, se-IVF, and de-IVF." Because the width of Cabannes spectrum approaches the FSR of the IR-FPI and is much broader than the Doppler shift of the LOS wind at 1064 nm, molecular scattering makes no contribution to wind sensitivity with the IR-FPI, leading to zero sensitivity at $R_b = 0$. The sensitivity of de-IVF is twice that of se-IVF as expected. Since the peak transmission of UV-FPI lies in the wing of the Cabannes spectrum where wind sensitivity is much larger, at $R_b = 0$ the UV-FPI has more than 1.5 (3) times higher sensitivity than the de-IVF (se-IVF). This advantage gradually diminishes as R_b increases. Indeed, "As R_b increases to 0.5, the de-IVF and the IR-FPI

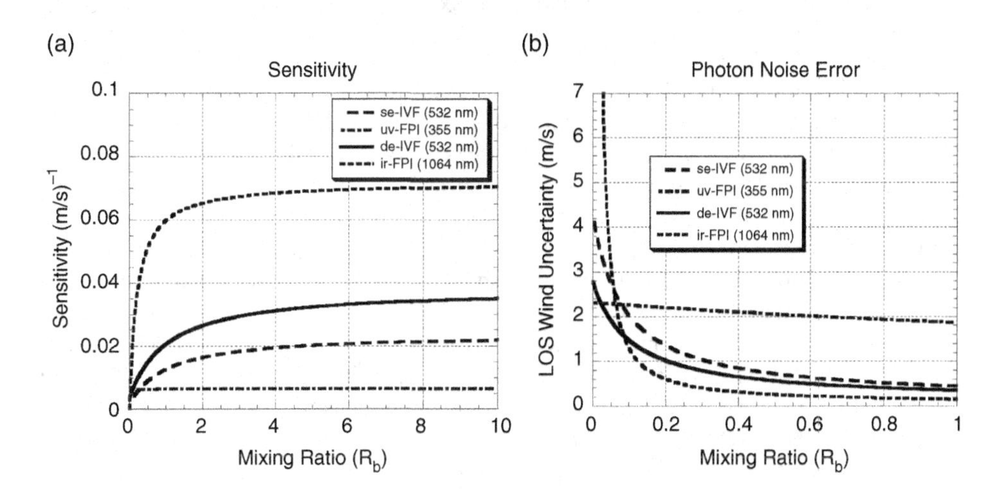

Fig. 7.24 (a) Measurement sensitivity for the four wind measurement scenarios as a function of aerosol mixing ratio, $0 < R_b < 10$. (b) LOS wind uncertainty based on 100,000 Cabannes photons (90,000 for de-IVF) received with an ideal photodetector for the four detection scenarios as a function of aerosol mixing ratio, $0 < R_b < 1.0$, for a clearer comparison in the lower R_b regime. They are, respectively, reproductions of Figures 3(a) and 4(a) of [7.37]. Reproduced with permission from Figs. 3(a) and 4(a) of [7.37]. © the Optical Society.

Table 7.4 *Parameters of the four analysis methods for the comparison study. Reproduced with permission from Table 1 of [7.37].* © *the Optical Society*

Analyzer	$\Delta\nu_{FSR}$†	$\Delta\nu$†a	$\mathcal{F}$	T_{max} [b]	f_{a1}	f_{a2}	f_{c1} [c]	f_{c2} [c]	η_1^{opt}	η_2^{opt}
UV-FPI	12	1.7	7.1	0.8	0.09	0.09	0.18	0.18	0.5	0.5
IR-FPI	3	0.1	30	0.8	0.4	0.4	0.08	0.08	0.5	0.5
de-IVF	–	1.92	–	~0.8	0.39	0.39	0.40	0.40	0.5	0.5
se-IVF	–	1.92	–	~0.8	0.39	1.0	0.40	1.0	0.6	0.4

† Frequency in GHz; $\mathcal{F}$= Finesse of the interferometer.
[a] This is the FWHM width of FPI or the well width of IVF.
[b] Maximum transmission of the four analyzers, which are nearly the same.
[c] Values based on Cabannes spectrum at 275 K and 0.75 atm.

already are, respectively, nearly 3 and 10 times more sensitive than the UV-FPI at 355 *nm*." "Since the slope of the right edge is a bit larger than the left edge of the IVF", see Figs. 7.22(c) and 7.22(d), "we expect the sensitivity of de-IVF to be higher than that of the se-IVF by less than a factor of 2. This difference, though negligible for $R_b = 0$, increases as R_b increases; as a result, the ratio of the sensitivities (de-IVF to se-IVF) is 1.99, 1.63, and 1.60 for $R_b = 0$, 1.0 and 10, respectively"; see Fig. 7.24(a). All quotes in this paragraph are from [7.37].

For the wind uncertainty comparison, we examine Fig. 7.24(b). Again, quoting from [7.37]:

"At $R_b = 0$ in the stratosphere and lower mesosphere, we see that the de-IVF and UV-FPI yielded comparable LOS wind uncertainties of 2.78 m/s and 2.33 m/s, respectively, while the se-IVF is higher with 4.31 m/s due to lower sensitivity. The LOS wind uncertainty of the IR-FPI at 1064 nm decreased from being the highest for $R_b = 0.06$ to being the lowest for $R_b \geq 0.09$. Our analysis suggests that the UV-FPI at 355 nm yields 16% (46%) lower LOS wind uncertainty compared to the de-IVF (se-IVF) at $R_b = 0$, while the latter outperforms the former when R_b is greater than 0.03 (0.08), i.e., under the condition with aerosol. When aerosol scattering dominates, in the planetary boundary layer for example, again assuming the same number of photons received, the use of IR-FPI is desirable; at $R_b = 0.7$, it yields 0.21 m/s, ~54% lower LOS wind uncertainty as compared to that for the de-IVF, 0.46 m/s. ... Generally speaking, with the direct-detection method for LOS wind measurements, the performance of the IR-FPI is the best for the planetary boundary layer and the least desirable for stratosphere and mesosphere, and the reverse is true for the UV-FPI. The de-IVF has a more consistent performance comparison throughout various atmospheric conditions with performance differences within ~50% of the best in the planetary boundary layer and in the stratosphere and mesosphere."

To facilitate comparison with the use of a potassium vapor filter (KVF) discussed below, we restate that at $R_b = 0$, the sensitivities are 0, 0.0064, 0.0019, and 0.0038, in $(\mathrm{m/s})^{-1}$, respectively, for the IR-FPI, UV-FPI, se-IVF, and de-IVF. At the same time, their corresponding LOS wind uncertainties for 100,000 Cabannes photons received are 0, 2.33 m/s, 4.31 m/s, and 2.64 m/s (scaled from 2.78 m/s for 90,000 photons). Using the optimum division between the two channels, with η_1^{opt} and η_2^{opt} listed in Table 7.4, we can calculate the single-photon uncertainty, ξ, and single-photon (intrinsic) velocity uncertainty, ξ_{Dir}^V, respectively to be

- $(4.71,\ 736\ \mathrm{m/s})$ for the UV-FPI,
- $(2.58,\ 1359\ \mathrm{m/s})$ for the se-IVF, and
- $(3.16,\ 832\ \mathrm{m/s})$ for de-IVF.

As a check, we divide these ξ_{Dir}^V by $\sqrt{100,000}$, and we indeed obtain the LOS wind uncertainties of 2.33, 2.63, 4.30 m/s at $R_b = 0$, 1.0, and 10, in agreement with Fig. 7.24(b).

It should be pointed out that the single-photon velocity uncertainties, ξ_{Dir}^V, for these scenarios are several times larger than the intrinsic $\xi_{Dir}^V = 281\ \mathrm{m/s}$ for the ~ 1 GHz-wide Cabannes-scattering spectrum as mentioned in the beginning of this Section. We wish to detect a Doppler shift, v_D, resulting from ~ 1 m/s LOS wind, a change of scattering frequency of ~ 1 MHz from the laser frequency (ca. 10^5 GHz or 10^8 MHz). Available optical detectors are simply too slow to follow the optical field, and successfully achieve 1 part in 10^8 experimental accuracy. With the help of a local oscillator field, optical heterodyning can lower the baseband optical field Doppler shift, v_D, to the RF (~ GHz) range (a manageable photocurrent frequency) – see (7.23.a) – and convert the 1 in 10^8 measurement problem to a 1 in 10^3 or 10^4 measurement problem. A fast optical detector can now measure GHz response and deduce the power spectral density of Cabannes scattering in the RF range. At the same time, it brings complications in the optical field (i.e., noise relating to spatial coherence and speckle modulation into the photocurrent).

By sensing optical intensity (photon counting) directly, direct detection has no such problem, but as pointed out, the detector is too slow to sense the Doppler shift in optical frequency. Optical filtering (such as the four scenarios considered or other similar techniques) allows determination of Doppler shift due to wind as well as random motions in the frequency domain and converts a frequency change to an intensity change, thus mitigating the detector speed problem (though it still needs to be fast enough to prevent photon pileup). The only price direct detection pays is that its single-photon velocity uncertainty, ξ_{Dir}^V, depending on the filtering technique used, can be several times larger than its intrinsic value. This problem is overcome by increasing the SNR of the experiment. If aerosols are responsible for the backscattering signal, the intrinsic rms width becomes $(\delta v_D)_{rms}^{532\mathrm{nm}} = 60$ MHz or $(\delta v_D)_{rms}^{2\mu\mathrm{m}} = 2.44$ MHz, about 17 or 410 times narrower than Cabannes scattering, thus increasing the accuracy of both direct and coherent scattering. However, since there are no aerosols in the upper atmosphere, direct detection is the only possible choice. Thus, we conclude that coherent detection is preferred for the lower atmosphere, while direct detection is preferred for the upper atmosphere.

7.4.4 A Proposed Direct Detection Potassium Filter–Based Temperature/Wind Lidar

Despite the presence of continuum absorption, Cabannes backscatter wind lidars employing iodine filters at 532 nm have been employed for LOS wind measurements for more than two decades [7.63, 7.64, 7.65, 7.66]. The iodine filter performs well in this case because it requires only one filter with moderate width. Alternatively, we can use the potassium vapor filter (KVF) previously proposed for temperature measurements to measure LOS wind. This option removes the concern of continuum absorption. It has the additional benefit of being a filter with a wider absorption band, increasing measurement sensitivity and reducing uncertainty somewhat. We can also explore the potential of the KVF lidar for simultaneous temperature and wind measurements. As pointed out in Section 7.3.4(2), setting the laser at –0.18 GHz, the KVF-2B, KVF-2A, and KVF-1 can respectively block aerosol scattering at levels of 20.4, 41.9, and 701 dB. Thus, we can perform the double-edge wind measurement with a left-edge frequency of $\nu_L = -0.180$ GHz and a right-edge frequency of ν_R. Due to the ground-state hyperfine splitting of 0.4618 GHz, the K D_1 absorption spectrum and its associated KVF transmission are asymmetric. For this reason, we set the right-edge frequency at $\nu_R = +0.208$ GHz, so that they are equally away from their respective 50% transmission point and enjoy equal capability for aerosol scattering rejection. By tuning the laser frequency cyclically to the right ν_R and the left ν_L of the D_1 absorption line of potassium, each filter (H.T. and L.T.) can perform an independent LOS wind measurement. In addition, the proposed two-filter CS/K lidar can be used for temperature measurement as well, leading to a device for simultaneous measurements of atmospheric temperature and LOS wind. This possibility has been analyzed and investigated in a recent article [7.46]. We refer the interested readers to this article without going into details, except to state here that this article outlines the procedure for data analysis and concludes that at the signal level of $N_0 = 1,000,000$ (not $N_0 = 10^8$), the uncertainties are respectively ~1K and 1.5 m/s. This LOS wind uncertainty is consistent with the result of 4.3 m/s for $R_b = 0$ shown in Fig. 7.24(b) at the signal level of $N_0 = 10^5$.

We note that a similar temperature/wind lidar with Cabannes scattering employing a Na double-edge magneto-optic filter was proposed [7.70] and demonstrated [7.71] in 2009. More recently, in 2014, by scanning the laser frequency, a Fabry–Perot interferometer has been used to measure atmospheric temperatures [7.72].

Incorporating the double-edge technique [7.57] with system-level optical frequency control [7.73], a scanning Fabry–Perot interferometer method has been used to measure stratospheric temperature and wind [7.74].

7.5 Mesopause Region Temperature and Wind Profiling with LIF

As the atmospheric density decreases exponentially with height, the signal from Cabannes scattering of air molecules quickly becomes dwarfed by background noise, even for a lidar with large PA product. This effectively precludes Cabannes lidar for practical measurements in the mesopause region (70–120 km). Fortunately, in this same region of the atmosphere, there exist stable concentrations of many metal atom species deposited by meteors, whose cross sections in the visible wavelengths can be 15 orders of magnitude stronger than that for Cabannes scattering. These present us with a resonance scattering tracer for atmospheric measurements. Assuming the availability of a laser source, a successful metal candidate is one whose product of metal concentration times absorption cross section is large [7.75]. A comparative study of metal lidar performances [7.76] suggests Na to be among the best due to its high absorption cross section and moderate atmospheric abundance. Neutral metal iron (Fe) is also very attractive; though it has much lower cross section, this is compensated for by its much higher atmospheric abundance. Since the Fe resonance occurs in the ultraviolet, there is much lower sky background interference, a definite advantage for lidar detection under sunlit conditions. Na LIF lidars have been widely implemented since the middle 1960s. Gibson et al. [7.77] was the first to report atmospheric temperature measurements with Na lidar in 1979, and the first practical system using a scanning narrowband dye laser was deployed by Fricke and von Zahn [7.78] in 1985. An ultra-narrowband system, employing an injection-seeded pulsed dye amplifier was deployed five years later by She et al. [7.79]. By locking the seed laser to Na D_2 peak and crossover frequencies, this group presented a practical system for temperature measurements. In 1994, it was upgraded to a system with three frequencies to make simultaneous temperature and wind measurements [7.80]. Further progress in temperature and wind measurements has been briefly reviewed [7.2]. In addition to the broadband Fe Boltzmann temperature lidar [7.81], considerable progress has been made in narrowband Fe lidar development at 386 nm [7.82] and 372 nm [7.83]. By using multiple Fabry–Perot étalons, Höffner and Lautenbach [7.82] were able to achieve an overall receiver bandwidth of ~ 2 GHz. Though the requirement of small field of view is a technical challenge, they achieved an impressive, detected signal-to-background ratio (S/B) of ~10 at noon. The 372 nm system for simultaneous temperature and wind measurements was developed by Chu et al. [7.83]. In what follows, we will discuss the salient features of resonance scattering with a simplified atomic system of one excited energy level. With these, we highlight the essential difference between Fe and Na atomic structure that suggests a simpler procedure for Fe lidar data processing than for Na. The main discussion of the Section is on the retrieval of temperature and wind from a narrowband Na lidar following the work of Krueger et al. [7.2], as this system has been widely deployed by researchers in the USA, Japan, and China.

7.5.1 The Salient Features of LIF as Illustrated by a Simple Atomic System and Iron Lidars

There are several differences between the process of nonresonant (Cabannes) scattering and resonant scattering (or laser-induced fluorescence, LIF). Firstly, the scattering cross section of LIF (at visible wavelengths) is about 15 orders of magnitude larger than nonresonant scattering. This means that for an incident laser frequency ν_L capable of inducing LIF from an atom (or molecule), its Cabannes scattering (though it exists) is safely ignored. Secondly, bulk movement of air along the beam (the wind velocity component along the beam) produces a different effect for Cabannes versus LIF scattering. In the case of Cabannes scattering, for an atom moving along the laser beam with speed υ, the backward scattering frequency is downshifted by $2\upsilon/\lambda$ according to (3.1.a). Atoms moving at different speeds backscatter light at different frequencies, producing an intensity distribution proportional to the Cabannes spectrum, which depends on ambient temperature and wind. In comparison, LIF consists of absorption followed immediately (within the lifetime of the excited state, ~ 10^{-8} s) by spontaneous emission. Only atoms whose beam-aligned velocity component Doppler shifts the laser frequency ν_L onto their atomic resonance transition frequency, ν_{Ff} (i.e., $\nu_L - \upsilon/\lambda = \nu_{Ff}$ in the atom's rest frame), can absorb laser photons. We begin with a simple atom with one excited state $|F>$ and two ground states $|f>$ and $|f'>$, as depicted in Fig. 7.25. According to (3.1.b), after excitation it will emit a photon at either of two discrete frequencies, $2\nu_{21} - \nu_L$ or $\nu_{21} + \nu'_{21} - \nu_L$, in the backward direction toward the ground-based detector. In doing so, the atom returns either to the initial ground state $|f>$ or to the other ground state $|f'>$. Since this provides no information on the atom's speed, analysis of LIF emission frequencies reveals no information on an atom's motion. Thus, unlike Cabannes scattering, frequency analysis cannot be employed to retrieve atmospheric temperature or wind. Fortunately, at a given laser frequency, ν_L, the absorption process selects a group of atoms with velocity in resonance with the beam for interaction and fluorescence emission. Since the probability of inducing absorption by the selected

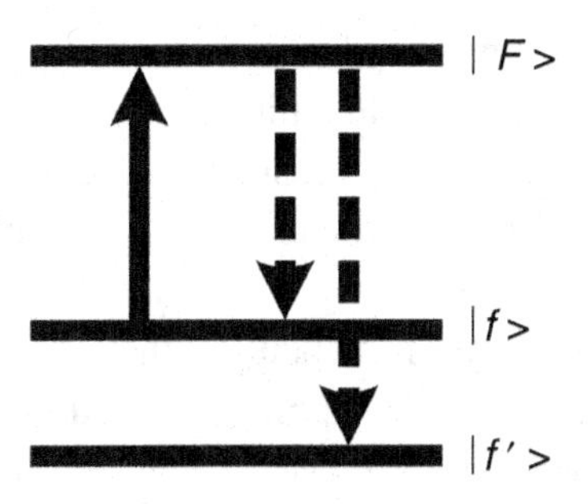

Fig. 7.25 The energy levels of a simple atom, showing excitation from $|f>$ to $|F>$, followed by emission to either $|f>$ or $|f'>$. See text for more information.

speed group of atoms depends on the atom's thermal velocity distribution, which in turn depends on atmospheric temperature and LOS wind, by analyzing the scattered signal strength properly, one can retrieve atmospheric temperature and wind. Thirdly, for one exciting laser frequency, ν_L, the LIF emission from one excited state $|F>$ occurs at more than one discrete frequency if there is more than one allowed lower state, such as $|f>$ and $|f'>$ shown in Fig. 7.25.

The existence of multiple allowed emission pathways (and thus wavelengths) for lidars of interest can be determined from the associated energy diagrams. For iron lidars, (energy diagram shown in Fig. A.2), the excitation at 372 nm, $(a\,^2D_4 \rightarrow z^5F^0_5)$ can emit only at the same (exciting) wavelength, while the 374 nm excitation to the excited state, $(z^5F^0_4)$, can emit either at 374 nm $(z^5F^0_4 \rightarrow a\,^2D_3)$ or at 368 nm $(z^5F^0_4 \rightarrow a\,^2D_4)$, and the excited state of the 386 nm excitation, $(z^5D^0_4)$, can emit to both lower states $a\,^5D_3$ and $a\,^5D_4$. The LIF backward differential cross section from initial lower state $|f>$ to excited state $|F>$ and back to either $|f>$ or $|f'>$ was derived in (3.21) as the sum of two terms, each corresponding to an allowed lower state as (7.28.a). The associated absorption cross section is given in (7.28.b).

$$\frac{d\sigma^{\pi}_{fF\Sigma}}{d\Omega} = \frac{\lambda^2_{Ff}}{32\pi^2}\frac{g_F}{g_f} A_{Ff}\boldsymbol{n}_f \mathbb{G}\left[(\nu_L - V/\lambda - \nu_{Ff}),\ T\right] \sum_{f'} \alpha^{\pi}_{fFf'}(0) \equiv \sum_{f'} \sigma^{\pi}_{fFf'}(\nu_L, T, V), \tag{7.28.a}$$

$$\text{and}\quad \sigma^{A}_{fF}(\nu_L, T, V) = \frac{\lambda^2_{Ff}}{8\pi}\frac{g_F}{g_f} A_{Ff}\boldsymbol{n}_f \mathbb{G}[(\nu_L - V/\lambda - \nu_{Ff}),\ T]\,. \tag{7.28.b}$$

Here, we have written out the contribution of each LIF pathway, $\sigma^{\pi}_{fFf'}(\nu_L, T, V)$, and note that the q-factor $q^{\pi}_{fF\Sigma}(0) = \sum_{f'} \alpha^{\pi}_{fFf'}(0)$ at zero magnetic field is the fluorescence rate sum of the two possible pathways, $\alpha^{\pi}_{fFf}(0) = \alpha_{Ff}(1 + 0.3g_F B^{(2)}_{fFf})$ and $\alpha^{\pi}_{fFf'}(0) = \alpha_{Ff'}(1 + 0.3g_F B^{(2)}_{fFf'})$, in which α_{Ff} and $\alpha_{Ff'}$ are the associated spontaneous emission rates from the excited state with degeneracy $g_F = 2F + 1$ and with the quantum structure coefficient $B^{(2)}_{fFf'}$ calculated from (3.10) via 6-j coefficients. $\mathfrak{G}[(\nu - V/\lambda - \nu_{Ff}),\ T]$ is a Voigt function, as given in (2.7.b), that results from the convolution of the natural linewidth with a Gaussian function depicting Doppler broadening and Doppler shift, and assuming excitation is by a truly monochromatic laser (ignoring laser lineshape for this purpose). The factor $\boldsymbol{n}_f$ is the fraction of the atomic population of the initial lower state $|f>$, which is in thermal equilibrium with the other lower state, $|f'>$ with fractional population $\boldsymbol{n}_{f'}$, given as:

$$\boldsymbol{n}_{f'} = \frac{g_{f'}}{g_{f'} + g_f \times \exp\left[-\ 0.04798 \times \left(\Delta\nu(\mathrm{GHz})/T\right)\right]};$$
$$\boldsymbol{n}_f = \frac{g_f \times \exp\left[-\ 0.04798 \times \left(\Delta\nu(\mathrm{GHz})/T\right)\right]}{g_{f'} + g_f \times \exp\left[-\ 0.04798 \times \left(\Delta\nu(\mathrm{GHz})/T\right)\right]} \quad (7.29)$$

where $\Delta\nu$ (GHz) is the frequency difference between the two lower states. For Na, $\Delta\nu = 1.772$ GHz, $\boldsymbol{n}_{f'} = 0.48$, and $\boldsymbol{n}_f = 0.52$ at $T = 200$ K, while for iron, $\Delta\nu = 415.4\ \mathrm{cm}^{-1} = 12{,}461.8$ GHz, $\boldsymbol{n}_{f'} = 0.96$, and $\boldsymbol{n}_f = 0.04$. Here, we justifiably assume only the two lowest states are thermally populated. For Na, the third state up (Fig. A.1), ${}^2P^o{}_{1/2}$, is at 16,956 cm^{-1}, much too high for thermal excitation at typical mesospheric temperatures near 200 K (or 139 cm^{-1}). For iron, the third state up, a^5D_2, is at 704 cm^{-1}, see Fig. A.2; not only is its thermal population negligibly small, selection rules prohibit this $J = 2$ ground state interacting with $J = 4$ or 5 excited states.

For iron lidar applications, we are interested in excitations at 372 nm, $(a^5D_4 \rightarrow z^5F^0_5)$, 374 nm, $(a^5D_3 \rightarrow z^5F^0_4)$, and 386 nm, $(a^5D_4 \rightarrow z^5D^0_4)$. The much stronger spin-orbit interaction in iron (than in sodium), as discussed in Section A.2.3, results in greater energy level splitting; see Fig. A.2. This large separation makes it impossible for a narrowband laser to reach more than one excited state and allows iron lidars to be treated as the simple atomic system with one excited state, such as shown in Fig. 7.25. For the three wavelengths of interest in iron (using J,j rather than F,f), the q-factors, $q^{\pi}_{jJ\Sigma}(0) = \sum_{j'} \alpha^{\pi}_{jJj'}(0)$, are, respectively, 1.087, 1.077, and 1.160 for 372 nm, 374 nm, and 386 nm due to the Hanle effect; see Eq. (3.23) and Table 3.2.b. Due to the existence of more than one emission pathway, we must calculate $\alpha^{\pi}_{jJj}(0) = \alpha_{Jj}(1 + 0.3g_J B^{(2)}_{jJj})$ and $\alpha^{\pi}_{jJj'}(0) = \alpha_{Jj'}(1 + 0.3g_J B^{(2)}_{jJj'})$ separately.

In this case, we have one upward and two downward attenuation factors, one for each emission pathway, given in (7.30.a). We define emission branching ratio and the effective downward attenuation in (7.30.b):

$$A_{\uparrow}(\nu_L, r) = \exp\left[-\int_{r_1}^{r} \mathcal{N}_{Fe}(r')\sigma^A(\nu_L, T, V)dr'\right], \text{ and}$$

$$A_{\downarrow}^{i}(\nu_i, r) = \exp\left[-\int_{r_1}^{r} \mathcal{N}_{Fe}(r')\sigma^A(\nu_i, T, V)dr'\right]; \tag{7.30.a}$$

$$\mathcal{B}^i(\nu_L, T, V) = \frac{\sigma_i^\pi(\nu_L, T, V)}{\sigma^\pi(\nu_L, T, V)}; \; A_{\downarrow eff}(\nu_L, T, V, r) = \sum_{i=jJj'}^{jJj} \mathcal{B}^i(\nu_L, T, V) A_{\downarrow}^i(\nu_i, T, V, r), \tag{7.30.b}$$

where $i = jJj$ and jJj' with $\sigma_i^\pi(\nu_L, T, V)$ and $\sigma^A(\nu_L, T, V)$ given in (7.28.a) and (7.28.b), respectively. Since there is only one excitation, the branching ratios of interest, $\mathcal{B}^i(\nu_L, T, V)$, are independent of (ν_L, T, V) with $\mathcal{B}^{jJj} = \alpha_{jJj}^\pi(0)/[\alpha_{jJj}^\pi(0) + \alpha_{jJj'}^\pi(0)]$ and $\mathcal{B}^{jJj'} = \alpha_{jJj'}^\pi(0)/[\alpha_{jJj}^\pi(0) + \alpha_{jJj'}^\pi(0)]$. As there is only one emission pathway for the 372 nm lidar, it has $q_{jJ\Sigma}^\pi(0) = \alpha_{jJj}^\pi(0) = 1.087$ with $\mathcal{B}^{jJj} = 1$ and $\boldsymbol{n}_j = 0.96$. For 374 nm and 386 nm lidars, the q-factors consist of two terms. Using the information in Table 3.1 for α_{Jj} and $B_{jJj'}^{(2)}$ along with $g_J = 2J + 1$, we can compute $\left(\alpha_{jJj}^\pi, \alpha_{jJj'}^\pi\right)$ as (1.001, 0.077) and (1.073, 0.086), giving rise to the branching ratio (92.9%, 7.1%) and (92.6%, 7.4%) for 374 nm and 386 nm, respectively, along with corresponding $\boldsymbol{n}_j = 0.04$ and $\boldsymbol{n}_{j'} = 0.96$.

Since there is only one emission pathway for the 372 nm lidar, the upward and downward attenuation for the laser and scattered light within the iron layer should be the same (i.e., $A_{\uparrow}(\nu_L, r) = A_{\downarrow}^{jJ}(\nu_L, r) = A_{\downarrow eff}(\nu_L, r)$ with $q_{jJ\Sigma}^\pi(0) = \alpha_{jJj}^\pi(0) = 1.087$). Even though for temperature/wind measurements using three different preset laser frequencies, ν_L [7.83], each still has only one LIF pathway. In this case, using the ratio technique for temperature/wind measurements incurs no error from the Hanle effect, though the retrieved metal density would be 8.7% too low [3.11]; the same is true for the scanning method [7.82]. For the 374 nm or 386 nm Fe lidar, each has two fluorescence paths, $A_{\uparrow}(\nu_L, r) \neq A_{\downarrow}^{fF}(\nu_L, r) \neq A_{\downarrow eff}(\nu_L, r)$. Thus, in principle, for these two cases, the attenuation of the two pathways should be included and treated separately with different effective downward attenuation factors, as described above. This subtle difference is nonetheless relatively unimportant because the emission rate to the other ground state is only about 10% or less.

For example, for the 386 nm case, using $\alpha^{\pi}_{jJj}(0) = \alpha_{Jj}(1 + 0.3g_J B^{(2)}_{jJj})$ and data in Table 3.1, we calculate $\alpha^{\pi}_{444}(0) = 1.073$ and $\alpha^{\pi}_{443}(0) = 0.086$. As part of the iron Boltzmann temperature lidar [7.81], since the separation between the ground state, a^5D_4, and the next state up, a^5D_3, is about 415 cm^{-1}, the use of a 1-nm (~ 70 cm^{-1}) bandpass filter at 386 nm prevents the other LIF emission from being detected. Whether the signal from the other pathway is detected (as with the broadband Boltzmann lidar) or not (as with the narrowband lidar), the emission of the other channel is part of the downward attenuation and should be included in data processing, somewhat complicating the evaluation. As explained below, the situation for a Na D_2 lidar is very different and necessitates development of the multiple pathway (multiple excited states and two ground states) approach to the lidar equation discussed above, generalized to handle the case of 10 LIF pathways.

7.5.2 LIF of the NaD$_2$ Transition and Temperature/Wind Lidar

The situation in a NaD$_2$ LIF lidar is much more complex than the simple case that applies to Fe. Due to the proximity of the four excited states, with |5>, |6>, |7>, and |8> separated by about 100 MHz as shown in Table A.1, and the fact that at the ambient temperature, ~ 200 K, both ground states, |1> and |2> separated by 1.772 GHz, are well populated (see Fig. A.1), all four excited states can be reached within the spectral width of a laser pulse with FWHM pulse width of 5 ns (Fourier-transform-limited FWHM frequency width of 88.2 MHz) tuned to NaD$_2$ resonance. Selection rules as shown in Fig. A.1 suggest that there are two allowed emission channels for states |6> and |7>, and one for |5> and |8>, leading to a total of ten emission channels at eight different frequencies. These are listed in Table 2 of [7.2] and repeated here in Table 7.5(a). Here the subscript fFf' indicates the pathway from initial state to excited state and back to the final state. They are relabeled by the pathway (or channel) number from $i = 0, 1, 2, \ldots 9$, for simplicity. The transition frequencies in Table 7.5(a) are listed in Table A.1.

Using the value of quantum structure factor $B^{(2)}_{fFf'}$ calculated from the 6-j coefficients in (3.10) – listed in Table 1 of [3.11] and Table 5 of [3.3] and repeated here as Table 7.5(b) – the LIF emission rate, $\alpha^{\pi}_{fFf'}(0) = \alpha_{Ff}$ $(1 + 0.3g_F B^{(2)}_{fFf'})$ (in units of Einstein coefficient A_0), for each pathway may be calculated separately. These are listed in Table 2 of [3.3] and Table 3 of [7.2], and they are reproduced here in Table 7.5(c).

7.5.2(1) The Differential Backscattering and Absorption Cross Sections of the NaD$_2$ Transition

With the information in Table 7.5(c), we can construct the differential backscattering cross section, $\sigma^{\pi}(\nu_L, T, V)$, from (7.28.a) by superposition of the contribution from the four excited states. The result is given in (7.31.a) below.

Table 7.5(a) *Ten pathways and eight emission LIF frequencies in the D_2 transitions. Reproduced with permission from Table 2 of [7.2].*

$\nu_i = \nu_{fFf'}$	Level 5 (F=0)	Level 6 (F=1)	Level 7(F=2)	Level 8 (F=3)
(1, 1)	$\nu_0 = \nu_{151} = 2\nu_{51} - \nu$	$\nu_1 = \nu_{161} = 2\,\nu_{61} - \nu$	$\nu_2 = \nu_{171} = 2\,\nu_{71} - \nu$	
(1, 2)		$\nu_3 = \nu_{162} = \nu_{61} + \nu_{62} - \nu$	$\nu_4 = \nu_{172} = \nu_{71} + \nu_{72} - \nu$	
(2, 2)		$\nu_5 = \nu_{262} = 2\,\nu_{62} - \nu$	$\nu_6 = \nu_{272} = 2\nu_{72} - \nu$	$\nu_7 = \nu_{282} = 2\nu_{82} - \nu$
(2, 1)		$\nu_8 = \nu_{261} = \nu_{62} + \nu_{61} - \nu$	$\nu_9 = \nu_{271} = \nu_{72} + \nu_{71} - \nu$	

Table 7.5(b) *The structure coefficient of NaD_2 transitions. Reproduced with permission from Table 1 of [3.11].*

$B^{(2)}_{fFf'}$	F=3 $(g_F = 7)$	F=2 $g_F = 5$	F=1 $g_F = 3$	F=0 $g_F = 1$
(1,1)	2/35	7/60	1/180	
(1.2)		–7/60	−1/36	
(2.2)		–7/60	−1/36	
(2,1)		7/60	5/36	

Table 7.5(c) *Emission rates of NaD_2 10 pathways. Reproduced with permission from Table 2 of [3.3].* $\alpha^{\pi}_{fFf'}(0)$

$\alpha^{\pi}_{fFf'}(0)$	|5> (F = 0)	|6> (F = 1)	|7> (F = 2)	|8> (F = 3)
(1, 1)	$\alpha_0 = 1$	$\alpha_1 = 5.625/6$	$\alpha_2 = 1.175/2$	0
(1, 2)	0	$\alpha_3 = 0.975/6$	$\alpha_4 = 0.825/2$	0
(2, 2)	0	$\alpha_5 = 1.005/6$	$\alpha_6 = 1.175/2$	$\alpha_7 = 1.12$
(2, 1)	0	$\alpha_8 = 4.875/6$	$\alpha_9 = 0.825/2$	0

$$\sigma^{\pi}(\nu_L, T, V) \equiv \sum_{i=0}^{9} \sigma_i^{\pi}(\nu_L, T, V)$$

$$= \frac{\lambda^2 A_0}{32\pi^2} \left\{ \begin{aligned} &\frac{n_1}{6}\Big[2\mathbf{G}_{51}(\nu_L - \frac{V}{\lambda} - \nu_{51}, T)\alpha_0^{\pi}(0) + 5\mathbf{G}_{61}(\nu_L - \frac{V}{\lambda} - \nu_{61}, T)\{\alpha_1^{\pi}(0) + \alpha_3^{\pi}(0)\} \\ &+ 5\mathbf{G}_{71}(\nu_L - \frac{V}{\lambda} - \nu_{71}, T)\{\alpha_2^{\pi}(0) + \alpha_4^{\pi}(0)\}\Big] + \frac{n_2}{10}\Big[\mathbf{G}_{62}(\nu_L - \frac{V}{\lambda} - \nu_{62}, T)\{\alpha_5^{\pi}(0) + \alpha_8^{\pi}(0)\} \\ &+ 5\mathbf{G}_{72}(\nu_L - \frac{V}{\lambda} - \nu_{72}, T)\{\alpha_6^{\pi}(0) + \alpha_9^{\pi}(0)\} + 14\mathbf{G}_{82}(\nu_L - \frac{V}{\lambda} - \nu_{82}, T)\alpha_7^{\pi}(0) \Big] \end{aligned} \right\}. \tag{7.31.a}$$

Here, the LIF transition rate for each pathway, given in (7.31.a) as $\alpha_i^\pi(0)$, with $i = 0, 1, 2, \ldots .9$, are the same as those in Eq. (3) of [7.2], with numerical values listed in Table 7.5(c) – replacing $\alpha_{fFf'}^\pi(0)$ with α_i to save space. The superscript π and the 0 in the bracket indicate, respectively, backward scattering direction and in zero magnetic field. Because, as shown in Appendix A of [7.2], the effect of the Earth's magnetic field (~ 50 μT) on the Na spectrum (thus on lidar temperature and wind retrieval) is small, we take it to be zero here for simplicity. We point out that if the backward LIF rate, $\alpha_{fFf'}^\pi(0)$, is replaced by the spontaneous emission rate, $\alpha_{Ff'}$ (i.e., the Hanle effect is spatially averaged out), we can obtain the absorption cross section by multiplying the resulting backscattering cross section by 4π. This leads to

$$\sigma^A(\nu, T, V) = \frac{\lambda^2 A_0}{8\pi} \times$$

$$\left\{ \begin{array}{l} \frac{n_1}{6}\left[2\mathbf{G}_{51}(\nu - \frac{V}{\lambda} - \nu_{51}, T) + 5\mathbf{G}_{61}(\nu - \frac{V}{\lambda} - \nu_{61}, T) + 5\mathbf{G}_{71}(\nu - \frac{V}{\lambda} - \nu_{71}, T)\right] \\ + \frac{n_2}{10}\left[\mathbf{G}_{62}(\nu - \frac{V}{\lambda} - \nu_{62}, T) + 5\mathbf{G}_{72}(\nu - \frac{V}{\lambda} - \nu_{72}, T) + 14\mathbf{G}_{82}(\nu - \frac{V}{\lambda} - \nu_{82}, T)\right] \end{array} \right\} \equiv \sigma_{D_2}^A . \tag{7.31.b}$$

Here, the frequency in the absorption cross section $\sigma^A(\nu, T, V)$ in (7.31.b) would be $\nu = \nu_L$, if it is used for the calculation of the upward attenuation $A_\uparrow(\nu_L, r)$ in (7.30.a); for the i-th pathway downward attenuation $A_\downarrow^i(\nu_i, r)$ calculation, it would be the LIF emission frequency, $\nu_i = \nu_{fFf'}$, in Table 7.5(a).

7.5.2(2) The 3-Frequency Technique: Count Ratio for (T, V) *Retrieval, Doppler-Free Spectrum*

As explained in Chapter 3 and detailed above, the LIF emission frequencies are different for different emission pathways, as given in Table 7.5(a). With the broadband receiver typically used for nocturnal observations, these cannot be distinguished for the different laser frequencies, ν_L, unless one employs a high-resolution spectral filter required for daylight observations for which one must apply a multiplicative function such as the frequency-dependent attenuations $\mathcal{F}^R(\nu_L)$ and $\mathcal{F}_{FF}^i(\nu_i, \nu_L)$ in (7.32.a) (to be discussed below). Without such a complication, the ten LIF transition rates $\alpha_i^\pi(0)$ in $\sigma^\pi(\nu_L, T, V)$ in (7.31.a) may be combined into six respective q-factors, one for each allowed excitation line as given in Table 1 of [7.2]: $q_{51}^\pi = \alpha_0^\pi(0) = 1$, $q_{61}^\pi = \alpha_1^\pi(0) + \alpha_3^\pi(0) = 1.1$, and $q_{71}^\pi = \alpha_2^\pi(0) + \alpha_4^\pi(0) = 1$ for excitations from $|1>$, and $q_{62}^\pi = \alpha_5^\pi(0) + \alpha_8^\pi(0) = 0.98$, $q_{72}^\pi = \alpha_6^\pi(0) + \alpha_9^\pi(0) = 1$, and $q_{82}^\pi = \alpha_7^\pi(0) = 1.12$ for excitations from $|2>$.

To show the temperature and wind dependence of the differential cross section, we plot the differential scattering cross section for the D_2 transition, $\sigma^{\pi}(\nu_L, T, V)$, as a function of laser frequency in Fig. 7.26(a). Here, we show $\sigma^{\pi}(\nu_L, T, V)$ at 200 K (and 150 K) and zero wind with thick (and thin) solid curves, along with $\sigma^{\pi}(\nu_L, T, V)$ at 200 K and 58.9 m/s LOS wind (100 MHz frequency shift) with the thick dashed curve. For the sake of simplicity, we ignore the natural linewidth, as in Fig. 2(a) of [7.2], and replace the Voigt function $\mathbb{G}_{Ff}(\nu_L - V/\lambda - \nu_{Ff}, T)$ with the associated Gaussian function $G^{RS}(\nu_L - V/\lambda - \nu_{Ff}, T)$; see (2.7.b). Notice that we have also marked four specific frequency locations (ν_-, ν_a, ν_c, ν_+) on the figure.

It is clear by inspection [7.2] that the cross section at ν_a (the offset frequency at the Na D_{2a} peak) increases as temperature decreases, while it decreases at ν_- (the red-shifted offset frequency for wind measurement) and ν_+ (the blue-shifted offset frequency for wind measurement), as well as at ν_c (the crossover frequency offset midway between D_{2a} and D_{2b}). Thus, using the four frequency locations ν_-, ν_a, ν_c, ν_+, the backscattering cross-section temperature ratio $R_T = \sigma^{\pi}(\nu_c, T, V=0)/\sigma^{\pi}(\nu_a, T, V=0)$ for the 2-frequency method – as described in [7.79] – or $R_T = \left[\sigma^{\pi}(\nu_+, T, V) + \sigma^{\pi}(\nu_-, T, V)\right] / 2\sigma^{\pi}(\nu_a, T, V)$ for the 3-frequency method – as described in [7.80] – are most sensitive to the ambient temperature changes. At 200 K and 58.9 m/s LOS wind, we see the spectrum shifted to the right by 100 MHz as shown by the dashed curve. This shift causes the cross section to increase at ν_+ and decrease at ν_-, and a ratio given by $R_V = \left[\sigma^{\pi}(\nu_+, T, V) - \sigma^{\pi}(\nu_-, T, V)\right]/\sigma^{\pi}(\nu_a, T, V)$ is then most sensitive to LOS winds. Since, generally speaking, lidar signal counts from range r, $N(r)$, are proportional to the scatterer's density multiplied by its cross section, $\mathcal{N}_{Na}(r)\sigma^{\pi}(\nu_L, T, V)$, these backscattering ratios can be determined directly from the signal count ratios, and thus can be used to retrieve temperature and wind.

An extensive sensitivity analysis of systems operating with two and three frequencies has been carried out by Papen et al. [7.84]. Consistent with this analysis, we opted to operate a 3-frequency system at $\nu_- = -1.2814$ GHz, $\nu_a = -0.6514$ GHz, and $\nu_+ = -0.0214$ GHz relative to the center of mass of the D_2 transition frequency at 508,848.7162 GHz (or wavelength 589.158 nm). This system began operations at Colorado State University in 1996. The University of Illinois group and the ALOMAR Observatory groups made the same choice for their 3-frequency systems. We therefore will discuss the procedure for retrieving T and V from the observed signals at these three frequencies.

Conceptually, using the scattering cross-section ratios to measure temperature and wind is all good, but since an accuracy of 1 MHz is required to determine wind velocity

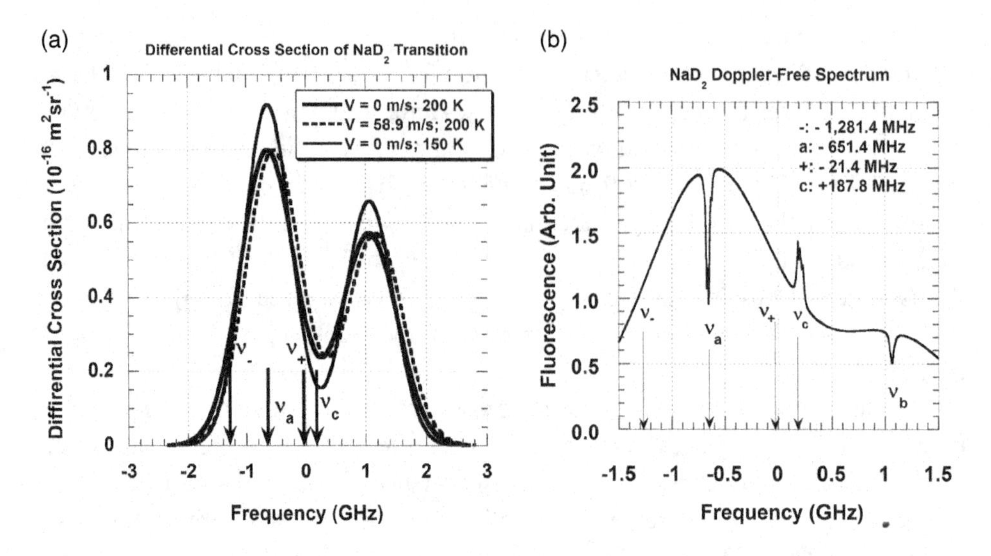

Fig. 7.26 (a) Differential backscattering cross sections of the Na D_2 transition at T = 150 K (thin solid) and 200 K with (thick dashed) and without (thick solid) a 100 MHz shift simulating a 58.9 m/s line-of-sight wind. (b) Doppler-free spectrum of the Na D_2 transition from a Na cell at 325 K. The frequencies relative to the spectral center of mass of the Doppler-free features at ν_a and ν_b are marked along with the acousto-optic switched frequencies ν_+ and ν_- by $\pm$630 MHz from ν_a (for details, see text). Reproduced with permission from Fig. 1 of [7.2].© the Optical Society.

to within 0.6 m/s, one must do much better than depend on the Doppler-broadened spectra at a GHz scale, as those shown in Fig. 7.26(a), to identify the absolute frequency. Fortunately, the technique of Doppler-free spectroscopy as described in She and Yu [7.8] can resolve the naturally occurring Na D_{2a} peak ν_a and crossover ν_c to an accuracy of about 1 MHz. Further, by locking the laser frequency at ν_a, we can modulate the laser beam with an acousto-optic modulator to up- and down-shift its frequency by a fixed radio frequency (chosen here to be ±630 MHz) to ν_+ and ν_- with very high precision. A Doppler-free spectrum for the Na D_2 transition is shown in Fig. 7.26(b), and the preselected frequencies (ν_-, ν_a, ν_c, and ν_+) are indicated. High-resolution (better than 1 MHz precision) spectral features are clearly seen at ν_a and ν_b, along with their crossover feature at $\nu_c = 0.5(\nu_a + \nu_b)$.

7.5.2(3) Lidar Equation, Theoretical Signal Counts, and A Priori Knowledge for Data Processing

For the LIF lidar equation, we consider an atmosphere with the following:

(1) A naturally occurring layer of metal atoms producing LIF from a range $r > r_1$,

(2) The metal atoms having more than one excited state (such as Na with four excited states and two ground states),
(3) Aerosols present (producing Mie scattering) below a reference range $r_R \ll r_1$, and
(4) Rayleigh/Cabannes scattering from clean atmosphere above r_R.

Since the integrals of the extinction coefficient in (5.3) in the exponent of the upward and downward attenuation coefficients are, respectively, additive between different ranges and different atmospheric species (or types of scattering processes), the total attenuation coefficient may be written as the product of the individual coefficients. A scattering channel photocount lidar profile for the described medium with range bin r_B at range $r > r_1 > r_R$ (or at altitude $z = r\cos\Theta$ for off-zenith beam angle of Θ) and dwell time τ_D, $N(\nu_L, r; r_B)$, may be derived from (5.3). For typical values for range and time resolutions $\left(r_B \sim 150\text{ m},\ \tau_D \sim 1\text{ min}\right)$, a representative photocount profile is quite noisy. In order to improve the signal-to-noise ratio, we sum many such profiles and then perform vertical smoothing, resulting in predetermined integration time τ_{int} and range (or altitude), Δr (or Δz), resolutions – for example, $\tau_{\text{int}} = 1$ hr, $\Delta z = 2$ km. Our common practice for vertical smoothing is to apply a running Hanning filter with FWHM = Δz. The resulting background-subtracted photon count profile, again derived from (5.3), is

$$N^{BS}(\nu_L, r; \Delta r, \tau_{\text{int}}) = N^{BS}(\nu_L, r_R; \Delta r, \tau_{\text{int}}) \frac{r_R^2}{r^2} \exp\left[-2\int_{r_R}^{r} \mathcal{N}_a(r')\sigma_T^R dr'\right]$$

$$\times \left[\mathcal{N}_a(r)\sigma_\pi^R \mathcal{F}^R(\nu_L) + \mathcal{N}_{Na}(r) A_\uparrow(\nu_L, T, V, r) \sum_{i=0}^{9} \sigma_i^\pi(\nu_L, T, V) \mathcal{F}_{FF}^i(\nu_i, \nu_L) A_\downarrow^i(\nu_i, r)\right], \text{ with} \tag{7.32.a}$$

$$N^{BS}(\nu_L, r_R; \Delta r, \tau_{\text{int}}) = \eta \frac{A_R \Delta r}{r_R^2} \mathcal{N}_a(r_R)\sigma_\pi^R \mathcal{F}^R(\nu_L) \exp\left[-2\int_0^{r_R} \left[\alpha_{aer}(r') + \mathcal{N}_a(r')\sigma_T^R\right] dr'\right] \frac{E_L}{h\nu_L}; \tag{7.32.b}$$

$$A_\uparrow(\nu_L, r) = \exp\left[-\int_{r_1}^{r} \mathcal{N}_{Na}(r')\sigma^A(\nu_L, T, V) dr'\right], \text{ and}$$

$$A_\downarrow^i(\nu_i, r) = \exp\left[-\int_{r_1}^{r} \mathcal{N}_{Na}(r')\sigma^A(\nu_i, T, V) dr'\right], \tag{7.32.c}$$

where $i = 0, 1, 2 \ldots, 9$ indicate the LIF pathways, $\alpha_{aer}(r')$, $\mathcal{N}_{Na}(r')$, σ_π^R, σ_T^R, $\mathcal{N}_{Na}(r')$, σ^A, σ_i^π, η and A_R are, respectively, aerosol extinction coefficient, atmospheric number density, backscattering Rayleigh cross section, Rayleigh extinction cross section, number density of metal (Na in this case) atoms, metal absorption cross section, metal LIF differential scattering cross section for the ith pathway, detector quantum efficiency, and the area of receiving telescope. The photon count at a reference altitude r_R (typically ~ 30 km) or range, $N^{BS}(\nu_L, r_R; \Delta r, \tau_{\text{int}})$, is the result of Rayleigh scattering. The exponential function in (7.32.a) represents the round-trip attenuation from Rayleigh scattering by air molecules between r_R and the range in question, r. For detection under sunlit conditions, a narrowband Faraday filter is used to reduce the sky background. This attenuates individual $\sigma_i^\pi(\nu_L, T, V)$ by a factor of $\mathcal{F}_{FF}^i(\nu_i, \nu_L)$ and Rayleigh scattering σ_π^R by a factor of $\mathcal{F}^R(\nu_L)$. These factors would be unity for observations at night when the Faraday filter is not used. Since the LIF emission frequencies $\nu_{fFf} = \nu_i$ are different from the incident frequency, ν_L, the upward and downward attenuation factors due to metal atoms, $A_\uparrow(\nu_L, r)$ and $A_\downarrow^i(\nu_i, r)$, must be computed separately and written as a product. For the latter, or downward, attenuation the individual $A_\downarrow^i(\nu_i, r)$ are multiplied by the associated σ_i^π. The result is summed over the ten LIF pathways. The complexity of the downward attenuation may by tidied up by defining branching ratios $\mathcal{B}^i(\nu_L, T, V)$, and the effective downward attenuation $A_{\downarrow eff}(\nu_L, T, V, r)$ derived for the 10 pathways for LIF Na D_2 emissions as in (7.33.a) and (7.33.b):

$$\mathcal{B}^i(\nu_L, T, V) \equiv \frac{\sigma_i^\pi(\nu_L, T, V)}{\sigma^\pi(\nu_L, T, V)}, \quad i = 0, 1, 2, \ldots, 9; \tag{7.33.a}$$

$$A_{\downarrow eff}(\nu_L, T, V, r) \equiv \sum_{i=0}^{9} \mathcal{B}^i(\nu_L, T, V) \mathcal{F}_{FF}^i(\nu_i) A_\downarrow^i(\nu_i, \mathrm{T}, \mathrm{V}, r). \tag{7.33.b}$$

Since the round-trip atmospheric attenuation below the reference altitude varies considerably, one normalizes the received counts to the counts at the reference altitude, shown in (7.32) as $N^{BS}(\nu_L, r_R)$, to yield $N^{nor}(\nu_L, r)$, given in (7.34.a). Note that from here on, for simplicity we ignore the range resolution and integration time $(\Delta r, \tau_{\text{int}})$ whenever it is understood. This leads to the total normalized signal counts, $N^{nor}(\nu_L, r)$, at the mesopause $(r > r_1)$, as the sum of the normalized Rayleigh counts, $N_{RS}^{nor}(\nu_L, r)$, between r_R and r in (7.34.b), and the normalized metal LIF counts, $N_{Na}^{nor}(\nu_L, r)$, between r_1 and r in (7.34.c) as

$$N^{nor}(\nu_L, r) = N_{RS}^{nor}(\nu_L, r) + N_{Na}^{nor}(\nu_L, r), \text{ with} \tag{7.34.a}$$

$$N_{RS}^{nor}(\nu_L, r) = \frac{r_R^2}{r^2}\frac{\mathcal{N}_a(r)}{\mathcal{N}_a(r_R)}\exp\left[-2\int_{r_R}^{r}\mathcal{N}_a(r')\sigma_T^R dr'\right], \text{ and} \tag{7.34.b}$$

$$N_{Na}^{nor}(\nu_L, r) = C\left[\mathcal{N}_{Na}(r)A_{\uparrow}(\nu_L, T, V, r)\sigma^{\pi}(\nu_L, T, V)A_{\downarrow eff}(\nu_L, T, V, r)\right], \text{ where}$$

$$C = \frac{r_R^2}{r^2\left(\mathcal{N}_a(r_R)\sigma_{\pi}^R\mathcal{F}^R(\nu_L)\right)}\exp\left[-2\int_{r_R}^{r}\mathcal{N}_a(r')\sigma_T^R dr'\right]. \tag{7.34.c}$$

For the 3-frequency method, the normalized counts, $N^{nor}(\nu_-, r)$, $N^{nor}(\nu_a, r)$, and $N^{nor}(\nu_+, r)$, are obtained from (7.34) by substituting for ν_L with (ν_-, ν_a, and ν_+).

Notice as expected, the normalized counts $N^{nor}(\nu_L, r)$ consist of two contributions, Rayleigh/Cabannes scattering from air molecules $N_{RS}^{nor}(\nu_L, r)$ and LIF from Na atoms $N_{Na}^{nor}(\nu_L, r)$ (ignoring the range resolution r_B for simplicity). In the mesopause region, the scattering from air molecules is negligible compared to the LIF from Na atoms – and in any case can be subtracted either by an empirical approach called "Rayleigh background subtraction" as discussed and shown in Fig. 4 of [7.2], or by using standard air molecular density, $\mathcal{N}_a(r)$, and normalized to Rayleigh counts, $N^{BS}(\nu_L, r_R; \Delta r, \tau_{\text{int}})$, at the reference range, r_R. From here on, we take $N_{Na}^{nor}(\nu_L, r)$ to be the experimentally deduced count profile by either equating it to $N_{Na}^{nor}(\nu_L, r)$ in (7.34.a) or to $N^{nor}(\nu_L, r)$ after its Rayleigh background subtraction.

The normalized Na counts, $N_{Na}^{nor}(\nu_L, r)$, in (7.34.c) are still not quite proportional to the product $\mathcal{N}_{Na}(r)\sigma^{\pi}(\nu_L, T, V)$. For this, we need to undo the effect of the upward and downward attenuations by defining the theoretical signal counts $N_{Na}^{Th}(\nu_L, r)$ at range r as

$$N_{Na}^{Th}(\nu_L, r) = A_{\uparrow}^{-1}(\nu_L, T, V, r)N_{Na}^{nor}(\nu_L, r)A_{\downarrow eff}^{-1}(\nu_L, T, V, r), \text{ leading to} \tag{7.35.a}$$

$$N_{Na}^{Th}(\nu_L, r) = \frac{r_R^2}{r^2\mathcal{N}_a(r_R)\sigma_{\pi}^R\mathcal{F}^R(\nu_L)}\exp\left[-2\int_{r_R}^{r}\mathcal{N}_a(r')\sigma_T^R dr'\right]\mathcal{N}_{Na}(r)\left[\sigma^{\pi}(\nu_L, T, V)\right], \text{ or} \tag{7.35.b}$$

$$\mathcal{N}_{Na}(r) = \frac{r^2}{r_R^2}\frac{\sigma_{\pi}^R\mathcal{F}^R(\nu_L)}{\sigma^{\pi}(\nu_L, T, V)}N_{Na}^{Th}(\nu_L, r)\exp\left[2\int_{r_R}^{r}\mathcal{N}_a(r')\sigma_T^R dr'\right]\mathcal{N}_a(r_R) \tag{7.35.c}$$

The theoretical signal counts $N_{Na}^{Th}(\nu_L, r)$ as seen in (7.35.b) are now directly proportional to $\mathcal{N}_{Na}(r)\sigma^{\pi}(\nu_L, T, V)$. In terms of these theoretical counts for the three frequencies (ν_-, ν_a and ν_+), the conceived 3-frequency temperature and wind ratios, R_T and R_V, can now be expressed in terms of the corresponding backscattering ratios as

$$R_T = \frac{\sigma^\pi(\nu_+, T, V) + \sigma^\pi(\nu_-, T, V)}{2\sigma^\pi(\nu_a, T, V)} = \frac{N_{Na}^{Th}(\nu_+, r) + N_{Na}^{Th}(\nu_-, r)}{2N_{Na}^{Th}(\nu_a, r)}, \text{ and} \tag{7.36.a}$$

$$R_V = \frac{\sigma^\pi(\nu_+, T, V) - \sigma^\pi(\nu_-, T, V)}{\sigma^\pi(\nu_a, T, V)} = \frac{N_{Na}^{Th}(\nu_+, r) - N_{Na}^{Th}(\nu_-, r)}{N_{Na}^{Th}(\nu_a, r)}. \tag{7.36.b}$$

These temperature and wind ratios, (R_T, R_V), are theoretically relatable to the three backscattering (T, V)-dependent cross-sections on the one hand, and to the three theoretical counts at a given altitude range on the other. The former leads to the calibration curves used to deduce (T, V) from (R_T, R_V), while the latter allows the two ratios to be relatable to $N_{Na}^{nor}(\nu_L, r; r_B)$ for $\nu_L = \nu_-$, ν_a and ν_+, which are in turn derivable from background subtracted experimental counts $N^{BS}(\nu_L, r; \Delta r, \tau_{\text{int}})$.

After replacing the Voigt function $\mathbb{G}_{Ff}(\nu_L)$ by the correlation between $\mathbf{L}(\nu_L)$ and $\mathbb{G}_{Ff}(\nu_\mathbf{L})$ in (7.31.a), where $\mathbf{L}(\nu_L)$ is the experimental laser lineshape function, as shown in Fig. 7.27(a), we calculate $\sigma^\pi(\nu_L, T, V)$ at each laser frequency (ν_-, ν_a, ν_+) for temperatures between 100 K and 300 K at 1 K intervals and for LOS winds between –100 and 100 m/s at 1 m/s intervals. From these calculations, we can produce 2-D calibration curves as shown in Fig. 7.27(b). Here, to avoid clutter, we show solid curves for constant V and T, at 20 m/s and 20 K intervals, respectively, as the calibration curves for a realistic Na lidar with the Hanle Effect included. Superimposed on these are ratio values for 200 K temperature and for 0 m/s wind in open circles and squares, respectively, ignoring the Hanle Effect. These ratios nearly coincide with their counterparts shown by the solid black curves. However, the small difference indicates that the Hanle Effect does indeed affect temperature and wind retrieval by a small amount. As stated on p. 9476 in [7.2], compared to the correct analysis results (solid curves), the retrieved temperature and line-of-sight wind from the temperature and wind ratios of these circles and squares ignoring "the Hanle effect are, respectively, 0.33 K cooler and 0.68 *m/s* faster." In order to calculate the theoretical signal counts at (ν_-, ν_a, ν_+) from $N_{Na}^{nor}(\nu_L, r)$, we need the information of $A_\uparrow^{-1}(\nu_L, T, V)$ and $A_{\downarrow eff}^{-1}(\nu_L, T, V)$. In addition to $\mathcal{N}_{Na}(r')$, we need to know $\sigma^A(\nu_L, T, V)$ to determine $A_\uparrow^{-1}(\nu_L, T, V)$; see (7.32.c). For $A_{\downarrow eff}^{-1}(\nu_L, T, V)$, we need to know the ten values for $\sigma^A(\nu_i, T, V)$, one for each $A_\downarrow^i(\nu_i, T, V, r)$ as well as the ten branching ratios, $\mathcal{B}^i(\nu_L, T, V)$, with $i = 0, 1, 2, \ldots, 9$; see (7.33.a), for the full range of (T, V) values. Along with $\sigma^\pi(\nu_L, T, V)$ and $A_\uparrow^{-1}(\nu_L, T, V)$, there are all together 22 (T, V)-dependent parameters for each of the three frequencies that are

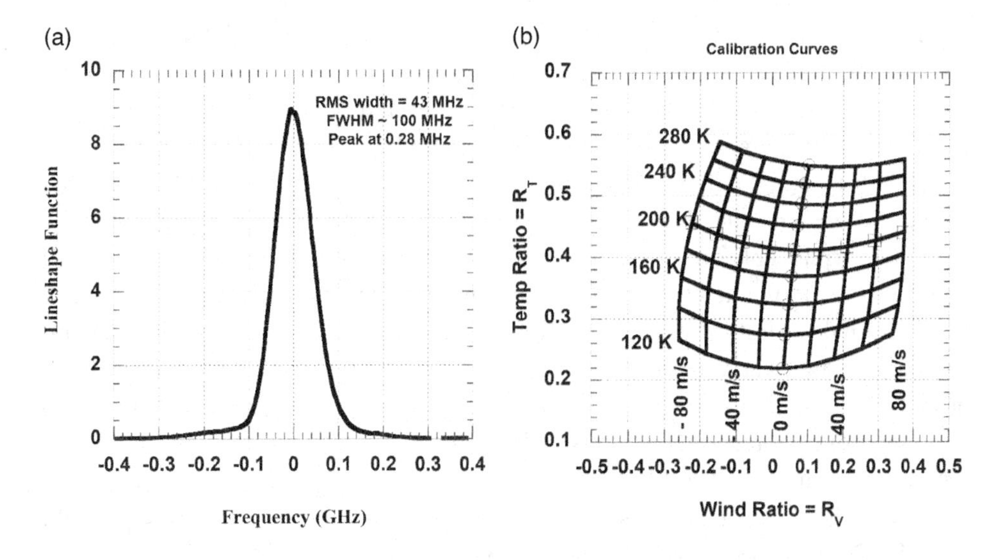

Fig. 7.27 (a) Pulsed laser lineshape function with measured FWHM and with Gaussian-fitted RMS width and peak location as shown in legend, and (b) the calibration curves relating the intensity ratios, (R_T, R_V) to retrieve temperature and LOS wind (T, V). (b) is reproduced with permission from Fig. 2(b) of [7.2]. © the Optical Society.

theoretically calculable for the range of temperature and LOS wind of interest, these being between 100 K and 300 K at 1 K intervals and between –100 and 100 m/s at 1 m/s intervals, respectively. These theoretical values for the three sets of 22 cross-section-related parameters are a priori knowledge required for data processing. We store the a priori knowledge of the 2D calibration curves and the three sets of 22 parameters at 1 K and 1 m/s intervals in text files to facilitate data processing. As a fast, accurate method of obtaining temperature and wind from values of R_T and R_V, we use a bilinear form with best-fit coefficients, as discussed in [7.2]. This method models all three sets of 22 cross-section-related parameters. At CSU, we called these text files, respectively, "TVfofRtRv3Frq" and "TVFCrossV22_3Frq." With a measured laser lineshape function, the readers can derive both files. These best-fit files produce temperature and LOS wind values continuously between a specified range of each.

One big advantage of the Na LIF lidar data processing is its ability to normalize the background subtracted photocount in the Na layer, $N^{BS}(\nu_L, r; \Delta r, \tau_{\text{int}})$ with $r > r_1$, as given in (7.32.a), to that at a lower reference altitude $N^{BS}(\nu_L, r_R; \Delta r, \tau_{\text{int}})$ at the same instant τ_{int} as given in (7.32.b) and hide the explicit information of the transmitter, such as the power-aperture product of the lidar system, $P_L A_R = E_L A_R / h\nu_L$, from the resultant normalized photocount, $N^{nor}(\nu_L, r)$, as given in (7.34.a). This lack of explicit $P_L A_R$ product carries through the intervening constructed theoretical count profiles, $N_{Na}^{Th}(\nu_+, r)$, $N_{Na}^{Th}(\nu_-, r)$, and $N_{Na}^{Th}(\nu_a, r)$, used for the deduction of Na density with (7.35.c), and of atmospheric temperature and wind via the temperature and wind ratios as given in (7.36.a) and (7.36.b), derived from the theoretical count files without the knowledge of P_L or A_R. All these may give the impression that the information of the transmitter power and the size of the receiver is irrelevant. This impression is, of course, not true. This information will be needed when we deduce the measurement uncertainties in Section 7.5.2(6), as the fractional theoretical count variance is directly proportional to the background subtracted count variance; see (7.38.a) and (7.38.b), which depends on the $P_L A_R$ product. The same is true when we examine the signal-to-background or signal-to-noise ratio of a photocount profile as in Section 7.5.2(4). For this reason, we provide the information of the CSU lidar below.

For the CSU CS/IVF lidar discussed in Section 7.3.2 and that following Eq. (7.20), or in [7.13], the laser transmits 5 W of power at 532 nm and the diameter of the receiving telescope is 35 cm (i.e., $A_R = 0.1\ \text{m}^2$ and $PA_R = 0.5\ \ \text{Wm}^2$). For the CSU Na lidar data presented here in Section 7.5 and in [7.2], the laser power is 1 W at 589 nm divided among two or three telescopes. The diameter of each telescope was 35 cm until the inclusion of two 70 cm-diameter telescopes in 2006. For current discussions, we take the nominal power of 0.5 W per beam received by a 35 cm telescope, or $PA_R = 0.05\ \text{Wm}^2$. Using this system and the received nighttime signal counts (averaged between 80 and 105 km) of 94.2 per 2 min in a 150 m bin (see Table 7.6 below), we deduce the received LIF signal from the Na layer within a slant range of 150 m to be about 0.8 photon/s. Since with 0.5 W at 589 nm we transmit 1.5×10^{18} photons/s, the detection sensitivity of atmospheric Na atoms within 150 m is about 1 part in 2×10^{18}. Though the signal of 1 photon is hardly enough to measure anything, fortunately, we can integrate over 1 hr and average over a 1 km vertical height for a meaningful atmospheric temperature measurement. For a more challenging problem like atmospheric turbulence that needs to get a measurement done in seconds, much more laser power and receiver area must then be used. The mesopause region is blessed to have naturally occurring Na atoms. Since the signal from Cabannes scattering is 5 orders weaker (10 orders higher in density, and 15 orders smaller in scattering cross section), if one needs to depend on Cabannes scattering for atmospheric measurement, a detection sensitivity of about one part in $2{\times}10^{23}$ would be required for the same vertical and temporal resolution (130 m and 2 min) using lidar with the same $P_L A_R$ product – an almost impossible proposition.

7.5.2(4) The Faraday Filter and Observations under Sunlit Conditions

In a nocturnal observation, a narrowband interference filter of, say, 1 nm (864 GHz or 28.8 cm^{-1}) bandwidth is wide enough to pass all six NaD_2 transitions and ten LIF emission lines, and narrow enough to block much of the night sky background. However, the sky background passing through a 1 nm bandpass filter under sunlit conditions will dwarf the signal and possibly cause severe damage to the photomultiplier. As shown in Fig. 7.28(a), a sodium Faraday filter consists of a Na vapor cell between two crossed polarizers in an axial magnetic field [7.85]. Light that is on-resonance with sodium and passes through the first polarizer (thus, initially linear polarized) has its polarization rotated by 90° in the cell with a suitable combination of Na vapor density, cell length, and axial magnetic field strength. This rotation is precisely that required to transmit optimally through the second polarizer. Light that is not on-resonance with sodium does not experience polarization rotation and is thus blocked by the second polarizer. The background extinction depends on the temperature-dependent birefringence of the cell windows, which was measured to be ~2×10^{-5} at the operation temperature in an early publication [7.86]. To obtain the desired vapor density, as described for the potassium cell discussed in Section 7.3.4(2), the cell consists of a Pyrex tube with optical-quality glass windows, constructed with a side reservoir tip for storing the metal. The temperatures of the reservoir and cell T_{res} and T_{cell}, are controlled so as to set vapor pressure and temperature separately, leading to the desired vapor density. We use Eq. 21 of [7.44], repeated here as (7.37), to determine the vapor pressure, p_{Na}, in solid and liquid phases for a given reservoir temperature, T_{res}:

$$\log_{10}(p_{Na}/\mathrm{Torr}) = 71.899 - 9217.2/T_{res} + 4.0693 \times 10^7/T_{res}^3 + 0.0061264 T_{res} - 9.6625 \ln(T_{res}). \quad (7.37)$$

To aid in setting correct values for (T_{res}/T_{cell}) given the cell length and axial magnetic field, we use a computer program following [7.44] to calculate the resulting Faraday filter transmission function. As an example, we show two transmission functions of a 4-cm-long Faraday filter with $(T_{res}/T_{cell}) = (440\ \mathrm{K}/443\ \mathrm{K})$ at two different axial magnetic fields 1,750 G and 2,200 G in Fig. 7.28(b). If small passive losses are ignored, they reach a peak transmission near 90% with a narrow FWHM bandwidth of about 2 GHz (1.2×10^{-3} nm), which rides over a nearly symmetric flat pedestal about 10 GHz wide with ~25% transmission. Background light outside this band is fully blocked by the crossed polarizer. The ~25% transmission occurs at frequencies when one circularly polarized component is nearly completely absorbed, while the other component is totally transmitted, as discussed in our early publication [7.86], where the agreement between experimental and theoretical filter transmissions is shown in Fig. 2. As the strength of magnetic field increases, the Zeeman split between the peak absorption frequencies of the two counter circularly polarized waves also increases, expanding the widths of both the

Table 7.6 *Counts from three photocount files shown in Fig. 7.29(a).*

Counts from three files	4:41 (without filter)	4:58 (with filter)	12:31 (with filter)
Mean* total counts, C	124.6	19.0	121.7
Background count, B	30.4	0.1	103
Mean* signal counts, S	94.2	18.9	18.7
Signal/Background, S/B	3.1	189	0.2

*Averaged between 80 and 105 km

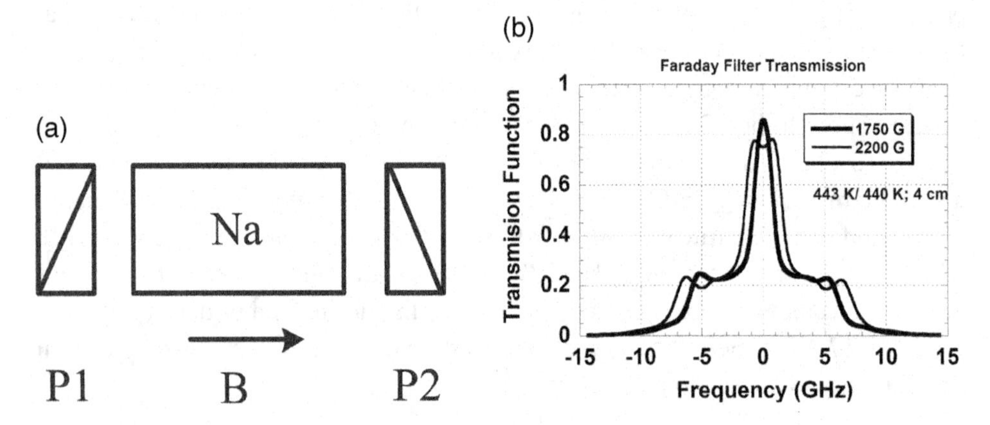

Fig. 7.28 (a) Schematic of a Faraday filter, a sodium cell between two crossed polarizers in an axial magnetic field. (b) Area-normalized transmission functions of a sodium Faraday filter in two magnetic fields.

pedestal and transmission peak. Continued increase in the magnetic field creates a valley in the center. These phenomena are seen clearly in Fig. 7.28(b).

The price for the very effective background light rejection, as shown in Fig. 5(a) of [7.2], lies in the different attenuations of the ten LIF emission rates, $\mathcal{F}^i_{FF}(\nu_i, \nu_L)$, and of the Rayleigh transmission, $\mathcal{F}^R(\nu_L)$, at each of the laser frequencies (ν_-, ν_a, and ν_+). This effect is shown graphically in Fig. 5(b) of [7.2]. We have recreated these figures as Fig. 7.29, (a) and (b), including a minor modification in (b) and a brief discussion. Raw photocount profiles collected under near-daybreak conditions without a Faraday filter (○), along with profiles at daybreak (•) and noon (×) with the Faraday filter installed, are shown in Fig. 7.29(a). From the raw photocount profiles, we compute the mean background counts B, averaged between 110 and 140 km, at just about daybreak without and with the Faraday filter, and at noon with the filter; the results are shown respectively in the third row of Table 7.6. Since most Na signal occurs between 80 and 105 km, we compute the respective mean counts, C, from which the mean signal counts, $S = C - B$, and listed them in the second and fourth rows of the Table and background counts B for the three instances in Fig. 7.29(a) in the third row. Comparing the background counts between 4:41 and 4:58 shows that the background rejection of this FF is a factor of about 300. Since at 589 nm, 1 nm is ~ 900 GHz, this rejection factor implies an effective FF transmission bandwidth of ~ 3 GHz.

The peak-normalized Faraday filter transmission is shown as the solid curve in Fig. 7.29(b). The relative transmissions, $\mathcal{F}^i_{FF}(\nu_i, \nu_L)$, at different laser frequencies ($\nu_- = -0.0214$ GHz, $\nu_a = -0.6514$ GHz, and $\nu_+ = -1.2814$ GHz), can be read from this. The three sets of relative filter transmissions in Fig, 7.29(b) are given in Table 7.7.

The data points represent groups of filtered LIF emission rates, which are the product of individual $\mathcal{F}^i_{FF}(\nu_i, \nu_L) \times \alpha^\pi_i(0)$ for ($\nu_L = \nu_-, \nu_a,$ and ν_+). Notice that though the filter transmission function is the same, the values for these points in Fig. 7.29(b) are somewhat different from those in Fig. 5(b) of [7.2]. This is because we have included the Hanle effect in this evaluation of $\alpha^\pi_i(0)$, which the authors of [7.2] ignored for that figure. Each group in Fig. 7.29(b) has nine different values, which is the result of the fact that while emission frequencies $\nu_3 = \nu_8$ and $\nu_4 = \nu_9$, only $\alpha^\pi_4(0) = \alpha^\pi_9(0)$ but $\alpha^\pi_3(0) \neq \alpha^\pi_8(0)$; see Table 7.5(c).

Assuming the same Na signal and sky background between 4:41 and 4:58, the measured Faraday filter gain reduction from Table 7.6 is a factor of about 94.2/ 18.8 = 5. This may be compared to the estimated gain reduction of 1/(0.5 × 0.8 × 0.76) = 3.3. Here the factor 0.76 is the average of the ten LIF channels' (associated with ν_a) normalized transmission, shown as solid points in Fig. 7.29(b). The factors 0.8 and 0.5 reflect, respectively, the peak transmission of the Faraday filter and the 50% light lost at the first polarizer. The difference between 3.3 and 5 is likely the result of scattering and reflection losses at the cell surfaces, along with some cell birefringence produced by the cell between the crossed polarizers. At noon, the layer mean *S/B* ratio achieved is only 0.2, but the *S/B* at the Na peak as can be seen from Fig. 7.29(a) is about 2. The PMT quantum efficiency in 2002 was 20%; its increase to 40% after 2005 increased signal but not the *S/B* ratio. Using two Faraday filters [7.87], one can recover the 50% loss at the first polarizer; this will increase the *S/B* ratio to 4, still much less than *S/B* = 10 achieved for a Fe lidar at 386 nm [7.82], showing the advantage of a UV system for observations under sunlit conditions.

Since the Rayleigh/Cabannes scattering spectrum is centered at the laser frequency, ν_L, with a width of about 2 to 3 GHz depending on atmospheric temperature, the fraction of Rayleigh-scattered light that passes through the Faraday filter at the reference altitude z_R will be different at each of the three laser frequencies (ν_-, ν_a, and ν_+). To account for this difference without knowing the temperature at z_R during the observation, we calculate the average value of $\mathcal{F}^R(\nu_L)$ for 150, 170,

Table 7.7 *Transmission of the Faraday filter shown in Fig. 7.28(b) for the 10 pathways*

ν_L (GHz)	$\mathcal{F}^0_{FF}$	$\mathcal{F}^1_{FF}$	$\mathcal{F}^2_{FF}$	$\mathcal{F}^3_{FF}$	$\mathcal{F}^4_{FF}$	$\mathcal{F}^5_{FF}$	$\mathcal{F}^6_{FF}$	$\mathcal{F}^7_{FF}$	$\mathcal{F}^8_{FF}$	$\mathcal{F}^9_{FF}$
ν_+	0.48761	0.48146	0.46921	0.99182	0.99415	0.65171	0.68141	0.73404	0.99182	0.99415
ν_a	0.40232	0.39961	0.39375	0.91393	0.88671	0.91662	0.93586	0.95853	0.91393	0.88671
ν_-	0.34624	0.34303	0.33597	0.62888	0.60263	0.98009	0.97957	0.98042	0.62888	0.60263

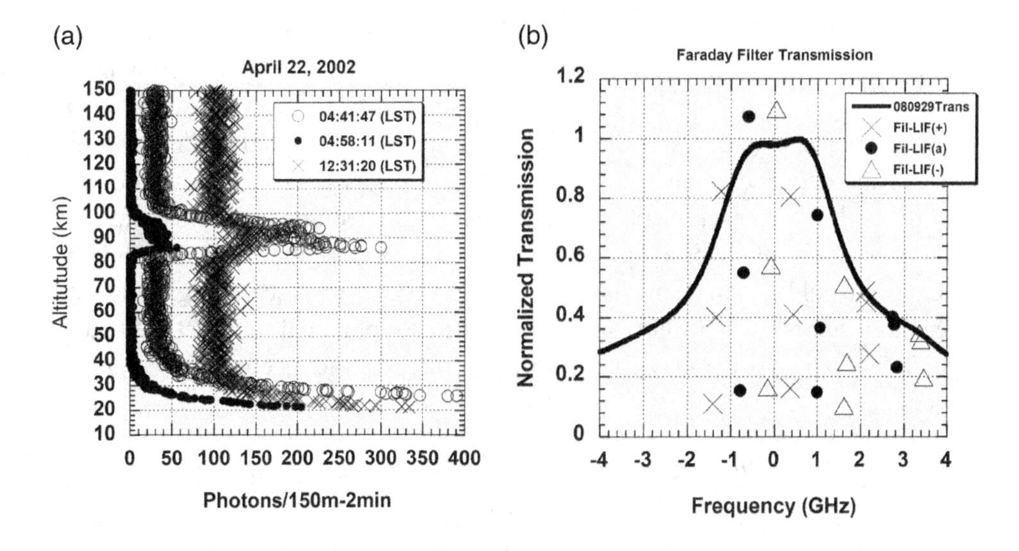

Fig. 7.29 (a) Photon profiles with $\nu_L = \nu_a$, 150 m radial resolution (30° from zenith) and 2-min integration at night (open circles), daybreak (filled circles), and local noon (×); a Faraday filter was used for daybreak and noon profiles. Reproduced with permission from Fig. 5(a) of [7.2]. © the Optical Society. (b) A peak-normalized Faraday filter transmission (black curve), along with transmissions of filtered relative LIF emission rates, including the Hanle effect, for the ten channels at eight frequencies induced by three laser frequencies, $\nu_L = \nu_+, \nu_a, \nu_-$, respectively as ×'s, solid circles, and open triangles.

190, 210, 230, and 250 K at (ν_-, ν_a, ν_+). With the peak-normalized Faraday transmission function shown in Fig. 7.29(b), we determine $\mathcal{F}^R(\nu_L)$ to be 0.976, 0.947, and 0.777, respectively, at the three laser frequencies [7.2]. For data processing under sunlit conditions, these three values for $\mathcal{F}^R(\nu_L)$ along with the 30 values for $\mathcal{F}^i_{FF}(\nu_i, \nu_L)$ shown in Table 7.7 are required. At CSU, we put this information into a text file called "TVfGenRayFiltB(. . . .)3Frq", along with links to "TVfofRtRv3Frq" and "TVFCrossV22_3Frq" as previously discussed. The information in parentheses denotes the particular Faraday filter used during the observation; for example, Filter B with transmission function measured on September 29, 2008, as "TVfGenRayFiltB(80929)3Frq".

We note that in Fig. 7.29(b), the peak-normalized Faraday filter transmission near the center is reasonably flat. Unfortunately, the flat region is too narrow to cover all the LIF emissions between –2 and +4 GHz. If the ground-state hyperfine splitting were much smaller, the frequency difference between LIF emission lines would, in turn, be much smaller, and a Faraday filter could then be built with a modest magnetic field and possess a wide enough flat top to encompass all emission frequencies. Were that the case, we could set $\mathcal{F}^i_{FF}(\nu_i, \nu_L)$ to be unity without spectral measurements; the potassium Faraday filter has this characteristic [7.87]. For a Na system, referring to Fig. 7.29(b), were it possible to make a narrowband top-hat filter between –2 GHz and +4 GHz relative to the Na D_2 transition frequency (16,973.36619 cm^{-1}) to an accuracy of say 10 MHz, one could decrease the gain reduction factor from 3.3 (using FF) to ~ 1. This would also simplify both the measurement and data analysis procedures for the daytime Na lidar observations. Perhaps a top-hat filter between –4 GHz and + 4 GHz would be better, as it would eliminate the filter's influence on Cabannes scattering in the data processing. Further, such filters might have a better potential to reject sky background than would a Faraday filter, to further increase the receiver's S/B ratio at 589 nm.

7.5.2(5) A Brief Outline of Data Processing and Retrieval of Temperature and LOS Wind Profiles

For a thorough step-by-step description of the data processing, the reader is referred to Section 6 of [7.2]. Here, we point out the basic features and give a brief summary.

First, we decide the temporal and vertical resolution ($\tau_{int}, \Delta r = \Delta z/\cos\Theta$) for the retrieval in question. With this in mind, we add enough photocount profiles and perform vertical smoothing with a Hanning window with full width Δz to create individual count profiles, $N(\nu_L, r; \Delta z, \tau_{int})$, and background subtracted count profiles, $N^{BS}(\nu_L, r; \Delta z, \tau_{int})$, which are then divided by the count value at a reference altitude, $N^{BS}(\nu_L, r_R)$, to obtain the (Rayleigh) normalized photocount profile, $N^{nor}(\nu_L, r)$.

Second, since the signal at the reference altitude, $N^{BS}(\nu_L, r_R)$, is still weak at the desired range/time resolution, the normalization process will introduce excess noise. To minimize this fluctuation, we consider two types of normalization counts with improved SNR:

(1) N_{Raysum} by summing all counts $N^{BS}(\nu_L, r)$ over a number of range bins centered at the reference range r_R ($z_R = r_R \cos\Theta$, based on altitude $z + H_0 = 30$ km, where H_0 is the altitude of the lidar station), for example, between 21 and 39 km, and
(2) N_{Rayfit} by fitting $N^{BS}(\nu_L, r)$ to a function of the form $C \exp(-br)$ over a short range, say, between 25 and 35 km, and determine the best fit value at r_R.

Since N_{Raysum} is larger than N_{Rayfit}, it is less noisy, and since $N_{Na}^{nor}(\nu_L, r)$ and $N_{Na}^{Th}(\nu_L, r)$ do not require absolute normalization, we use N_{Raysum} to compute $N_{Na}^{nor}(\nu_L, r)$ and the corresponding theoretical counts $N_{Na}^{Th}(\nu_L, r)$ via (7.35.a), followed by temperature and wind ratios via (7.36.a) and (7.36.b). To determine Na density, which requires absolute normalization, we would use N_{Rayfit} to obtain $N_{Na}^{nor}(\nu_L, r)$ and $N_{Na}^{Th}(\nu_L, r)$ for the evaluation of Na density, $\mathcal{N}_{Na}(r)$, via (7.35.c) with a model atmospheric density at, r_R, $\mathcal{N}_a(r_R)$, N_{Rayfit} being directly proportional to atmospheric density at z_R or r_R. One can determine a Na density profile from $N_{Na}^{nor}(\nu_L, r)$ for each laser frequency; though because they are equal to one another to within photon count statistics, we usually use that determined from $N_{Na}^{nor}(\nu_a, r)$.

Third, to determine $N_{Na}^{Th}(\nu_L, r)$ from $N_{Na}^{nor}(\nu_L, r)$, we need to account for the upward and downward attenuations due to the sodium layer utilizing the information in the "CrossV22 TVFCrossV22_3Frq" file. We calculate (*T*, *V*) using the Raysum N_{Raysum} normalized $N_{Na}^{nor}(\nu_L, r)$, and $\mathcal{N}_{Na}(r_R)$ using Rayfit N_{Rayfit} normalized $N_{Na}^{nor}(\nu_L, r)$. We begin below the sodium layer (e.g., 70 km) where there is no Na nor attenuation correction. Using these properties of the layer underside, we calculate the attenuation of photons passing through that layer. This attenuation is used to obtain the $N_{Na}^{Th}(\nu_L, r)$ in the layer above.

Using hourly means as example, the retrieved temperature, LOS wind, and Na density profiles at night (9.5 UT at the lidar site) and near noon local time (18.5 UT) of January 19, 2009 were analyzed and presented in Figs. 8(a), 8(b), and 8(c) of [7.2]. Since all photocounts are normalized to the modeled atmospheric density at the reference altitude, the authors also consider the effect on the retrieved quantities if the reference atmospheric density is 10% higher, or if the attenuation due to the Na layer is omitted. These results are also shown in the same figures. In addition, this comprehensive reference [7.2] also offers tips for real-time monitoring of observations and techniques of numerical analysis for lidar data processing; it is worthy of the lidar practitioner's attention.

The signal from Rayleigh scattering from the Na lidar, though not mentioned, can be used to retrieve atmospheric temperature via the integration technique (see Section 7.2). However, for a lidar with modest PA product of 0.05 Wm^2, the Rayleigh signal becomes negligibly weak at ~60 km; see Fig. 4 of [7.2]. Because of the initial guess of temperature and pressure at the top, the altitude range for valid temperature retrieval is likely limited to between 30 and 50 km. For a lidar with much larger PA product, a Na lidar temperature measurement can provide a measured temperature at the top altitude for Rayleigh lidar temperature retrieval. This was first reported in 2018 [7.88].

7.5.2(6) Photon Noise–Induced Uncertainty and Intrinsic Stability of a Narrowband LIF Lidar

The estimation of measurement uncertainty in retrieved temperature and LOS wind due to photon noise, though tedious, is straightforward. As given in Eqs. (12a) and (12b) of [7.2], the variance of observed temperature and LOS wind may be related to measurement sensitivities, as derivatives of (*T*, *R*) with respect R_T and R_V, derivable from the calibration curves, multiplied by the associated fractional variances of theoretical photocounts, $<(\Delta N_{\pm}^{Th})^2>/(N_a^{Th})^2$, very much like the ratio techniques discussed in Sections 7.3 and 7.4. These relations are reproduced here as (7.37.a) and (7.37.b):

$$\left\langle(\Delta T)^2\right\rangle = \frac{\left\{\left[\frac{1}{2}\frac{\partial T}{\partial R_T}+\frac{\partial T}{\partial R_V}\right]^2\left\langle(\Delta N_+^{Th})^2\right\rangle+\left[\frac{1}{2}\frac{\partial T}{\partial R_T}-\frac{\partial T}{\partial R_V}\right]^2\left\langle(\Delta N_-^{Th})^2\right\rangle+\left[R_T\frac{\partial T}{\partial R_T}+R_V\frac{\partial T}{\partial R_V}\right]^2\left\langle(\Delta N_a^{Th})^2\right\rangle\right\}}{\left(N_a^{Th}\right)^2}; \tag{7.37.a}$$

$$\left\langle(\Delta V)^2\right\rangle = \frac{\left\{\left[\frac{1}{2}\frac{\partial V}{\partial R_T}+\frac{\partial V}{\partial R_V}\right]^2\left\langle(\Delta N_+^{Th})^2\right\rangle+\left[\frac{1}{2}\frac{\partial V}{\partial R_T}-\frac{\partial V}{\partial R_V}\right]^2\left\langle(\Delta N_-^{Th})^2\right\rangle+\left[R_T\frac{\partial V}{\partial R_T}+R_V\frac{\partial V}{\partial R_V}\right]^2\left\langle(\Delta N_a^{Th})^2\right\rangle\right\}}{\left(N_a^{Th}\right)^2}. \tag{7.37.b}$$

The fractional variances of theoretical photocounts may in turn be related to the fractional variances of the background subtracted photocounts, $<(\Delta N_\pm^{BS})^2>/(N_a^{BS})^2$, via their respective inverse attenuation coefficient, $Ext^{-1}(\nu_L) = A_\uparrow^{-1}(\nu_L, T, V)A_{\downarrow eff}^{-1}(\nu_L, T, V)$, whose uncertainty can be deemed negligible, as given in (13a) of [7.2]. In the shot-noise limit, the variance-squared of the background-subtracted photocount, $<(\Delta N_{\nu_L}^{BS})^2>$, equals the background subtracted photocount, $N_{\nu_L}^{BS}$, plus the associated background counts, B_{ν_L}. Including the fluctuations in the background counts, the fractional variances of theoretical photocounts, $(\Delta N_\pm^{Th})^2/(N_a^{Th})^2$, are then given in (13b) of [7.2]. We reproduce these relations as (7.38.a) and (7.38.b):

$$\frac{(\Delta N_\pm^{Th})^2}{(N_a^{Th})^2} = \frac{Ext^{-2}(\nu_\pm)}{Ext^{-2}(\nu_a)}\frac{(\Delta N_\pm^{nor})^2}{(N_a^{nor})^2} = \left(\frac{Ext^{-1}(\nu_\pm)N_{Raysum}^a}{Ext^{-1}(\nu_a)N_{Raysum}^\pm}\right)^2\frac{(\Delta N_\pm^{BS})^2}{(N_a^{BS})^2}, \text{ or} \tag{7.38.a}$$

$$\frac{(\Delta N_\pm^{Th})^2}{(N_a^{Th})^2} = \left(\frac{Ext^{-1}(\nu_\pm)N_{Raysum}^a}{Ext^{-1}(\nu_a)N_{Raysum}^\pm}\right)^2\frac{N_\pm^{BS}+B_\pm}{(N_a^{BS})^2} \text{ and } \frac{(\Delta N_a^{Th})^2}{(N_a^{Th})^2} = \frac{(\Delta N_a^{BS})^2}{(N_a^{BS})^2} = \frac{N_a^{BS}+B_a}{(N_a^{BS})^2}. \tag{7.38.b}$$

With (7.37) and (7.38), the temperature and LOS wind uncertainties, ΔT and ΔV, can be calculated along with the retrieving of T and V as the output of data analysis. The photon noise uncertainty of $\mathcal{N}_{Na}(r)$ may be directly evaluated from (7.35.c) by noting that $\Delta N_{Na}^{Th} = \Delta N_{Na}^{BS}/N_{Rayfit}^{\nu_L} = \sqrt{N_{Na}^{BS}+B^{\nu_L}}/N_{Rayfit}^{\nu_L}$. Using these formulae, the measured uncertainties for the same example at 9.5 UT were also presented in Figs. 8(a), 8(b), and 8(c) of [7.2]; there, at 91 km, the deduced temperature, LOS wind, and Na density were 198.7 ± 0.39 K, –25.0 ± 0.34 m/s, and $2.81 \times 10^9 \pm 3.96 \times 10^6$ m^{-3} with $PA = 0.05$ Wm^2.

After many years of data collection, lidars have become an important instrument for monitoring long-term changes in the atmosphere. As such, in addition to photon noise, the stability of the measurements is a potential issue of concern. Since precisely determining the absolute lidar receiving sensitivity, which depends on PMT quantum efficiency, alignment, and optics, is a difficult challenge, the received photon files in Na lidar are normalized to atmospheric density $\mathcal{N}_a(r_R)$ at a lower reference height. Most Na lidars that measure Na density use only a broadband laser system, whose backscattered ratio of atmospheric Rayleigh scattering, σ_π^R, to Na resonance scattering spectrum $\sigma^\pi(\nu_L)$ changes from one laser pulse to the next, resulting in an unreliable retrieval of absolute Na density or Na abundance [7.89]. This is not the case for a narrowband Na lidar, since its laser frequency ν_L is locked to within ~ 1 MHz in the fluorescence linewidth of ~ 1 GHz; since σ_π^R is insensitive to frequency jitter, the ratio $\sigma_\pi^R/\sigma^\pi(\nu_L)$ is then known to be better than one part in one thousand. Therefore, the narrowband Na lidar, which takes full advantage of intensity ratios with its frequencies locked in real time to an absolute frequency standard – i.e., Doppler-free spectroscopy – is free from calibration issues. Its measurements over the years can be employed with confidence for long-term change studies [7.90]. This is particularly true for the CSU/USU Na lidar, as $\sigma^\pi(\nu_L)$ depends on temperature T and LOS wind V, which the lidar measures at all altitudes at the same time. To further discuss the systematic errors of the Na lidar measurements, we point out that (T, V) are retrieved from the ratios of three photon files, each at a preselected frequency ν_L within the Na D_2 transition via the well-known, accurate backscattering cross-section $\sigma^\pi(\nu_L, T, V)$. Assuming the lidar receiving sensitivity remains the same during the dwell time of a set of the frequencies (60 ms in a 50 Hz system), the ratios used in the retrieval are independent of $\mathcal{N}_a(r_R)$ and of the possible day-to-day changes in lidar receiving sensitivity – the only systematic error of the (T, V) measurements comes from frequency locking accuracy. By locking the absolute laser frequency to within ~1 MHz, the systematic errors in the retrieved (T, V) have been shown to be 0.005 K and 0.589 m/s, respectively [7.2]. The retrieved Na density, however, in addition to statistical error due to photon noise has an additional systematic error associated with the uncertainty in $\mathcal{N}_a(r_R)$, unless its value is measured simultaneously.

7.5.3 Scientific Contributions and Future Challenges of Narrowband LIF Lidars

We have not discussed potassium LIF lidar mainly because potassium and sodium have the same energy-level structure, thus all discussions of Na LIF lidar apply to the K LIF lidar. Because of potassium's much lower abundance compared to sodium, most lidar improvements and deployments have been for sodium measurements [7.42, 7.91, and 7.92]. However, early potassium lidar observations have made significant contributions to MLT science. These include establishing the global two-level nature of the mesopause thermal structure [7.93], and temperature climatology in both polar [7.94] and equatorial [7.95] regions.

In the past three-plus decades, LIF lidars have greatly advanced MLT science. Among the most interesting is the two-level mesopause thermal structure previously mentioned, which demonstrates the powerful influences of atmospheric buoyancy waves that produce the counterintuitive temperature structure between 65 km and 100 km, with summer colder than winter. A conclusive report on this interesting temperature and wind climatology at a fixed midlatitude location was done with the CSU narrowband Na lidar [7.96, 7.97]. With the full-diurnal-cycle (24-hr continuous) observations, midlatitude climatology of temperatures along with zonal and meridional winds were carried out with four years of data collection [7.98]. It is gratifying to see the observed seasonal results in qualitative agreement with our current understanding of the thermal and dynamical structure of the midlatitude mesopause region: reproducing salient features, such as the summer-to-winter two-level mesopause, zonal wind reversal, and the "residual" meridional flow from summer pole to winter pole. Another interesting result came out of a record nine days of 24-hr continuous observations, which showed correlated tidal 24-hr and 12-hr variations. The resulting variations in raw temperature along with zonal and meridional wind profiles were so graphically clear [7.99] as to be amazing. Many subtle and important phenomena in the MLT relating to atmospheric gravity (buoyancy) waves, long-term climatic changes, and the physics and chemistry of normal, sporadic, and thermospheric metal layers were enabled by LIF lidar studies. These achievements have recently been summarized in a book chapter [7.100].

As the recognition of the upper atmosphere as an integral part of the global weather system, the need to monitor temperature and wind in the mesosphere by lidars from many ground-based stations is imperative. This, of course, needs to be done automatically with maintenance-free lidar systems, that is, all-solid-state systems (Na or Fe) employing diode laser pumping with much better thermal stability, minimum maintenance, and longer lifetime (i.e., years) compared to the classic flashlamp-pumped systems. Diode-pumped Nd:YAG-based [7.42] and fiber-coupled [7.92] systems arrive timely for this purpose. A new addition in this connection is the diode-pumped alexandrite laser [7.43]. At the current level of 1 W of available power at 770 nm, a diode-pumped alexandrite ring laser is being designed for the VAHCOLI (Vertical And Horizontal COverage by LIdar) network intended for worldwide measurement campaigns [7.101]. Like the sum frequency generation from two Nd: YAG lasers (at 1064 nm and 1319 nm) that gives rise to 589 nm coherent light [7.42, 7.103] for probing atmospheric Na atoms, the second harmonic of the alexandrite laser at 744 nm that generates 372 nm light can be used to probe Fe atoms in the mesopause region. Realizing that lasing has been achieved at 1116 nm in Nd:YAG, whose third harmonic generates 372 nm light, a new lidar with a diode-pumped third harmonic YAG system has also been built and tested for the detection of fluorescence from Fe atoms [7.103]. To minimize pulse-to-pulse jitter, recent research to generate high-power single-frequency laser pulses using a hybrid master oscillator power amplifier (MOPA) configuration should be noted [7.104]. In such a system, one starts from a CW source modulated by an AOM for pulse shaping followed by sometimes challenging multiple stages of hybrid crystal and fiber amplifiers. This process ultimately results in pulses of desired shape and power without the use of a Q-switch, and thus minimizes jitter. All these activities will certainly lead to lidar systems playing the major role in weather forecasting by supplying temperature and wind data of the upper atmosphere (from the stratosphere to the mesopause).

Another exciting area of scientific research is to take up the challenging investigations of atmospheric turbulence, which are believed to have timescales between 1 s and 1 hr, and length scales between millimeters and kilometers, corresponding to inner and outer ranges (ℓ_0 and L_0, respectively) of the Kolmogorov turbulence spectrum. For many years, in situ (with rockets) and ground-based radar measurements have explored turbulence at scales down to 0.1 hr and 0.1 km, being limited by instrument resolution [7.105]. These have seen decreasing power toward smaller sizes and shorter periods; see Fig. 3 of [7.106], where it shows $\ell_0 \sim 27$ m and $L_0 \sim 1670$ m in the MLT region. Using a Na lidar with temporal and vertical resolution of 6 s and 25 m, turbulence spectra of temperature variations have been measured down to a period and wavelength of 12 s and 50 m [7.107]. One "advantage of this high-resolution lidar technique is that both turbulence and gravity wave perturbations can be measured at the same time so that the transport induced by gravity waves and turbulence can be easily compared." Due to considerable photon noise with this moderate lidar system, the result was derived from 150 hours of observations. Lidars with larger PA product would be necessary to assess atmospheric turbulence within a short time duration, and this is a challenge. These are two of the many reasons for desiring a lidar with much larger PA product for the study of atmospheric turbulence. For atmospheric physics interest, we would like to resolve and identify fine structures created by atmospheric interactions and to characterize turbulent heating associated with transient events, such as the breaking of an atmospheric gravity wave. For astronomy, we would like to characterize atmospheric turbulence in real time to negate the wavefront distortions it induces in integrating adaptive optics with LGS (laser guide star) in all modern giant telescopes.

An example of this dual benefit to both atmospheric and astronomical research is in order. In LGS systems, rapid fluctuations in the Na layer centroid altitude induce focusing errors on the adaptive optics. Motivated by the need to monitor and correct for this effect, researchers developed a high-resolution lidar with PA ~ 130 Wm^2 (4–5 W power and 6 m diameter telescope) and used it to probe the atmospheric Na layer structure, measuring the temporal power spectrum of the Na centroid altitude down to 1 Hz [7.108]. The same observations also reveal highly structured multiple Na layers that vary in density and altitude on timescales ranging from minutes to hours, along with direct observations of periodic oscillations with a 4-min apparent period, large-scale instabilities, and Kelvin–Helmholtz billows with an overall downward propagation, implicating the presence of atmospheric waves [7.109].

The availability of a high-power (up to 50 W), single-mode, narrowband, frequency-doubled, and Raman amplified fiber laser at 589 nm [7.110], in support of Na LGS research, is celebrated in the astronomy community. A lower-power version (~ 2 W) of this laser has already been used to replace the notorious CW ring laser in the many existing pulsed-dye-amplifier-based Na lidar systems. Using a 20 W version along with the pseudorandom code modulation technique [7.111, 7.112] – despite the power penalty in a CW system – observations of high-resolution Na density profiles were demonstrated in 2020 [7.113]. These suggest that this technique could be implemented by diverting a small portion of the returned laser light and allow the determination of Na centroid altitude every 5 s. Since the position (the centroid altitude) of a Na LGS moves around in the Na layer, a natural guide star (NGS) is used for focus and tip-tilt sensing in current AO systems. "If the sodium density profile can be measured with sufficient spatial and temporal resolution, the focus term could be sensed using LGS alone. If tip/tilt could also be sensed using LGS, no NGS would be required, and AO systems would be able to achieve 100 per cent sky coverage," as stated in [7.113].

References

7.1 Hair, J. W., C. A. Hostetler, A. L. Cook et al. (2008). Airborne high spectral resolution lidar for profiling aerosol optical properties. *Appl. Opt.* **47**(36), 6734–6753.

7.2 Krueger, D. A., C.-Y. She, and T. Yuan. (2015). Retrieving mesopause temperature and line-of-sight wind from full-diurnal-cycle Na lidar observations. *Appl. Opt.* **54**(32), 9469–9489.

7.3 Hansch, T. W., I. S. Shahin, and A. L. Schawlow. (1971). High resolution saturation spectroscopy of the sodium *D* lines with a pulsed tunable dye laser. *Phys. Rev. Lett.* **27** (11), 707–710.

7.4 Kuchta, E., R. J. Alvarez, II, Y. H. Li, D. A. Krueger, and C. Y. She. (1990). Collisional broadening of BaI line (553.5 *nm*) by He or Ar. *Appl. Phys. B*, **50**(2), 129–132.
7.5 Arie, A., S. Schiller, E. K. Gustafson, and R. L. Byer. (1992). Absolute frequency stabilization of diode-laser-pumped Nd:YAG lasers to hyperfine transitions in molecular iodine. *Opt. Lett.* **17**(17), 1204–1206.
7.6 Arie, A. and R. L. Byer. (1993). Frequency stabilization of the 1064-*nm* Nd: YAG lasers to Doppler-broadened lines of iodine. *Appl. Opt.* **32**(36), 7382–7386.
7.7 Fiedler, J., and G. von Cossart. (1999). Automated lidar transmitter for multiparameter investigations within the Arctic atmosphere. *IEEE Trans. Geosci. Remote Sensing*, **37** (2), 748–755.
7.8 She, C. Y. and J. R. Yu. (1995). Doppler-free saturation fluorescence spectroscopy of Na atoms for atmospheric applications. *Appl. Opt.*, **34**(6), 1063–1075.
7.9 Fiocco, G., G. Beneditti-Michelangeli, K. Maischberger, and E. Madonna. (1971). Measurement of temperature and aerosol to molecule ratio in the troposphere by optical radar. *Nature, Phys. Sci.*, **229**(3), 78–79.
7.10 Schwiesow, R. L. and L. Lading. (1981). Temperature profiling by Rayleigh-scattering lidar. *Appl. Opt.*, **20**(11), 1972–1979.
7.11 Shimizu, H., S. A. Lee, and C. Y. She. (1983). High spectral resolution lidar system with atomic blocking filters for measuring atmospheric parameters. *Appl. Opt.*, **22**(9), 1373–1381.
7.12 She, C. Y., R. J. Alvarez II, L. M. Caldwell, and D. A. Krueger. (1992). High-spectral-resolution Rayleigh–Mie lidar measurement of aerosol and atmospheric profiles. *Opt. Lett.*, **17**(7), 541–543.
7.13 Hair, J. W., L. M. Caldwell, D. A. Krueger, and C.-Y. She. (2001). High-spectral-resolution lidar with iodine-vapor filters: measurement of atmospheric-state and aerosol profiles. *Appl. Opt.*, **40**(30), 5280–5294.
7.14 Ansmann, A., M. Riebesell, and C. Weitkamp. (1990). Measurement of atmospheric aerosol extinction profiles with a Raman lidar. *Opt. Lett.*, **15**(13), 746–748.
7.15 Eloranta, E. (2005). High spectral resolution lidar. Chapter 5 in *Lidar Range-Resolved Optical Remote Sensing of the Atmosphere*, C. Weitkamp, ed., Springer.
7.16 Ansmann, A., U. Wandinger, M. Riebesell, C. Weitkamp, and W. Michaelis. (1992). Independent measurement of extinction and backscatter profiles in cirrus clouds by using a combined Raman elastic-backscatter lidar. *Appl. Opt.* **31**(33), 7113–7131.
7.17 Ansmann, A. and Müller, D. (2005). Lidar and atmospheric aerosol particles. Chapter 4 in *Lidar Range-Resolved Optical Remote Sensing of the Atmosphere*, C. Weitkamp, ed., Springer.
7.18 Shipley, S. T., D. H. Tracy, E. W. Eloranta et al. (1983). High spectral resolution lidar to measure optical scattering properties of atmospheric aerosols. 1: Theory and instrumentation. *Appl. Opt.*, **22**(23), 3716–3724.
7.19 Alvarez II, R. J., L. M. Caldwell, Y. H. Li, D. A. Krueger, and C. Y. She. (1990). High spectral resolution lidar measurement of tropospheric backscatter-ratio with barium atomic blocking filters. *J. Atmos. Oceanic Technol.*, **7**(6), 876–881.
7.20 Piironen, P. and E. W. Eloranta. (1994). Demonstration of a high-spectral-resolution lidar based on an iodine absorption filter. *Opt. Lett.*, **19**(3), 234–236.
7.21 Gerstenkorn S. and P. Luc. (1978). Atlas du Spectre d'Absorption de la Molecule d'Iode 14800–20000 cm^{-1}. Paris: Editions du Centre National de la Recherche Scientifique (CNRS).
7.22. Forkey, J. N., W. R. Lempert, and R. B. Miles. (1997). Corrected and calibrated I_2 absorption model at frequency-doubled Nd:YAG laser wavelengths, *Appl. Opt.*, 36 (27), 6729–6738.

7.23 Zhang, Y.-P., D. Liu, X. Shen et al. (2017). Design of iodine absorption cell for high-spectral-resolution lidar. *Opt. Ex.*, **25**(14), 15913–15926.
7.24 Eloranta, E., I. Razenkov, and J. Garcia. (2018). HSRL measurements of Lidar ratios in the presence of oriented ice crystals. Abstract. American Meteorological Society website. Accessed at https://ams.confex.com/ams/2019Annual/webprogram/Paper351863.html
7.25 Eloranta, E. (1998). Practical model for the calculation of multiply scattered lidar returns. *Appl. Opt.* **37**(12), 2464–2472.
7.26 Elterman, L. (1954). Seasonal trends of temperature, density, and pressure to 66 km obtained with the searchlight probing technique. *Jour. Geophys. Res.* **59**(3), 351–358.
7.27 Hauchecorne, A., and M. L. Chanin. (1980). Density and temperature profiles obtained by lidar between 35 and 70 km, *Geophys. Res. Letters* **7**(8), 565–568.
7.28 Chanin, M.-L., and A. Hauchecorne. (1981). Lidar observation of gravity and tidal waves in the stratosphere and mesosphere, *J. Geophys. Res.* **86**(C10), 9715–9721.
7.29 Nedeljkovic, D., A. Hauchecorne, and M. L. Chanin. (1993). Rotational Raman lidar to measure the atmospheric temperature from the ground to 30 km, *IEEE Trans. Geosci. Remote Sens.* **31**(1), 90–101.
7.30 Behrendt, A. and J. Reichardt. (2000). Atmospheric temperature profiling in the presence of clouds with a pure rotational Raman lidar by use of an interference-filter-based polychromator. *Appl. Opt.* **39**(9), 1372–1378.
7.31 Behrendt, A., T. Nakamura, and T. Tsuda. (2004). Combined temperature lidar for measurements in the troposphere, stratosphere, and mesosphere. *Appl. Opt.* **43**(14), 2930–2939.
7.32 Arshinov, Yu. F., S. M. Bobrovnikov, V. E. Zuev, and V. M. Mitev. (1983). Atmospheric temperature measurement using a pure rotational Raman lidar. *Appl. Opt.* **22**(19), 2984–2990.
7.33 Cooney, J. A. (1984). Atmospheric temperature measurement using a pure rotational Raman lidar: comment. *Appl. Opt.* **23**(5), 653–654.
7.34 Alvarez, R. J., II. (1991). *Measurement of tropospheric temperature and aerosol extinction using high-spectral-resolution lidar.* Ph.D. thesis, Colorado State University. University Microfilms International, Ann Arbor, MI.
7.35 Krueger, D. A., L. M. Caldwell, R. J. Alvarez II, and C. Y. She. (1993). Self-consistent method for determining vertical profiles of aerosol and atmospheric properties using high-spectral-resolution Rayleigh–Mie lidar. *J. Atm. Oceanic Tech.* **10**(4), 533–545.
7.36 Alvarez II, R. J., L. M. Caldwell, P. G. Wolyn et al. (1993). Profiling temperature, pressure, and aerosol properties using a high spectral resolution lidar employing atomic blocking filters. *J. Atm. Oceanic Tech.* **10**(4), 546–556.
7.37 She, C.-Y., J. Yue, Z.-A. Yan et al. (2007). Direct-detection Doppler wind measurements with a Cabannes–Mie lidar: A. Comparison between iodine vapor filter and Fabry–Perot interferometer methods. *Appl. Opt.*, **46**(20), 4434–4443.
7.38 Tellinghuisen, J. (1973). Resolution of the visible-infrared absorption spectrum of I_2 into three contributing transitions. *J. Chem. Phys.* **58**(7), 2821–2834.
7.39 Forkey, J. N. (1996). *Development and demonstration of filtered Rayleigh scattering – a laser based flow diagnostic for planar measurement of velocity, temperature and pressure.* Ph.D. dissertation, Department of Mechanical and Aerospace Engineering, Princeton University.
7.40 Hair, J. W. (1998). *A high spectral resolution lidar at 532 nm for simultaneous measurement of atmospheric state and aerosol profiles using iodine vapor filters.* Ph.D. dissertation, Department of Physics, Colorado State University.

7.41 Hua, D.-X., M. Uchida, and T. Kobayashi. (2004). Ultraviolet high-spectral-resolution Rayleigh–Mie lidar with a dual-pass Fabry–Perot etalon for measuring atmospheric temperature profiles of the troposphere. *Opt. Lett.* **29**(10), 1063–1065.
7.42 Kawahara, T. D., S. Nozawa, N. Saito et al. (2017). Sodium temperature/wind lidar based on laser-diode-pumped Nd:YAG lasers deployed at Tromsø, Norway (69.6°N, 19.2°E). *Opt. Express* **25**(12), A491–A501.
7.43 Munk, A., B. Jungbluth, M. Strotkamp et al. (2018). Diode-pumped alexandrite ring laser in single-longitudinal mode operation for atmospheric lidar measurements. *Opt. Express* **26**(12), 14,928–14,935.
7.44 Harrell, S. D., C.-Y. She, T. Yuan et al. (2009). Sodium and potassium vapor Faraday filters revisited: Theory and applications. *J. Opt. Soc. Amer. B*, **26**(4), 659–670.
7.45 Voss, E., and C. Weitkamp. (1992). Investigations on atomic-vapor filter high-spectral-resolution lidar for temperature measurements. *Proc. 16th. Int. Laser Radar Conf.*, Cambridge, MA, NASA Conf. Pub. 3158, Part 2, 699–702.
7.46 She, C.-Y., D. A. Krueger, Z.-A. Yan, and X. Hu. (2021). Atomic vapor filter revisited: a Cabannes scattering temperature/wind lidar at 770 nm. *Opt. Express* **29**(3), 4338–4362.
7.47 Werner, C. (2005). Doppler wind lidar. Chapter 12 in *Lidar Range-Resolved Optical Remote Sensing of the Atmosphere*, C. Weitkamp, ed., Springer.
7.48 Post, M. J. and R. E. Cupp. (1990). Optimizing a pulsed Doppler lidar. *Appl. Opt.* **29** (28), 4145–4158.
7.49 Kavaya, M. J., J. Y. Beyon, G. J. Koch et al. (2014). The Doppler Aerosol Wind (DAWN) airborne, wind-profiling coherent-detection lidar system: overview and preliminary flight results. *J. Atm. Oceanic Tech.* **31**(4), 826–842. doi: https://doi.org/10.1175/JTECH-D-12-00274.1.
7.50 Yu, J., U. Singh, N. Barnes, and M. Petros. (1998). 125-*mJ* diode-pumped injection-seeded Ho:Tm:YLF laser. *Opt. Lett.*, **23**(10), 780–782. doi: https://doi.org/10.1364/OL.23.000780.
7.51 Henderson, S. W., P. Gatt, D. Rees, and R. M. Huffaker. (2005). Wind lidar. Chapter 7 in *Laser Remote Sensing*, T. Fujii, & T. Fukuchi, eds., CRC Press, Taylor and Francis Group.
7.52 Kingston, R. H. (1995). *Optical Sources, Detectors, and Systems*. Academic Press. doi: https://doi.org/10.1016/B978-0-12-408655-5.X5000-8.
7.53 Goodman, J. W. (1985). *Statistical Optics*. John Wiley and Sons, New York.
7.54 Henderson, S. W. (2013). Review of Fundamental Characteristics of Coherent and Direct Detection Doppler Receivers and Implications to Wind Lidar System Design, *Proc. 17th Coherent Laser Radar Conference* (CLRC 2013), 45–49, ISBN: 9781629931494.
7.55 She, C. Y., J. R. Yu, H. Latifi, and R. E. Bills. (1992). High-spectral-resolution fluorescence light detection and ranging for mesospheric sodium temperature measurements. *Appl. Opt.*, **31**(12), 2095–2106.
7.56 Chanin, M. L., A. Garnier, A. Hauchecorne, and J. Porteneuve. (1989). A Doppler lidar for measuring winds in the middle atmosphere. *Geophys. Res. Lett.* **16**(11), 1273–1276, doi: https://doi.org/10.1029/GL016i011p01273.
7.57 Korb, C. L., B. Gentry, and C. Weng. (1992). Edge technique: Theory and application to the lidar measurement of atmospheric wind. *Appl. Opt.* **31**(21), 4202–4213.
7.58 Korb, C. L., B. M. Gentry, S. X. Li, and C. Flesia. (1998). Theory of the double-edge technique for Doppler lidar wind measurement. *Appl. Opt.* **37**(15), 3097–3104.
7.59 Flesia, C., and C. L. Korb. (1999). Theory of the double-edge molecular technique for Doppler lidar wind measurement. *Appl. Opt.* **38**(3), 432–440.

7.60 Korb, C. L., B. Gentry, and X. Li. (1997). Edge technique Doppler lidar wind measurements with high vertical resolution. *Appl. Opt.* **36**(24), 5976–5983.
7.61 Gentry, B.M., H. Chen, and S. X. Li. (2000). Wind measurements with 355 *nm* molecular Doppler lidar. *Opt. Lett.* **25**(17), 1231–1233.
7.62 Liu, Z. S., W. B. Chen, T. L. Zhang, J. W. Hair, and C. Y. She. (1996). An incoherent Doppler lidar for ground-based atmospheric wind profiling. *Appl. Phys. B* **64**(5), 561–566.
7.63 Friedman, J. S., C. A. Tepley, P. A. Castleberg, and H. Roe. (1997). Middle-atmospheric Doppler lidar using an iodine-vapor edge filter. *Opt. Lett.* **22**(21), 1648–1650.
7.64 Liu, Z.-S., D. Wu, J.-T. Liu et al. (2002). Low-altitude atmospheric wind measurement from the combined Mie and Rayleigh backscattering by Doppler lidar with an iodine filter. *Appl. Opt.* **41**(33), 7079–7086.
7.65 Wang, Z., Z. Liu, L. Liu et al. (2010). Iodine-filter-based mobile Doppler lidar to make continuous and full-azimuth-scanned measurements: Data acquisition and analysis system, data retrieval methods, and error analysis. *Appl. Opt.*, **49**(36), 6960–6978.
7.66 Yan, Z., X. Hu, W. Guo et al. (2017). Development of a mobile Doppler lidar system for wind and temperature measurements at 30–70 km. *J. Quant. Spectrosc. Radiat. Transf.*, **188**, 52–59.
7.67 Baumgarten, G. (2010). Doppler Rayleigh/Mie/Raman lidar for wind and temperature measurements in the middle atmosphere up to 80 km. *Atmos. Meas. Tech.*, **3**(6), 1509–1518, doi: https://doi.org/10.5194/amt-3-1509-2010.
7.68 von Zahn, U., G. von Cossart, J. Fiedler et al. (2000). The ALOMAR Rayleigh/Mie/Raman lidar: Objectives, configuration, and performance. *Ann. Geophys.* **18**(7), 815–833.
7.69 Hecht, E. (1998). *Optics*. 3rd ed., Addison-Wesley, pp. 413–417.
7.70 Huang, W., X. Chu, B. P. Williams et al. (2009). Na double-edge magneto-optic filter for Na lidar profiling of wind and temperature in the lower atmosphere. *Opt. Lett.*, **34** (2), 199–201.
7.71 Huang, W., X. Chu, J. Wiig et al. (2009). Field demonstration of simultaneous wind and temperature measurements from 5 to 50 km with a Na double-edge magneto-optic filter in a multi-frequency Doppler lidar. *Opt. Lett.*, **34**(10), 1552–1554.
7.72 Witschas, B., C. Lemmerz, and O. Reitebuch. (2014). Daytime measurements of atmospheric temperature profiles (2–15 km) by lidar utilizing Rayleigh–Brillouin scattering. *Opt. Lett.*, **39**(7), 1972–1975.
7.73 Xia, H., X. Dou, D. Sun et al. (2012). Mid-altitude wind measurements with mobile Rayleigh Doppler lidar incorporating system-level optical frequency control method. *Opt. Express*, **20**(14), 15286–15300.
7.74 Xia, H., X. Dou, M. Shangguan et al. (2014). Stratospheric temperature measurement with scanning Fabry–Perot interferometer for wind retrieval from mobile Rayleigh Doppler lidar. *Opt. Express*, **22**(18), 21775–21789. doi: https://doi.org/10.1364/OE.22.021775.
7.75 She, C.-Y. (2005). On atmospheric lidar performance comparison: from power aperture product to power aperture mixing ratio scattering cross-section product. *Modern Optics*, **52**(18), 2723–2729, doi: https://doi.org/10.1080/09500340500352618.
7.76 Gardner, C. S. (2004). Performance capabilities of middle-atmosphere temperature lidars: comparison of Na, Fe, K, Ca, Ca^+, and Rayleigh systems. *Appl. Opt.*, **43**(25), 4941–4956.

7.77 Gibson, A. J., L. Thomas, and S. K. Bhattachacharyya. (1979). Laser observation of the ground-state hyperfine structure of sodium and of temperature in the upper atmosphere. *Nature* **281**(5727), 131–132.
7.78 Fricke, K. H. and U. von Zahn. (1985). Mesopause temperature derived from probing the hyperfine structure of the D_2 resonance line of sodium by lidar. *J. Atmos. Terr. Phys.* **47**(5), 499–512.
7.79 She, C. Y., H. Latifi, J. R. Yu et al. (1990). Two frequency lidar techniques for mesospheric Na temperature measurements. *Geophys. Res. Lett.* **17**(7), 929–932.
7.80 She, C. Y. and J. R. Yu. (1994). Simultaneous three-frequency Na lidar measurements of radial wind and temperature in the mesopause region. *Geophys. Res. Lett.* **21**(17), 1771–1774.
7.81 Chu, X., W. Pan, G. C. Papen, C. S. Gardner, and J. A. Gelbwachs. (2002). Fe Boltzmann temperature lidar: design, error analysis and initial results at the North and South Poles. *Appl. Opt.* **41**(21), 4400–4410.
7.82 Höffner, J., and J. Lautenbach. (2009). Daylight measurements of mesopause temperature and vertical wind with the mobile scanning iron lidar. *Opt. Lett.* **34**(9), 1351–1353.
7.83 Chu, X., W. Huang, J. P. Thayer, Z. Wang, and J. A. Smith. (2010). Progress in MRI Fe-Resonance/Rayleigh/Mie Doppler Lidar. In *Proceedings of the 25th International Laser Radar Conference*, Saint Petersburg, Russia, 947–950.
7.84 Papen, G. C., W. M. Pfenninger, and D. M. Simonich. (1995). Sensitivity analysis of Na narrowband wind-temperature lidar systems. *Appl. Opt.* **34**(3), 480–498.
7.85 Chen, H., M. A. White, D. A. Krueger, and C. Y. She. (1996). Daytime mesopause temperature measurements using a sodium-vapor dispersive Faraday filter in lidar receiver. *Opt. Lett.* **21**(15), 1093–1095.
7.86 Chen, H., C. Y. She, and E. Korevaar. (1993). Sodium vapor dispersive Faraday filter. *Opt. Lett.* **18**(12), 1019–1021.
7.87 Fricke-Begemann, C., M. Alpers, and J. Höffner. (2002). Daylight rejection with a new receiver for potassium resonance temperature lidars. *Opt. Lett.* **27**(21), 1932–1934.
7.88 Sox, L., V. B. Wickwar, T. Yuan, and N. R. Criddle. (2018). Simultaneous Rayleigh-scatter and sodium resonance lidar temperature comparisons in the mesosphere-lower thermosphere. *J. Geophys. Res. Atmos.*, **123**(18), 10,688–10,706, doi: https://doi.org/10.1029/2018jd029438.
7.89 Clemesha, B. R., Simonich, D. M., Batista, P. P., Vondrak, T., and Plane, J. M. C. (2004). Negligible long-term temperature trend in the upper atmosphere at 23°S. *J. Geophys. Res.*, **109**(D5), D05302, https://doi.org/10.1029/2003JD004243.
7.90 She, C.-Y., U. Berger, Z.-A. Yan et al. (2019). Solar response and long-term trend of midlatitude mesopause region temperature based on 28 years (1990–2017) of Na lidar observations. *J. Geophys. Res.*, **124**(8), 7140–7156, https://doi.org/10.1029/2019JA026759.
7.91 Li, T., X. Fang, W. Liu, S.-Y. Gu, and X. Dou. (2012). Narrowband sodium lidar for the measurements of mesopause region temperature and wind. *Appl. Opt.*, **51**(22), 5401–5411, doi: https://doi.org/10.1364/AO.51.005401.
7.92 Xia, Y., L.-F. Du, X.-W. Cheng et al. (2017). Development of a solid-state sodium Doppler lidar using an all-fiber-coupled injection seeding unit for simultaneous temperature and wind measurements in the mesopause region. *Opt. Express*, **25**(5), 5264–5278, doi: https://doi.org/10.1364/OE.25.005264.
7.93 von Zahn, U., and J. Hőffner. (1996). Mesopause temperature profiling by potassium lidar. *Geophys. Res. Lett.*, **23**(2), 141–144, doi: https://doi.org/10.1029/95GL03688.

7.94 Lübken, F.-J., J. Lautenbach, J. Hőffner, M. Rapp, and M. Zecha. (2009). First continuous temperature measurements within polar mesosphere summer echoes. *J. Atmos. Sol.-Terr. Phys.*, **71**(3–4), 453–463, doi: https://doi.org/10.1016/j.jastp.2008.06.001.

7.95 Friedman, J. S., and X. Chu. (2007). Nocturnal temperature structure in the mesopause region over the Arecibo Observatory (18.35°N, 66.75°W): Seasonal variations. *J. Geophys. Res.*, **112**(D14), D14107, doi: https://doi.org/10.1029/2006JD008220.

7.96 She, C. Y. (1994). New lidar reveals seasonal temperature variations in a midlatitude mesopause region. *Optics & Photonics News*, December issue, pp. 23–24.

7.97 She, C. Y., S. S. Chen, Z. L. Hu et al. (2000). Eight-year climatology of nocturnal temperature and sodium density in the mesopause region (80 to 105 km) over Fort Collins, CO (41°N, 105°W). *Geophys. Res. Lett.*, **27**(20), 3289–3292.

7.98 Yuan, T., C.-Y. She, D. A. Krueger et al. (2008). Climatology of mesopause region temperature, zonal wind, and meridional wind over Fort Collins, Colorado (41°N, 105°W), and comparison with model simulations. *J. Geophys. Res.*, **113**(D3), D03105, doi: https://doi.org/0.1029/2007JD008697.

7.99 She, C.-Y. and D. A. Krueger. (2007). Laser-induced fluorescence: Spectroscopy in the sky. *Optics & Photonic News*, **18**(9), 35–41.

7.100 She, C. Y., A. Z. Liu, T. Yuan et al. (2021). MLT science enabled by atmospheric lidars. In *Upper Atmospheric Dynamics and Energetics*, Space Physics and Aeronomy Collection, vol. 4, ed. Wenbin Wang, Yongliang Zhang, Larry J. Paxton, chap. 20. John Wiley and Sons. doi: https://doi.org/10.1002.9781119507512.ch20.

7.101 Lübken, F.-J. and Höffner, J. (2021). VAHCOLI, a new concept for lidars: technical setup, science applications, and first measurements. *Atmos. Meas. Tech.*, **14**(5), 3815–3836. https://doi.org/10.5194/amt-14-3815-2021.

7.102 Vance, J. D., C. Y. She, and H. Moosmüller. (1998). Continuous-wave, all-solid-state, single-frequency 400-mW source at 589 *nm* based on doubly resonant sum-frequency mixing in a monolithic lithium niobate resonator. *Appl. Opt.* **37**(21), 4891–4896.

7.103 Kaifler, B., C. Büdenbender, P. Mahnke et al. (2017). Demonstration of an iron fluorescence lidar operating at 372 *nm* wavelength using a newly developed Nd: YAG laser. *Opt. Lett.*, **42**(15), 2858–2861. doi: https://doi.org/10.1364/OL.42.002858.

7.104 Fu, S., Shi, W., Feng, Y., et al. (2017). Review of recent progress on single-frequency fiber lasers [Invited]. *J. Opt. Soc. Am. B*, **34**(3), A49–A62. doi: https://doi.org/10.1364/JOSAB.34.000A49.

7.105 Strelnikov, B., A. Szewczyk, I. Strelnikova et al. (2017). Spatial and temporal variability in MLT turbulence inferred from in situ and ground-based observations during the WADIS-1 sounding rocket campaign. *Ann. Geophys.*, **35**(3), 547–565. doi: https://doi.org/10.5194/angeo-35-547-2017.

7.106 Lübken, F.-J. (1997). Seasonal variation of turbulent energy dissipation rates at high latitudes as determined by in situ measurements of neutral density fluctuations. *J. Geophys. Res.*, **102**(D12), 13,441–13,456. doi: https://doi.org/10.1029/97JD00853.

7.107 Guo, Y., A. Z. Liu, and C. S. Gardner. (2017). First Na lidar measurements of turbulence heat flux, thermal diffusivity, and energy dissipation rate in the mesopause region. *Geophys. Res. Lett.*, **44**(11), 5782–5790. doi: https://doi.org/10.1002/2017GL073807.

7.108 Pfrommer, T. and P. Hickson. (2010). High-resolution lidar observations of mesospheric sodium and implications for adaptive optics. *J. Opt. Soc. Am. A*, **27**(11), A97–A105.

7.109 Pfrommer, T., P. Hickson, and C.-Y. She. (2009). A large-aperture sodium fluorescence lidar with very high resolution for mesopause dynamics and adaptive optics studies. *Geophys. Res. Lett.*, **36**(15), L15831. doi: https://doi.org/10.1029/2009GL038802.

7.110 Taylor, L. R., Y. Feng, and D. Bonaccini Calia. (2010). 50 W CW visible laser source at 589 *nm* obtained via frequency doubling of three coherently combined narrow-band Raman fibre amplifiers. *Opt. Express* **18**(8), 8540–8555.

7.111 She, C.-Y., M. Abo, J. Yue et al. (2011). Mesopause-region temperature and wind measurements with pseudorandom modulation continuous-wave (PMCW) lidar at 589 nm. *Appl. Opt.* **50**(18), 2916–2926.

7.112 Butler, D. J., R. I. Davies, R. M. Redfern et al. (2003). Measuring the absolute height and profile of the mesospheric sodium layer using a continuous wave laser. *Astron. Astrophys.* **403**(2), 775–785. doi: https://doi.org/10.1051/0004-6361:20030379.

7.113 Hellemeier, J. A., D. Bonaccini Calia, P. Hickson, A. Otarola, and T. Pfrommer. (2020). Measuring line-of-sight sodium density structure using laser guide stars. *Monthly Notice of the Royal Astronomical Society* MNRAS **494**(2), 2798–2808.

8
Transmitting and Receiving Optics

In this chapter, we begin with a discussion of the concept of the light beam parameter product and conservation of étendue in transmitting and receiving optics. We then use these concepts to describe the telescopes and detectors relevant to atmospheric lidars. We conclude with a brief conceptual discussion on the effects of atmospheric turbulence, how these reduce the effective receiver aperture (for coherent detection lidar), and their mitigation for astronomical telescopes.

8.1 Conservation of Étendue in Transmission and Reception Optics

Ideally, the laser transmitter and optical receiver beams have cylindrical symmetry. The beam quality may be quantified by the beam parameter product (BPP), defined as the product of the beam radius at its beam waist, w_0, and its half divergence angle θ_d [8.1, 8.2]. The best quality laser beam has a TEM_{00} spatial mode and is diffraction limited. As is well-known and shown in Fig. 8.1(a), the beam radius of a Gaussian beam, described by $w(z) = w_0\sqrt{1 + (z/z_R)^2}$, in which $z_R = \pi w_0^2/\lambda$ (λ being the wavelength) and z is the position along the beam, has its minimum beam radius (waist), w_0, where $z = 0$. The parameter z_R is known as the Rayleigh range; thus, at $z = \pm z_R$, or $z/z_R = \pm 1$, the beam radius is $\sqrt{2}w_0$. The divergence (half) angle of a Gaussian beam is diffraction limited and is given by $\theta_d = \lambda/(\pi w_0)$. As shown in Fig. 8.1(b), the peak normalized amplitude, A (dashed), and intensity, I (solid), of a Gaussian beam are respectively $A(r, z) = \exp[-(r/w)^2]$ and $I(z) = \exp[-2(r/w)^2]$. Here, the beam radius, $w(z)$, equals twice the standard deviation of the intensity profile of a Gaussian beam (i.e., $w(z) = 2\sigma(z)$). The percentage of the total power of a Gaussian beam contained within an aperture defined by this beam radius is ~95%. Thus, the half divergence angle, θ_d, and the minimum beam radius, w_0, are both measured in the far-field on the 5% intensity contour.

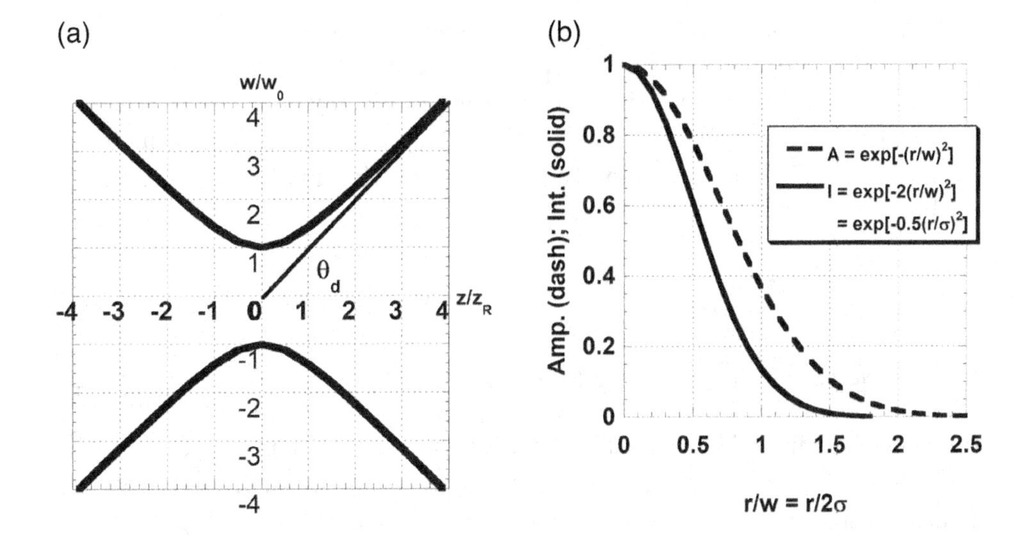

Fig. 8.1 (a) Beam radius of a Gaussian beam, $w(z)$, with the waist, w_0, at $z = 0$; z_R is the Rayleigh range. (b) Peak normalized amplitude and intensity of a Gaussian beam, A (dashed) and I (solid). Note, the beam radius equals twice the standard deviation of the intensity profile of a Gaussian beam (i.e., $w = 2\sigma$).

For a Gaussian beam profile as just described, its variance or second moment at z, $\sigma^2(z)$, is related to its value at $z = 0$, σ_0^2, and the square of the angular spread (intensity) from its waist $\theta_\sigma^2 = (2\lambda/\pi w_0)^2 = (\lambda/\pi\sigma_0)^2$ is given as

$$w^2(z) = w_0^2 + \left(\frac{w_0}{z_R}\right)^2 z^2 = w_0^2 + \left(\frac{\lambda}{\pi w_0}\right)^2 z^2 \to \sigma^2(z) = \sigma_0^2 + \theta_\sigma^2 z^2. \qquad (8.1)$$

Equation (8.1), above, relates the respective equalities, width, and variance of a Gaussian beam profile, as the beam propagates from its waist through free space. Remarkably, this quadratic free-space propagation rule, as stated in Siegman [8.1], "holds for any arbitrary real laser beam, whether it be Gaussian or non-Gaussian, fully coherent or partially incoherent, single mode or multiple transverse mode in character." Thus, by defining the beam radius or intensity variance of a real beam in terms of its second moment, these beam parameters will propagate with distance in free space exactly like an ideal Gaussian beam. For a non-diffraction limited beam, we require a multiplication factor, $M^2 = \Theta/\theta_d$, the ratio of the measured divergence (half angle), Θ, to that of a Gaussian beam (diffraction limited), θ_d, with the same minimum beam radius, w_0, to account for the difference in the far-field beam divergence between real and Gaussian beams.

The aforementioned multiplication factor quantifies the quality of a real laser beam. Unlike the BPP, M^2 as given in (8.2.a) is not only dimensionless but also wavelength independent. In optics, one describes the quality of an optical beam by the quantity *étendue*, which is defined as the product of the minimum beam cross-sectional area and its subtended solid angle; it is thus related to the BPP, as given in (8.2.b):

$$M^2 = \frac{\Theta}{\theta_d} = \Theta \frac{\pi w_0}{\lambda} \rightarrow BPP = w_0 \Theta = \frac{\lambda}{\pi} M^2 \text{ , and} \tag{8.2.a}$$

$$\textit{étendue} = \left(\pi w_0^2\right) \Delta\Omega = \left(\pi w_0^2\right)\left(\pi \Theta^2\right) = \pi^2 (BPP)^2 = \lambda^2 M^4. \tag{8.2.b}$$

Notice that both the BPP and étendue depend only on beam parameters (λ and M^2), thus they are conserved quantities as the beam propagates through passive optical elements, which do not alter the quality of the beam. Since the beam power remains the same as it propagates through passive elements, the *brightness*, defined as power per unit area–solid angle, is then also conserved. We can thus use the beam waist radius, $w_0 = 2\sigma_0$, and the corresponding (half) divergence angle, Θ, for a real (partially coherent) beam and $\theta_d = \lambda/(\pi w_0)$ for a Gaussian beam, in which $\Theta > \theta_d$, for the real beam BPP calculation. The cross-sectional diameter of an aperture of that passes 95% of the total beam power is $d = 2w = 4\sigma$, a parameter sometimes called $D4\sigma$. On a practical note, a well-centered aperture with diameter $d \approx 3w$ will transmit more than 99% of the beam energy. In the case of a real beam without cylindrical symmetry, (8.1), (8.2), and BPP conservation apply separately to each principal cross-axis.

8.2 Laser Beam Expander and Its Use

Background stray light is a major contributor to noise on the lidar signal and is a major challenge for daytime operations. Reducing the field of view is the most effective way to mitigate this. The spot size of the far-field coherent or partially coherent laser beam may be controlled and reduced using a laser beam expander, which is simply a small telescope with the placement of its objective and image lenses reversed. By enlarging the beam size, the transmitting beam divergence angle is accordingly reduced due to conservation of BPP. This, in turn, produces a smaller receiving field of view for the reception of the scattered light signal. There are two types of design for the beam expander, Keplerian and Galilean. The former design employs a positive image lens that focuses the collimated input beam to a spot before the objective lens, while the latter employs a diverging/converging lens combination, thus having no internal foci. For high-pulse-energy lasers, the

Galilean design is preferred, as it avoids nonlinear effects caused by extremely high irradiance in the air gap between the lenses. The input lens is oriented with a flat or convex entry surface so that any reflection from that surface does not converge, thus avoiding potential damage or safety concerns from the (generally small) reflection. The specifications of a laser beam expander will include its expansion ratio, input aperture diameter, diffraction-limited input diameter, output aperture diameter, and applicable wavelengths; as, for example, 3x, 7.0 mm, 5.0 mm, 15 mm, and 400–650 nm.

To utilize a given laser beam expander to get the desired expansion, one needs to ensure that the beam diameter of the laser output is much less than the specified maximum input diameter, D, of the beam expander. For these purposes, we adopt $D = 3w_0$ to allow 99% of the beam power to transmit through the system. For the specifications given above, $w_0 \leq 2.3$ mm, the output beam diameter (similarly defined) will be enlarged by a factor of three and the beam divergence simultaneously reduced by the same factor, nominally from 1 mrad to 0.33 mrad. For lidar applications, 2- or 3-times expansion is usually adequate, since a 3x expansion reduces the far-field beam cross-sectional area, and thus background noise, by a factor of 9 without penalty in the detected signal. Expanding the beam beyond this may come at a penalty, as the reduction of divergence angle brought about by expansion increases beam intensity in the scattering region, which will eventually lead to undesirable effects. In the case of resonance lidars, high beam intensity produces saturation effects in the scattering from the metal layer, as discussed in Chapter 3, distorting the desired measurements of wind and temperature. This is particularly acute for lidars detecting atoms with large scattering cross sections, like Na, Fe, and K.

8.2.1 Beam Expander Design

In Fig. 8.2(a), we show a Galilean beam expander design with a magnification of 3.3. The optics are all low-cost and off-the-shelf, so the cost is nominal. Cage system mounting provides reliable concentricity and easy adjustment. Note that the input lens orientation is the opposite of that recommended to minimize spherical aberrations. This eliminates a converging back reflection, which, although it might represent 0.1% or less of the beam power, can still have kilowatts of peak power and present a danger to upstream optics. This design accepts a slight penalty in spherical aberration in return for a safety margin. Nonetheless, output divergence angle induced by this expander is under 0.02 mrad for collimated input, as demonstrated by the y axis scale in Fig. 8.2(b), so spherical aberration has an insignificant effect on the divergence of the output, and we achieve the desired field-of-view reduction of 3.3 through beam expansion.

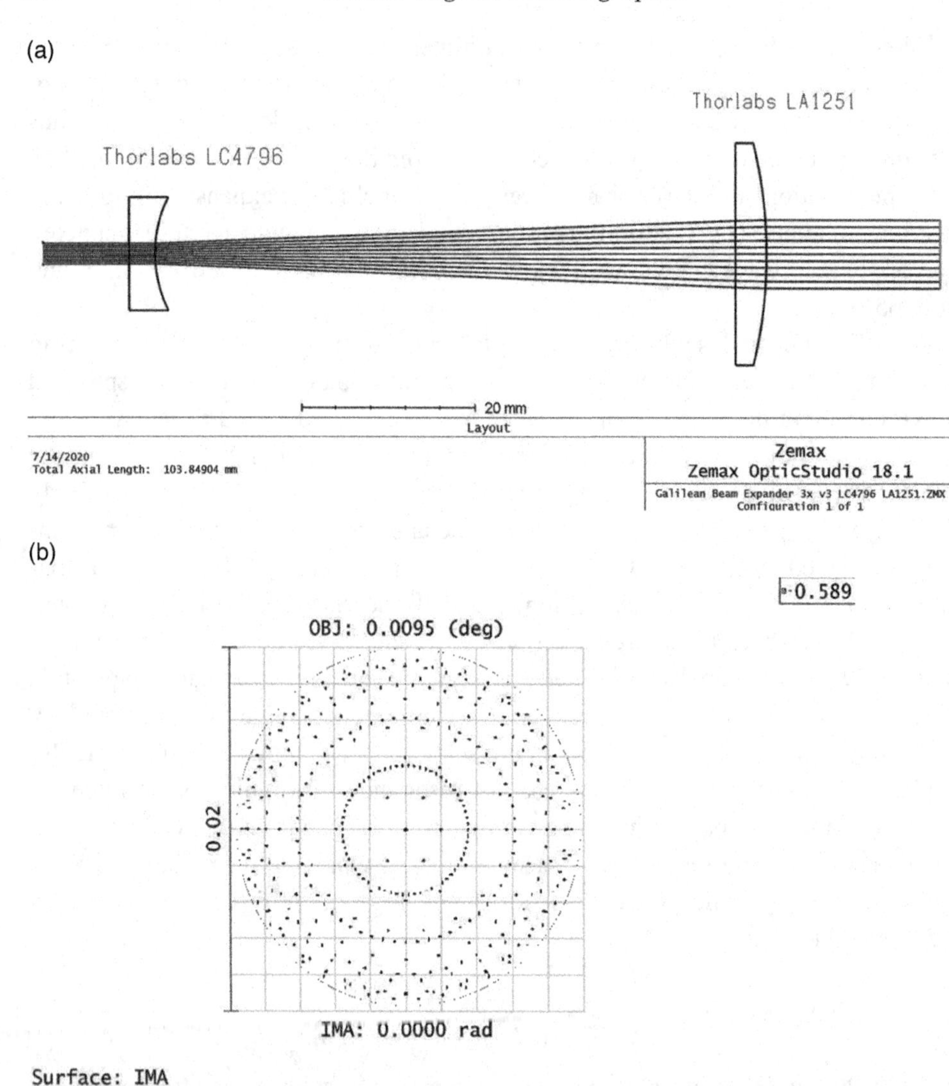

Fig. 8.2 (a) Zemax® OpticStudio design of a high-power Galilean beam expander using low-cost, off-the-shelf lenses. The beam trajectory is left-to-right. Note that the input plano-concave lens is placed in reverse, meaning this design induces slight spherical aberration in return for avoiding a converging back reflection from the input lens. (b) Output spot diagram. The scale is in mrad.

8.3 Telescope Optics

For many years, lidar telescopes were considered simple light buckets – to collect the scattered return and deliver it to a photon-counting detector – where image quality was unimportant. Today, thanks to computer modeling of receiver systems and more stringent requirements for daylight observations, we are more careful when specifying these optics. Principal considerations for telescope selection include the most obvious, such as mirror size, reflectivity, and *f*-number, along with others whose importance is more ambiguous, such as exit pupil size, field of view, detector photocathode size, and mirror surface figure. Telescope parameters sometimes interact in unexpected ways, and one must make trade-offs. For example, it is now common to couple the light from the telescope into an optical fiber. The optical fiber limits the telescope field of view, which in turn places constraints on the physical length of the telescope and affects its *f*-number (or numerical aperture, *NA*). The telescope *NA* must be equal to or less than the fiber *NA*, which limits the primary mirror diameter. At the output side of the optical fiber, the detector photocathode size may constrain both the fiber core diameter and the numerical aperture. For this reason (as well as cost), 10-m class telescopes proposed for community lidar facilities always consist of an array of smaller telescopes. In this section, we consider these factors to develop a selection space from which one can determine an optimum choice for a system telescope.

8.3.1 Size Considerations

Size is always the first consideration in specifying the receiver telescope. This may be obvious; after all, lidars generally detect weak signals and look to collect as many photons as possible, and as all astronomers know, there is no substitute for size. Based on the above interactions, what are the limits? We start by considering the system field of view. In the envisioned lidar system, the field of view is 0.3 mrad, as determined by a laser beam divergence of 1 mrad, further reduced by the 3.3x beam expander. Furthermore, we will project the beam onto an optical fiber of 1 mm diameter with $NA = 0.22$. To start, we will consider a prime-focus parabolic mirror that matches the fiber *NA*. Fig. 8.3 is a sketch of the geometry and guides the subsequent discussion. The fiber core being the output pupil, its diameter limits the telescope field of view (FOV) of 0.3 mrad, as previously determined by the laser-output and beam expander. By projecting the fiber face onto the telescope primary mirror, we determine the maximum telescope focal length of $1\ \text{mm}/0.3\ \text{mrad} = 3.3\ \text{m}$. From here, the fiber *NA* determines the maximum diameter of the telescope primary mirror, as $d_{telescope} = 2 \times f_{telescope} \times \tan(\arcsin(NA)) = 1.49\ \text{m}$.

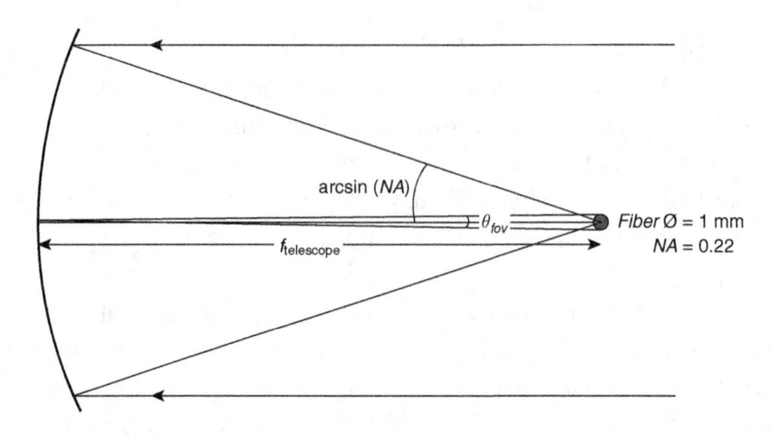

Fig. 8.3 Sketch of a prime-focus lidar telescope.

Another size factor is the physical space limitation. A 3.3-m focal length prime focus telescope requires a large space, as the length of the scope alone will likely exceed 4 m; and if it is steerable, the steering mechanics will add to the overall length. For this reason, it is worth considering alternative configurations, such as Newtonian (prime focus with a turning mirror to extract the light) or Cassegrain (which includes a convex secondary mirror, slowing the telescope for higher magnification and greatly reducing the physical size).

8.3.2 Configurations

As mentioned at the end of the previous section, it is generally convenient to use telescope configurations other than prime focus so as to make the size of the system more manageable. There are two common configurations, Newtonian and Cassegrain.

A Newtonian telescope places a secondary reflector oriented at 45° in front of the primary, to direct collected light to an output pupil outside the telescope bore. The effects of the obstruction of the secondary mirror include a usually small (< 10%) blockage of light entering the telescope and diffraction that impacts image quality. For lidar purposes, the Newtonian secondary does not change the results derived in the previous section for a prime focus system.

A Cassegrain is somewhat different, as these telescopes are generally designed for imaging with high magnification, so the system *f*-number will be much slower. We will consider the Celestron C-14 telescope utilized, for example, by the pioneering work of the CSU Na lidar in the 1990s [8.3] in Section 8.3.5.

8.3.3 Fiber-Coupled vs. Free-Space

Lidar system designers have to consider the costs and benefits of fiber coupling. Fiber-coupled telescopes offer big advantages to ease of operation. If the lidar scans the sky, for example to make Doppler measurements or map the horizontal structure of the metal layer, fiber coupling greatly simplifies the problem of transferring the return signal to the back-end optics. It also places rigid limits on FOV and *NA*, as mentioned previously. On the negative side, Fresnel reflections from the fused silica core at 589 nm cost the system about 7% of the light collected by the telescope, and dust, insects, and other contaminants can deposit on the fiber, further reducing its transmission.

Free-space systems avoid the fiber Fresnel reflections, but they introduce complexity if the designer intends to include pointing. An additional advantage for a free-space telescope pupil is that the exit pupil of the system can be adjustable. This aids initial alignment of the system by using a wide exit pupil while working to overlap the output beam and the receiver axis. Once the initial signal is established, the exit pupil is reduced in order to fine-tune alignment and diminish background noise. However, when having to move the telescope for pointing, receiving optics have to be directly mounted to the telescope in order to maintain the through path. As the telescope is exposed to the atmosphere, this exposes the sensitive and expensive dielectric-coated downstream optics, which may include a Faraday filter, and the detector – which is sometimes cooled to near-cryogenic temperatures. Thus, such an arrangement increases expense and complexity.

8.3.4 Mirror Quality

The view of a lidar telescope as a light bucket has evolved, thanks to optical modeling. While it is true that a lidar telescope need not resolve an image of the sampled area in the sky, it is still vital to ensure that the light couple through the telescope output pupil. This throughput is determined by the Strehl ratio, which is the fraction of the irradiance of an aberrated to an unaberrated focus. For optics of RMS wavefront error < 0.1, the Strehl ratio, S, is given by [8.4] as

$$S = e^{-(2\pi\Phi)^2} \approx 1 - (2\pi\Phi)^2, \tag{8.3}$$

where Φ is the RMS wavefront error in number of waves. The RMS wavefront error for a mirror with peak-to-valley wavefront error of $\lambda/4$, assuming the difference between the mirror surface and the ideal reference surface varies linearly across its diameter, is $\Phi = 0.073$. The resulting Strehl ratio is 0.81, which is generally considered to be "acceptable image quality" for astronomical applications. Applying this approach to peak-to-valley mirror wavefront errors of $\lambda/8$, $\lambda/4$, $\lambda/3$, and $\lambda/2$, we build Table 8.1.

Table 8.1 *Relation between surface figure and Airy disk*

Peak-valley wavefront error	RMS wavefront error, Φ	Strehl Ratio, S
$\lambda/8$	0.036	0.95
$\lambda/4$	0.073	0.81
$\lambda/3$	0.097	0.69
$\lambda/2$	0.15	$< 0.43^{\dagger}$

$^{\dagger}S$ for $\lambda/2$ peak-valley wavefront error will be less than the value given by (8.3) due to the RMS exceeding 0.1.

From Table 8.1, the image quality falls below 0.81, and in fact quite rapidly, when the mirror peak-valley wavefront error is more than $\lambda/4$, so this is an important design consideration for photon-starved lidar receivers. Collecting light that falls outside a diffraction-limited pupil means accepting higher background noise, which impacts error budget, especially for daytime operations. Thus, although incoherent lidar designs often do not assume that diffraction-limited optics are required, the rapid falloff of the Strehl ratio with wavefront error should not be ignored.

8.3.5 Coupling the Received Light into an Optical Fiber

In many modern lidars, the receiver telescope condenses scattered light into an optical fiber located at its image plane. From there, the fiber delivers the light to downstream optics, where it is gated, collimated, spectrally filtered, and detected by a photomultiplier tube (PMT) or avalanche photodiode (APD). To conserve BPP, at the entrance of the optical fiber the product of optical image diameter and convergence angle must be compatible with the product of the core diameter and *NA* of the optical fiber. *NA* is defined in terms of the refractive index n_i of the incident material (in this case, air) and the maximum half acceptance angle θ_a of the fiber, so that $NA = n_i \sin\theta_a$. Since the acceptance angle is small and $n_i \approx 1$, $NA \approx \theta_a$. Depending on the extent of object illumination, to make this possible one may need to use a short-focal-length lens to condense the received light into the optical fiber, as shown in the arrangement in Fig. 8.4(b), while being mindful of the need to match fiber-core diameter and *NA* while conserving BPP. To understand this arrangement, we first employ the laws for optical imaging by a telescope, as shown in Fig. 8.4(a).

We point out that the diameters of the telescope and the short-focal-length lens, D_T and D_L, define light propagating through these optical components. The diameters d_o and d_i are respectively the $2w_0$ diameter (containing 95% of the input light) of the lidar viewing volume and its image at the telescope image plane; they share the same beam radius w_T $(= 0.5D_T)$ at the primary mirror.

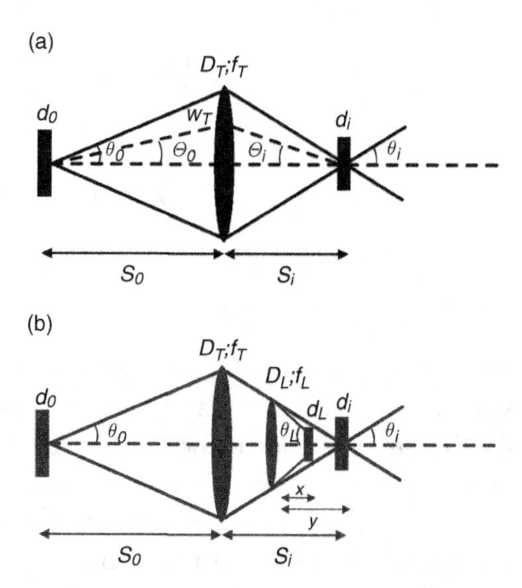

Fig. 8.4 Schematic of the receiver optics. (a) Imaging by a telescope with diameter and focal length of D_T and f_T. The object and image diameters d_o and d_i are $2w_0$ diameters, satisfying the BPP conservation law: $d_o\Theta_o = d_i\Theta_i$. The objective and image distances are, respectively, s_o and s_i. (b) The imaging system modified by insertion of a short-focal-length coupling lens with $2w_0$ diameter and focal length, D_L and f_L. Finally, d_L and θ_L are, respectively, the $2w_0$ diameter and half convergence angle tailored to be compatible with the fiber diameter, $D \geq 1.5\, d_L$ and numerical aperture, $\theta_a \geq \theta_L$. The object and image distances of the coupling lens are, respectively, y and x; here, the object is virtual with respect to the coupling lens.

Equations (8.4) provide a mathematical description using the simple lens law and associated image magnification of the geometry in Fig. 8.4(a). Equations (8.4.a) and (8.4.b) relate image diameter d_i to

(1) object distance (i.e., altitude) s_o
(2) full-width divergence of transmitting laser beam, Θ_{FW}, and
(3) telescope focal length, f_T.

The results of these plus BPP conservation applied to the coupling lens give (8.4.c) and (8.4.d), which determine the coupling lens location, x and y, the focusing half-angle θ_L, and the minimum coupling lens diameter, D_L, to optimize light transmission into the fiber in terms of the coupling lens focal length f_L and the fiber entrance diameter $D = d_L$. These equations are:

$$\frac{1}{s_o} + \frac{1}{s_i} = \frac{1}{f_T} \rightarrow \frac{f_T}{s_i} = 1 - \frac{f_T}{s_o};\ \frac{s_i}{s_o} = \frac{s_i}{f_T} - 1 = \frac{1}{(s_o/f_T) - 1}, \tag{8.4.a}$$

$$\frac{d_i}{d_o} = \frac{s_i}{s_o} \rightarrow d_i = \frac{s_i}{s_o} d_o = \frac{s_o \Theta_{FW}}{(s_o/f_T) - 1}; \; \theta_i \approx \frac{D_T}{2f_T}, \tag{8.4.b}$$

$$\frac{1}{-y} + \frac{1}{x} = \frac{1}{f_L} \rightarrow \frac{y}{x} = 1 + \frac{y}{f_L}; \; \frac{y}{x} = \frac{d_i}{d_L} \rightarrow \frac{y}{f_L} = \frac{d_i}{d_L} - 1 \text{ , and} \tag{8.4.c}$$

$$\frac{y}{x} = \frac{d_i}{d_L}; \; \theta_L = \frac{\theta_i d_i}{d_L} = \frac{D_T}{2 f_T} \frac{d_i}{d_L}; D_L = (2x)\theta_L. \tag{8.4.d}$$

To gain some physical insight, we consider a numerical example using a simple monoaxial lidar with the transmitting laser beam full-angle divergence $\Theta_{FW} = 1.0$ mrad and a Celestron-14 telescope with $D_T = 0.35$ m and $f_T = 4$ m as the receiver. The distance (altitude) of the illuminated atmosphere of interest is $s_o = 80 - 105$ km. The image diameter, d_i, shown in Fig. 8.4(a), is calculated to be 4.0 mm over the entire object range. If we wish to couple the received light into an optical fiber with a core diameter of $D = 1.0$ mm, we employ a coupling lens to reduce d_i to a $(2w_0)$ diameter of $d_L = 1.0$ mm. We choose the coupling lens focal length to be 60 mm, to both match *NA* and minimize inducing spherical aberration, and employ (8.4.c) and (8.4.d) to calculate the image distance, the (virtual) object distance, the focusing half-angle, and the minimum lens diameter (for passing all received light) of the coupling lens with the results shown in Table 8.2.

In Fig. 8.5 we show a model of a Celestron C14 telescope, with a commercial, off-the-shelf $f = 60$-mm (e.g., Thorlabs LA1134) "coupling" lens used to focus light onto an optical fiber. The Celestron C-14 components comprise the primary and secondary mirrors along with two field lenses located behind the primary mirror. The coupling lens is the "small" optic (25 mm diameter) following the field lenses. For the simulation, the spacing between the second field lens and the 60-mm focusing lens was optimized for best coupling into a 1-mm fiber core diameter with $NA = 0.22$. The spot image diagram within a 1 mm × 1 mm square shows 97% of the light within the fiber aperture. Using a plano-convex lens to refocus converging light is not optimal, as it adds spherical aberration. An appropriate focal

Table 8.2 *Parameters for focusing light into a 1 mm, NA = 0.22 fiber*

Coupling lens focal length	60 mm
Image distance, x	45 mm
Virtual object distance, y	180 mm
Focusing half angle, θ_L	0.175 rad
Minimum lens diameter clear aperture	15.8 mm

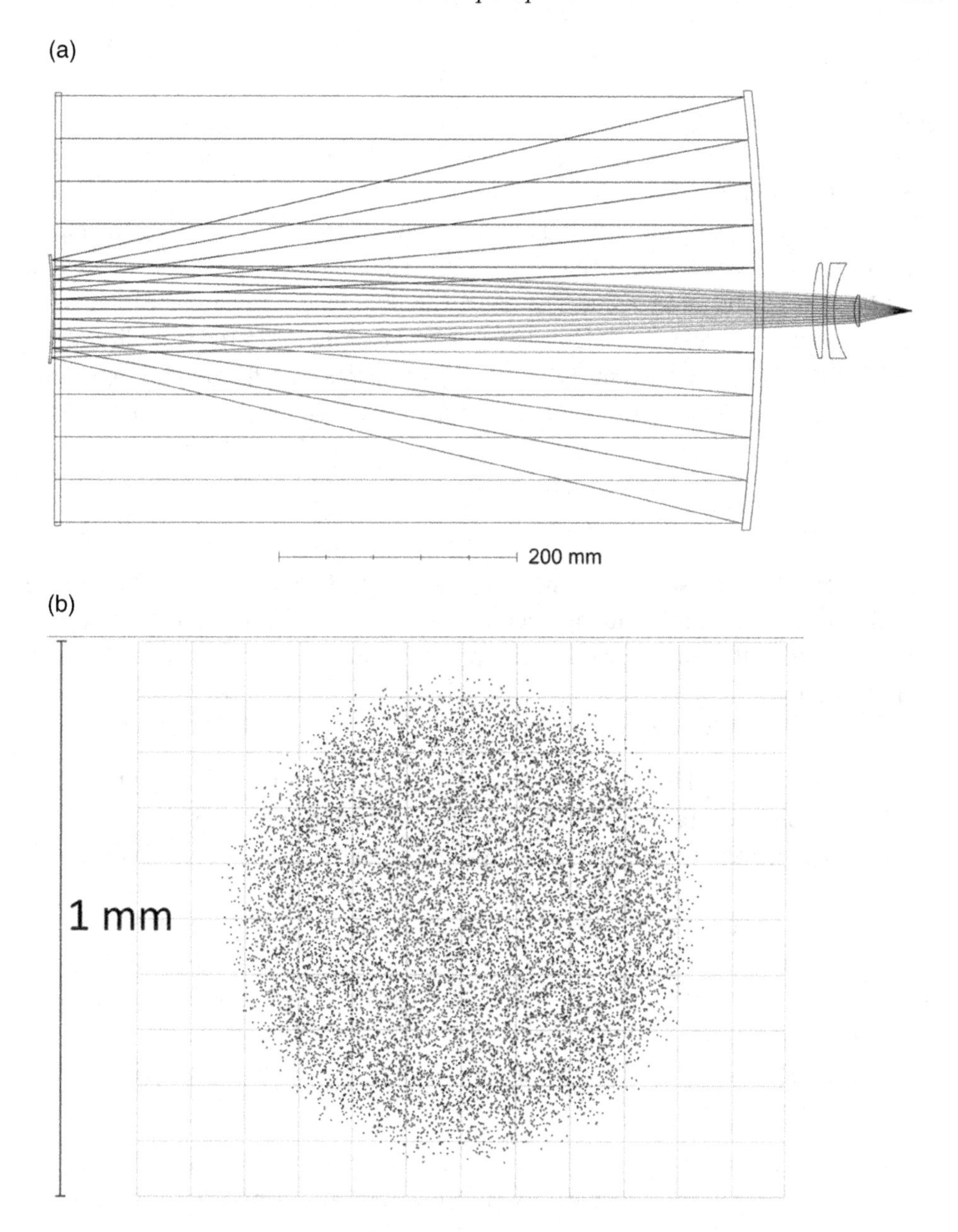

Fig. 8.5 (a) Model of a Celestron C14 telescope with a commercial off-the-shelf 60-mm-focal-length coupling lens positioned to match the fiber NA and optimize coupling efficiency. (b) Spot diagram for this arrangement. The scale bar is 1 mm.

length (in this case, 60 mm) meniscus lens would improve the performance. As mentioned above, these are valid for the full range of object distances between 80 and 105 km. The fact that x and y are independent of the object distance makes the positioning of the short-focal-length lens straightforward. The calculated focusing half-angle of $\theta_L = 0.175$ rad requires us to choose a commercially available fiber with $D = 1$ mm and $NA = 0.22$, which is the smallest available NA greater than θ_L (the next smaller $NA = 0.116$). Since $D_L = 15.8$ mm, a common off-the-shelf focusing lens with $f_L = 60$ mm and 25 mm diameter serves our purpose well.

We close by pointing out that if the transmitting laser is transmitted through a 4x beam expander, its full-angle divergence can be optimized to $\Theta_{FW} = 0.25$ mrad, and the $(2w_0)$ image diameter reduces to $d_i = 1.00005$ mm, a value compatible with $d_L = 1.0$ mm. The telescope imaging half-angle of $\theta_i = 0.04375$ rad is readily adaptable to a fiber with $D = 1.0$ mm and $NA = 0.116$. In this case, the coupling lens is no longer necessary, and we can simply center the fiber directly in the focal plane of the telescope.

8.4 Detectors

Atmospheric lidar presents challenges to detector technology. While close-range signals are strong enough to be seen visually (and damage photomultipliers), scattering from the stratosphere–mesosphere is so weak as to require photon counting, ultra-low-noise detectors. In selecting a detector, the lidar operator must balance the following requirements:

1. Dynamic range and linearity – to manage a wide range of signal strength and be able to make spectral measurements without detector-induced biases.
2. Photon detection efficiency – to capture as many of the return photons as possible.
3. Noise – to see weak background signals, detector and spurious noise should be minimized.
4. Bandwidth – the detector bandwidth should be sufficient to resolve the minimum required range bin.
5. Photocathode size – to make sure that the full breadth of the field of view can be imaged onto the detector.

In this section, we discuss these five detector selection considerations. Finally, we will examine the value of computer modeling to optimize system performance, and as such, signal. For purposes of comparison and example, we will consider two detectors in this section: the Hamamatsu® model H7422P-40 photomultiplier (PMT) and the Excelitas® SPCM-AQRH Geiger-mode avalanche photodiode. These are selected for their respective photon detection efficiency in the visible and near-infrared parts of the spectrum. The H7422P-40 spectral response range is from 300 to 720 nm, peaking at 580 nm, which is close to optimum for Na lidar. The SPCM-AQRH spectral response range is about 400–1050 nm, peaking near 660 nm, with sensitivity ~ 95% of that peak at 770 nm for K lidar.

8.4.1 Dynamic Range and Linearity

Perhaps the greatest challenge lidar presents to detector systems is the requirement for wide dynamic range. Consider that, to detect metal atom layers in the thermosphere, abundances below 1 atom/cc must be resolved. In a 0.15 km range bin (1 μs range gate) and 1,000 laser pulses, the signal may be only one or two photons over the background detector noise. Thus, signal photon rates are on the order of 1 kHz. This sets a noise floor (which we discuss in Section 8.4.3). At the same time, in sporadic layers lidar signals can easily exceed 100 MHz. Thus, a dynamic range of over 50 dB is a minimum acceptable for atmospheric lidar. In fact, >60 dB (100 Hz – >100 MHz) is highly desirable, particularly with ongoing initiatives for lidars with power-aperture products of 1,000 W-m^2.

In the case of our sample detectors, we have the following, as summarized in Table 8.3.

According to the data sheet, the SPCM-AQRH noise floor is about 100 counts per second (cps) (it can be as low as 25 cps). Its saturation level is about 10^7 cps. This gives a dynamic range of 50 dB of linear performance. The H7422P-40 also has a noise floor of about 100 cps. From its gain and maximum output signal current specifications, it also has a saturation level of about 10^7 cps, and thus 50 dB of linear performance.

Linearity is extremely important in resonance lidar detectors, as temperature and wind measurements involve sounding the spectrum of the metal atom resonance line. Since the line center produces much more signal than measurements on the spectral wings, any saturation will show up as a positive temperature offset for the Na lidar. Nevertheless, effective nonlinearity or detector saturation correction has been demonstrated [8.5, 8.6, 8.7], but it is advisable to avoid nonlinearities whenever possible.

8.4.2 Quantum (or Photon Detection) Efficiency

Quantum (QE) or photon detection (PDE) efficiency is often the first thing one thinks of when selecting a low-signal-level detection system. This is not surprising, as it behooves us to detect every possible photon that falls on the photocathode. Let us consider this question.

First, we clarify the difference between quantum and photon detection efficiencies. Quantum efficiency (QE) is the probability of an incident photon generates a photoelectron (or, electron–hole pair in a semiconductor). However, generating the photoelectron does not guarantee generating a detection pulse. The generated photoelectron needs to trigger the generation of a pulse in the system. This probability is contained in the triggering probability factor, P_{tr}. In a conventional photomultiplier (PMT), the photoelectron must overcome the low barrier energy (electron affinity) of the photocathode to be emitted to the vacuum and arrive at the first dynode to be amplified. The use of III-V GaAsP(Cs) photocathodes for the high QE H7422P-40 PMT, in which cesium activation induces negative electron affinity, leads to unity P_{tr} for these PMTs. The only noise in these PMTs is thermal (dark) electrons, whose effect can be mitigated by cooling and by proper setting of the threshold voltage in the pulse-height discrimination in the pulse detection system. In a semiconductor detector, P_{tr} is the probability that the generated carrier (be it an electron or hole) will traverse the high-electric-field region and trigger an avalanche. From here on, we will focus where possible on photon detection efficiency (PDE), which is equal to $P_{tr} \times$ QE.

Initially, we seek the highest possible PDE. However, if high PDE requires sacrifice of one of the other parameters, such as dark noise, dynamic range, or photocathode size, a lower PDE may be preferred. PDE is wavelength dependent. For example, silicon typically has maximum PDE between 500 and 600 nm, and this falls off quickly in the blue, more slowly in the red. Its PDE can be enhanced in the red, and in fact, the SPCM-AQRH, with a Si photocathode, reaches 70% at 660 nm (~ 66% at 770 nm). Silicon photomultipliers typically have lower PDE of around 15%, even with QE in excess of 90%. The Hamamatsu H7422P-40 photomultiplier tube achieves 40% QE at 589 nm, which is the same as its PDE since P_{tr} is practically 1.

Semiconductor devices, such as multi-pixel photon counters, may reach 90% PDE at some wavelengths. However, these cannot perform channeled range-profiling detection. They also often suffer from high noise and low dynamic range. These devices are often used for "first photon" detection, where the target is a solid object at a single range. On detecting the first photon, the detector is then blind for the period of time it takes to transfer the information from the pixel to the back-end electronics. This detection scheme is not suitable for atmospheric lidar.

8.4.3 Noise

We have mentioned noise previously. Photon counting detectors such as those used in lidar generally have only one appreciable intrinsic noise source, which is their dark counts. As detailed in Table 8.3, the detectors we use for comparison both have dark count rates of about 100 cps. Considering a lidar operation where we integrate signal

Table 8.3 *Detector Photon Counting Specifications*

Detector	H7422P-40	SPCM-AQRH
Noise floor (cps)	100	100
Saturation level (cps)	10^7	10^7
Dynamic range (dB)	50	50

for 1 min from a system that operates at 50 pulses/second, that implies an integration time of 3,000 laser pulses. If the range bin size is 150 m (1 μs channels), that means that over 3,000 pulses, 3 ms of data are accumulated in each channel. A dark count of 100 cps means that a dark photon is detected every 10 ms, so a dark photon will appear for approximately every three channels in a 1-min integration, which will be mostly removed by the pulse-height discrimination process in photon counting.

In addition to the random noise of dark counts, the lidar also sees the background sky as a dominating noise source. Nevertheless, careful spectral and spatial filtering can reduce this background noise to a level comparable with the dark noise. This fact is evident in the photon count data recorded by modern resonance and Rayleigh lidars.

8.4.4 Bandwidth

Bandwidth in photon counting detectors is effectively the upper limit of the dynamic range. For photon counting detectors, we can use the relation between the output pulse width and instrument bandwidth, given as $BW = 0.375/\Delta\tau$, in which BW is the detector bandwidth and $\Delta\tau$ is the width of the current pulse from the detector, produced by detection of a photon. For the detectors under consideration, the risetime of the pulses output by the H7422P-40 is 1 ns, leading to a bandwidth of 375 MHz, and similarly the SPCM-AQRH output pulses are 35 ns wide, leading to a bandwidth of about 11 MHz. These are particularly important for the strong signal regime (not usually a problem for atmospheric lidar measurements) as well as for making high-range-resolution measurements. For middle-atmosphere lidar, range bins smaller than 75 m are rarely used, but even at 10 m, the bandwidth required is only about 30 MHz, well within the bandwidth requirements of many photon counting detectors.

8.4.5 Photocathode Size

Photocathode size is mainly an issue of the difference between using a PMT versus an APD. Typical PMT photocathodes are large, several to tens of mm^2. However,

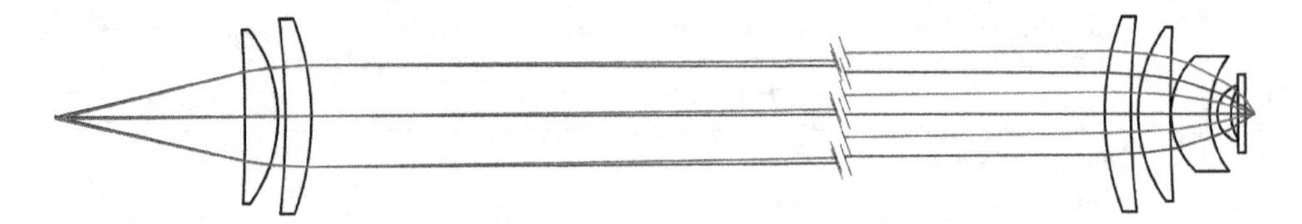

Fig. 8.6 Lens system for projecting light from the telescope pupil fiber (left) onto the APD sensor (right). The system consists of conjugate pairs made up of meniscus lenses followed by two meniscus lenses to further concentrate the image onto the APD. The last element is the entrance window of the APD. Both paraxial and chief rays are shown.

an APD, as a semiconductor device, has a trade-off between bandwidth and photocathode size. The H7422P-40 photocathode is relatively large, with a 5 mm diameter, so the downstream optics that project the image of the telescope exit pupil through the narrowband filter onto the photocathode are not critical. The SPCM-AQRH model, on the other hand, is only about 0.2 mm diameter. For this reason, the downstream optics are more critical, as is the telescope exit pupil (i.e. fiber) diameter.

For the potassium lidar at the Arecibo Observatory, a 0.55 mm, $NA = 0.22$ fiber is used at the telescope focus, and custom optics were developed to image the fiber face onto the photocathode of the APD. These are shown in Fig. 8.6. In order to efficiently image the fiber onto a detector one-third its diameter, the Arecibo group developed a lens system to speed $NA = 0.22$ to $NA = 0.5$. It consists of conjugate collimating-refocusing lenses, with a separation of 30 cm for location of a Faraday filter or Fabry–Perot étalon plus interference bandpass filters. Following the second conjugate lens pair are two lenses that speed the system as required to image the fiber face onto the APD. All lenses are meniscus in order to minimize spherical aberration, and high-quality antireflective coatings are required to minimize Fresnel reflection losses from the 14 surfaces. Optionally, two aspheric lenses can replace the six spherical lenses.

8.4.6 Importance of Computer Modeling for Optimum Efficiency of the Receiver Path

The above sections provide an ample demonstration of the value of using modern optical design software to optimize the performance of the receiver system. In evaluating telescope-to-fiber coupling designs, such as implemented in Fig. 8.4 and Fig. 8.5, we considered a range of options for the coupling lens. We found that, given the fact that the entrance and exit rays for the coupling lens are not paraxial, we had to consider the effect of spherical aberration, and in so doing, we determined that the 60-mm focal length plano-convex lens led to the best performance for the system requirements. At the exit end of the fiber, the lens system shown in Fig. 8.6 required extensive computer design, simulation, and optimization to both effectively collimate the fiber output and then condense the image of its 0.5-mm face onto a photocathode less than 0.2 mm in diameter. This resulted in a successful design employing spherical surfaces, which are less susceptible to alignment-induced coma, easier to fabricate, and less costly. In a one-to-one comparison between the lens system built from the design in Fig. 8.6 and a system employing aspheric lenses, the Fig. 8.6 system performed notably better.

8.5 The Effects of Atmospheric Turbulence on Receiver Aperture and Na Laser Guide Star

To reduce background noise, particularly for daytime measurements, lidar engineers' first efforts involve reducing the observing field of view (including both transmitter divergence and the receiving telescope). Reducing field of view (FOV) reduces background noise by the square of the reduction in angular FOV without penalizing signal. However, atmospheric turbulence restricts the minimum field of view, just as it prevents the formation of diffraction-limited images for astronomy and other applications. This atmospheric turbulence-limited angular resolution of a telescope is commonly referred to as atmospheric "seeing" [8.8]. Given that direct detection atmospheric lidar does not require image formation (or seeing), the receiver need only be a photon bucket, but as mentioned, it does impact the effectiveness of methods to reduce sky background noise. In contrast, coherent lidar requires maintaining spatial coherence across the receiver aperture for optical heterodyne detection. As a result, atmospheric seeing limits the effective area of the receiving telescope of a coherent detection lidar. Fortunately, coherent detection lidar typically investigates the lower troposphere where the aerosol and molecular signals are strong, so a small telescope is usually adequate. Where the effects of atmospheric seeing are most profound is on ground-based astronomical telescopes. Astronomers must be able to correct the phase-front distortions resulting from atmospheric turbulence in order to produce near-diffraction-limited images of celestial bodies. For the remainder of this section, we provide a very brief overview of (a) atmospheric turbulence, as pointed out in Section 7.5.3, which impacts not only astronomers but also atmospheric dynamicists who investigate turbulent mixing and heating, and

(b) progress on Adaptive Optics using Na LGS (see Section 3.5) to enable ground-based telescopes to approach and even exceed the resolution of the Hubble Space Telescope.

8.5.1 Atmospheric Turbulence and the Fried Parameter

Instabilities in the atmosphere caused by small temperature lapse rate and large windshear create random inhomogeneous temperature variations (or eddies) of different sizes that interact with atmospheric waves and lead to turbulent mixing and heating. The accompanying random variations in atmospheric density lead to random inhomogeneities in optical refractive index, causing distortions in the shape and variations in intensity of a wavefront propagating through the medium [8.9]. These effects may be quantified in a statistical sense by the structure functions of temperature $D_T(\vec{r})$ and refractive index $D_n(\vec{r})$ fluctuations, where $\vec{r}$ is the displacement between two points in space. The theory of atmospheric turbulence was developed by Kolmogorov and Obukhov and summarized concisely and clearly by Boyd in [8.8]. They showed, as is now widely known, that turbulent flow can be described statistically by eddies of different sizes, ranging from the largest dimension L_0 to the smallest ℓ_0. These energies cascade downward (adiabatically) from the large to the small and are eventually damped (energy dissipated) by eddies with sizes smaller than ℓ_0. For eddies within the inertial subrange (i.e., with sizes between L_0 and ℓ_0), the turbulence is isotropic, and its intensity depends only on the single parameter ε, the per unit mass turbulent energy dissipation rate. In this line of thinking, the structure function (or variance) of turbulent velocities is taken (or assumed) to be directly proportional to a power law of εr. The power in question must be 2/3 since it is the only power, p, by which $(\varepsilon r)^p$ with $p = 2/3$ has dimensions of speed squared. This is the "two-thirds law" of Kolmogorov and Obukhov, as given in (8.5.a). Once the fundamental statistics of turbulent velocity, $D_{\text{v}}(\vec{r})$, are figured out, the structure functions of the derived quantities, such as temperature eddies and refractive index eddies, which preserve the $D_{\text{v}}(\vec{r})$ dependence on the distance between any two points, may be given as in (8.5.b):

$$D_{\text{v}}(\vec{r}) = \left\langle \left[\vec{\text{v}}\left(\vec{R} + \vec{r}\right) - \vec{\text{v}}\left(\vec{R}\right) \right]^2 \right\rangle \rightarrow D_{\text{v}}(r) = C(\varepsilon r)^{2/3} \equiv C_{\text{v}}^2 r^{2/3}. \qquad (8.5.\text{a})$$

$$D_T(\vec{r}) \rightarrow D_T(r) = C_T^2 r^{2/3}, \quad \text{and} \quad D_n(\vec{r}) = D_n(r) = C_n^2 r^{2/3}. \qquad (8.5.\text{b})$$

Here, C_x^2 is the fluctuation structure coefficient of the relevant parameter, x, where x is v, T, or n. It is often convenient to work in the spectral domain. For isotropic turbulence, the structure function, $D_n(r)$, is transformed into its power spectral density (PSD) of the refractive index fluctuations, $\Phi_n(\kappa)$, where κ is the modulus of the 3D spatial frequency vector $\vec{\kappa}$; they are related by [8.10]:

$$D_n(r) = 8\pi \int \Phi_n(\kappa) \left[1 - \frac{\sin(\kappa r)}{(\kappa r)} \right] \kappa^2 d\kappa. \tag{8.5.c}$$

The PSD for the refractive index structure function $D_n(r) = C_n^2 r^{2/3}$ is [8.8, 8.10]:

$$\Phi_n(\kappa) = 0.0330\, C_n^2 \kappa^{-11/3} \quad \text{or} \quad \Phi_n(\kappa, z) = 0.0330\, C_n^2(z) \kappa^{-11/3}, \tag{8.6}$$

where the refractive-index structure coefficient, C_n^2, is obviously altitude dependent (i.e., $C_n^2 = C_n^2(z)$ and $\Phi_n(\kappa) = \Phi_n(\kappa, z)$). The numerical constant in (8.6) is $\Gamma(8/3)\sin(\pi/3)/(4\pi^2)$ [8.11], where $\Gamma(8/3)$ is a Gamma function. To consider the effect of atmospheric turbulence on wave propagation, we follow Hickson [8.10] and write the wave function in question in terms of its log-amplitude $\chi(\vec{r})$ and phase $\varphi(\vec{r})$ as

$$\Psi(\vec{r}) = \Psi_0 \exp\left[\chi(\vec{r}) + i\varphi(\vec{r}) \right]. \tag{8.7.a}$$

Above the atmosphere, $\Psi(\vec{r}) = \Psi_0$. As the light wave propagates down through the atmosphere, it experiences random variations in both $\chi(r)$ and $\varphi(r)$, which induce phase-front distortions and intensity variations, affecting the angular resolution and the quality of optical imaging [8.9]. The structure functions of log-amplitude $D_\chi(r)$ and phase fluctuation $D_\varphi(r)$ are obviously related to the refractive index structure function $D_n(r) = C_n^2 r^{2/3}$, and in turn to its PSD $\Phi_n(\kappa)$. Thus, the sum of log-amplitude and phase structure functions, termed the "wave-structure function" $D_n(r)$ [8.9, 8.10], can be expressed in terms of integration of $\Phi_n(\kappa)$ over vertical height z as given in Eq. (14) of [8.10] and repeated here:

$$D_n(r) \equiv D_\chi(r) + D_\varphi(r) = 8\pi^2 k^2 \int_0^\infty dz \int_0^\infty \Phi_n(\kappa, z) \left[1 - J_0(\kappa, z) \right] \kappa d\kappa, \tag{8.7.b}$$

where $k = 2\pi/\lambda$ and κ are, respectively, the wave numbers of light wave and structure function, and J_0 is the 0th-order Bessel function. For the isotropic Kolmogorov turbulence spectrum, with $\Phi_n(\kappa, z)$ given by (8.6), the integral of (8.7.b) may be carried out; its result is given in Eqs. (24) and (25) of [8.10] and repeated here:

$$D_n(r) \equiv D_\chi(r) + D_\varphi(r) = 6.884\,(r/r_0)^{5/3}, \quad \text{with } r_0^{-5/3} = 0.423\,k^2 \int_0^\infty C_n^2(z)dz. \tag{8.7.c}$$

For astronomical work, $D_\chi(r) \ll D_\varphi(r)$, r_0, the "Fried parameter," given above represents the characteristic length of the structure function. It is the transverse scale at which the RMS phase difference between two points in the wavefront has reached $\sqrt{6.884} = 2.634$ radians [8.10]. In practical terms, it is the scale beyond which fluctuations become uncorrelated and begin to affect astronomical seeing. The Fried parameter is wavelength dependent [8.8], that is, $r_0^{5/3}\,\alpha\,\lambda^2$ from (8.7.c), or $r_0\,\alpha\,\lambda^{6/5}$; thus, seeing improves as wavelength increases. For infrared-visible wavelengths, the Fried parameter ranges between 10 and 20 cm. The phase-front distortion caused by atmospheric turbulence is the main cause of atmospheric seeing and image blurring.

8.5.2 Atmospheric Turbulence Correction and LGS Telescope

The exciting and timely discussion of the adaptive optics for LGS telescopes is obviously too big a topic and outside the scope of this book. We refer interested readers to a recent review article by Hickson [8.10] for authoritative information. Here, we shall give only a very brief description of how a LGS telescope works and its interesting history and bright prospects.

If the distorted phase front of light from a reference star can be sensed by a wavefront sensor (WFS), a deformable mirror can in principle be adjusted to correct this distortion caused by atmospheric turbulence and restore its phase front to that near a plane wave. If this can be done periodically at a rate faster than that of turbulence fluctuations, the image of the reference star as well as those nearby will be sharp, and diffraction limited. This is, in essence, adaptive optics (AO). For good AO performance, a reference star must be bright enough and exist within a few arcsec of a science target. Unfortunately, not many natural stars are available to cover enough sky for even a small fraction of interesting objects.

To overcome the rarity of suitable natural guide stars (NGS), AO systems use a laser beam to create an artificial laser guide star (LGS) either in the atmosphere via Rayleigh scattering or in Earth's Na layer via laser-induced fluorescence (LIF). To create a star-like source via Rayleigh scattering (or Rayleigh beacon), a pulsed laser along with electronic gating must be employed. Since Earth's Na layer exists between 80 and 110 km and is sharply peaked at ~ 90 km, the Na LGS (or sodium beacon) created by either a CW or a pulsed laser tuned to the Na D_2 transition, is a point-like source. Unlike NGS, as they form within the atmosphere, either LGS option is too close to completely sample the atmospheric turbulence experienced by light from astronomical objects. This is due to the so-called "cone effect"; see Fig. 6 of [8.10]. In this respect, the sodium beacon, being located at ~ 90 km, does a much better job than the Rayleigh beacon at ~ 10 km.

Despite the advantage of high brightness, LGS suffers tilt degeneracy [8.10]. In tilt degeneracy, a turbulent inhomogeneity deflects the downward LIF emission by an angle exactly opposite that of the upward laser beam, and, as a result, the LGS image does not move on the image plane. Thus, a LGS is incapable of sensing the tip-tilt component of the atmospheric turbulence [8.10], and a natural guide star is required for that first-order correction. The combination of the LGS and one (or preferably more) NGS enables the AO system to correct the mean phase tilt across the telescope aperture. Furthermore, for Na LGS, its exact range suffers variations caused by fluctuations of the Na layer centroid. The centroid altitude can be measured with high accuracy and precision in real time with a pulsed laser. Notwithstanding, due to the use of optical pumping discussed in Section 3.5 to enhance the Na LGS efficiency, LGS systems employ high-power, narrowband (~ 1 MHz), circularly polarized CW laser systems. These are unable to provide a direct real-time measurement of the Na layer centroid. As pointed out in Section 7.5.3, in parallel with the construction of very large AO systems, much ongoing research is characterizing the Na layer, measuring its PSD and centroid altitude fluctuations to provide a real-time feedback to correct for the focus variations. By monitoring and correcting both atmospheric tilt and focus in real time, Na LGS can operate without needing a NGS and potentially provide 100% sky coverage.

In Section 3.5, we briefly discussed the quantum physics of the CW laser system required for the formation of a Na LGS, and in Section 7.5.3 we mentioned relevant measurement work on Na density fluctuations and power spectrum. The technological developments in adaptive optics for devices like deformable mirrors and wavefront sensors (WFS) are beyond the scope of this book. For a simple and understandable description of LGS adaptive optics, in addition to the review articles, such as Hickson [8.10], readers are encouraged to examine a presentation in the 197th meeting of American Astronomical Society by Claire Max, entitled "Introduction to adaptive optics and its history"; its transcript [8.12] can be downloaded from the Internet. Here, she describes how adaptive optics (deformable mirrors and WFS) and LGS work and explains the complementary nature between the Hubble Space Telescope and ground-based adaptive optics telescopes in 2000s. For example, she demonstrates how an array of 2-D lenses can be configured to sense wavefront distortion (shown in the figure on p. 6 of [8.12] and in Fig. 4 of

[8.10]) and how a deformable mirror can be shaped to make phase-front corrections (upper figure on p. 5 of [8.12]).

The history of LGS research and development is both interesting and inspiring [8.12, 8.13]. The concept of adaptive optics was first proposed in 1953 by Babcock [8.14]. The research then disappeared from open literature, though it continued under JASON, an independent group of elite scientists who advised the US government on sensitive science and technology issues. According to an account by Wright [8.13], "A brief history of laser adaptive optics," posted on a blog entitled "Astronomy and meta-Astronomy" on January 17, 2015, many scientists in the group were in favor of declassifying their research. Once his research project was declassified, Freeman Dyson published a theoretical paper in 1975 entitled "Photon noise and atmospheric noise in active optical systems" [8.15], which explains the promise and limitations of adaptive optics. The idea of using lasers to create an artificial reference star was proposed independently by Harper in classified literature in 1983 and by Foy and Labeyrie in 1985 in open literature [8.16]. This was later tested by Thompson and Gardner in 1987 [8.17]. Both the military and aerospace communities played significant roles in the initial technology development in adaptive optics between the 1960s and 1970s, which led to a more sustained and sophisticated Air Force effort in adaptive optics technology. After the technology declassification in 1991 [8.13], the results of the successful two generations of LGS deployments using Rayleigh beacons at the Air Force Starfire Optical Range (SOR) in Albuquerque, NM were published in 1994 [8.18]. Summarizing the continued efforts of his group on sodium beacons at SOR, Robert Fugate gave one of the AFRL Inspire talks on LGS in 2016, entitled "Fire in the Sky." It is excellent and can be downloaded from YouTube® (https://youtu.be/dIRG2J7nrxw). Adaptive optics research and technology development have come a long way, from the military interest in improving satellite tracking to uncracking the secrets of the universe in infrared and optical astronomy. Adaptive optics telescopes deployed in 2008 have already achieved or exceeded the resolution of Hubble Space Telescope at IR wavelengths; see Fig. 1 of [8.10]. Today, LGS adaptive optics is an essential technology for large telescopes worldwide. The next generation of extremely large telescopes, including the Giant Magellan Telescope (22 m aperture, on Cerro Las Campanas, Chile), the European Extremely Large Telescope (39 m, on Cerro Armazones, Chile), and the Thirty Meter Telescope (30 m, on Mauna Kea, Hawaii), are all designed to have resolution a few tenths that of the Hubble Space Telescope, allowing astronomers to obtain diffraction-limited quality images in the visible and IR from any part of the sky.

References

8.1 Siegman, A. E. (1997). How to (Maybe) Measure Laser Beam Quality. Tutorial presentation at the Optical Society of America Annual Meeting, Long Beach, CA, October 1997. Accessed at https://en.wikipedia.org/wiki/Beam_parameter_product.

8.2 Kogelnik, H. and T. Li (1966). Laser beams and resonators. *Applied Optics*, **5**(10), 1550–1567, doi: https://doi.org/10.1364/AO.5.001550.

8.3 She, C. Y., H. Latifi, J. R. Yu et al. (1990). Two-frequency lidar technique for mesospheric Na temperature measurements. *Geophys. Res. Lett.*, **17**(7), 929–932.

8.4 Mahajan, V. N. (1982). Strehl ratio for primary aberrations: some analytical results for circular and annular pupils. *J. Opt. Soc. Am.*, **72**(9), 1258–1266.

8.5 Donovan, D. P., J. A. Whiteway, and A. I. Carswell. (1993). Correction of nonlinear photon-counting effects in lidar systems. *Appl. Opt.*, **32**(33), 6742–6753.

8.6 Smith, J. A. and X. Chu. (2015). High-efficiency receiver architecture for resonance-fluorescence and Doppler lidars. *Appl. Opt.*, **54**(11), 3137–3184. doi: https://doi.org/10.1364/AO.54.003173.

8.7 Liu, A. and Y. Guo. (2016). Photomultiplier tube calibration based on Na lidar observation and its effect on heat flux bias. *Appl. Opt.*, **55**(33), 9467–9475. doi: https://doi.org/10.1364/AO.55.009467.

8.8 Boyd, R. (1978). The wavelength dependence of seeing. *J. Opt. Soc. Am.*, **68**(7), 877–883.

8.9 Fried, D. (1966). Optical resolution through a randomly inhomogeneous medium for very long and very short exposures. *J. Opt. Soc. Am.*, **56**(10), 1372–1379.

8.10 Hickson, P. (2014). Atmospheric and adaptive optics. *Astron. Astrophys. Rev.*, **22**(1), 76. doi: https://doi.org/10.1007/s00159-014-0076-9.

8.11 Roddier, F. (1981). The effect of atmospheric turbulence in optical astronomy. *Prog. Opt.*, **XIX**, 281–376. doi: https://doi.org/10.1016/S0079-6638(08)70204-X.

8.12 Max, C. (2001). Introduction to adaptive optics and its history. The 197th Annual Meeting of the American Astronomical Society, jointly with the American Association of Physics Teachers, 7–11 January 2001, San Diego, California. Accessed at www.researchgate.net/publication/228761974.

8.13 Wright, J. (2015). A brief history of laser adaptive optics. Posted in AstroWright under science, Uncategorized and tagged science on January 17, 2015. Accessed at https://sites.psu.edu/astrowright/2015/01/17.

8.14 Babcock, H. (1953). The possibility of compensating astronomical seeing. *Publ. Astron. Soc. Pac.*, **65**(386), 229–23.

8.15 Dyson, F. (1975). Photon noise and atmospheric noise in active optical systems. *J. Opt. Soc. Am.*, **65**(5), 551–558.

8.16 Foy, R., and Labeyrie, A. (1985). Feasibility of adaptive telescope with laser probe. *Astron. Astrophys.*, **152**(2), L29–L31.

8.17 Thompson L., and Gardner C. (1987). Experiments on laser guide stars at Mauna Kea Observatory for adaptive imaging in astronomy. *Nature* **328**(6127), 229–231.

8.18 Fugate, R. (1994). Two generations of laser-guide-star adaptive-optics experiments at the Starfire Optical Range. *J. Opt. Soc. Am. A*, **11**(1), 310–324.

Appendix A

Electric Dipole Interaction and Structures of Atoms and Linear Molecules

We adopt semiclassical (quantum) theory for treating light–matter interaction, in which atoms (and molecules) are treated quantum mechanically, and the impinging and scattered light classically. The quantum state of a single atom (or molecule) is described by a wave function, $|\Psi(t)>$, which satisfies the time-dependent Schrödinger equation (A.1a) with a Hamiltonian operator $\mathscr{H}(t) = \mathscr{H}_0 + \mathscr{H}_I(t)$, where the time-dependent (or perturbation) Hamiltonian, $\mathscr{H}_I(t)$, represents the interaction energy with an external (classical) electromagnetic field. The associated time-independent (or unperturbed) Hamiltonian, $\mathscr{H}_0$, which represents the energy of the atomic system, consisting of the dominating attraction from the nucleus, $\mathscr{H}_{e-n}$, electron–electron repulsion, $\mathscr{H}_{e-e}$, electronic spin-orbit coupling (leading to the fine structure), $\mathscr{H}_{SO}$, the alignment energy to nuclear magnetic moment (leading to the hyperfine structure), $\mathscr{H}_{HF}$, and that due to a D.C. external magnetic field $\mathscr{H}_{MG}$, when present, gives rise to the time-independent Schrödinger equation with eigen-energy E_n and eigenstate $|\psi_n>$ as solution in (A.1.b).

$$i\hbar\frac{\partial|\Psi(t)>}{\partial t} = \mathscr{H}(t)|\Psi(t)>, \quad \text{with } \mathscr{H}(t) = \mathscr{H}_0 + \mathscr{H}_I(t), \text{ and}$$

$$\mathscr{H}_0 = \mathscr{H}_{e-n} + \mathscr{H}_{e-e} + \mathscr{H}_{SO} + \mathscr{H}_{HF} + \mathscr{H}_{MG}. \tag{A.1.a}$$

$$\text{For } \mathscr{H}_I(t) = 0, \quad i\hbar\frac{\partial|\Psi(t)>}{\partial t} = \mathscr{H}_0|\Psi(t)> \rightarrow |\Psi_n(t)> = |\psi_n>e^{-iE_nt/\hbar} \text{ with}$$

$$\mathscr{H}_0|\psi_n> = E_n|\psi_n> \tag{A.1.b}$$

In the absence of optical perturbation (i.e., $\mathscr{H}_I(t) = 0$), the evolution of the system's wave function is governed by the time-independent Schrödinger equation, $\mathscr{H}_0|\psi_n> = E_n|\psi_n>$, whose solution gives rise to stationary states of the atom, with each being a product of a time-independent eigenfunction $|\psi_n>$ and an exponential oscillatory function of time at angular frequency $E_n/\hbar$, where E_n is the associated eigen-energy as depicted in (A.1.b). The eigenstate index n specifies a set of quantized stationary motions, each denoted by a quantum number. Since $\mathscr{H}_0 \gg \mathscr{H}_I(t)$, the time-dependent Schrödinger equation (A.1.a) may be solved by time-dependent perturbation theory to yield the transition probabilities between the stationary eigenstates of the atom, from an initial eigenstate $|i\rangle$ to a final eigenstate $|k\rangle$, from which we can determine useful quantities like absorption cross section; see Section A.1.

For atoms (or molecules) in the atmosphere, the system consists of an ensemble of weakly colliding particles, independently interacting with an external optical field, and its quantum state is more appropriately described by a density matrix, ρ, whose time development, including a mild damping term, is governed by the Liouville equation:

$$\frac{\partial \rho}{\partial t} = \frac{1}{i\hbar}\left[\mathscr{H}_0 + \mathscr{H}_I(t), \rho\right] + \frac{1}{i\hbar}\frac{d\rho}{dt}. \tag{A.2}$$

As stated, the perturbation Hamiltonian generates the interaction energy with the impinging light. Since the bound electrons are nonrelativistic, the electric dipole interaction dominates, which is then assumed to be the only term in the perturbation Hamiltonian. The laser electric field, $\vec{E}(t) = 0.5(\vec{\mathscr{E}} e^{-i\omega t} + \vec{\mathscr{E}}^* e^{i\omega t})$, external to the system, is treated classically. It interacts with the dipole moment, $\vec{p}(t) = 0.5$ $(\vec{p}e^{-i\omega t} + \vec{p}^{\dagger} e^{i\omega t})$, represented by a quantum Hermitian operator (i.e., $\vec{p} = \vec{p}^{\dagger}$). Since $\mathscr{H}_I(t) \ll \mathscr{H}_0$, the effect of the interaction $\mathscr{H}_I(t)$ on the system imposes a temporal dependence on the density matrix, and its matrix elements $\rho_{\alpha\beta}(t)$ may be deduced from perturbation theory with a hierarchy of time-dependent Liouville equations:

$$\frac{\partial \rho^{(n)}_{\alpha\beta}(t)}{\partial t} = \frac{1}{i\hbar}\left\{\hbar\omega_{\alpha\beta}\rho^{(n)}_{\alpha\beta}(t) + \left[\mathscr{H}_I(t), \rho^{(n-1)}(t)\right]_{\alpha\beta}\right\} - \Gamma_{\alpha\beta}\rho^{(n)}_{\alpha\beta}(t), \tag{A.3}$$

where $n = 1, 2, \ldots$, $\mathscr{H}_I(t) = 0.5\left[\mathscr{H}_I e^{-i\omega t} + \mathscr{H}_I^{\dagger} e^{i\omega t}\right]$, $\mathscr{H}_I = -\left(\vec{p} \bullet \vec{\mathscr{E}}\right)$, $\mathscr{H}_I^{\dagger} = -\left(\vec{p} \bullet \vec{\mathscr{E}}^*\right)$. The factor $\Gamma_{\alpha\beta}$ is the damping rate of the matrix element at indices (α, β), resulting from interactions with other atoms as well as the atom's spontaneous emission induced by impinging radiation.

In this Appendix, we first discuss the basics of the electric dipole transition in Section A.1. This provides the characteristic parameters associated with the transitions in a two-level system, leading to the relationships between the Einstein A coefficient, oscillator strength, line-strength and absorption cross section, and the formal solution to the Liouville equation to the first and second orders. The solution to the time-independent Schrödinger equation, which leads to stationary eigenstates and the structures of selected atoms, will be discussed conceptually in Section A.2, followed by the discussion of the structure of linear molecules in A.3 and relevant thermodynamics in A.4.

A.1 Basics of the Electric Dipole Transition

Since the electric dipole interaction dominates the interactions between light and matter, to understand the interaction between a laser beam and the atoms and molecules of interest, we can neglect the magnetic dipole and electric quadrupole interactions. Because we choose to study light–matter interactions by the semiclassical (quantum) theory, which cannot handle spontaneous emission, we are forced, as were other authors like Corney [A.1], to install spontaneous emission by invoking classical radiation theory along with some heuristic arguments; see Subsection A.1.1. Luckily, it leads to results that agree with experiments. To do so, we recall Eq. (2.1.b), which gives the radiated power per unit solid angle from an $\hat{e}$-polarized dipole moment located at the origin, $\vec{p} = |p|\hat{e}$, into the observation direction $\hat{k}$ with two transverse components polarized along $\hat{e}_a{}'$ and $\hat{e}_b{}'$, or generally along the unit vector $\hat{e}'$ on the $\hat{e}_a{}' - \hat{e}_b{}'$ plane (see Fig. B.1):

$$\frac{dP}{d\Omega} = \frac{\omega^4|p|^2}{32\pi^2\mathrm{e}_0c^3}\left[(\hat{e}\bullet\hat{e}_a{}')^2 + (\hat{e}\bullet\hat{e}_b{}')^2\right] = \frac{\omega^4|p|^2}{32\pi^2\mathrm{e}_0c^3}(\hat{e}\bullet\hat{e}')^2\,. \tag{A.4.a}$$

The radiated power, or the rate of system energy loss, W per unit time, may be computed by integrating (A.4.a) over all directions (4π solid-angle), resulting in:

$$-\frac{dW}{dt} = \left(\frac{\omega^4|p|^2}{32\pi^2\mathrm{e}_0c^3}\right)\left(\frac{8\pi}{3}\right) = \frac{\omega^4|p|^2}{12\pi\mathrm{e}_0c^3}\,. \tag{A.4.b}$$

The simplest way to see this is to choose the textbook example of letting $\hat{e} = \hat{z} \rightarrow \hat{e}\bullet\hat{e}_a{}' = 0$, and $\hat{e}\bullet\hat{e}_b{}' = -\sin\theta$, and $\int \sin^2\theta d\Omega = 8\pi/3$.

A.1.1 Definitions of and Relations between Einstein Coefficient, Line-Strength and f-Values

Now, we follow the development in Chapter 4 of Corney [A.1] and imagine that the radiating dipole moment in question, $\vec{\mathrm{p}}(t) = \frac{|p|}{2}(\hat{e}e^{-i\omega t} + \hat{e}^* e^{i\omega t})$, is due to the excitation of a two-level atom with the energy difference between the excited and ground states, $E_k - E_i = \hbar\omega_{ki} \approx \hbar\omega$. Therefore, we identify:

$$\left(\frac{|p|}{2}\hat{e}e^{-i\omega t}\right)_{cl} \quad \text{as} \quad \int \psi_i^* \vec{p}\,\psi_k dV e^{-i\omega_{ki}t} = <i\,|\vec{p}|k> e^{-i\omega_{ki}t}\,. \tag{A.5}$$

Here, $\vec{p}$ is a quantum mechanical vector operator. Therefore, we can obtain the rate of spontaneous emission, or Einstein A coefficient from state $|k\rangle$ to state $|i\rangle$ from (A.4.b) by the substitution of $|p|/2$ with $<i|\vec{p}|k>$ from (A5), giving

$$-\frac{dW}{dt} = \frac{\omega^4|p|^2}{12\pi\mathrm{e}_0c^3} \equiv \hbar\omega_{ki}A_{ki} \rightarrow A_{ki} = \frac{\omega_{ki}^3}{3\pi\hbar\mathrm{e}_0c^3}|<i|\vec{p}|k>|^2. \tag{A.6}$$

In quantum systems, the energy levels often have degeneracies, specified by magnetic quantum numbers, for example. In this case, the total decay (spontaneous emission) rate from an upper state $|k, m_k\rangle$ to a degenerate lower level $|i\rangle$ is computed by summing over all possible values of m_i. Since the result is (or should be) independent of the upper state index m_k, we may also sum it over and divide by the degeneracy of the upper level, leading to:

$$A_{ki} = \frac{\omega_{ki}^3}{3\pi\hbar\epsilon_0 c^3} |< i|\vec{p}|k>|^2 \rightarrow$$

$$\frac{\omega_{ki}^3}{3\pi\hbar\epsilon_0 c^3} \sum_{m_i} |< i,\ m_i|\vec{p}|k, m_k>|^2 = \frac{\omega_{ki}^3}{3\pi\hbar\epsilon_0 c^3} \frac{1}{g_k} \sum_{m_k, m_i} |< i,\ m_i|\vec{p}|k,\ m_k>|^2. \tag{A.7}$$

The square of the dipole transition-matrix element, $|< i,\ m_i|\,\vec{p}\,|k, m_k>|^2$ (which is proportional to the transition probability) is of capital importance, and we define its sum over all degenerate substates as the line-strength of the dipole transition:

$$S_{ik} = S_{ki} = \sum_{m_k, m_i} |< i,\ m_i|\,\vec{p}\,|k,\ m_k>|^2 \rightarrow A_{ki} = \frac{\omega_{ki}^3}{3\pi\hbar\epsilon_0 c^3} \frac{S_{ki}}{g_k}. \tag{A.8.a}$$

Thus, the line-strength, S_{ki}, and Einstein coefficient, A_{ki}, are related by (A.8.a); they both are provided in the NIST reference data with examples shown in Tables A.2 and A.3. We point out that the line-strength, $S_{ik} = S_{ki}$, defined here is identical to the square of the reduced matrix element $< i\|\,p\,\|k>$, defined by the Wigner–Eckart theorem, as given in p. 75 of Edmonds [A.2], which expresses the matrix element of a tensor operator $< i, m_i|T_q^\kappa|k, m_k>$ as the product of a reduced matrix element, $< i\|T^\kappa\|k>$, and a relevant 3–j coefficient, as repeated in (A.8.b), with the former reflecting the dynamics and the latter the geometry of the transition. Squaring (A.8.b) and summing over three component indices (m_i, q, m_k) and using the identity given in the first equation of (A.8.c), which is to be derived in (A.22.e) later, we have the sum rule leading to the line-strength $S_{ik} = S_{ki}$ given in the second equation of (A.8.c).

$$< i, m_i|T_q^\kappa|k, m_k> = (-1)^{i-m_i} \begin{pmatrix} i & \kappa & k \\ -m_i & q & m_k \end{pmatrix} < i\|T^\kappa\|k>. \tag{A.8.b}$$

$$\sum_{m_1, m_2, m_3} \begin{pmatrix} j_1 & j_2 & j_3 \\ m_1 & m_2 & m_3 \end{pmatrix}^2 = 1 \rightarrow S_{ki} = \sum_{m_i, q, m_k} \left|< i, m_i|T_q^\kappa|k, m_k>\right|^2 = \left|< i\|T^\kappa\|k>\right|^2. \tag{A.8.c}$$

Another related parameter is the oscillator strength f, or absorption f-value, which is defined in relation to a one-dimensional classical electronic harmonic oscillator with dipole moment ey, and whose time-averaged energy is $W = m_e\omega^2|y|^2/2$ with m_e being the mass of the electron. Using the classical radiation rate (A.4.b), we can calculate the classical energy decay rate, γ_{cl} or 2Γ used in Section 3.2, as the fractional energy decrease per unit time, in agreement with (4.10) of [A.1]:

$$\gamma_{cl} = 2\Gamma = \frac{1}{W}\left(-\frac{dW}{dt}\right) = \frac{1}{m_e\omega^2|y|^2/2}\left(\frac{\omega^4|ey|^2}{12\pi\epsilon_0 c^3}\right) = \frac{e^2\omega^2}{6\pi\epsilon_0 m_e c^3}. \quad \text{(A.9)}$$

Some find it convenient to define the emission oscillator strength (f-value), f_{ki}, for the quantum oscillator (which emits in three dimensions) as the ratio of quantum emission rate, A_{ki}, to its classical counterpart, (γ_{cl}):

$$f_{ki} = -\frac{A_{ki}}{3\gamma_{cl}} = -\left(\frac{2\pi\epsilon_0 m_e c^3}{e^2\omega_{ki}^2}\right)A_{ki} = -\frac{2m_e\omega_{ki}}{3\hbar g_k}S_{ki} \text{ and } g_i f_{ik} = -g_k f_{ki}. \quad \text{(A.10)}$$

Although the absorption f-value, f_{ik}, is given in the NIST tables, and since it is related to the Einstein A coefficient and the line-strength S, it will not be necessary for quantum mechanical calculations in this book, except in occasional use for the connection between quantum and classical oscillators. Equations (A.9) and (A.10) can also be found in Chapter 4 of Corney [A.1].

Another useful parameter is the absorption cross section of a two-level transition. This can be derived from the time-dependent, first-order harmonic perturbation calculation with (A.1.a). Here we assume the system is in its initial state $|i\rangle$ at $t = 0$ when the interaction is initiated by turning on $\mathscr{H}_I(t) = 0.5\left[\mathscr{H}_I e^{-i\omega t} + \mathscr{H}_I^\dagger e^{i\omega t}\right]$. At a later time t, assuming $\omega_{ki} \approx \omega$, the transition amplitude $a_k(t)$ from $|i\rangle$ to the state $|k\rangle$ and the associated transition probability $|a_k(t)|^2$ may be calculated by first-order, time-dependent perturbation theory, which is treated in many elementary quantum mechanics texts; it is not repeated here. The result can be found in Chapter 9 of Corney [A.1] or in Appendix A of Selveto [A.3] as

$$\left|a_k(t)\right|^2 = \frac{\left|<k|\mathscr{H}_I|i>\right|^2}{\hbar^2}\frac{\sin^2\left[(\omega - \omega_{ki})t/2\right]}{(\omega - \omega_{ki})^2}. \quad \text{(A.11.a)}$$

Since $\int_{-\infty}^{\infty}\left[\sin(\Delta\omega t/2)/(\Delta\omega)\right]^2 d(\Delta\omega) = \pi t/2$, the integrand in the limit of $t \gg 2\pi/|\omega - \omega_{ki}|$, may be expressed in terms of the delta function as $\lim_{t\gg 2\pi/(\Delta\omega)}\left[\sin(\Delta\omega t/2)/(\Delta\omega)\right]^2 = (\pi t/2)\delta(\Delta\omega)$. Therefore, $|a_k(t)|^2$ and its associated rate of absorption transition probability, w_{ik}, may be reduced to (A.11.b) as follows:

$$|a_k(t)|^2 = \frac{|<k|\mathscr{H}_I|i>|^2}{\hbar^2}\frac{\pi t}{2}\delta(\omega - \omega_{ki}); \; w_{ik} = \frac{\pi|<k|\mathscr{H}_I|i>|^2}{2\hbar^2}\delta(\omega - \omega_{ki}), \quad \text{(A.11.b)}$$

where the delta function, $\delta(\omega - \omega_{ki})$, is a statement of energy conservation. Since the energy level of the excited state is not infinitely sharp due to spontaneous emission, nor is the incident light truly monochromatic, the delta function may be replaced by a Lorentzian lineshape function, $g(\omega - \omega_{ki})$, given in (A.12). If the ground and excited states are degenerate, we sum over the degenerate states as in the treatment in Chapter 9 of (A.1). Thus, the rate of the (induced) absorption and (stimulated) emission transition probabilities, w_{ik} and w_{ki}, become

$$w_{ik} = \frac{\pi}{2\hbar^2}\frac{1}{g_i}\sum_{m_k,m_i}|<k,m_k|\mathscr{H}_I|i,m_i>|^2 g(\omega-\omega_{ki}),\ \text{with}\ g(\omega-\omega_{ki}) = \frac{\Gamma/\pi}{(\omega-\omega_{ki})^2+\Gamma^2},\ \text{and}$$
$$w_{ki} = \frac{\pi}{2\hbar^2}\frac{1}{g_k}\sum_{m_k,m_i}|<k,m_k|\mathscr{H}_I|i,m_i>|^2 g(\omega-\omega_{ki}),\ \text{with}\ g_i w_{ik} = g_k w_{ki} = \frac{\pi}{2\hbar^2}S_{ik}\,g(\omega-\omega_{ki})\,. \tag{A.12}$$

Since the atomic dipole moment is randomly oriented with respect to the polarization of the external field, when summing over all degeneracies (or orientations), the magnitude square of its projection on the external field is one-third of the product of their magnitudes squared, as given below. Noting that the intensity of the impinging light field is $I = \mathrm{e}_0 c|\mathscr{E}|^2/2$, we can derive the absorption cross section, σ_{ik}, or stimulated emission cross section, σ_{ki}, in relation to line-strength S in (A.13), much like the relation between S and Einstein A coefficient, as in (A.8.a):

$$\sum_{m_k,m_i}\left|<k,m_k|\mathscr{H}_I|i,m_i>\right|^2 = \sum_{m_k,m_i}\left|<k,m_k|-\left(\vec{p}\cdot\vec{\mathscr{E}}\right)|i,m_i>\right|^2$$
$$= \sum_{m_k,m_i}\left|<k,m_k|\vec{p}|i,m_i>\right|^2\frac{\left|\mathscr{E}\right|^2}{3} = \frac{S_{ik}\left|\mathscr{E}\right|^2}{3},$$

$$\sigma_{ik} = \frac{\hbar\omega_{ki}w_{ik}}{I} = \frac{\pi\omega_{ki}}{2\hbar}\frac{1}{g_i}\frac{2S_{ik}\left|\mathscr{E}\right|^2}{3\mathrm{e}_0 c\left|\mathscr{E}\right|^2}g(\omega-\omega_{ki}) = \frac{\pi\omega_{ki}}{3\mathrm{e}_0\hbar c}\frac{S_{ik}}{g_i}g(\omega-\omega_{ki})$$
$$= \frac{4\pi^3}{3\mathrm{e}_0 h\lambda}\frac{S_{ki}}{g_i}g(\nu-\nu_{ki}),\ \text{and}$$

$$g_i w_{ik} = g_k w_{ki} \rightarrow g_k\sigma_{ki} = g_i\sigma_{ik} \rightarrow \sigma_{ki} = \frac{\pi\omega_{ki}}{3\mathrm{e}_0\hbar c}\frac{S_{ki}}{g_k}g(\omega-\omega_{ki}) = \frac{4\pi^3}{3\mathrm{e}_0 h\lambda}\frac{S_{ki}}{g_k}g(\nu-\nu_{ki}). \tag{A.13}$$

Note that (A.13) is identical to (3.6) derived from density matrix formulation. Relating A_{ki} to the emission oscillator strength f_{ki}, the last expression is consistent with (9.11) of [A.1].

A.1.2 Formal Solutions to the First- and Second-Order Liouville Equation

We next solve the time-dependent Liouville equation to the first and second order. To evaluate the transition between an initial ground state $|f,\mu>$ and an excited state $|F,m>$ of a metal atom in the atmosphere with damping rate between substates, $\Gamma_{m\mu}$, we solve (A.3) in the first order by assuming the ground-state density matrix is given as $\rho^{(0)}_{\mu\mu'} = \delta_{\mu\mu'}\rho^{(0)}_{\mu\mu}$ with $\rho^{(0)}_{\mu\mu} = 1/g_f$, where $g_f = (2f+1)$ is the ground-state degeneracy. Since both the excited state |F> and the ground state $|f>$ are degenerate (Zeeman splitting in energy due to the weak terrestrial magnetic field is ignored in comparison with the optical transition energy), we set $\Gamma_{m\mu} = \Gamma_{Ff}$ and $\omega_{m\mu} = \omega_{Ff}$ independent of m and μ. The steady-state first-order solution of (A.3) can then be written as

$$\rho_{\mu m}{}^{(1)}(t) = \frac{1}{2}\left[\rho^{(1)}_{\mu m}(\omega)e^{-i\omega t} + \rho^{(1)}_{\mu m}(-\omega)e^{i\omega t}\right],$$

$$\text{where, } \rho^{(1)}_{\mu m}(\omega) = \frac{1}{\hbar}\frac{<f\mu|\left(\vec{p}\cdot\vec{\mathscr{E}}\right)|Fm>}{(\omega+\omega_{Ff})+i\Gamma_{Ff}}\frac{1}{g_f} \text{ and } \rho^{(1)}_{\mu m}(-\omega) = \rho^{(1)*}_{m\mu}(\omega)\,. \tag{A.14}$$

The second-order equation of (A.3) is

$$\frac{\partial\rho^{(2)}_{mm'}(t)}{\partial t} = \frac{1}{i\hbar}\left\{\hbar\omega_{mm'}\rho^{(2)}_{mm'}(t) + \left[H_I(t),\rho^{(1)}(t)\right]_{mm'}\right\} - \Gamma_{mm'}\rho^{(2)}_{mm'}(t),$$

$$\text{with } \omega_{mm'} = \omega_B(m-m'), \text{ where } \omega_B = g_F\mu_B B/\hbar, \tag{A.15}$$

where g_F (not to be confused with $\mathcal{g}_F$), μ_B, B, and ω_B are hyperfine Landé-g factor, Bohr magneton, magnetic field (or induction) in Wbm^{-2}, and Larmor frequency, respectively. When the system has Hertzian (radio-frequency) coherence (see discussions in 3.2.1), the off-diagonal matrix element, $\rho^{(2)}_{mm'}(t)$, is nonzero. Though the substates in both excited and ground states are split under a weak magnetic field, we ignore the Zeeman splitting in comparison to the energy difference between $|f>$ and $|F>$ in carrying out the first-order perturbation calculation. On the other hand, since the frequency difference between the Zeeman split levels of the excited substates is comparable to those resulting from the lifetime broadening of the excited state and the spectral width of Hertzian coherence, we explicitly include the Zeeman splitting of the excited level in the second-order perturbation treatment.

To solve (A.15), we first calculate the commutator in it. The result is expressed in (A.16) in terms of the factor $F_{mm'}$, defined below it in terms of the product of dipole matrix elements summing over the substates of $|f>$ and representing the connection between $|F,m>$ and $|F,m'>$, both coherently excited from $|f>$. Assuming the incident light has a constant intensity $\mathscr{I} = \varepsilon_0 c|\mathscr{E}|^2/2$ between $t'=0$ and $t'=t$, the solution for $\rho^{(2)}_{mm'}(t)$, following She et al. [A.4], is given in (A.17) along with the steady-state solution under the condition $\Gamma_{mm'}t \gg 1$.

$$\left[H_I(t),\rho^{(1)}(t)\right]_{mm'} = \left[i\pi/(\hbar g_f)\right]F_{mm'}|\mathscr{E}|^2 g(\omega-\omega_{m\mu}),$$
$$\text{where } F_{mm'} \equiv \sum_{\mu} \langle Fm|(\vec{p}\cdot\hat{e})|f\mu\rangle \langle f\mu|(\vec{p}\cdot\hat{e}^*)|Fm'\rangle. \tag{A.16}$$

$$\rho^{(2)}_{mm'}(t) = \frac{\pi}{\hbar^2}F_{mm'}|\mathscr{E}|^2\frac{1}{g_f}g(\omega-\omega_{m\mu})\frac{1-e^{-[\Gamma_{mm'}+i\omega_B(m-m')]t}}{\left[\Gamma_{mm'}+i\omega_L(m-m')\right]}$$

$$\xrightarrow{\Gamma_{mm'}t\gg 1} \rho^{(2)}_{mm'}(t) = \frac{\pi}{\hbar^2}\frac{F_{mm'}|\mathscr{E}|^2}{\left[\Gamma_{mm'}+i\omega_B(m-m')\right]}\frac{1}{g_f}g(\omega-\omega_{m\mu}). \tag{A.17}$$

A.2 Energy Structure of Na, K, and Fe Atoms and of Nd^{3+}, Cr^{3+} in Crystals

We consider an isolated atom with several electrons moving around a positively charged nucleus. The moving electrons, each with spin quantum number s = ½, experience Coulomb attraction from the nucleus as well as Coulomb repulsion from other electrons. For an atom with one electron (i.e., the hydrogen atom), the allowed stationary states are known to be quantized with four quantum numbers: the principal quantum number, $n = 1, 2, 3 \ldots$; the orbital angular momentum quantum number, ℓ and its magnetic quantum number m; and the spin quantum number, s = ½ and its associated magnetic quantum numbers, $m_s = \pm 1/2$. The values of n, ℓ, m_ℓ, and m_s depend, respectively, on the distance of the electron from the nucleus, the magnitude and orientation of its orbital angular momentum, and of the orientation of the electron spin. The boundary conditions for the solution to the Schrödinger equation (A.1.b) impose a limit on the allowed values of ℓ for a given n to be $\ell = 0,\ 1, \cdots,\ n-1$. For a given value of angular momentum, ℓ, the requirement of a single value in azimuth (or the magnetic quantization condition) limits the magnetic quantum numbers m_ℓ, representing the orientation of the electron orbit with same magnitude of angular momentum, to $m_\ell = -\ell, -\ell+1, .., \ell-1$, and ℓ. In the absence of an external magnetic field, different orientations give rise to the same energy, leading to a $(2\ell+1)$ degeneracy for a configuration (n,ℓ), with both the principal quantum number n and orbital angular momentum quantum number ℓ specified. In the absence of electron repulsion, the energy for states corresponding to different values of ℓ for a given value of n are the same due to the $1/r$ dependence of Coulomb potential, where r is the distance from the electron to the nucleus. This energy degeneracy is called the "accidental degeneracy." Thus, for a one-electron atom, its energy levels are specified by the principal quantum number, n, with $2n(2\ell+1)$ degeneracy the first factor of 2 coming from the two equivalent (degenerate) spin orientations $m_s = \pm 1/2$.

We now consider a basic model of multiple-electron atoms where a moving electron sees both the attractive potential from the nucleus and the repulsive interactions from other electrons, $\mathscr{H}_0 = \mathscr{H}_{e-n} + \mathscr{H}_{e-e}$, and the kinetic energy of the electron is included in $\mathscr{H}_{e-n}$. The electrons still move in a spherically symmetric potential, though having a radial dependence more complex than $1/r$, the accidental degeneracy is thus lifted. Since the total angular momentum of the electrons is still conserved, the energies of the states now depend upon two quantum numbers (n, ℓ), termed the electron configuration, each with a degeneracy of $2(2\ell+1)$. For multiple-electron atoms, the Pauli Exclusion Principle, which states that for fermions (including electrons) no two identical particles can occupy states with the same quantum numbers, must be obeyed. Thus, as the number of electrons increases, the configurations of higher energies will be successively occupied to form the atom's ground state. This is the underlying principle of the periodic table. For the interests of metal resonance lidars, we consider only sodium (Na), potassium (K), and iron (Fe) atoms, having, respectively, 11, 19, and 26 electrons. These electrons fill subshells according to their rising energies, conforming to the Pauli Exclusion Principle. The ground-state configurations are then $(1s)^2(2s)^2(2p)^6(3s)^1$ for Na, $(1s)^2(2s)^2(2p)^6(3s)^2(3p)^6(4s)^1$ for K, and $(1s)^2(2s)^2(2p)^6(3s)^2(3p)^6(4s)^2(4p)^6(4d)^6(5s)^2$ for Fe. Due to the importance of Nd:YAG ($Y_3Al_5O_{12}$) lasers for lidar applications, we shall also investigate the structure of the ground-state configuration of Nd^{3+} ions in the crystalline field of a YAG crystal. In addition, since Alexandrite lasers are often used for K and Fe resonance fluorescence lidars, we treat the energy structure of Cr^{3+} in $BeAl_2O_4$, together with that of ruby (Cr^{3+} in Al_2O_3), due to their intimate relationships.

A.2.1 The Term Symbol, Atomic Configuration, and Hund's Rules

Since the total angular momentum of a fully occupied configuration is zero, the possible eigenfunctions of the atom depend on the combined motion of those electrons in the partially filled subshell. Subject to the Pauli Exclusion Principle, their combination leads to several allowed values of total angular momentum L and total spin S, denoted by a term symbol as ${}^{2S+1}L$. It is customary to use capital letters S, P, D, F, … to denote L = 0, 1, 2, 3, … respectively. When the spin-orbit interaction, $\mathscr{H}_{SO}$, is included, the total angular momentum (orbital plus spin), J, is conserved, its stationary (eigen-) level is then denoted by J added to the lower right corner and denoted as ${}^{2S+1}L_J$. When the magnetic quantum number of J, M_J, is also specified, we then have a set of quantum numbers, called good quantum numbers, L^2, S^2, J^2, M_J, representing squared magnitudes of L, S, J and the projection of J on the z-axis. These quantum numbers define the state (eigenfunction) of the atom. In the absence of an external magnetic field, the z-axis is arbitrary, and the states with different M_J have the same energy (i.e., the ${}^{2S+1}L_J$ term has $g_J = 2J + 1$ degeneracy).

It is straightforward to determine the possible terms for electrons in different subshells, for they are not limited by the Pauli Exclusion Principle. For example, for two electrons in the configuration (1p) (2p), the possible values of L and S are, respectively, 0, 1, 2 and 0, 1, leading to three singlet terms ${}^1S_0$, ${}^1P_1$, ${}^1D_2$ (9 states) and three triplet terms ${}^3S_1$, ${}^3P_{0,1,2}$, ${}^3D_{1,2,3}$ (27 states), a total of 36 states. This is what we expect, as there are three values of m_ℓ and two values of m_s, leading to six choices for one electron and giving rise to 6×6=36 states. For two electrons in the same p-subshell, some choices will be forbidden by the Pauli Exclusion Principle. Thus, the first electron has six choices and the second only five, and since electrons are indistinguishable, we cannot count the same choice in reverse order as a valid choice, so the total number of allowed choices is 6×5/2! = 15. A careful tabulated examination will show that the allowed terms for 2p electrons in the same subshell, or in the configuration $(p)^2$ are ${}^1S_0$, ${}^1D_2$, ${}^3P_{0,1,2}$, consisting of 1, 5, and 9, or a total of 15 states. The same meticulous method can be applied to determine the possible terms for four (or six) electrons in the same d subshell, which has 10 states. The resulting allowed terms for $(d)^4$ or $(d)^6$ configurations, following Table A.1 in Appendix A of [A.5], are 2^1S, 2^1D, ${}^1F$, 2^1G, ${}^1I$, 2^3P, ${}^3D$, 2^3F, ${}^3G$, ${}^3H$, ${}^5D$. The total number of choices should be 10×9×8×7×6×5/6! = 10×9×8×7/4! = 210, in agreement with the number of states summing over all allowed terms, that is, 2×1+2×5+7+2×9+13 +2×(5+3+1) +(7+5+3) +2×(9+7+5) +(11+9+7)+(13+11+9) +(9+7+5+3+1) = 210. These are then the possible configurations of the Fe atom; it is much more complicated. These terms will have different energies, and we will discuss which one is lowest below. If the configuration in question has an odd number of electrons with odd parity (p-electron or f-electron, for example), the parity of all the resulting terms is then odd, and we label it by a right superscript "o". Since $(p)^2$ and $(f)^4$ configurations consist of an even number of electrons with odd parity, their resulting terms have even parity; thus, no superscript "o" appears in the resulting term symbols.

It is of interest to know which term among the possible ones in each configuration has the lowest energy. Without invoking quantum calculations, one can see that if we could turn off the electron–electron repulsion and spin-orbit coupling, all terms arising from a given configuration would have the same energy. When the electron–electron repulsion is turned on, different spatial distributions of electrons will correspond to different energies. Since the Pauli Exclusion Principle states that electrons with the same spin orientation cannot occupy the same orbital, electrons with parallel spins must reside in different spatial orbitals, which on average reduces their repulsive interactions. Though making no prediction about the energy ordering of different terms, the well-known Hund's rules predict which term will be the lowest in energy in a given configuration, and when spin-orbit interaction is turned on, which level in each term will have lowest energy. Hund's rules, which predict the lowest energy term in each configuration, and the order of energy levels in a given term, are as follows:

(a) For terms in each configuration, that with highest total spin S will lie lowest in energy.
(b) For terms having the highest possible values of spin, S, the one having highest total angular momentum, L, will lie lowest in energy.
(c) For terms with the same L and S and split by spin-orbit interaction, the energy levels will be ordered as follows:
   i. If the responsible subshell is less than half-filled, then the level with the lowest J value lies lowest in energy.
   ii. If the subshell is more than half-filled, the order in J values is reversed.

A.2.2 The Terms and Energy Levels for the D*-Line Transitions of Na and K*

We follow She and Yu [A.6] and consider the energy-level diagram of Na for the D-line transitions in terms of a hierarchy of three related physical models. As a basic model we first consider the Na atom as a two-state system labeled by orbital angular momentum ℓ (i.e., $3s$ and $3p$, with respective degeneracies of 1 and 3). In the term symbol, the ground and excited levels are respectively $(2p)^6(3s) - {}^2S$ and $(2p)^6(3p) - {}^2P^o$; the associated line-strengths and the spontaneous emission rates (or Einstein A coefficients) are denoted, respectively, as S_0 and A_0. Since the configuration for the excited level, $(1s)^2(2s)^2(2p)^6(3p)^1$, has an odd number of electrons with odd parity (p-electron in this case), the resulting term has odd parity, and we label it by a right superscript "o" as ${}^2P^o$. We then consider the next model by turning on the electron spin-orbit coupling. In this intermediate model (fine structure), the degeneracies are doubled, and the excited terms split into doublet and quadruplet levels, labeled by ${}^2P^o{}_{1/2}$ and ${}^2P^o{}_{3/2}$ with the $J = 3/2$ level lying above the $J = 1/2$ level, according to Hund's rule (c); when they transit to the ground level, ${}^2S_{1/2}$, they respectively emit the D_1 and D_2 lines at 589.6 nm and 589.0 nm. As the final model (hyperfine structure), we also include the nuclear spin. Since ^{23}Na is the only stable (and the only primordial) isotope and has nuclear spin $I = 3/2$, this leads to an energy-level structure with hyperfine splitting. In this final model, the degeneracy increases by another factor of 4, and the levels ${}^2S_{1/2}$, ${}^2P^o{}_{1/2}$, and ${}^2P^o{}_{3/2}$ split into two, two, and four states, respectively, with different energies labeled by the total angular momenta of the atom, $F = 1, 2$; $F = 1, 2$ and $F = 0, 1, 2, 3$. The energy-level diagram showing the three models of Na D-line transitions is given in Fig. A.1, taken from Fig. 1 of She and Yu [A.6].

A potassium atom has 19 electrons; the configurations of its ground and excited states are, respectively, $(1s)^2(2s)^2(2p)^6(3s)^2(3p)^6(4s)^1$ and $(1s)^2(2s)^2(2p)^6(3s)^2(3p)^6(4p)^1$. Fortunately, the most abundant (93%) naturally occurring isotope of potassium is ^{39}K, which also has nuclear spin of 3/2. Therefore, other than the fact that the ground and excited states in the basic model are 4s and 4p, the structure of the three models and respective electron terms and degeneracies in Fig. A.1 also describe ^{39}K. The numerical values for the energy levels lead to D_1 and D_2 transitions at 770.1 nm and 766.7 nm. The energy levels [A.7] for the intermediate model are 0

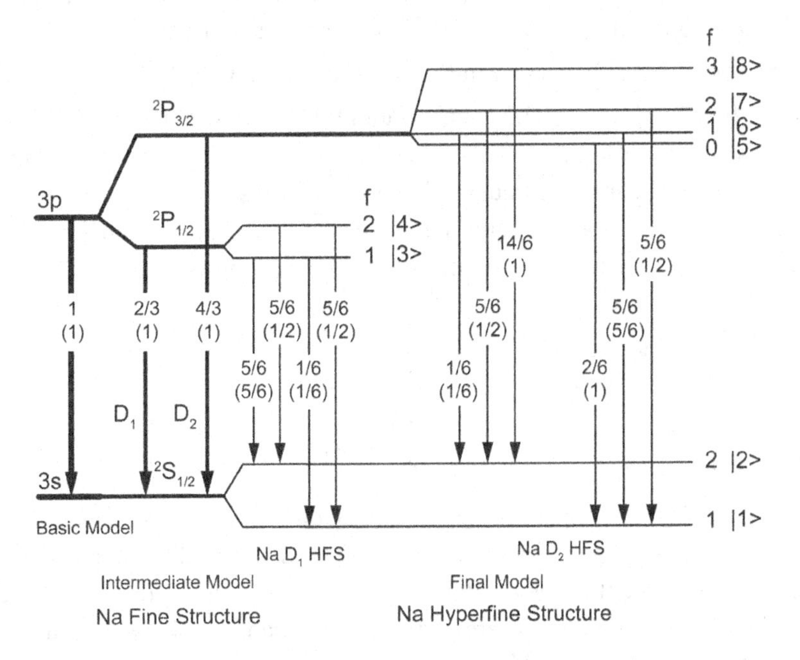

Fig. A.1 Energy-level diagram of a Na atom described by three related models. The basic model is a two-level system (3s and 3p with respective degeneracies of 1 and 3). In the intermediate model, the electron spin $S = 1/2$ is included, which doubles the degeneracies. The spin-orbit coupling splits the excited state into a doublet and a quadruplet. In the final model, the nuclear spin $I = 3/2$ and the associated hyperfine interaction lead to the energy level structure with hyperfine splitting. The numbers given in a transition arrow of each model are the transition line strength S (not bracketed) and the Einstein coefficient A (in the bracket) of that transition in units of the transition line-strength S_0 and the spontaneous emission rate A_0 of the basic model computed from (A.18.a) and (A.18.b). As discussed in the text, other than the fact that the ground and excited states in the basic model are 4s and 4p, this 3-model structure also describes the most abundant potassium isotope, ^{39}K. Reprinted with permission from [A.6]. © The Optical Society.

for $^2S_{1/2}$, 16956.172 cm^{-1} (Na) and 12985.186 cm^{-1} (K) for $^2P^o{}_{1/2}$, and 16973.368 cm^{-1} (Na) and 13042.896 cm^{-1} (K) for $^2P^o{}_{3/2}$.

We note that the numbers given in the transition arrows of each model are the line-strength S (not bracketed) and the Einstein A coefficient (in the brackets) of that transition for that model, in terms of the values of the basic model, S_0 and A_0. As noted, S and A coefficients of the three models are simply related. This is because the spin-orbit and hyperfine interactions that lead to the spin-orbit and hyperfine splitting are negligibly small relative to the energy difference between $3p$ and $3s$ levels for Na and $4s$ and $4p$ levels for K atoms. For the ground configuration of Na, ($3s$), there is no spin-orbit splitting, and its hyperfine splitting is only 1.772 GHz, or 0.059 cm^{-1} (0.015 cm^{-1} for K); its excited spin-orbit splitting, responsible for the D-doublet, is only 17.2 cm^{-1} (57.72 cm^{-1} for K), about 0.1% (0.4%) of the s-p transition energy. In comparison, the spin-orbit splitting between Fe ground levels $(3d)^6(4s)^2a^5D_3$ and $(3d)^6(4s)^2a^5D_4$ is 415.393 cm^{-1}, about 1.5% of the a^5D to z^5F transition energy, and that between the excited levels $3d^6(^5D)4s4p(^3P^0)z^5F^0_4$ and $3d^6(^5D)4s4p(^3P^0)z^5F^0_5$ is 292.27 cm^{-1}, about 1.0% of the a^5D to z^5F transition energy. For this reason, there is no simple connection between the line-strengths from the basic to the intermediate models for Fe atoms.

For Na or K, the line-strengths between different models can be related by respective 6-j coefficients as given in Eq. (49.9) of [A.8] as

$$|<J\|p\|j>|^2 = g_j g_J \begin{bmatrix} J & 1 & j \\ \ell & S & L \end{bmatrix}^2 |<L\|p\|\ell>|^2; \; S = \frac{1}{2}, \text{ and} \tag{A.18.a}$$

$$|<F\|p\|f>|^2 = g_f g_F \begin{bmatrix} F & 1 & f \\ j & I & J \end{bmatrix}^2 |<J\|p\|j>|^2; \; I = \frac{3}{2}. \tag{A.18.b}$$

Since the electric dipole operator depends only on orbital coordinates of the atom, the line-strength of the intermediate and final models should depend only on the reduced matrix element of the orbital state and 6-j coefficients (assuming the spin-orbit and hyperfine interactions are negligible), as can be achieved by the successive use of (A.18.a) and (A.18.b), as well as (A.8.a), to relate the line-strength, S_{ki}, to the A coefficient, A_{ki}, for each model, with the assumption that the frequency ω_{ki} is approximately independent of the models. Thus, we can express A_{Ff} of the final model in terms of the basic model A_0 as

$$A_{Ff} = g_f g_J \begin{bmatrix} F & 1 & f \\ j & I & J \end{bmatrix}^2 A_{Jj} = g_f g_J g_j g_L \begin{bmatrix} F & 1 & f \\ j & I & J \end{bmatrix}^2 \begin{bmatrix} J & 1 & j \\ \ell & S & L \end{bmatrix}^2 A_0 \, . \tag{A.18.c}$$

The numerical values of S and A for the final model in terms of S_0 and A_0 are shown in Fig. A.1; they were also "checked with long-hand calculation

Table A.1 *The* D*-line transition frequencies relative to the* D*-line doublets* of Na and K*

Transition frequency Offset (GHz) to NaD_2	Transition frequency Offset (GHz) to NaD_1	Transition frequency Offset (GHz) to KD_2	Transition frequency Offset (GHz) to KD_1
Na		K	
$\vert 7>\rightarrow\vert 1>$ 1.0911	$\vert 4>\rightarrow\vert 1>$ 1.1780	$\vert 7>\rightarrow\vert 1>$ 0.2822	$\vert 4>\rightarrow\vert 1>$ 0.3100
$\vert 6>\rightarrow\vert 1>$ 1.0566	$\vert 3>\rightarrow\vert 1>$ 0.9894	$\vert 6>\rightarrow\vert 1>$ 0.2725	$\vert 3>\rightarrow\vert 1>$ 0.2528
$\vert 5>\rightarrow\vert 1>$ 1.0408	$\vert 4>\rightarrow\vert 2>$ –0.5936	$\vert 5>\rightarrow\vert 1>$ 0.2688	$\vert 4>\rightarrow\vert 2>$ –0.1517
$\vert 8>\rightarrow\vert 2>$ –0.6216	$\vert 3>\rightarrow\vert 2>$ –0.7822	$\vert 8>\rightarrow\vert 2>$ –0.1589	$\vert 3>\rightarrow\vert 2>$ –0.2089
$\vert 7>\rightarrow\vert 2>$ –0.6806		$\vert 7>\rightarrow\vert 2>$ –0.1795	
$\vert 6>\rightarrow\vert 2>$ –0.7150		$\vert 6>\rightarrow\vert 2>$ –0.1892	

* *D*-lines: 16956.172 cm^{-1} (NaD_1), 16973.368 cm^{-1} (NaD_2), 12985.186 cm^{-1} (KD_1), and 13042.896 cm^{-1} (KD_2).

Table A.2 *The Einstein A coefficient, oscillator strength, line strengths of Na D transitions*

Obs λ (nm)	A_{ki} (s^{-1})	f_{ik}	S_{ik} (a.u.)	Lower Level Energy (cm^{-1})	Upper Level Energy (cm^{-1})	$g_i - g_k$
588.995	6.16×10^7	6.41×10^{-1}	2.49×10^1	$(2p)^6(3s)^2S_{1/2}$ 0	$(2p)^6(3p)^2P^0_{3/2}$ 16 973.36619	2 – 4
589.592	6.14×10^7	3.20×10^{-1}	1.24×10^1	$(2p)^6(3s)^2S_{1/2}$ 0	$(2p)^6(3p)^2P^0_{1/2}$ 16 956.17025	2 – 2

sequentially, using the relevant Clebsch–Gordan coefficients" [A.6]. Convincingly, the results make physical sense, as the total line-strengths for the basic, intermediate (spin-orbit), and final (hyperfine) models are, respectively, S_0, $2S_0$, and $8S_0$, and the total transition rate from a given upper state is not only independent of the model, but also equal to the Einstein coefficient of the $(3s) - (3p)$ transition, A_0.

Unlike Na, K has three naturally occurring isotopes, two stable ^{39}K (93.3%) and ^{41}K (6.7%), and one long-lived radioactive ^{40}K (0.012%); their nuclear spins are, respectively, 3/2, 4, and 3/2. Though ^{41}K, which has a more complex hyperfine structure due to the higher nuclear spin of $I = 4$, represents only 6.7% of the total, it should not be ignored for potassium lidar applications [A.9]. The majority isotope ^{39}K (93.3%) has nuclear spin, I = 3/2, leading to identical hyperfine structure as ^{23}Na but with different values for excited level spin-orbit splitting and ground-level hyperfine splitting. Thus, the energy-level diagram for ^{23}Na in Fig. A.1 may be used to depict the three models of ^{39}K, including the relationships of S and A between different models in terms of those of the basic model. In Table A.1 we list the frequencies for six D_2 transitions and four D_1 transitions for sodium Na and potassium K, relative to their respective D_1 and D_2 transition frequencies for the atoms' fine structure (intermediate model). These are based on the experimentally determined nuclear magnetic dipole and electric quadrupole interaction strengths, responsible for hyperfine splitting in the alkali atoms presented by Arimodo et al. [A.10]. For reference, we also list useful parameters for Na D transitions in Table A.2, taken from [A.11], showing $2S_0 = 37.3\ a.u.$, or $S_0 = 18.7(ea_0)^2$ where e and a_0 are electron charge and Bohr radius, respectively.

A.2.3 The Terms and Partial Energy Levels of Fe for the 386, 374, and 372 nm Transitions

Naturally occurring iron (Fe) with 26 electrons comprises four stable isotopes: 5.845% ^{54}Fe (possibly radioactive with a half-life over 10^{22} years), 91.754% ^{56}Fe, 2.119% ^{57}Fe, and 0.282% ^{58}Fe. The nuclear spins of the three dominant species, ^{54}Fe, ^{56}Fe, and ^{58}Fe, are all zero. The only exception is ^{57}Fe with ~2% abundance. We shall consider only the major isotope ^{56}Fe (91.8%) here. Its zero-nuclear spin means no hyperfine splitting, so we only need consider the basic and intermediate models. However, its partially filled ($3d$) subshell with six electrons makes the ground configuration, $(1s)^2(2s)^2(2p)^6(3s)^2(3p)^6\ (3d)^6(4s)^2$, quite complex, with these possible terms, 2^1S, 2^1D, ^1F, 2^1G, ^1I, 2^3P, ^3D, 2^3F, ^3G, ^3H, ^5D. Fortunately, we can apply the Hund's rules (a) and (b) to determine that the ^5D term has the lowest energy. For the transitions of interest, only the $(1s)^2(2s)^2(2p)^6(3s)^2(3p)^6$ $(3d)^6(4s)(4p)$ excited configuration with two different terms therein needs to be considered. It is then sufficient to use simplified term symbols, $(3d)^6(4s)^2\ a^5D$ for the ground term, and $3d^6(^5D)4s4p(^3P^0)\ z^5D$ and $3d^6(^5D)4s4p(^3P^0)\ z^5F^0$ for the excited terms. For reference, we list useful parameters for the three transitions in Fe at 386, 374, and 372 nm in Table A.3, taken from the NIST atomic spectra database [A.11]. Notice that the three transitions involved are between the J levels in the respective excited and ground terms. They are, respectively, $z^5F_5^0 \to a^5D_4$ for 372 nm, $z^5F_4^0 \to a^5D_3$ for 374 nm, and $z^5D_4 \to a^5D_4$ for 386 nm in the respective ground and excited configurations mentioned. Relevant partial energy levels of Fe are shown in Fig. A.2. Upon excitation by 372 nm radiation, the atom can fluoresce back only to the same ground level. However, upon excitation by 374 nm or 386 nm radiation, the atom can fluoresce back either to the same ground level or to another ground level (a^5D_4 at 368 nm for the former case and a^5D_3 at 392 nm for the latter case) with their respective different A coefficients and line-strengths given in Table A.3.

To appreciate the energy level diagram of the Fe atom shown in Fig. A.2, additional discussion is in order. The configuration with lowest energy is $(3d)^6(4s)^2$. As mentioned, among the allowed terms in this configuration, ^5D has lowest energy since this is the term with the highest total spin, $S = 2$. In the ^5D term, $J = 4$ level has the lowest energy per Hund's rule (c) since the (3*d*) subshell is more than half filled. As a result, there exist higher-energy terms, like ^{3}H in the same configuration, $(3d)^6(4s)^2$. Since the energies of the (3*d*) and (4*s*) orbitals are comparable, one expects that configurations like $(3d)^7(4s)$ would have energy only slightly higher than that of $(3d)^6(4s)^2$. The $(3d)^7$ (4*s*) configuration contains many possible terms, one of which is also ^3H. In order to express these lower energy levels without writing out the details of their configurations, we distinguish them with a prefix, *a*, *b*, *c*, . . ., arranged in order of increasing energy. Thus, we have a^3H in $(3d)^6(4s)^2$, and b^3H in $(3d)^7(4s)$ . . . . To distinguish the higher

Table A.3 *The Einstein A coefficient, oscillator strength, line strengths of three Fe transitions*

Obs λ (nm)	A_{ki} (s^{-1})	f_{ik}	S_{ik} (*a.u.*)	Lower Level Energy (cm^{-1})	Upper Level Energy (cm^{-1})	$g_i - g_k$
367.99131	1.38×10^6	2.80×10^{-3}	3.05×10^{-1}	$(3d)^6(4s)^2a^2D_4$ 0	$3d^6(^5D)4s4p(^3P^o)z^5F_4^0$ 27166.82	9 – 9
371.99345	1.62×10^7	4.11×10^{-2}	4.53	$(3d)^6(4s)^2a^2D_4$ 0	$3d^6(^5D)4s4p(^3P^o)z^5F_5^0$ 26874.55	9 – 11
373.7133	1.41×10^7	3.81×10^{-2}	3.28	$(3d)^6(4s)^2a^2D_3$ 415.393	$3d^6(^5D)4s4p(^3P^o)z^5F_4^0$ 27166.82	7 – 9
385.9911	9.69×10^6	2.17×10^{-2}	2.48	$(3d)^6(4s)^2a^2D_4$ 0	$3d^6(^5D)4s4p(^3P^o)z^5D_4$ 25899.989	9 – 9
392.29115	1.08×10^6	3.19×10^{-3}	2.98×10^{-1}	$(3d)^6(4s)^2a^2D_3$ 415.393	$3d^6(^5D)4s4p(^3P^o)z^5D_4$ 25899.989	7 – 9

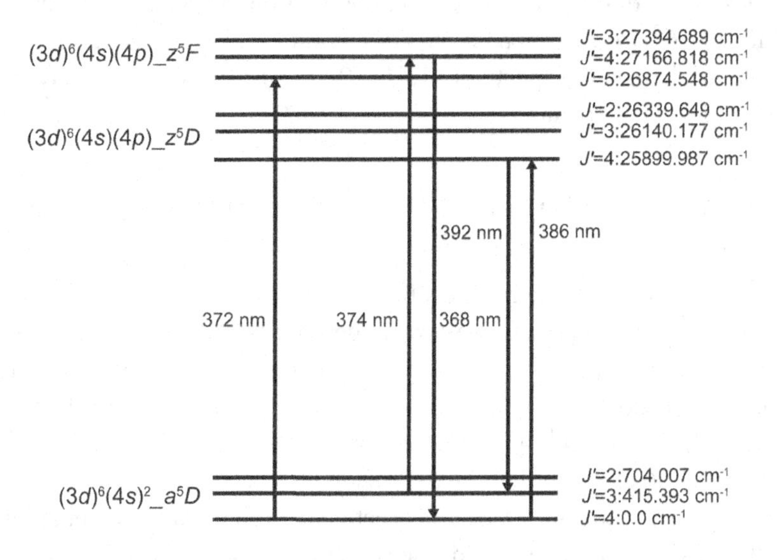

Fig. A.2 Partial energy levels of Fe, showing resonant excitation and fluorescence at 372 nm, 374 nm, and 386 nm. For 374 nm and 386 nm excitations, in addition to 374 and 386 nm, fluorescence also occurs at 368 nm and 392 nm, respectively.

energy (or excited) levels, which have a partially filled (p) or (sp) subshell in addition to the partially filled (d) subshell, by term symbols without specifying their parent configurations, we use a prefix in reverse, that is, with z, y, x, ... arranged in the order of increasing energy ($z < y$, etc.). For example, for Fe levels, we have $z\,^5D_4$ in the $3d^6(^5D)4s4p(^3P^0)$ configuration, and $y\,^5D_3^0$ in $3d^7(^4F)4p$, and so on. Another example is that the terms for both upper and lower levels of the 386 nm transition are $^5D_4$; it is imperative to separate them as $z^5D_4 \rightarrow a^5D_4$, if the term notation is to be used without confusion.

Some discussion on the term symbol in the excited configuration, like $3d^6(^5D)$ $4s4p(^3P^0)$, is also worthwhile. Here the configuration is $3d^64s4p$, which consists of two partially filled subshells, namely $(3d)^6$ and $(4s4p)$. The lowest energy terms with respect to the partially filled subshells are ^5D and $^3P^0$ as per Hund's rules (a) and (b). Since these two terms are in two different subshells, their respective electrons are independent from one other. Thus, the possible total angular and spin momenta are $L = 3, 2, 1$, and $S = 3, 2, 1$. Within these choices, the term $z\,^7F^0$ has the lowest energy, as per Hund's rules. However, the Rules make no prediction about the energy ordering of the other terms; indeed, as shown in Table A.3, experimentally, the term z^5F^0 has higher energy than the term z^5D.

Due to the large spin-orbit interaction, the values of line-strengths, S, and Einstein coefficients, A, are not simply related between the basic model and intermediate models. Fortunately, the values of S and A have been measured for each spin-orbit level and are listed in Table A.3, and thus, we do not need to know their values in the basic model. For example, two fluorescence channels from the excited level $z^5F_4^0$ are $z^5F_4^0 \rightarrow a^5D_3$ at 374 nm, and $z^5F_4^0 \rightarrow a^5D_4$ at 368 nm. Their Einstein coefficients are respectively 1.41×10^7 s^{-1} and 1.38×10^6 s^{-1}, giving rise to branching ratios of 91% and 9%.

A.2.4 Partial Energy Levels of Nd:YAG Relevant to the 0.94, 1.06, and 1.32 μm Transitions

The Nd:YAG is no doubt the most useful laser material for lidar applications. The absorption bands and emission lines are all derived from the Nd^{3+} ion in a crystalline lattice of $Y_3Al_5O_3$ (YAG). The ground state configuration of a Nd atom, with its 60 electrons, consists of the filled first three shells of 28 electrons plus $(4s)^2(4p)^6(4d)^{10}(5s)^2(5p)^6(4f)^4(6s)^2$. It is interesting to note that the ground state of Ba with 56 electrons has the $(6s)^2$ shell filled with both $(4f)$ and $(5d)$ shells empty, implying that the $(4f)$ shell is energetically higher than or comparable to the $(6s)$ shell. Thus, the Nd atom would lose two $(6s)$ and one $(4f)$ electrons to become the Nd^{3+} ion in question. Then, the configuration for the ground term of Nd^{3+} is $(4f)^3$, leading to a possible total electron spin of $S = 3/2$ or $1/2$, with the quadruplet having lower energy per Hund's rule (a). The method for figuring out the allowed terms in the subshell $(4f)^3$ is straightforward but very challenging. Fortunately, we only need to know the lowest term in this subshell. It is clear that within the quadruplet manifold ($S = 3/2$), the highest value for total orbital angular momentum is $L = 6$, because for all three electrons to have values of $m_s = +1/2$, their m_ℓ values must be different, resulting in the largest M value of $3+2+1 = 6$, or the term ^4I. This term with the highest possible L value ($L = 6$) has the lowest energy per Hund's rule (b). The possible total angular momenta for this term are then $J = 15/2$, $13/2$, $11/2$, and $9/2$ with the $J = 9/2$ level lying lowest in energy per Hund's rule (c). As a laser host, other terms of relevance are $^4S_{3/2}, ^4F_{7/2}$; $^4F_{5/2}, ^2H_{9/2}$, respectively forming two absorption bands at ~730 and ~800 nm, along with the term $^4F_{3/2}$, serving as the upper state for laser emission. Since the electrons in the $(4f)^3$ subshell are well protected from the crystalline field, the spin-orbit interaction is stronger than the splitting caused by the crystalline field. A diagram of the relevant energy levels for Nd:YAG is given in Fig. A.3, where under the crystalline field, the terms $^4F_{3/2}(R \text{ levels}), ^4I_{13/2}(X \text{ levels})$, $^4I_{11/2}(Y \text{ levels})$, and $^4I_{9/2}(Z \text{ levels})$ are split, respectively, into two, seven, six, and five doubly degenerate levels. The energy levels for these terms are denoted, respectively, as (R_1, R_2), $(X_1, X_2, \ldots X_6, X_7)$, $(Y_1, Y_2, \ldots Y_5, Y_6)$, and $(Z_1, Z_2, \ldots Z_4, Z_5)$, with the value of subscript increasing as energy increases.

The transitions from $^4F_{3/2}$ to $^4I_{13/2}, ^4I_{11/2}, ^4I_{9/2}$ give rise to the emissions at 1.32, 1.06, and 0.94 μm, respectively. There are five doubly degenerate emission lines for $^4F_{3/2} \rightarrow {^4I_{9/2}}$ transitions, six lines for $^4F_{3/2}, \rightarrow {^4I_{11/2}}$ transitions, and seven lines for $^4F_{3/2} \rightarrow {^4I_{13/2}}$ transitions. The center wavelength, half linewidth, α, emission cross section, σ_e, and branching ratio, β, of all observed fluorescence lines at room temperatures were tabulated in Table I of [A.12]; this Table listing the potential

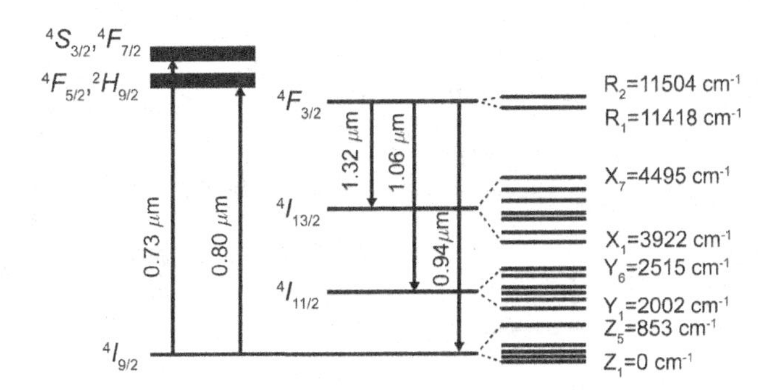

Fig. A.3 Relevant energy levels of Nd:YAG for laser operations [A.3]. The pump bands, and the energy levels of Nd^{3+} under crystal fields of Al_2O_3 for lasing transitions are given in cm^{-1} and grouped into emissions around 1.32, 1.06, and 0.94 μm.

laser transitions in YAG is highly relevant to new lidar development and mentioned in Section 7.5.3.

A.2.5 Partial Energy Levels of Ruby and Alexandrite

The host material of the first laser made to operate in 1960 by T. H. Maiman was ruby, in which some of the Al^{3+} ions in Al_2O_3 (corundum) are replaced by Cr^{3+} ions (about 0.1% Cr_2O_3 by weight). The ground configuration of a chromium (Cr) atom with its 24 electrons consists of the filled first two shells of 18 electrons plus $(3s)^2(3p)^6(3d)^5(4s)$. To form Cr^{3+} in Cr_2O_3, the $(4s)$ electron and two $(3d)$ electrons in the Cr atom are freed. Thus, the configuration for the ground term of Cr^{3+} is $(3d)^3$, following Table A.1 in Appendix A of [A.5], leading to allowed terms of ${}^2P$, ${}^4P$, 2^2D, ${}^2F$, ${}^4F$, ${}^2G$, ${}^2H$ with possible total electron spin of $S = 3/2$, or $1/2$; the quadruplet ${}^4F$ has the lowest energy per Hund's rules (a) and (b). Since the wavefunction of the $(3d)$ orbital extends out much farther than the $(4f)$ electrons, the crystalline field has a bigger effect on the energy of Cr^{3+} than its own spin-orbit interaction. As a result, the energy levels of ruby are labeled by the irreducible representations of its crystalline point group, A_1 and A_2 (singlet), E (doublet), and T_1 and T_2 (triplet). The ground term quadruplet ${}^4F$ of the Cr^{3+} ion splits into ${}^4A_2 \oplus {}^4T_1 \oplus {}^4T_2$, where the superscript indicates its spin degeneracy. The lowest doublet term of the free Cr^{3+} ion is ${}^2G$, which splits into ${}^2A_1 \oplus {}^2E \oplus {}^2T_1 \oplus {}^2T_2$. A simplified energy-level diagram relevant to ruby laser transitions is shown in the left panel of Fig. A.4. Here, the ground state, ${}^4A_2$, and the blue/green absorption bands, ${}^4T_1$ and ${}^4T_2$, are mainly derived from the quadruplet ${}^4F$ term. In the literature these energy bands are sometimes called ${}^4F_1$ and ${}^4F_2$ bands. The doublet ${}^2E$

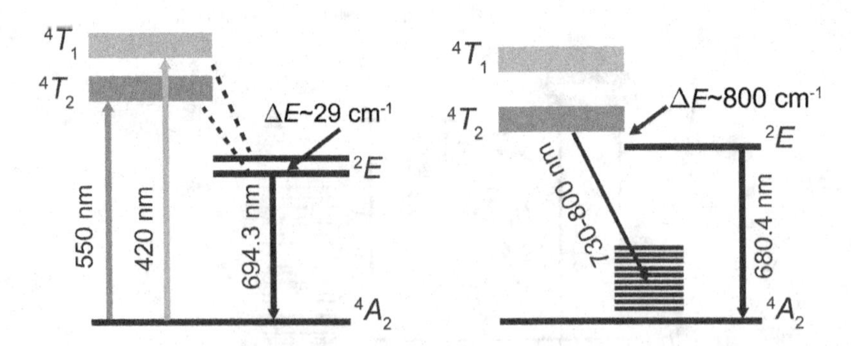

Fig. A.4 Simplified energy diagram of Cr^{3+} in Al_2O_3 (ruby), left panel, and in $BeAl_2O_4$ (alexandrite), right panel [A.3]. Rapid non-radiative transitions between the ${}^4T_1$ and ${}^4T_2$ bands to the metastable state ${}^2E$ (indicated by dashed arrows on the left) exist in both cases. Due to the close coupling between the ${}^2T_2$ and ${}^2E$ states (~800 cm^{-1}) in alexandrite, the vibronic states of the ${}^4T_2 - {}^2E$ manifold are populated, and their transition to the empty vibrational levels of the ${}^4A_2$ state makes alexandrite a phonon-terminated tunable laser.

term is derived from the ^2G term and splits into two levels via spin-orbit interaction, historically denoted as $2A$ and $\overline{E}$ with $2A$ higher by ~29 cm^{-1} (about 20 K); they are, respectively, the upper states of the R_2 and R_1 laser emissions. These states are metastable because the electric dipole transition between quadruplet and doublet states is forbidden. Since the ^2E state is easily populated by rapid non-radiative transitions from the $^4T_1$ and $^4T_2$ bands, in spite of it being a 3-level laser, it is not difficult to achieve population inversion between the ^2E and $^4A_2$ states in ruby. We present the energy levels of the ruby laser, not due to its status as the first laser ever made, but rather because its energy structure is tightly related to that of the solid-state tunable alexandrite ($Cr{:}BeAl_2O_4$), which is an important laser for potassium and iron fluorescence lidars.

Alexandrite is a crystal of $BeAl_2O_4$ with some of the Al^{3+} ions replaced by Cr^{3+}ions (~ 0.1% by weight). Its basic energy structure is similar to that of ruby. Since the equilibrium distance between Cr^{3+} and the neighboring O^{2-} ions is larger in alexandrite than in ruby, the crystalline field is weaker, which pushes the ^2E level higher and the $^4T_2$ level lower as shown in right panel of Fig. A.4. Their separation is only about $\Delta E \sim 800$ cm^{-1} with potential curves partially overlapping at internuclear distance slightly larger than R_e. The presence of the long-lived ^2E level nearby serves as a reservoir for and increases the lifetime of the vibronic states in $^4T_2$. Thus, in addition to the R_1-line transition (like the emission in ruby but at 680.4 nm), the transitions from the longer-lived vibronic states in $^4T_2$ to the empty vibrational levels of the $^4A_2$ state make alexandrite a preferred 4-level phonon-terminated laser, tunable between 730 and 800 nm. The D_1 transition for potassium at 770 nm falls within this region, as do 744 and 772 nm, whose second harmonics at 372 nm and 386 nm are appropriate for a Fe fluorescence wind/temperature lidar.

A.3 Structure of Linear Molecules

In a molecule, the motion of nuclei is negligible in comparison with that of the valence electrons. This makes it possible to take the nuclei as frozen in a fixed configuration when electronic energy and wave functions are considered. For a diatomic molecule in an electronic state $|n\rangle$, the average motion of the valence electrons yields an effective potential energy $U_{n,\Lambda}(R)$, a function of the internuclear separation, R, under which the two nuclei move relative to each other. This function $U_{n,\Lambda}(R)$ is regarded as the potential curve for molecular motion; for a bound state, $U_{n,\Lambda}(R)$ has a minimum at the equilibrium separation, R_e. Another significant difference between atoms and molecules is that the (valence) electrons no longer move in a centrosymmetric potential, thus the total orbital angular momentum L of the electrons is no longer conserved. However, for a linear molecule, the absolute value of the projection of the orbital angular momentum along the symmetry axis of the molecule (called the figure axis), Λ, is conserved; we can use it to classify the electron terms of the molecule with the Greek letters, $\Sigma, \Pi, \Delta, \ldots$ for $\Lambda = 0, 1, 2, 3, \ldots$, respectively. Each (electronic) state of the molecule is also characterized by its total spin, S, giving $2S+1$ multiplicity for each term, written for example as ${}^{2S+1}\Sigma$, ${}^{2S+1}\Pi$, ${}^{2S+1}\Delta, \ldots$.

Due to the axial symmetry of a linear molecule, the Hamiltonian, $\mathscr{H}_0(r, R)$, commutes with the axial projection of the orbital angular momentum Λ. The electronic state energy is then unchanged upon reflection of any plane passing through the figure axis; see p. 301 [A.13]. However, reflection about this axis changes the sign of the angular momentum (an axial vector) on this axis, leading to a double degeneracy if $\Lambda \neq 0$. If $\Lambda = 0$, the sign-changing degeneracy does not exist; in that case, the wave function of a Σ term after a reflection can only be changed by a constant multiplier. Notwithstanding, since a double reflection on the same plane returns the wavefunction to its original state, this constant must be ± 1; we need to distinguish the wave function unaltered by reflection from that which changes sign upon reflection by denoting them, respectively, by a plus superscript, Σ^+, or a minus superscript, Σ^-. For the linear molecules with inversion symmetry, such as N_2, O_2, and CO_2, the angular momentum is invariant with respect to inversion, so we classify terms with a given value of Λ into those with even (gerade) parity by a right subscript g and those with odd (ungerade) parity by a u, such as $\Sigma_g{}^+$, Π_g or $\Sigma_u{}^+$, Π_u. When two electrons, one from each atom, are shared to form a molecular bond, their spins may be parallel (triplet) or antiparallel (singlet), as indicated on the upper left of the term symbol, and the associated electronic orbital functions will be symmetric or antisymmetric upon exchange, respectively, giving rise to bonding or antibonding molecular orbitals. We state that the ground electronic terms of N_2, O_2, and CO_2 are respectively ${}^1\Sigma_g{}^+$, ${}^3\Sigma_g{}^-$, and ${}^1\Sigma_g{}^+$. In the following section, we describe molecular orbitals briefly and explain the meaning of molecular term symbols, and why ${}^1\Sigma_g{}^+$ is the ground electronic term of N_2 while ${}^3\Sigma_g{}^-$ is the ground term of O_2.

A.3.2 Molecular Orbitals and the Ground Terms of Nitrogen and Oxygen Molecules

Nitrogen and oxygen are both homogeneous diatomic molecules. The former consists of two nitrogen atoms each with seven electrons in the ground-state configuration of $(1s)^2(2s)^2(2p)^3$, and the latter two oxygen atoms each with eight electrons in the configuration of $(1s)^2(2s)^2(2p)^4$. Two electrons in $(1s)^2$ of each atom are core electrons, which have nothing to do with the formation of a molecule. The electrons in $(2s)$ and $(2p)$ atomic orbitals are valence electrons, so they may be shared by the two atoms that form a molecule. To first order of approximation, we may assume these electrons in the orbitals $(2s)$, $(2p_x)$, $(2p_y)$, and $(2p_z)$ remain independent of one another. The valence electrons in each molecular orbital can move under the influence of both nuclei and form molecular states. There are two possible molecular states formed by the valence electrons of the same type of atomic orbital (AO): one with parallel electron spins (triplet) and the other with antiparallel spins (singlet). Their corresponding molecular orbitals (MO) may be constructed by the linear superposition of AOs of A and B atoms as

$$\Psi_{\pm}(r_A, r_B) = \frac{1}{\sqrt{2}}\left[\Phi_{AO}(r_A, R) \pm \Phi_{AO}(r_B, R)\right], \tag{A.19}$$

with r_A and r_B being the locations of the electron relative to the nuclei of atoms A and B, respectively, and R the internuclear distance. The two electrons in the symmetric MO (+ combination, constructive interference) are more likely to be found between the nuclei and experience attractive forces from both; they are bonding electrons having antiparallel spins, according to the Pauli Exclusion Principle, and form a singlet bonding MO. The two electrons in the antisymmetric molecular state (combination) are more likely to be outside the internuclear region; they are nonbonding (or lone) electrons, forming a triplet antibonding MO with higher energy. Therefore, the atomic orbitals ($2s$), ($2p_x$), and ($2p_y$) of each atom superimpose to form bonding MOs, σ_{2s}, π_{2p_x}, and π_{2p_y} and antibonding MOs, σ^*_{2s}, $\pi^*_{2p_x}$, and $\pi^*_{2p_y}$; the two ($2s$) orbitals approach each other end to end to form sigma bonds, and the ($2p_x$) and ($2p_y$) orbitals approach each other sideways to form pi bonds. Since we choose the z-axis to be the figure axis of the molecule, the ($2p_z$) orbitals also approach each other end to end, thus forming sigma bonds, σ_{2p_z} and $\sigma^*_{2p_z}$. The two orbitals constructed from ($2s$) electrons, σ_{2s} below σ^*_{2s}, have lower energies. The molecular orbitals constructed from three ($2p$) AOs with the axial projection of total molecular orbital angular momentum $\Lambda = 0$, as shown in Fig. A.5, have higher energies. From their electron distributions, one can see that the bonding orbitals, σ_{2p_z}, π_{2p_x} and π_{2p_y} have lower energy than the antibonding orbitals, $\pi^*_{2p_x}$ and $\pi^*_{2p_y}$, and $\sigma^*_{2p_z}$, but whether σ_{2p_z} or π_{2p_x} and π_{2p_y} have lower energy is a tough call. As we have pointed out, all the bonding orbitals are singlet and antibonding orbitals are triplet. Since the reflection about a plane containing the figure axis preserves the σ_{2p_z} and $\sigma^*_{2p_z}$ orbitals and changes the sign of π_{2p_x}, π_{2p_y}, $\pi^*_{2p_x}$, and $\pi^*_{2p_y}$ orbitals, we could add a plus sign to the right superscript of the former and a minus sign to that of the latter. In addition, inversion of a homogeneous diatomic molecule preserves σ_{2p_z}, $\pi^*_{2p_x}$ and $\pi^*_{2p_y}$ orbitals as they have even (gerade) parity, while it changes the sign of π_{2p_x} π_{2p_y}, and $\sigma^*_{2p_z}$ as they have odd (ungerade) parity.

Armed with this basic understanding of molecular orbital theory, we can now construct the ground electronic state (or electronic term) of the oxygen molecule. An oxygen molecule has 12 valence electrons; we fill them into the molecular orbitals as shown in the right panel of Fig. A.6. Here, the individual spins are

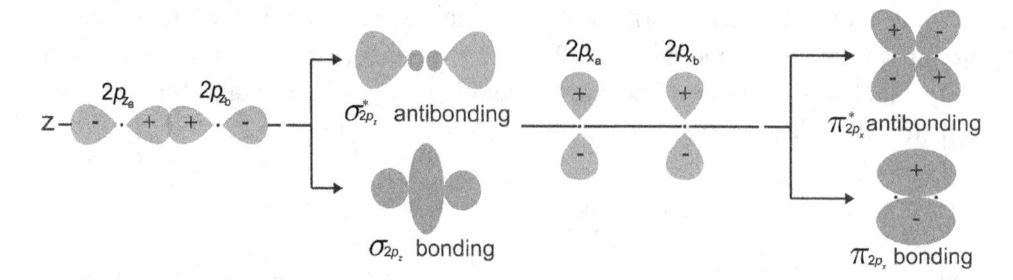

Fig. A.5 The molecular orbitals formed by atomic orbitals (AO) ($2p_x$), ($2p_y$) and ($2p_z$). Since the z-axis is chosen to be the figure axis of the molecule, the electrons of the ($2p_z$) AOs approach one another head-on, and they form sigma orbitals, $\sigma_{2p_z} - {}^1\Sigma_g^+$ and $\sigma^*_{2p_z} - {}^3\Sigma_u^+$. The electrons of the ($2p_x$) AOs form $\pi_{2p_x} - \Sigma_u^-$ and $\pi^*_{2p_x} - {}^3\Sigma_g^-$ MOs.

marked by arrows (up or down). Since the last two electrons are unpaired, respectively occupying $\pi^*_{2p_x}$ and $\pi^*_{2p_y}$ orbitals, oxygen is paramagnetic, and its ground electronic term may be designated as ${}^3\Sigma_g^-$ – see the antibonding orbital on the top right of Fig. A.5. For the nitrogen molecule, there are 10 valence electrons; if its molecular orbital (MO) energy diagram were the same, the last filled orbitals would be π_{2p_x} and π_{2p_y}, leading to the ground electronic term ${}^1\Sigma_u^-$. This is, however, incorrect because in nitrogen, the energy of the orbital σ_{2p_z} is higher than that of π_{2p_x} and π_{2p_y}, leading to the filling of the molecular orbitals as shown in the left panel of Fig. A.6. Ultimately, this leads to the filling of the last two electrons into the σ_{2p_z} MO, whose ground electronic term is designated as ${}^1\Sigma_g^+$ – see the bonding orbital on the bottom of the second column from the left of Fig. A.5. Therefore, ${}^1\Sigma_g{}^+$ is the ground electronic term of N_2, while ${}^3\Sigma_g^-$ is the ground term of O_2.

Like the constituent atomic orbitals ($2p$) and ($2s$), the MOs formed by them remain orthogonal. As shown in the right panel of Fig. A.6, the MO energy increases as σ_{2p_z}, π_{2p_x} and π_{2p_y}, $\pi^*_{2p_x}$ and $\pi^*_{2p_y}$, to $\sigma^*_{2p_z}$. However, this is not the case for nitrogen, because as the ($2p$) AO is filling up from boron $[(2s)^2(2p)]$ to fluorine $[(2s)^2(2p)^5]$, the interaction between the ($2s$) AO of one atom and the ($2p$) AO of the other atom – termed $2s$-$2p$ interaction – remains significant to a varied degree. As more electrons occupy the ($2p$) orbital, the energy difference between ($2s$) and ($2p$) AOs in the constituent atoms increases and the $2s$-$2p$ interaction (between two atoms) decreases, since the $2s$-$2p$ interaction will mix the two MOs with even parity, σ_{2s} and σ_{2p_z}, and push the energy of the higher level σ_{2p_z} higher and the lower level σ_{2s} lower. As the $2s$-$2p$ interaction decreases from N_2 to O_2, the reverse happens. The result is that while the level of σ_{2p_z} is below those of π_{2p_x} and π_{2p_y} in oxygen, as shown in the right panel of Fig. A.6, it is above in nitrogen as shown in the left panel of Fig. A.6. The effect on MOs with odd parity, σ^*_{2s} and $\sigma^*_{2p_z}$, is similar, but much smaller in magnitude. The two pairs of degenerate molecular orbitals, π_{2p_x} and π_{2p_y} and $\pi^*_{2p_z}$ and $\pi^*_{2p_z}$, may be considered

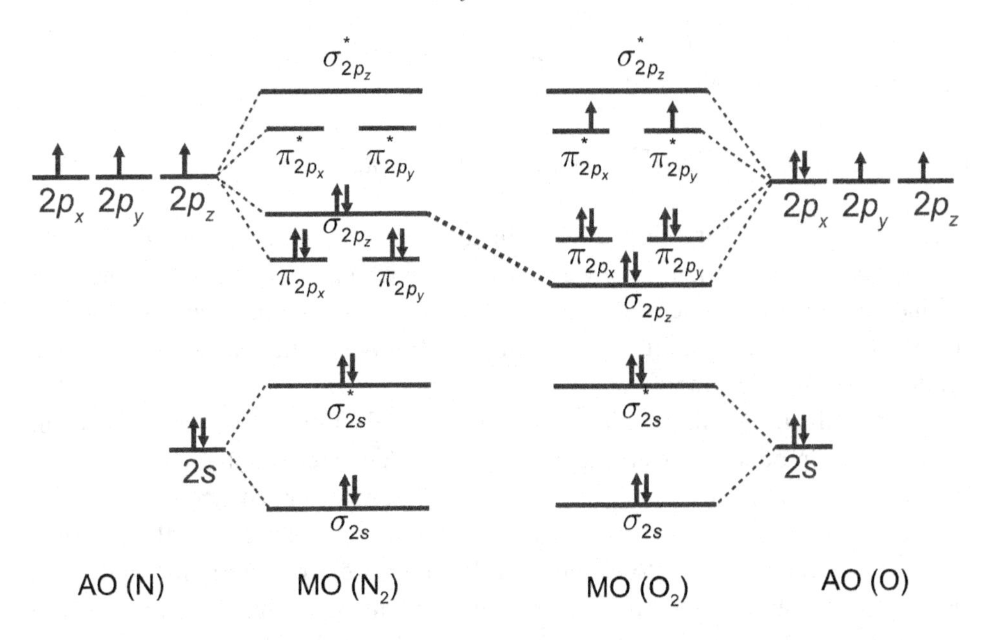

Fig. A.6 The formation of molecular orbitals from their constituent atomic orbitals (AO). The left panel for nitrogen and the right panel for oxygen. The heavy dashed line shows how the 2*s*-2*p* interaction pushes the σ_{2p_z} orbital to a higher level for nitrogen.

independent from the (2*s*) AOs. A more detailed treatment on the formation of MOs from the mixing (2*s*)-(2*p*) AOs can be found in Chapter 8 of Atkins and Friedman [A.14]. To account for their Fig. 8.18, on which our Fig. A.6 is based, the authors state: "It is hard to predict the order of energy levels, particularly the relative ordering of the σ and π sets, but it is found experimentally and confirmed by more detailed calculations that the order shown on the left of Fig. A.6 applies from Li_2 to N_2, whereas the order shown on the right applies to O_2 and F_2."

A.3.2 Vibrational and Rotational Motion of a Molecule

Once the electronic state (term) of a molecule is determined, the motion of a linear molecule in its ground electronic state $|e^g>$ may be described by a hierarchy of mechanical models. We give a brief account of the development in Chapter III of Herzberg [A.15] here. The simplest model is that of a rigid rotator. It disregards the motion of electrons and models the bond by a massless rod that connects the nuclei. Or, equivalently, the motion of the electron around the nuclei is such as to produce a massless spring with infinite stiffness, so that the (equilibrium) internuclear distance R_e does not change as the molecule rotates. The rigid rotator can only rotate around an axis through its center of mass and perpendicular to the internuclear axis (called the figure axis) of the molecule with angular momentum (of the nuclear frame) $\vec{N}$. Its rotational energy E_{rot}, as given in (III.7) p. 68 of [A.15], is the centrifugal energy of the rotation:

$$E_{rot} = hcB(R_e)\left(\vec{N}\bullet\vec{N}\right) = hcB(R_e)(N+1);\ B(R_e) = \frac{h}{8\pi^2 cI_\perp}\,. \tag{A.20.a}$$

Here, $h, c,$ and $B(R_e)$ are, respectively, Planck's constant, the speed of light, and the rotational coefficient (in units of wavenumber), in which $I_\perp$ is the rotator's moment of inertia. For a diatomic molecule, the constant R_e is the distance between two nuclei; the moment of inertia is $I_\perp = \mu R_e^2$, with μ being the reduced mass of the molecule (or nuclear frame).

More realistically, the motion of the valence electrons around the nuclei in $|e^g\rangle = |n, \Lambda\rangle$ produces a restoring potential $U_{n,\Lambda}(R)$ under which the nuclei move relative to each other to maintain an equilibrium internuclear distance R_e. The effect of the electronic motion then produces a massless spring that tries to pull a mass μ in a one-dimensional motion to counter the deviation from equilibrium $\xi = R - R_e$. For small deviation ξ, the relative motion is harmonic with vibrational energy $(1/2)\mu\omega_e^2\xi^2$, where $\omega_e = 2\pi\nu_e = \sqrt{k/\mu}$ with ω_e (ν_e) being the angular frequency of a simple harmonic oscillator. This leads to equal-distance quantized eigenstates described by the vibrational quantum number, v = 1, 2,3, ... As ξ increases, the vibration becomes anharmonic. In the anharmonic oscillator model, its energy E_{vib} may be obtained by expanding in a power series of ξ; this is given in (III.73), p. 93 of [A.15] as

$$E_{vib} = hc\nu_e(\mathrm{v}+1/2) - hc\nu_e x_e(\mathrm{v}+1/2)^2 + hc\nu_e y_e(\mathrm{v}+1/2)^3 + \ldots\,. \tag{A.20.b}$$

Here, the constants are derived from the second, third, and fourth derivatives of the potential function $U_{n,\Lambda}(R)$ with $\nu_e \gg \nu_e x_e \gg \nu_e y_e \gg \ldots$ representing harmonic and successively higher orders of anharmonicity. From Table 39 of [A.15], the values for $\nu_e, \nu_e x_e,$ and $\nu_e y_e$ are 2360, 14.5, and 0.0075 cm^{-1} for N_2 and 1580, 12.1, and 0.055 cm^{-1} for O_2.

A linear molecule also rotates in the space-fixed coordinates with respect to its center of mass. To allow vibration between nuclei, we model this rotational motion by a nonrigid rotator. In this case, the restoring force of the effective spring will act to balance the centrifugal force due to molecular rotation, settling down to a new equilibrium internuclear distance R_c, which is slightly larger than R_e. R_c depends on the angular momentum of molecular rotation $\vec{N}$. In other words, the new equilibrium distance R_c, or moment of inertia, increases (rotational energy decreases) as the rotational quantum number N increases. It can be shown that the first-order correction to R_e, $(R_c - R_e)$ is proportional to R_e^{-3}, corresponding to a correction to rotational energy term proportional to R_e^{-6}. The anharmonicity in rotational energy may be given in a series expansion as in (A.20.c), following (III.110), p. 104 of [A.15]:

$$E_{rot} = hcB_eN(N+1) - hcD_eN^2(N+1)^2 + \ldots;\ B_e = \frac{h}{8\pi^2\mu cR_e^2};\ D_e = \frac{h^3}{32\pi^4\mu^2 cR_e^6}\ . \tag{A.20.c}$$

Here, the rotational constants are in terms of equilibrium distance, R_e, and reduced mass, μ. The effect of molecular vibration will change the mean values of $<R^{-2}>$ and $<R^{-6}>$ from R_e^{-2} and R_e^{-6}, making the rotational constants dependent on the vibrational quantum number. The leading correction term results from harmonic oscillation and is proportional to the mean value of ξ^2, or to $(\mathrm{v} + 1/2)$, effectively replacing B_e and D_e with B_v and D_v, due to rotation-vibration coupling. The combined effects yield the vibrating rotator model with energy given in (A.20.d):

$$\begin{aligned} E_{vib} + E_{rot} &= hc\nu_e(\mathrm{v}+1/2) - hc\nu_e x_e(\mathrm{v}+1/2)^2 + \ldots \\ &+hcB_\mathrm{v}N(N+1) - hcD_\mathrm{v}N^2(N+1)^2 + \ldots, \text{where} \\ B_\mathrm{v} = B_e - \alpha_e\,(\mathrm{v}+1/2) &+ \ldots;\ D_\mathrm{v} = D_e + \beta_e(\mathrm{v}+1/2) - \ldots\ . \end{aligned} \tag{A.20.d}$$

A.3.3 Vibrational Modes, and Infrared and Raman Activity of Linear Molecules

It is well known that for linear molecules, the number of normal vibrational modes (vibrations that preserve the center of mass) is $3n - 5$, with n being the number of nuclei in the molecule, equaling the total number of degrees of freedom minus three parameters that would define the displacement of the center of mass and three angles (two for a linear molecule) that define rotation of the molecule with respect to the center of mass. For the linear molecules of interest, diatomic N_2 and O_2 have only one normal mode, while for CO_2 there are four. The vibrational normal modes for these linear molecules are shown pictorially in Fig. A.7. The vibrational frequencies for N_2 and O_2 from Table 4.2 and those for CO_2 from [A.16] are given respectively in the figure caption.

As pointed out, the spring that connects the nuclei is formed by the motion of the electrons in a stationary electronic state. Thus, the electronic polarizability of the molecule may be altered by the relative motion of the nuclei. Since each nuclear core may possess different charge, the vibration of nuclei collectively may also alter the (or induce an) electric dipole moment of the molecule. By visual observation, it is obvious that the symmetric stretching (ν_1) of a N_2, O_2, or CO_2 molecule, which has Σ_g^+ symmetry, alters its electronic polarizability, but it does not induce a dipole moment; therefore, the symmetric stretch mode of these molecules is Raman but

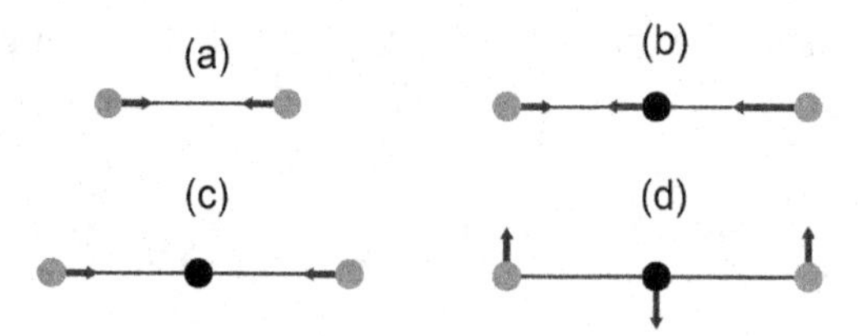

Fig. A.7 Vibrational normal modes of symmetric linear molecules, (a) symmetric stretch of N_2 or O_2, and for CO_2 (b) asymmetric stretch, (c) symmetric stretch, and (d) bending (doubly degenerate). The vibration frequency of N_2, O_2, and CO_2 are, respectively, 2331 cm^{-1}, 1556 cm^{-1}, and 1340 cm^{-1} (symmetric stretch, ν_1), 667 cm^{-1} (CO_2 bending, ν_2) and 2349 cm^{-1} (CO_2 antisymmetric stretch, ν_3) .

not infrared active. On the other hand, either bending (ν_2) or asymmetric stretching (ν_3) of a CO_2 molecule, both exhibiting Π_u symmetry, do not alter its electronic polarizability; rather, they induce an electric dipole moment; they are infrared active and Raman inactive. However, their overtones ($2\nu_2$ and $2\nu_3$), having $\Pi_u \otimes \Pi_u = \Sigma_g^+ + \Delta_g$ symmetry, are Raman active. When the frequencies of an overtone and a Raman active mode are close, the weak overtone can siphon intensity from the strong fundamental, and both can be observed by Raman scattering with frequency slightly shifted away from each other via Fermi resonance interaction. The combination of the ν_1 fundamental at 2340 cm^{-1} and $2\nu_2$ at $2\times667=1334$ cm^{-1} of CO_2 presents a good example; the observed Raman shifts become 2388.2 cm^{-1} and 1285.4 cm^{-1}, as given in Table 1 of [A.17].

A.3.4 Linear Molecules as a Nonrigid Symmetric Top

The nuclei in the molecule are, of course, not connected by massless springs; they are connected by the motions of electrons yielding a binding potential energy $U_{n,\Lambda}(R)$, under which the nuclei can move relative to one another, leading to vibrational modes as mentioned. The much faster motions of the electrons yield a small moment of additional inertia to molecular rotation with respect to the figure axis, $I_{\|}$; meanwhile, their contribution to $I_{\perp}$ is negligible in comparison to that of the nuclear frame. This contributes to the model of a (nonrigid) symmetric top; see p. 115 of [A.15]. Since $I_{\|} \ll I_{\perp}$, we also call it a symmetric prolate top. With the addition of the minute electronic contribution, $I_{\perp}$ is slightly larger than the moment of inertia of the nuclear frame. We consider a linear molecule in the molecular coordinates (X, Y, Z), with Z being the figure axis. The electrons are in the state $|n, \Lambda>$, where n is the principal electronic quantum number, and $\hbar\Lambda$ is the projection of electronic orbital angular momentum on the figure axis Z. In this case, R is the internuclear distance, and the relative motion between nuclei along the Z axis amounts to a change of internuclear distance R from its equilibrium value R_e, yielding a vibration amplitude, $\xi = R - R_e$, with associated potential energy $U_{n,\Lambda}(R)$. The total energy of the molecule is then

$$E = E_{ele,\text{vib}} + E_{rot} = U_{n,\Lambda}(R) + hcB_\perp(J_X^2 + J_Y^2) + hcB_\parallel J_Z^2;$$
$$B_\perp = \frac{h}{4\pi^2 cI_\perp} \text{ and } B_\parallel = \frac{h}{4\pi^2 cI_\parallel} \quad \text{(A.21.a)}$$

where $E_{ele,\text{vib}} = U_{n,\Lambda}(R)$ is the electronic and vibrational energy of the molecule, and $B_\perp$ and $B_\parallel$ are rotational constants, which strictly speaking are functions of R and thus dependent on the vibrational coordinate, ξ. The total angular momentum of the molecule is $\hbar\vec{J}$, which includes the contribution from the rotations of the nuclear frame and of electrons in the state $|n, \Lambda>$ (i.e., $\vec{J} = \vec{L} + \vec{N}$ or its component on the figure axis, $J_Z = \Lambda$). In other words, J_X and J_Y are mainly the result of rotation of the nuclear frame, and $J_Z = \Lambda$ is the Z-projection of the electronic angular momentum of electrons in the state $|n, \Lambda>$. Expanding $U_{n,\Lambda}(R)$ in a power series of $\xi = R - R_e$ and including its effect on the nonrigid rotator, we generalize the model of the vibrating rotator given in (A.20.d) into the rotational and vibrational motion of a symmetric top in the electronic state $|n, \Lambda>$, and we express its energy up to the leading anharmonic term as

$$E = E_{n,\Lambda} + \hbar\omega_e(\text{v} + \frac{1}{2}) - x_e\hbar\omega_e(\text{v} + \frac{1}{2})^2 + hcB_\text{v} J(J+1)$$
$$-hcD_\text{v} J^2(J+1)^2 + hc\left(B_\parallel - B_\perp\right)\Lambda^2 . \quad \text{(A.21.b)}$$

The last term in (A.21.b) is independent of J and is zero for the ground electronic state of N_2, O_2, and CO_2. In fact the last term and part of the second and third terms are independent of vibrational and rotational quantum numbers. Because of this mutual independence, we are able to lump these terms together into the electronic energy term. If in addition, we collect the terms proportional to the vibrational quantum number v in the second and third terms of (A.21.b), we may express the energy of $|n\Lambda, \text{v}, J>$, $T_{n\Lambda,\text{v},J}$, using (A.21.c) in units of wave numbers:

$$T_{n\Lambda,\text{v},J} = E_{n\Lambda,\text{v},J}/hc = T_{n,\Lambda} + \nu_0(\text{v}) - x_e\nu_e(\text{v}^2) + hcB_\text{v}J(J+1) - hcD_\text{v}J^2(J+1)^2 ,$$
$$\text{with } \nu_0 = (1 - x_e)\nu_e,\ B_\text{v} = B_e - \alpha_e(\text{v} + 1/2);\ D_\text{v} = D_e + \beta_e(\text{v} + 1/2) . \quad \text{(A.21.c)}$$

where $T_{n,\Lambda}$ is the electronic term, or electronic energy in units of cm^{-1}. The values for the above constants, $T_{n,\Lambda}$, ν_e, $\nu_e x_e$, B_e, α_e, D_e, have been tabulated in Table 39 of [A15] in units of cm^{-1} for most diatomic molecules. For the ground and first excited states of O_2 and N_2, they are reproduced in Table A.4; for details, see Table

Table A.4 *Vibrational and rotational constants of the ground and first excited electronic states of* O_2 *and* N_2 *(in units of* cm^{-1}*).*

Mol/State	$T_{n\Lambda}$	ν_e	B_e	α_e	$\nu_e x_e$	R_e (Å)	Comment
$O_2/a^1\Delta_g$	7,918.1	1,509	1.426	0.0171	12	1.216	Excited state
$O_2/X^3\Sigma^-_g$	0.0	1,580	1.446	0.0158	12.073	1.207	Ground state
$N_2/a^1\Pi_g$	69,290	1,692	1.637	0.022	12.791	1.213	Excited state
$N_2/X^1\Sigma^+_g$	0.0	2,360	2.010	0.0187	14.456	1.094	Ground state

39 of [A.15]. Note that the calculated vibrational energies ν_0, 2,345.5 cm^{-1} for N_2 and 1,567.9 cm^{-1} for O_2 in the ground electronic state, are somewhat different from those measured. When compared to the associated rotational transition B_0, about 2 cm^{-1} for N_2 and 1.5 cm^{-1} for O_2 and compared to the electronic transition of about 70,000 cm^{-1} and 8,000 cm^{-1}, respectively, the spectra resulting from electronic, molecular vibrational, and rotational motions can clearly be very different, with their spectral features well separated.

There is only one vibrational mode in a diatomic molecule. In the case of a linear molecule with more than two atoms, like CO_2, the vibrational energy is the sum over all vibrational modes. Ignoring the spin-axis interaction [A.15] of electrons (analogous to the spin-orbit interaction in atoms), the electron spins do not appear in (A.21), as they do not affect the dynamics in this approximation. The value of the total electron spin, however, leads to a (2*S*+1) degeneracy of vibration and rotation states. The spin-axis interaction is zero for the ground electronic state of the molecules of interest, such as N_2, O_2, and CO_2, since they correspond to the Σ (i.e., $\Lambda = 0$) terms. Since the anharmonic terms in (A.21.c) are small, they will be ignored in the quasi-harmonic approximation for this book, except that we keep two leading effects: (1) making the rotational constant dependent on vibrational quantum number (i.e., $B_1 \neq B_0$), and (2) modifying the vibrational frequency slightly, replacing ν_e by $\nu_0 = \nu_e(1 - x_e)$. Generally, the difference in rotational constants between the ground vibrational state, B_0, and the first excited vibrational state, $B_1 = B_0 - \alpha_e$, is the result of vibration-rotation coupling. It lifts the degeneracy of the Q-branch vibrational-rotational Raman scattering lines, a topic discussed in Section 4.3.4. For atmospheric Raman scattering, we address only nonresonant scattering (i.e., between rotational and vibrational states in the ground electronic state) for linear molecules of interest O_2, N_2, and CO_2. Their electronic term is Σ, and its vibrational and rotational energy expressed in units of wavenumber is

$$T_{eg,\mathrm{v},J} \equiv \frac{E_{eg,\mathrm{v},J}}{hc} = \nu_0\left(\mathrm{v} + \frac{1}{2}\right) + B_\mathrm{v} J(J+1);\ \nu_0 = \nu_e(1 - 0.5x_e),\ B_\mathrm{v} = B_0 - \alpha_e \mathrm{v}\,. \tag{A.21.d}$$

Here $T_{eg} = 0$ for the ground electronic term. The above expression is employed in Section 4.3.1 when we discuss Rayleigh and vibrational Raman scattering.

A.3.5 The Eigenstates of a Symmetric Top and Associated Integrals

The stationary state of a molecule may be conceptually deduced by first solving the time-independent Schrödinger equation (A.1.b) for electrons in a configuration of nuclei, giving rise to electronic terms. For a given electronic term, the nuclei may vibrate from their equilibrium positions in a normal mode oscillatory pattern with vibration quantum number, v. A linear molecule may also rotate with respect to the space-fixed coordinate system with a rotational wave function depending on the total angular momentum J ($\vec{J} = \vec{L} + \vec{N}$) of the molecule and its projections on the space-fixed z-axis and on the figure axis (Z-axis), denoted respectively by the quantum numbers, M and Λ. An eigenstate of the molecule may then be characterized by its eigenfunction $|\Psi_{n\Lambda,\mathrm{v},J\Lambda M}>$, and its associated eigen-energy $E_{n\Lambda,\mathrm{v},J\Lambda}$, (A.21.c) under the harmonic approximation becomes

$$|\Psi_{n\Lambda,\mathrm{v},J\Lambda M}> = |ele>|\mathrm{vib}>|rot>, \text{with } |ele> = |n,\Lambda>, |\mathrm{vib}> = |\mathrm{v}>, |rot> = |J,\Lambda,M>, \tag{A.22.a}$$

$$\text{and } E_{n\Lambda,\mathrm{v},J\Lambda} = E_{ele} + \hbar\omega_0\left(\mathrm{v} + \frac{1}{2}\right) + hcB_{\mathrm{v}}J(J+1) + hc\left(B_{e,\parallel} - B_{e,\perp}\right)\Lambda^2, \tag{A.22.b}$$

where $\omega_0 = \omega_e(1 - 0.5x_e)$, and $B_{\mathrm{v}} = B_{e,\perp} - \alpha_e(\mathrm{v} + 0.5) = B_0 - \alpha_e\mathrm{v}$. Since the electronic state (of the Λ term) assumes the equilibrium configuration of nuclei (R_e), vibration is the result of relative motion of nuclei from equilibrium positions along the figure axis, and rotations are motions without altering the figure axis (or the rotation of molecular coordinates, X-Y-Z with respect to space-fixed coordinates, x-y-z). The three modes, electronic + vibrational + rotational, of motion may be considered to be independent of one another, at least approximately. As such, the molecular wave function, according to the Born–Oppenheimer approximation, is the product of electronic, vibrational, and rotational components, $|ele>|\mathrm{vib}>|rot>$, as in (A.22.a). The rotational wave function $|rot>$ in a space-fixed coordinate results in the well-known spherical harmonics, $Y_{J,M}(\theta,\phi)$; see Section 2.5 of [A.2], where (θ,ϕ) are polar and azimuthal angles; see Fig. B.1. The principal quantum number of a rotational state is the total angular momentum J of the molecule. As explained earlier, its projection on the figure (molecular) axis is a conserved quantity, and it defines the electronic term Λ, which takes values of $\Lambda = 0$, 1, 2, 3, ...; it is doubly degenerate for $\Lambda \geq 1$ because there are two possible directions for any given Λ. Thus, we may distinguish the direction of projection by employing the quantum number $K = \pm\Lambda = 0, \pm1, \pm2, \ldots$. The total angular momentum can also be projected onto the z-axis of a space-fixed coordinate, yielding quantum number $M = 0, \pm1, \pm2, \ldots$. An alternative form for the eigenfunction of angular momentum, as discussed in Section 4.7 of [A.2], is the matrix element of the finite rotation operator, $D(\alpha,\beta,\gamma)$, as a function of Euler angles

(α, β, γ); see (B.4.a). In its irreducible representation, $D^J_{K,M}(\alpha,\ \beta,\ \gamma) \equiv D^J_{K,M}$, the eigenfunction of operator J^2 with eigenvalue of $J(J+1)$, and of J_z (in space-fixed coordinates) with eigenvalue M, and simultaneously, of J_Z (in molecular coordinates) with eigenvalue of K. This alternative presentation is preferred as the transformation from the space-fixed z-axis to the molecular Z-axis defines the rotation of x-y-z to X-Y-Z via Euler angles, (α, β, γ). In this view, the rotational quantum state then depends on three quantum numbers as $|rot> \ = \ |J, \Lambda, M> \equiv |J, K, M>$; it is proportional to the function $D^J_{K,M}(\alpha, \beta, \gamma) \equiv D^J_{K,M}$ as

$$|rot> \ = \ |J, \Lambda, M> \equiv |J, K, M> \ = \sqrt{\frac{2J+1}{8\pi^2}} D^J_{K,M}; <J, K, M| \ = \sqrt{\frac{2J+1}{8\pi^2}} D^J_{-K,-M}. \tag{A.22.c}$$

The corresponding normalization condition, as given in (4.6.1) of [A.2], is:

$$\frac{1}{8\pi^2} \iiint D^{j_1}_{-m_1',-m_1}(\alpha\beta\gamma) D^{j_2}_{m_2' m_2}(\alpha\beta\gamma)\ d\alpha\ \ \sin\beta\ d\beta\ d\gamma$$

$$= \delta_{j_1 j_2} \delta_{m_1' m_2'} \delta_{m_1 m_2} \frac{1}{2j_1+1} \quad \rightarrow <J^f K^f M^f \,|\, J^i K^i M^i> = \delta_{J^f J^i} \delta_{K^f K^i} \delta_{M^f M^i} \ .$$

Employing (4.6.1) and (4.3.2) of [A.2], the integral of the product of three D's may be expressed as the matrix element of one D invoking (A.22.c), resulting in the product of two 3-j coefficients, as given in (4.6.2) of [A.2]. This, in turn, may be cast as (A.22.d), the formula used for the computation of Placzek–Teller coefficients in Chapter 4:

$$\frac{<J^f K^f M^f | D^{(j)}_{m'm} | J^i K^i M^i>}{\sqrt{(2J^i+1)(2J^f+1)}} = \begin{pmatrix} J^f & j & J^i \\ -M^f & m & M^i \end{pmatrix} \begin{pmatrix} J^f & j & J^i \\ -K^f & m' & K^i \end{pmatrix}. \tag{A.22.d}$$

Another formula we use for the computation of Placzek–Teller coefficients is summarized here. We recall the definition of the 3-j coefficient in terms of Clebsch–Gordon coefficients as given, for example, in (3.7.3) of [A.2], in terms of the eigenstates of angular momentum addition, $\vec{j_1} + \vec{j_2} = \vec{j_3}$, leading to the normalization condition of the eigenstate $|j_1 j_2 j_3 - m_3>$:

$$\begin{pmatrix} j_1 & j_2 & j_3 \\ m_1 & m_2 & m_3 \end{pmatrix} = (-1)^{j_1 - j_2 - m_2} (2j_3+1)^{-0.5} <j_1 m_1 j_2 m_2 | j_1 j_2 j_3 - m_3>$$

$$\rightarrow \sum_{m_1, m_2} \begin{pmatrix} j_1 & j_2 & j_3 \\ m_1 & m_2 & m_3 \end{pmatrix}^2 = \frac{1}{2j_3+1} \sum_{m_1, m_2} \left| <j_1 m_1 j_2 m_2 | j_1 j_2 j_3 - m_3> \right|^2 = \frac{1}{2j_3+1} \ .$$

Since the selection rule for the 3-j coefficient selects nonzero values of m_3 for given values of m_1 and m_2, and there are $2j_3 + 1$ values of m_3, we then get unity when we sum over m_3. Invoking the symmetry of the 3-j coefficients, we thus have these useful formulae:

$$\sum_{m_1,m_2} \begin{pmatrix} j_1 & j_2 & j_3 \\ m_1 & m_2 & m_3 \end{pmatrix}^2 = \frac{1}{2j_3+1}; \sum_{m_1,m_3} \begin{pmatrix} j_1 & j_2 & j_3 \\ m_1 & m_2 & m_3 \end{pmatrix}^2 = \frac{1}{2j_2+1};$$

$$\sum_{m_2,m_3} \begin{pmatrix} j_1 & j_2 & j_3 \\ m_1 & m_2 & m_3 \end{pmatrix}^2 = \frac{1}{2j_1+1} \rightarrow \sum_{m_1,m_2,m_3} \begin{pmatrix} j_1 & j_2 & j_3 \\ m_1 & m_2 & m_3 \end{pmatrix}^2 = 1 . \quad \text{(A.22.e)}$$

Since for nonzero values of the 3-j coefficient, one of the m-indices is selected by the other two, and thus one of the m-indices under the summation sign is often removed for simplicity. This causes confusion, since when two indices were given under the summation sign, it makes a difference whether the third m-index is given or summed over; the result is $(2j_k + 1)^{-1}$ for a given value of m_k (k is the index not summed over), and it is unity when summing over all three m-indices.

What is treated above should be valid for the ground electronic state of molecules of interest, N_2, O_2, and CO_2 with $K^f = K^i = 0$ for their electronic ground states in the Σ term. Taking the Z-axis to be the figure axis of a linear molecule, two rotations are sufficient to define an arbitrary finite rotation; in terms of the Euler angles, this corresponds to setting the Euler angles of Fig. B.2 to $\alpha = \Phi$, $\beta = \Theta$, and $\gamma = 0$.

A.4 A Brief Account of Thermodynamics and Transport of an Ideal Molecular Gas

To understand light scattering from an ensemble of molecules, we start from a description of the thermodynamics and transport properties of an ideal gas. Thus, we consider an ensemble (system) of identical spherical particles in a volume V at equilibrium, with temperature, T, and number density, $\mathcal{N}$. Such a system nicely represents an atomic ensemble. It can also represent the translational (center-of-mass) motions of a molecular ensemble free from rotation or vibration. Its equation of state (the Ideal Gas Law), which relates pressure p, temperature T, and number density, $\mathcal{N}$ (in m^{-3}) or n (in kmol^{-1}), is $p = \mathcal{N} k_B T$ or $p = nRT$, where $k_B = 1.38 \times 10^{-23}$ J K^{-1} and $R = 8.31 \times 10^3$ J K^{-1} kmol^{-1} are, respectively, the Boltzmann constant and the universal gas constant. The first law of thermodynamics states the increase of system's internal energy dU equals heat energy intake δQ (equals TdS for reversible processes) minus mechanical work done by the system $\delta W = pdV$, as given in (A.23.a). For a gas system, it is useful to define an additional state variable, the enthalpy $H = U + pV$, as the sum of the internal energy (energy required to create the system) and the product of the pressure p and volume V

(energy required to make room for the system and establishing its volume and pressure from its environment). Constrained to the first law, the change in enthalpy is related to the changes in (perturbations of) entropy and in pressure (i.e., dS and dp), as given in (A.23.b). See, for example, (2.17) and (2.18) of [A.18]:

$$dU = \delta Q - \delta W = TdS - pdV \text{ , and} \tag{A.23.a}$$

$$H = U + pV \rightarrow dH = TdS + Vdp. \tag{A.23.b}$$

Like enthalpy, it is convenient to express other state variables and thermodynamic parameters as a function of entropy and pressure. In thermal equilibrium, the probability distribution of an ideal gas ensemble is described by the well-known Boltzmann function, the ratio of the particle's kinetic energy, $0.5\ m\mathrm{v}^2$, to thermal energy, k_BT; here m and v are, respectively, the mass and speed of the particle. From the associated Maxwellian speed probability density function, we can determine three useful mean values; they are the most probable speed, $\mathrm{v}_0 = \sqrt{2k_BT/m}$, the average speed, $\overline{\mathrm{v}} = \sqrt{8k_BT/\pi m}$, and the root-mean-square speed, $\mathrm{v}_{rms} = \sqrt{3k_BT/m}$.

Since it is the collisions between molecules that produce perturbations from the mean, the mean free path, $\overline{\ell}$, between collisions (mean distance a particle can travel before encountering a collision), and the collision time (the mean time between collisions), which is the inverse of collision frequency λ_c, with $\overline{\ell} = \overline{\mathrm{v}}/\lambda_c$, may be related to the collision cross section σ and number density as $\overline{\ell} = \left(\sqrt{2}\mathcal{N}\sigma\right)^{-1}$. The effective particle radius for collision cross section is known to be twice that of physical particle radius, R, leading to $\sigma = \pi(2R)^2$. Following the common procedure from thermodynamics and statistical mechanics textbooks (e.g., Chapter 16 of Stowe [A.19]), we discuss the transport properties of a gas system by first considering the transport of a macroscopic quantity $Q(x)$ in one dimension, with the density (per unit volume) of this quantity $Q(x)$, $q(x)$ at $(x + \Delta x/2)$ and $(x - \Delta x/2)$ respectively being $q(x + \Delta x/2)$ and $q(x - \Delta x/2)$. We investigate the transport of $Q(x)$ carried by a group of gas particles moving in the x-direction with speed v into a box centered at x with unit cross-sectional area. Half of the particles at $(x - \Delta x/2)$ will move into the box (with +v), and half will move out of the box; the same is true for particles at $(x + \Delta x/2)$, with half of the particles moving into the box (with –v). Thus, the change in the quantity $Q(x)$ per unit area per unit time, or the flux of the quantity $Q(x)$, is

$$\begin{aligned} &\frac{\mathrm{v}}{2}q(x-\Delta x/2) - \frac{\mathrm{v}}{2}q(x+\Delta x/2) \approx \frac{\mathrm{v}}{2}\left(q(x) - \frac{dq}{dx}\frac{\Delta x}{2}\right) - \frac{\mathrm{v}}{2}\left(q(x) + \frac{dq}{dx}\frac{\Delta x}{2}\right) \\ &\rightarrow \frac{dq(x)}{dt} = -\frac{\mathrm{v}}{2}\frac{dq(x)}{dx}(\Delta x) \xrightarrow{\text{Averaged over 3 dimensions}} \frac{dq(x)}{dt} = -\frac{1}{2}\left(\frac{\overline{\mathrm{v}}}{3}\right)\frac{dq(x)}{dx}(n\overline{\ell}) . \end{aligned} \tag{A.24.a}$$

To arrive at the last identity, we have averaged over the 3-dimensional motions (the isotropic average of $\cos^2\theta$ is 1/3), and over the Maxwellian speed distribution. We then set (Δx) equal to a number (integer) times the mean free path $\overline{\ell}$, implying that it takes a number of n collisions for a particle to come to terms with its new environment, that is, for the values $q(x+\Delta x/2)$ and $q(x-\Delta x/2)$ to settle down to the value in the box, $q(x)$. Applying (4.26.a) to the transport of particles, thermal energy, and the y-component of linear momentum with their respective densities, $\mathcal{N}$, $\mathcal{N}c_{\mathrm{v}}T$, and $\mathcal{N}m\mathrm{v}_y$, (A.24.a) gives the rate of the associated density (in the box) increase in relation to their respective flux as

$$\begin{aligned}
\frac{d\mathcal{N}}{dt} &= -\frac{1}{2}\left(\frac{\overline{\mathrm{v}}}{3}\right)\frac{d\mathcal{N}}{dx}\left(n\overline{\ell}\right) = -\frac{\text{integer}\times\overline{\ell}\overline{\mathrm{v}}}{6}\frac{d\mathcal{N}}{dx} \rightarrow D = \frac{\text{integer}\times\overline{\ell}\overline{\mathrm{v}}}{6};\\
\frac{d(\mathcal{N}c_{\mathrm{v}}T)}{dt} &= -\frac{1}{2}\left(\frac{\overline{\mathrm{v}}}{3}\right)\frac{d(\mathcal{N}c_{\mathrm{v}}T)}{dx}\left(n\overline{\ell}\right) \approx -\frac{\text{integer}\times\overline{\ell}\overline{\mathrm{v}}}{6}(\mathcal{N}c_{\mathrm{v}})\frac{d(T)}{dx} \rightarrow \kappa = D(\mathcal{N}c_{\mathrm{v}});\\
\frac{d(\mathcal{N}m\mathrm{v}_y)}{dt} &= -\frac{1}{2}\left(\frac{\overline{\mathrm{v}}}{3}\right)\frac{d(\mathcal{N}m\mathrm{v}_y)}{dx}\left(n\overline{\ell}\right) \approx -\frac{\text{integer}\times\overline{\ell}\overline{\mathrm{v}}}{6}(\mathcal{N}m)\frac{d(\mathrm{v}_y)}{dx} \rightarrow \mu = D(\mathcal{N}m).
\end{aligned} \tag{A.24.b}$$

Diffusion is the basic property of the transport; its coefficient, D, depends on the product of the mean free path and the average speed in thermal equilibrium. The absolute values of the three transport coefficients depend on the choice of the "integer", not explicitly shown in (A.24.b). Often, this integer is taken to be 2, implying that only two collisions are needed for a particle to settle in its new local values; in this case, we have $D = \left(\overline{\ell}\overline{\mathrm{v}}\right)/3$. The thermal conductivity, κ, depends, in addition to $\left(\overline{\ell}\overline{\mathrm{v}}\right)$, on the product of $\mathcal{N}$ and c_{v}; the latter is $1.5\,k_BT$ for a monoatomic gas and $2.5\,k_BT$ for a diatomic molecular gas, when both rotational degrees of freedom are fully activated at ambient temperature. The shear (dynamic) viscosity, μ, is the coefficient for the transport of linear momentum; it depends on the mass of an individual molecule, m, or the mass density, $\rho = \mathcal{N}m$. The coefficients along with the kinematic viscosity η are related in (A.24.c):

$$D = \frac{\overline{\mathrm{v}}^2}{\lambda_c};\quad \kappa = D\mathcal{N}c_{\mathrm{v}};\quad \mu = D\rho;\ \eta = \frac{\mu}{\rho} = D;\quad \frac{p}{\mu} = \frac{k_BT}{Dm} \tag{A.24.c}$$

This brief review is useful for our discussion of the Cabannes scattering spectrum in Section 4.4.

References

A.1 Corney, A. (1977). *Atomic and Laser Spectroscopy.* Oxford University Press.

A.2 Edmonds, A. R. (1957). *Angular Momentum in Quantum Mechanics.* Princeton University Press.

A.3 Svelto, O. (1976). *Principles of Lasers.* 4th ed. Plenum Press.

A.4 She, C.-Y., H. Chen, and D. A. Krueger. (2015). Optical processes for middle atmospheric Doppler lidars: Cabannes scattering and laser induced resonance fluorescence. *Jour. Opt. Soc. Am.*, **B32**(8), 1575–1592.

A.5 Capitelli, M., G. Colonna, and A. D'Angola (2012). *Fundamental Aspects of Plasma Chemical Physics: Thermodynamics*. Springer. Appendix A, "Spectral Terms for Atoms and Molecules." Accessed at https://cds.cern.ch/record/1433720/files/978-1-4419-8182-0_BookBackMatter.pdf.

A.6 She, C. Y., and J. R. Yu (1995). Doppler-free saturation fluorescence spectroscopy of Na atoms for atmospheric applications. *Appl. Opt.*, **34**(6), 1063–1075.

A.7 Taken from *NIST Atomic Spectra Database Levels Form*: https://physics.nist.gov/PhysRefData/ASD/levels_form.html

A.8 Berestetski, V. B., E. M. Lifshitz, and L. P. Pitevski (1971). *Relativistic Quantum Theory: Part I.* Addison-Wesley.

A.9 von Zahn, U., J. Höffner, V. Eska, and M. Alpers (1996). The mesopause altitude: Only two distinctive levels worldwide? *Geophys. Res. Lett.*, **23**(22), 3231–3234.

A.10 Arimondo, E., M. Inguscio, and P. Violino (1977). Experimental determinations of the hyperfine structure in the alkali atoms, *Rev. Mod. Phys.*, **47**(1), 31–76.

A.11 Taken from *NIST Atomic Spectra Database Lines Form*: https://physics.nist.gov/PhysRefData/ASD/lines_form.htm

A.12 Singh, S., R. G. Smith, and L. G. Van Uitert. (1974). Stimulated-emission cross section and fluorescent quantum efficiency of Nd^{3+} in yttrium aluminum garnet at room temperature. *Phys. Rev. B*, **10**(6), 2566–2572.

A.13 Landau, L. D. and E. M. Lifshitz. (1997). *Quantum Mechanics, Non-relativistic Theory.* 3rd ed. Pergamon Press.

A.14 Atkins, P., and R. Friedman. (2005). *Molecular Quantum Mechanics*. 4th ed. Oxford University Press.

A.15 Herzberg, G. (1950). *Molecular Spectra and Molecular Structure: I. Spectra of Diatomic Molecules*. 2nd ed. Van Nostrand Reinhold Company.

A.16 Günzler, H. and H.-U. Gremlich. (2002). *IR Spectroscopy: An Introduction*. Wiley-VCH.

A.17 Lemus, R., M. Sánchez-Castellanos, F. Pérez-Bernal, J. M. Fernández, and M. Carvajal. (2014). Simulation of the Raman spectra of CO_2: Bridging the gap between algebraic models and experimental spectra. *Jour. Chem. Phys.*, **141**(5), 054306.

A.18 Andrews, D. G. (2000). *An Introduction to Atmospheric Physics*. Cambridge University Press.

A.19 Stowe, K. (2007). *An Introduction to Thermodynamics and Statistical Mechanics*. 2nd ed. Cambridge University Press.

Appendix B

Coordinate Systems and Mueller Matrices

To describe laser light scattering, we need two Cartesian/polar coordinate systems with a common origin, one a space-fixed (or laboratory) system with (x, y, z) axes and the other a rotating system, that fixes to a molecule with (X, Y, Z) axes. For the space-fixed Cartesian coordinate system with unit vectors $(\hat{x}, \hat{y}, \hat{z})$, one may use equivalent spherical (or polar) coordinates with unit vectors $(\hat{r}, \hat{\theta}, \hat{\phi})$, or the scattering Cartesian coordinates with unit vectors $(\hat{e}'_a, \hat{e}'_b, \hat{k})$, where $\hat{k} = \hat{r}$ points along the direction of the scattered light. The $\hat{k}$ unit vector is the $\hat{z}$ axis rotated into the polar and azimuthal directions, $\hat{\theta}$ and $\hat{\phi}$. As shown in Fig. B.1, these three sets of unit vectors are all real vectors and mutually perpendicular to one another within each system.

These unit vectors are related, see, for example, [B.1]. Those of the spherical and the scattering coordinates may be expressed in terms of their direction cosines in Cartesian coordinates $(\hat{x}, \hat{y}, \hat{z})$ as follows:

$$\hat{k} = (\sin\theta\cos\phi \quad \sin\theta\sin\phi \quad \cos\theta)\,, \tag{B.1.a}$$

$$\hat{\theta} = (\cos\theta\cos\phi \quad \cos\theta\sin\phi \quad -\sin\theta), \tag{B.1.b}$$

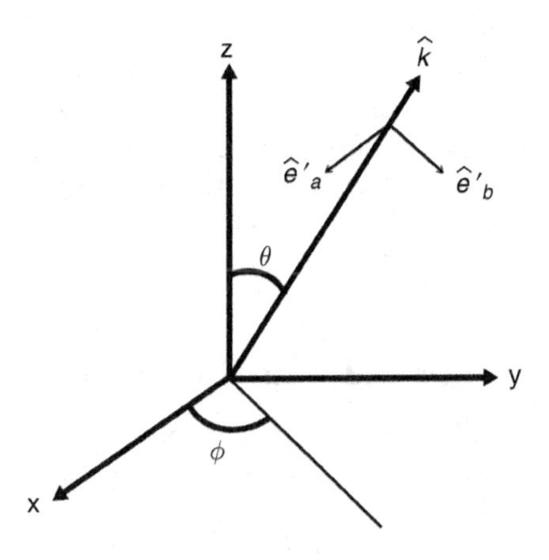

Fig. B.1 Cartesian, spherical, and scattering coordinates with respective unit vectors, $(\hat{x}, \hat{y}, \hat{z})$, $(\hat{r}, \hat{\theta}, \hat{\phi})$, and $(\hat{k}, \hat{e}'_a, \hat{e}'_b)$.

$$\hat{e}_a{}' = -\hat{\phi} = (\sin\phi \quad -\cos\phi \quad 0), \text{ and} \tag{B.1.c}$$

$$\hat{e}_b{}' = \hat{\theta} = (\cos\theta\cos\phi \quad \cos\theta\sin\phi \quad -\sin\theta). \tag{B.1.d}$$

To facilitate the circular polarization of light and the discussion of angular momenta quantum mechanically, we introduce a complex coordinate system with the unit vectors of $\hat{\varepsilon}_q$ (q = 1, 0, −1). They are related to the Cartesian coordinates and vice versa, as in (B.2.a) and (B.2.b):

$$\hat{\varepsilon}_1 = -\frac{1}{\sqrt{2}}(\hat{x} + i\hat{y}),\ \hat{\varepsilon}_0 = \hat{z}, \hat{\varepsilon}_{-1} = \frac{1}{\sqrt{2}}(\hat{x} - i\hat{y}), \text{ and} \tag{B.2.a}$$

$$\hat{x} = -\frac{\hat{\varepsilon}_1 - \hat{\varepsilon}_{-1}}{\sqrt{2}},\ \hat{y} = i\frac{\hat{\varepsilon}_1 + \hat{\varepsilon}_{-1}}{\sqrt{2}},\ \hat{z} = \hat{\varepsilon}_0. \tag{B.2.b}$$

We note that the complex unit vectors are equal to the negatives of their respective complex conjugates, as $\hat{\varepsilon}_{\pm1} = -\hat{\varepsilon}_{\mp1}{}^*$, $\hat{\varepsilon}_{\pm1}\bullet\hat{\varepsilon}_{\pm1}{}^* = 1$, and $\hat{\varepsilon}_{\pm1}\bullet\hat{\varepsilon}_{\mp1} = -1$, and in terms of these, $\hat{x}$ and $\hat{y}$ remain real. In terms of the unit vectors of the complex coordinate system, $(\hat{\varepsilon}_{+1}, \hat{\varepsilon}_0, \hat{\varepsilon}_{-1})$, a vector may be expressed as $\vec{V} = \sum_{q=-1}^{+1} V_{-q}\hat{\varepsilon}_q = (V_{-1}, V_0, V_{+1})$. Thus, the direction of received radiation and polarizations may be expressed as

$$\hat{k} = (-\frac{1}{\sqrt{2}}\sin\theta e^{-i\phi},\ \cos\theta,\ \frac{1}{\sqrt{2}}\sin\theta), \tag{B.3.a}$$

$$\hat{e}_a{}' = \left(-\frac{i}{\sqrt{2}}e^{-i\phi},\ 0,\ -\frac{i}{\sqrt{2}}e^{i\phi}\right), \text{ and} \tag{B.3.b}$$

$$\hat{e}_b{}' = \left(-\frac{\cos\theta}{\sqrt{2}}e^{-i\phi},\ -\sin\theta,\ \frac{\cos\theta}{\sqrt{2}}e^{i\phi}\right). \tag{B.3.c}$$

Because the incident direction of the laser beam may be different from the scattering direction, we also need incident Cartesian coordinates, $(\hat{k}_0, \hat{e}_a, \hat{e}_b)$, with $\hat{k}_0$ being the incident beam propagation direction, specified by the angles θ_0 and ϕ_0 with respect to a space-fixed system. The direction cosines of these unit vectors and their projections on the complex spherical unit vectors are the same, respectively, as (B.1.a), (B.1.c), and (B.1.d), and as (B.3.a), (B.3.b), and (B.3.c), except that θ and ϕ are replaced by the polar and azimuthal angles of the incident beam, θ_0 and ϕ_0.

B.1 Rotation of an Object and Transformation of Coordinate Systems

For simplicity, we first consider the rotation of a point 1, or vector $\overrightarrow{O1}$, from the origin of a two-dimensional space to point 2, or vector $\overrightarrow{O2}$, as given in Fig. B.2(a); this is called active rotation (of an object). In this case, the coordinates of point 1 ($x = OA$ and $y = OB$) can be transformed into the coordinates of point 2 ($x' = OA'$ and $y' = OB'$) in the same coordinate system, as seen in Fig. B.2(a). Equivalently, we can rotate the coordinate system from (x, y) to the system (x', y'), as given in Fig. B.2(b);

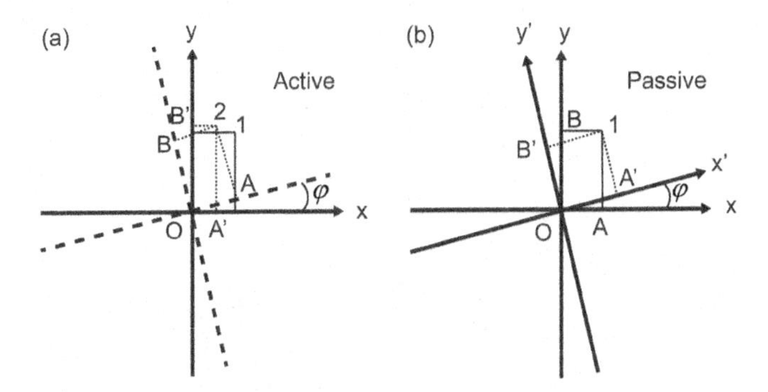

Fig. B.2 (a) Active rotation of $\overrightarrow{O1}$, whose projections are $x = OA$ and $y = OB$, by an angle φ to $\overrightarrow{O2}$, whose projections on the same axes are $x' = OA'$ and $y' = OB'$; their relations are given in (B.4.a). (b) Passive rotation of the coordinates (x, y) by an angle φ to the coordinates (x' y'); their relations are given in (B.4.b).

this is called passive rotation (of coordinate systems). In this case, the coordinates of the same point (x, y) and (x', y') projected in two different coordinate systems may be related as (B.4.b).

Thus, the transformation between coordinates in active rotation of an object by angle φ is given in (B.4.a), and that between two coordinates of the same point in two coordinate systems related by rotating angle φ is given in (B.4.b). Or, alternatively related in matrix form in (B.4.c) and (B.4.d),

$$\begin{aligned} x' &= x\cos\varphi - y\sin\varphi \\ y' &= x\sin\varphi + y\cos\varphi, \text{ or} \end{aligned} \tag{B.4.a}$$

$$\begin{aligned} x' &= x\cos\varphi + y\sin\varphi \\ y' &= -x\sin\varphi + y\cos\varphi; \end{aligned} \tag{B.4.b}$$

$$\begin{bmatrix} x' \\ y' \end{bmatrix} = \begin{bmatrix} \cos\varphi & -\sin\varphi \\ \sin\varphi & \cos\varphi \end{bmatrix} \begin{bmatrix} x \\ y \end{bmatrix}, \text{ or} \tag{B.4.c}$$

$$\begin{bmatrix} x' \\ y' \end{bmatrix} = \begin{bmatrix} \cos\varphi & \sin\varphi \\ -\sin\varphi & \cos\varphi \end{bmatrix} \begin{bmatrix} x \\ y \end{bmatrix}. \tag{B.4.d}$$

Notice the transformation matrix of the passive rotation (of coordinate systems) is the transpose of that of the active rotation (of points in the same coordinate system).

B.2 Rotations in Three-Dimensional Space and Euler Angles

Since atoms and molecules live in a three-dimensional space, the rotating (molecular) Cartesian coordinate system (X, Y, Z) may be obtained from the fixed Cartesian coordinate system (x, y, z) by means of three successive rotations (of coordinate systems):

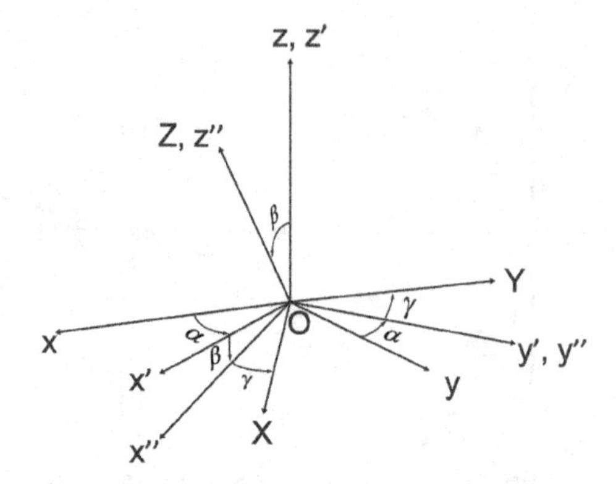

Fig. B.3 Rotation from the fixed (*x, y, z*) axes to the rotating (molecule) system (*X, Y, Z*). For details, see Fig. 1.1 of [B.2].

1. With respect to the *z*-axis by an angle α to ($x', y', z' = z$);
2. With respect to the new y'-axis by β to ($x'', y'' = y', z''$);
3. With respect to the new z''-axis by γ to ($X, Y, z'' = Z$).

These angles (α, β, γ) are known as the Euler angles, as indicated in Fig. B.3. The corresponding rotation matrices are denoted as $D_z(\alpha)$, $D_{y'}(\beta)$, and $D_{z''}(\gamma)$, or as $D(\alpha, \beta, \gamma) = D_{z''}(\gamma) D_{y'}(\beta) D_z(\alpha)$.

The three transformation matrices that rotate the fixed Cartesian system into the new coordinate system (passive rotations) are given in (B.5.a). By inspection of Fig. B.3, we can see that these three operations with respect to successive new axes are equivalent to the three rotations in the reverse order with respect to the corresponding original axes; see (1.3.1) of [B.2]. A more detailed discussion on this point may be found in Section A5.2 of [B.3]. Therefore, by adopting the passive rotation of the coordinate system, the three-dimensional transformation matrix $D(\alpha, \beta, \gamma)$ may be obtained by multiplication of matrices in (B.5.a), resulting in (B.5.b).

$$D_z(\alpha) = \begin{bmatrix} \cos\alpha & \sin\alpha & 0 \\ -\sin\alpha & \cos\alpha & 0 \\ 0 & 0 & 1 \end{bmatrix},$$

$$D_{y'}(\beta) = \begin{bmatrix} \cos\beta & 0 & -\sin\beta \\ 0 & 1 & 0 \\ \sin\beta & 0 & \cos\beta \end{bmatrix}, D_{z''}(\gamma) = \begin{bmatrix} \cos\gamma & \sin\gamma & 0 \\ -\sin\gamma & \cos\gamma & 0 \\ 0 & 0 & 1 \end{bmatrix}; \qquad \text{(B.5.a)}$$

$$D(\alpha, \beta, \gamma) = D_{z''}(\gamma) D_{y'}(\beta) D_z(\alpha) = D_z(\alpha) D_y(\beta) D_z(\gamma)$$
$$= \begin{pmatrix} \cos\alpha\cos\beta\cos\gamma - \sin\alpha\sin\gamma & \sin\alpha\cos\beta\cos\gamma + \cos\alpha\sin\gamma & -\sin\beta\cos\gamma \\ -\cos\alpha\cos\beta\sin\gamma - \sin\alpha\cos\gamma & -\sin\alpha\cos\beta\sin\gamma + \cos\alpha\cos\gamma & \sin\beta\sin\gamma \\ \cos\alpha\sin\beta & \sin\alpha\sin\beta & \cos\beta \end{pmatrix}. \qquad \text{(B.5.b)}$$

The resulting matrix, (B.5.b) is identical to Table A5.1, [B.3] if we identify (α, β, γ) as (ϕ, θ, χ). Note that $D^T(\alpha, \beta, \gamma)\ D(\alpha, \beta, \gamma) = 1$, where $D^T(\alpha, \beta, \gamma)$ is the transpose of

$D(\alpha, \beta, \gamma)$. The coordinates of a point (or projection of a vector with tail on the origin) in the rotating coordinate system may be obtained by multiplying the finite rotation operator $D^T(\alpha, \beta, \gamma)$ by its coordinates in the fixed system.

B.3 Specification of Selected Sets of Unit Vectors Associated with Finite Rotations

As an example of finite rotation, we notice that the Euler angles (α, β) so defined are identical to (ϕ, θ) of the spherical coordinates. This identification allows the derivation of the unit vectors $(\hat{e}'_a, \hat{e}'_b, \hat{k})$ in the Cartesian system $(\hat{x}, \hat{y}, \hat{z})$, as given in (B.1). We can see this by realizing that the unit vectors $(\hat{e}'_a, \hat{e}'_b, \hat{k})$ may be obtained from the unit vectors $(\hat{x}, \hat{y}, \hat{z})$ by active rotation with Euler angles $(\alpha = \phi, \beta = \theta, \gamma = 0)$. Or, in general, we consider a point (x, y, z), or a vector from the origin to this point, after active rotation through specified Euler angles to become a new point (X, Y, Z), or a new vector with its distance from the origin preserved. The matrix that takes (x, y, z) to (X, Y, Z) is then the transpose of (B.5.b), which (passively) relates the coordinates. By setting $(\alpha = \phi, \beta = \theta, \gamma = 0)$, the transformation matrix that transforms (x, y, z) to (X, Y, Z), $D^T(\phi, \theta, 0)$, is given in (B.6.a). The unit vectors $(\hat{X}, \hat{Y}, \hat{Z})$ are derived from $(\hat{x}, \hat{y}, \hat{z})$, and their results are given in (B.6.b).

$$\begin{bmatrix} X \\ Y \\ Z \end{bmatrix} = \begin{bmatrix} \cos\phi\cos\theta & -\sin\phi & \cos\phi\sin\theta \\ \sin\phi\cos\theta & \cos\phi & \sin\phi\sin\theta \\ -\sin\theta & 0 & \cos\theta \end{bmatrix} \begin{bmatrix} x \\ y \\ z \end{bmatrix},$$

$$D^T(\phi, \theta, 0) = \begin{bmatrix} \cos\phi\cos\theta & -\sin\phi & \cos\phi\sin\theta \\ \sin\phi\cos\theta & \cos\phi & \sin\phi\sin\theta \\ -\sin\theta & 0 & \cos\theta \end{bmatrix}; \tag{B.6.a}$$

$$\begin{bmatrix} 0 \\ 0 \\ 1 \end{bmatrix} \rightarrow \hat{Z} = \begin{bmatrix} \cos\phi\sin\theta \\ \sin\phi\sin\theta \\ \cos\theta \end{bmatrix} = \hat{k}, \quad \begin{bmatrix} 0 \\ 1 \\ 0 \end{bmatrix} \rightarrow \hat{Y} = \begin{bmatrix} -\sin\phi \\ \cos\phi \\ 0 \end{bmatrix}$$

$$= -\hat{e}'_a; \quad \begin{bmatrix} 1 \\ 0 \\ 0 \end{bmatrix} \rightarrow \hat{X} = \begin{bmatrix} \cos\phi\cos\theta \\ \sin\phi\cos\theta \\ -\sin\theta \end{bmatrix} = \hat{e}'_b. \tag{B.6.b}$$

By renaming $(\hat{X} \rightarrow \hat{e}'_b, \hat{Y} \rightarrow -\hat{e}'_a, \hat{Z} \rightarrow \hat{k})$, we obtain the unit vectors for the scattered beam, $(\hat{e}'_a, \hat{e}'_b, \hat{k})$, in agreement with (B.1.a), (B.1.c), and (B.1.d). If we replace θ and ϕ by θ_0 and ϕ_0 in (B.6.b), we obtain the Cartesian vectors for the incident beam $(\hat{k}_0, \hat{e}_a, \hat{e}_b)$. The unit vectors $(\hat{X}, \hat{Y}, \hat{Z})$ may be attached to a molecule as its principal axes; they then become the frame for the molecular (rotating) system. To avoid confusion, when transforming a molecule from fixed coordinates to molecular coordinates, we replace θ and ϕ by Θ and Φ in (B.6.a) and (B.6.b).

B.4 Atoms in the Earth's Magnetic Field

Another situation needing the use of a different coordinate system arises when an ensemble of atoms is under the Earth's magnetic field with declination angle $\mathscr{D}$ and inclination angle $\mathscr{I}$. In an external magnetic field, the magnetic axes of all atoms will align with it, and the atomic *Z*-axis will be parallel to the Earth's magnetic field. Since, in an external magnetic field, the magnetic axis (*z*-axis) of a quantum mechanical atom is now parallel to Earth's magnetic field, one needs to determine the coordinates (*X, Y, Z*) of a point (or vector) defined in the geometric (fixed) coordinates (*x, y, z*), with *z*-axis zenith pointing, and *x*-axis and *y*-axis for pointing at east and north. To do so, we rotate the coordinate system (*x, y, z*) by an angle $-\mathscr{D}$ (clockwise) with respect to the *z*-axis so that the *y'*-axis is pointing to the magnetic north (MN); this is then followed by a rotation of an angle $(-\mathscr{I})$ with respect to the new $x'=X$ axis so that the *y'*-axis becomes the *Z*-axis parallel to the Earth's magnetic field, and the *z'*-axis becomes –*Y*-axis (not shown), as shown in Fig. B.4 [B.4].

The transformation matrices for the passive rotations, as we wish to relate the same vectors in the old and new coordinates, are

$$D_z(-\mathscr{D}) = \begin{bmatrix} \cos\mathscr{D} & -\sin\mathscr{D} & 0 \\ \sin\mathscr{D} & \cos\mathscr{D} & 0 \\ 0 & 0 & 1 \end{bmatrix}, \text{ and } D_{x'}(-\mathscr{I}) = \begin{bmatrix} 1 & 0 & 0 \\ 0 & -\sin\mathscr{I} & -\cos\mathscr{I} \\ 0 & \cos\mathscr{I} & -\sin\mathscr{I} \end{bmatrix}, \tag{B.7.a}$$

resulting in the transformation matrix of

$$D(\mathscr{D},\mathscr{I}) = D_{x'}(-\mathscr{I})D_z(-\mathscr{D}) \rightarrow$$
$$\begin{bmatrix} X \\ Y \\ Z \end{bmatrix} = \begin{bmatrix} \cos\mathscr{D} & -\sin\mathscr{D} & 0 \\ -\sin\mathscr{I}\sin\mathscr{D} & -\sin\mathscr{I}\cos\mathscr{D} & -\cos\mathscr{I} \\ \cos\mathscr{I}\sin\mathscr{D} & \cos\mathscr{I}\cos\mathscr{D} & -\sin\mathscr{I} \end{bmatrix} \begin{bmatrix} x \\ y \\ z \end{bmatrix}. \tag{B.7.b}$$

The above is the same as (A11) of Papen et al. [B.5] and in agreement with (A3) of Krueger et al. [B.4], though their notations for the coordinates are reversed from those used in Fig. B.4.

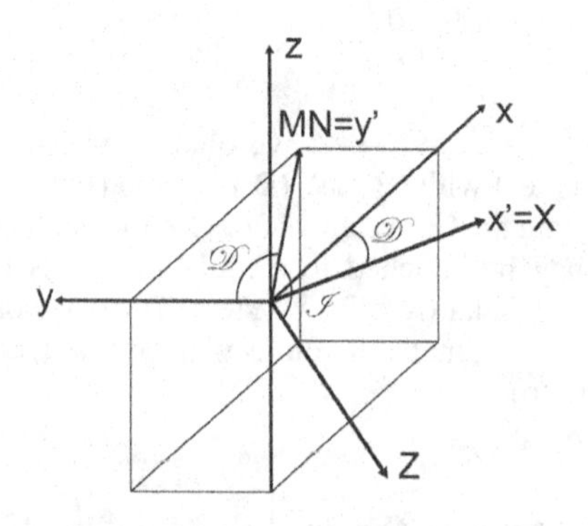

Fig. B.4 Geometric coordinates: *x, y, z* pointing east, north, and zenith with Earth's magnetic field ($\mathscr{D}$ = Declination angle; $\mathscr{I}$ = Inclination angle) pointing toward the new Z-axis.

B.5 Jones Vectors, Jones Matrices, and Nondepolarizing Optical Devices

The electric field of coherent (polarized) light propagating along the z-axis may be expressed as $E_x = E_{0x}\cos(kz - \omega t) = \text{Re}\left[E_{0x}e^{i(kz-\omega t)}\right]$, $E_y = E_{0y}\cos(kz - \omega t + \delta) = \text{Re}\left[E_{0y}e^{i(kz-\omega t)+i\delta}\right]$; $|E| = \sqrt{E_{0x}^2 + E_{0y}^2}$, where E_{0x} and E_{0y} are the magnitudes of the field along the x- and y-axes, respectively, with $|E|$ and δ being the magnitude and relative phase (with y-component leading) between the two components. It is convenient to represent this field by vector on the x-y plane or a complex two-element matrix known as a Jones matrix,

$$\vec{E} = \text{Re}\left[\left(E_{0x}\hat{x} + E_{0y}e^{i\delta}\hat{y}\right)e^{i(kz-\omega t)}\right] \rightarrow \vec{E} = \begin{bmatrix} a\cos\chi \\ a\sin\chi e^{i\delta} \end{bmatrix}, \text{ with } a = |E| \text{ and } \tan\chi = \frac{E_{0y}}{E_{0x}}, \tag{B.8.a}$$

by setting the phase as $\delta = 90°$. This represents a right-handed (with the point of view of the source, or with thumb pointing in the wave propagation direction) elliptically polarized field of magnitude a and ellipticity angle χ; the principal axes $a\cos\chi$ and $a\sin\chi$ ($0° \le \chi \le 90°$) are parallel to the x- and y-axes, respectively. For $\chi = 45°$, the light wave is then right-hand circularly polarized. For $\chi = 0°$ and $90°$, the wave is linearly polarized with polarization axis parallel to the x- and y-axes, respectively. To orient the principal axes in other directions, we rotate the ellipse by an angle φ, as depicted in Fig. B.2(a); its Jones vector in the same coordinate system is then transformed by the 2×2 matrix, as in (B.4.c), into

$$\vec{E} = a\begin{bmatrix} \cos\varphi & -\sin\varphi \\ \sin\varphi & \cos\varphi \end{bmatrix}\begin{bmatrix} \cos\chi \\ i\sin\chi \end{bmatrix} = a\begin{bmatrix} \cos\chi\cos\varphi - i\sin\chi\sin\varphi \\ \cos\chi\sin\varphi + i\sin\chi\cos\varphi \end{bmatrix}. \tag{B.8.b}$$

The Jones vector (B.8.b) then represents an elliptically polarized field with the same characteristics, except with principal axes rotated from the x- and y-axes by angle φ.

A polarization optical element alters the polarization state of light by changing the amplitudes and/or the relative phase of the electric field vector components. Two types of nondepolarizing elements, diattenuators and retarders, are of interest. The diattenuator changes only the amplitudes, while the retarder changes only the phases of the electric-field components. The diattenuation D of a diattenuator and the retardance R of a retarder are expressed [B.6] in terms, respectively, of the difference of transmittances and phases between the principal eigenstates, (q, r), of the optical element, and are given in (B.8.c):

$$D = \frac{|T_q - T_r|}{T_q + T_r};\quad R = |\delta_q - \delta_r|, \tag{B.8.c}$$

where T_q and T_r are the transmittances and δ_q and δ_r are the phase changes for (orthogonal) eigen-polarizations (associated with the eigenstates q, r). Polarizers and wave plates are examples of diattenuators and retarders, respectively. Jones matrices (and vectors) for selected polarization elements and states are given in Table B.1.

B.6 Stokes Vectors, Mueller Matrices, and Atmospheric Scattering

As shown above, the polarized (coherent) field and optical polarization components can be represented by Jones vectors and matrices, respectively. They can also be represented by

Table B.1 *Jones matrices and vectors for polarizing elements and polarization states*

Optical Element	Jones Matrix
linear polarizer (horizontal)	$\begin{pmatrix} 1 & 0 \\ 0 & 0 \end{pmatrix}$
linear polarizer (at +45°)	$\frac{1}{2}\begin{pmatrix} 1 & 1 \\ 1 & 1 \end{pmatrix}$
quarter-wave plate (fast axis-horizontal)	$e^{i\pi/4}\begin{pmatrix} 1 & 0 \\ 0 & i \end{pmatrix}$
half-wave plate	$\begin{pmatrix} 1 & 0 \\ 0 & -1 \end{pmatrix}$
homogeneous circular polarizer (right-handed)	$\frac{1}{2}\begin{pmatrix} 1 & -i \\ i & 1 \end{pmatrix}$
horizontal linear polarization	$\begin{bmatrix} 1 \\ 0 \end{bmatrix}$
vertical linear polarization	$\begin{bmatrix} 0 \\ 1 \end{bmatrix}$
45° linear polarization	$\frac{1}{\sqrt{2}}\begin{bmatrix} 1 \\ 1 \end{bmatrix}$
–45° linear polarization	$\frac{1}{\sqrt{2}}\begin{bmatrix} 1 \\ -1 \end{bmatrix}$
circular polarization (+ for right-handed)	$\frac{1}{\sqrt{2}}\begin{bmatrix} 1 \\ \pm i \end{bmatrix}$

four-element Stokes vectors and 4 × 4 Mueller matrices. Despite the additional complication, there are two advantages of the latter representations: (a) they can be used to handle and describe both natural and scattered light, including both unpolarized and polarized light, and (b) the four elements in the Stokes vector are intensities, which unlike optical amplitudes and phases are measurable quantities. The elements of the 4 × 4 Mueller matrix then relate the input and output intensities, and thus describe an atmospheric scattering medium with known symmetry [B.7] in addition to simpler optical elements. The four elements of Stokes vector are the intensity of the light beam and, respectively, after transmitting through the specified filters/polarizers, as follows:

$$\begin{bmatrix} S_0 \\ S_1 \\ S_2 \\ S_3 \end{bmatrix} = \begin{bmatrix} I_0 \\ 2I_1 - I_0 \\ 2I_2 - I_0 \\ 2I_3 - I_0 \end{bmatrix}, \text{ with } \begin{bmatrix} I_0 = \text{the total intensity of light transmitted} \\ I_1 = \text{intensity of light transmitted by a horizontal polarizer} \\ I_2 = \text{intensity of light transmitted by a } 45^\circ \text{ polarizer} \\ I_3 = \text{intensity of light transmitted by a RCP filter} \end{bmatrix}. \tag{B.9.a}$$

Here, we note, the total intensity, $S_0 = I_0$, includes both polarized and unpolarized light, and the other three elements, S_1, S_2, and S_3, are zero for unpolarized light, thus providing a definition for the degree of polarization as $p = \sqrt{S_1^2 + S_2^2 + S_3^2}/S_0$ (not to be confused with dipole moment).

To deduce the Stokes vector for the right-handed elliptically polarized field given in (B.8.b), we first calculate the polarized light intensity components I_1, I_2, and I_3, as follows. We first project its associated electric field vector, $\vec{E}$, whose E_x and E_y are given by (B.8.b), then multiply by $\sqrt{p}$, respectively onto the horizontal and 45° linearly, and right-handed circularly polarized analyzers, whose Jones matrices are given in Table B.1, and then calculate I_1, I_2, and I_3. Together with its total intensity $I_0 = a^2$, the resulting matrix is

$$\begin{pmatrix} S_0 \\ S_1 \\ S_2 \\ S_3 \end{pmatrix} = a^2 \begin{pmatrix} 1 \\ p\cos(2\chi)\cos(2\varphi) \\ p\cos(2\chi)\sin(2\varphi) \\ p\sin(2\chi) \end{pmatrix}, \tag{B.9.b}$$

where $a^2 = S_0$ is the total light intensity, $a^2 p$ its polarized portion, and $a^2(1 - p)$ its unpolarized portion. A light wave whose polarization changes randomly in a time longer than many periods but shorter than the time you can perform an intensity measurement is deemed unpolarized. For unpolarized light, only S_0 is nonzero, while for completely polarized light, $p = 1$.

The atmosphere under different conditions may be represented by Mueller matrices. According to Lu et al. [B.6], in general, a Mueller matrix that characterizes the atmosphere may be decomposed into factors with different attributes: a diattenuator, a retarder, and a depolarizer. Like those representing optical elements, the diattenuator and retarder are nondepolarizing polarization elements that convert polarized light into polarized light; these matrices can also be described by equivalent Jones matrices as previously described. In contrast, the depolarizer converts completely polarized light into partially polarized light, and there is no Jones matrix equivalent to the depolarizing Mueller matrices. The form of the Mueller matrices representing disordered media (such as the atmosphere) depends on the symmetry of the medium [B.7; B.8; B.9], which we discuss further, as needed, when we address polarization lidars. Here, we follow Eqs. (3) to (7) of [B.9] and point out that a scattering matrix F, satisfying the law of reciprocity and the law of isotropic mirror symmetry, respectively, satisfies the relations $F = \Delta_3 F^T \Delta_3$ and $F = \Delta_{3,4} F \Delta_{3,4}$. A scattering matrix that satisfies both laws is a matrix with six independent elements $F(6)$. The matrices Δ_3, $\Delta_{3,4}$, and $F(6)$ are given in (B.10):

$$\Delta_3 = \begin{bmatrix} 1 & 0 & 0 & 0 \\ 0 & 1 & 0 & 0 \\ 0 & 0 & -1 & 0 \\ 0 & 0 & 0 & 1 \end{bmatrix}; \quad \Delta_{3,4} = \begin{bmatrix} 1 & 0 & 0 & 0 \\ 0 & 1 & 0 & 0 \\ 0 & 0 & -1 & 0 \\ 0 & 0 & 0 & -1 \end{bmatrix}; \quad F(6) = \begin{bmatrix} a_1 & b_1 & 0 & 0 \\ b_1 & a_2 & 0 & 0 \\ 0 & 0 & a_3 & b_2 \\ 0 & 0 & -b_2 & a_4 \end{bmatrix}. \tag{B.10}$$

Unless otherwise stated, the only Mueller matrix of interest for this book is the type depicted by $F(6)$.

References

B.1 She, C.-Y., H. Chen, and D. A. Krueger. (2015). Optical processes for middle atmospheric Doppler lidars: Cabannes scattering and laser induced resonance fluorescence. *J. Opt. Soc. Am.* **B32**(8), 1575–1592.

B.2 Edmonds, A. R. (1957). *Angular Momentum in Quantum Mechanics*. Princeton University Press.

B.3 Long, D. A. (2002). *The Raman Effect: A Unified Treatment of the Theory of Raman Scattering by Molecules*. John Wiley & Sons, Ltd.

B.4 Krueger, D. A., C.-Y. She, and T. Yuan (2015). Retrieving mesopause temperature and line-of-sight wind from full-diurnal-cycle Na lidar observations. *Appl. Opt.* **54**(32), 9469–9489.

B.5 Papen, G. C., W. M. Pfenninger, and D. M. Simonich. (1995). Sensitivity analysis of Na narrowband wind-temperature lidar systems. *Appl. Opt.* **34**(3), 480–498.

B.6 Lu, S. Y. and R. A. Chipman. (1996). Interpretation of Mueller matrices based on polar decomposition. *J. Opt. Soc. Am. A* **13**(5), 1106–1113.

B.7 van de Hulst, H. (1981). *Light Scattering by Small Particles*. Wiley.

B.8 Kaul, V., I. V. Samokhvalov, and S. N. Volkov. (2004). Investigating particle orientation in cirrus clouds by measuring backscattering phase matrices with lidar. *Appl. Opt.* **43**(36), 6620–6628.

B.9 Brown, A. J. (2014). Equivalence relations and symmetries for laboratory, LIDAR, and planetary Müeller matrix scattering geometries. *J. Opt. Soc. Am. A* **31**(12), 2789–2794.

Appendix C

Chiao-Yao She's Research Career at Colorado State University: An Autobiographical Acknowledgment

My four-decade research career at Colorado State University (CSU) may be divided roughly in 1985 into before-lidar days and lidar days. All my activities have directly or indirectly influenced the writing of this book. Hired in 1968 as a junior faculty member without a startup package (typical in those days), I gratefully recall Dave Edwards, who invited me to work in a project he initiated on laser-induced damage in transparent solids. In the process, I learned laser Raman spectroscopy and published papers on Raman investigation of ferroelectric phase transitions with Ted Broberg, whom I consider to be my first Ph.D. student. The credit for helping me to finally establish an independent research project at CSU goes to the Army Research Office (ARO) program directors, who suggested the redirection of my laser beam toward atmospheric applications and studies. Their office later provided a series of two small ARO grants that stretched out from 1974 to 1983 and enabled the Ph.D. dissertations of Keith Bartlett and Rich Kelley. Many interesting nightly trips to Boulder Table Mountain in 1975–6 with Keith for laser aerosol scattering and horizontal wind measurements are still vivid in my memory. Near the end of this period, I had an opportunity to work with colleague Jim Sites and secure a grant (1983–6) from the Air Force to resume my Raman spectroscopy work, this time on thin films and supporting the Ph.D. work of L. S. Hsu. At the same time, we welcomed Hiroshi Shimizu from Japanese National Institute of Environmental Studies for a year-long sabbatical, and from whom I learned about Rayleigh–Mie lidar. This was timely, as I was investigating the use of atomic vapor filters to block aerosol scattering. Together, we published a paper in 1983 establishing a laser atmospheric temperature measurement technique in the troposphere that was effective despite the presence of aerosols. Based on this paper, we obtained another round of ARO support, this time for laboratory atmospheric temperature measurements, which supported the Ph.D. work of Frank Lehmann.

Needing summer support for myself, in 1973 I took a summer job at Lawrence Livermore National Laboratory and in 1974 was awarded an ASEE fellowship at NASA/Ames Research Center. In both I was engaged in nonlinear optics research, which complemented nicely my earlier interests (1966–8 in Minnesota) in the stimulated Raman effect; I was grateful that my NASA host, Ken Billman, continued to support me with seed funding in late 1974, which initiated my nonlinear optics research at CSU. These efforts led to a very successful sabbatical year (1976–7) at the Naval Research Laboratories, where I worked in a team led by

John Reintjes and generated what was then the world's shortest-wavelength coherent radiation at 53 nm and 38 nm via 5th and 7th harmonic generation. Returning from NRL, I learned from a younger colleague, Bill Fairbank, the technique for single atom detection with laser-induced fluorescence (LIF) and, with Ken Billman's help, obtained a grant in 1977–80 for velocity measurements using LIF. This funding supported the Ph.D. work of my student, C. L. Pan, along with Bill's student, John Prodan. More significantly, in a visit with Bill to NASA/Langley to explore the LIF technique for practical flow measurements, our host, Reggie Exton, described their need for wind-tunnel supersonic flow measurements. To solve this problem, I proposed employing stimulated Raman gain techniques. It is my great fortune that Reggie was able to provide funding, including capital equipment to set up a nonlinear optics laboratory at CSU, along with research support from 1979 to 1986 for this endeavor. He invited me to give a week-long short course at Langley on coherent laser light scattering; from my lecture notes I developed a semester course in nonlinear optics at CSU in the 1980s. During this period, I was also fortunate to be able to work with a young colleague, SiuAu Lee, and learned laser precision measurement techniques from her. While establishing their own laboratories, I appreciate that both SiuAu and Bill took time to help guide my graduate students. This period of measuring temperature and wind in supersonic flows turned out to be very productive for my two students, Greg Herring and Hans Moosmüller. Not only did we develop a robust system that Reggie could duplicate at Langley and improve upon for use in NASA wind tunnels, but we also developed the new stimulated Rayleigh–Brillouin gain spectroscopy for investigating kinetic theory in gases. This book's coauthor, Jonathan Friedman, came into this program near the end of this period; he and fellow graduate student Cheryl Gratias developed stimulated Raman and Brillouin gain spectroscopy to study orientational and compressional dynamics in simple organic liquids at both ambient and high pressure for their dissertations. Since the lidar days began in 1985, Jonathan had the opportunity to simultaneously gain knowledge in both coherent light-scattering spectroscopy and incoherent detection atmospheric lidar.

In the middle of the last round of laboratory atmospheric temperature measurements in 1985, I had the good fortune to be informed by the then ARO program director, Walter Bach, of the news of the CSU Geoscience Center proposal. He suggested that I join the team, and I proposed to build a lidar observatory for tropospheric temperature-profiling measurements. This was the period when the Defense Department realized the need to support University Research Equipment, so the proposal PI, Tom von der Haar of the CSU Atmospheric Science Department, allocated $300,000 via the Geoscience Center for a pulsed lidar observatory, along with a five-year research grant (1986–91) followed by Phase II funding (1995–7). These funds partially supported three Ph.D. students, Raul Alvarez, Max Caldwell, and John Hair, who successfully measured atmospheric temperature profiles, respectively using barium filters at 554 nm and iodine filters at 589 nm and 532 nm. As is altogether too common and yet somehow unexpected, the investigation turned out to be more challenging than anticipated, and it took a total of twelve years to develop a practical system at 532 nm sufficiently robust to demonstrate field temperature measurements from the ground to the tropopause. This was finally described in John Hair's dissertation in 1998.

To complete this work in the face of funding gaps, I needed bridge money to support the students. Fortunately, I had had another stroke of fortune in 1987 when I learned about mesospheric sodium atoms in a remote sensing meeting. I immediately realized that I could tune the narrowband lidar system in the Lidar Observatory to 589 nm to induce fluorescence from these atoms, and not only that, but to simultaneously measure atmospheric temperatures from near 100 km in height. Chet Gardner of the University of Illinois was also at the meeting and provided me some literature on the recent success of Ulf von Zahn's pulsed dye laser system (of Bonn University, Germany) for upper mesospheric temperature measurements, along with the name of the relevant Program Director at the NSF. With this information, I was able to obtain an initial two-year NSF grant for temperature measurements in the upper mesosphere and lower thermosphere (MLT). It turns out that the upper atmosphere is, in fact, a clean laboratory; thus, one can predict its behavior more easily than that of the troposphere. In the middle of August 1989, Richard Bills, Chet's graduate student, came to CSU with a truck loaded with a 1.22-m Fresnel lens and the photon-counting electronics (not yet commercially available) that he designed. Together with my student Jay Yu and postdoc Hamid Latifi, who were getting the Na lidar transmitter ready, we deployed a narrowband Na lidar in the CSU lidar observatory. In about a week's time, they performed the first MLT temperature measurements in North America with three nights of observation (in the company of Chet and myself) ending on August 25, 1989, still in the first year of the two-year grant. Chet's guidance and collaboration continue to this day. With new and continuing NSF grant support, the CSU Na lidar facility continued in operation for the pursuit of different MLT science goals through March 2010, when my former Ph.D. student and postdoc, Titus Yuan, relocated the lidar to Utah State University (USU) in anticipation of my retirement in January 2011. There, he and his students continue MLT lidar research. Credit to Titus that our regular MLT nocturnal temperature observations, begun in 1991, have (as of this writing) continued for 30 plus years with no end in sight. This long data set is a credit to the generations of my CSU students and visitors (named below) along with those of Titus at USU since 2010, who continue to contribute to investigations of atmospheric waves and long-term change in the MLT.

The initial temperature measurements were performed by tuning the laser cyclically between the Na D_{2a} peak and D_2 crossover (midway between the D_{2a} and D_{2b} peaks) frequencies. It was upgraded to measure both temperature and wind in 1994 by precise cyclical frequency switching among the D_{2a} peak and fixed frequencies on either side. In 1996, we developed a Na Faraday filter (FF) – or dispersive Faraday filter (DFF) – and demonstrated measurements under sunlit conditions. Employing this filter, we achieved 24-hour continuous observations in 2002. In addition to MLT science studies, early Na lidar students Jay Yu, H.-L. Chen, Mike White, and Sam Chen made vital contributions to lidar system developments and upgrades. During the development period, we were fortunate to have visiting faculty M. C. Lee from Chiao-Tung University in Taiwan and Chikao Nagasawa from Tokyo Metropolitan University in Japan. They spent, respectively, one year and three-month sabbaticals with us. Later, we hosted Z. S. Liu of Ocean University of China and his Ph.D. student Weibiao Chen for over a year. In 2004, Richard Collins of the University of Alaska spent a semester sabbatical between CSU and National Center for Atmospheric Research (NCAR) in Boulder.

The six Ph.D. students following 2000 (Jim Sherman, Titus Yuan, Tao Li, Phil Acott, Jia Yue, and Sean Harrell) concentrated principally on using the observational data in different aspects of MLT science studies for their dissertations. These include the determination of the counterintuitive MLT temperature climatology (colder during summer than in winter), and groundbreaking studies of tidal and buoyancy waves and their interactions. Between 2001 and 2006, our Na lidar served as a ground truth station for NASA-TIMED satellite observations of the mesosphere and thermosphere. Additional funding has helped the group to engage additional members, including undergraduate students and visitors. In this connection, I am grateful to have had Kam Arnold join us as a part-time undergraduate for two years and then stay for one more year after graduation as well as visiting faculty Taku Kawahara from Shinshu University in Japan for two years, contributing to different aspects of research and laboratory operation. This additional manpower made possible a long 14-day campaign in September 2003 with a record-setting nine days of 24-hour continuous observation. Later, in 2006, we hosted Zhao-Ai Yan as a resident visiting graduate student from Ocean University of China for one year.

My dear friend and colleague Dave Krueger has been my co-PI for all lidar work since 1985 and continues to this day to be my closest research collaborator. I am grateful for his great efforts all these years, sharing the responsibility of guiding graduate students in the group. Special credit goes to our wives, Lucy She and Minnie Krueger, who treated our group members as part of the family, often keeping them fed and entertained. Their contributions to maintaining the well-being of a happy CSU lidar group cannot be over exaggerated.

The idea of using laser-induced fluorescence of atmospheric Na for temperature measurements at altitudes from 80 to 100+ km was considered sufficiently novel to be noted by famed physicists. In 1992, while I was on sabbatical in Tsing Hua University in Taiwan, Arthur Shawlow came to CSU for a public talk and Physics Department seminar, and he visited our lidar observatory. He later told me in his visit to Tsing Hua that to his knowledge, ours is the first application of the "crossover" resonances. When Jan Hall gave a Physics Colloquium at CSU in January 2009, he was interested in the lidar system and spent considerable time talking to my graduate students at the site. Later, in 2018, he gave a colloquium at Utah State University and was surprised and delighted to see the lidar still in operation at its new home.

Continuing the second half of my sabbatical in Spring 1993, I spent three months with M.-L. Chanin's group at Service d'Aeronomie du CNRS, France, where I learned how to treat and analyze a large quantity of lidar data. While in Europe, I also visited Ulf von Zahn, who became the Director of the Institute of Atmospheric Physics in Germany, and he arranged for me a visit to the Arctic Lidar Observatory for Middle Atmosphere Research (ALOMAR) in Andøya, Norway, in 1994. Also, in his role as the ALOMAR director, Ulf gave us a verbal open invitation to deploy a Na lidar at the facility. As a result, we become collaborators for more than a decade. I have learned much MLT science from Ulf, overlapping with him frequently at ALOMAR throughout the years. The Na lidar wish finally became reality in 1999 when Paul Bellaire of AFOSR came through with a Defense University Research Instrumentation Program grant based on my collaborative proposal, together with Dave Fritts of Colorado Research Associates who developed the science justifications for the proposal to deploy a Na lidar at an Arctic site. To enhance the chances of success, I applied for and fortunately was granted a Fulbright scholarship to work with Eivind Thrane and Ulf Hoppe at the Norwegian Defence Research Institute in Oslo for five months. Since the first observation in August 2000 (story below), the ALOMAR Na lidar (later named Weber lidar) has provided research opportunities to many young international scientists. At CSU, we also hosted numerous visits from scientists and graduate students associated with ALOMAR.

The CW seed laser of the ALOMAR Na lidar transmitter was a sum-frequency-generator (SFG), totally different from previous seeders; it was developed and constructed mainly by Joe Vance under the guidance of Hans Moosmüller at the Desert Research Institute and University of Nevada at Reno; after receiving his MS, Joe became my graduate student and received his Ph.D. at CSU in 2004. In addition to Joe, our initial team for lidar installation and deployment at ALOMAR consisted of postdocs Z. L. Hu and Biff Williams and me. The interesting story of the lidar's "First Light" has been published in *EOS, Transactions, American Geophysical Union*, in the July 2, 2002, issue:

"The new lidar was first installed and tested in the summer of 2000. Two scientists were on site between mid-July and mid-August 2000, along with two additional scientists present for 3 weeks each. Gorgeous weather prevailed for the first 2 weeks, but typical deployment delays and various electrical noise problems prevented us from acquiring data during the first 4 days in August as scheduled when all four participants were present. By the time the system was ready on 10 August, the weather had deteriorated; it was cloudy and raining at times. The first atmospheric Na signal was detected at 09:50:15 UT on 13 August 2000 under daylight and partly cloudy conditions. That we could detect the first light with only a half hour of searching under these conditions testifies, among other things, to the effectiveness of our DFF. Marginal weather turned worse before it improved a day later. Fair weather prevailed at ALOMAR the following night, with all three lidars operating between 21:00 on 14 August and 01:00 on 15 August. The resulting data quality clearly demonstrated the potential of clustered lidar observation at ALOMAR. The first 24-hour observation period with the Weber lidar was completed ~1:30 UT, on 16 August 2000."

This was just in time for our scheduled flight to leave Andøya in the same day. After my retirement, the responsibility for the Weber lidar was transferred to Dave Fritts and Biff Williams. In 2018, it was relocated to Poker Flats Research Range, Alaska, another Arctic site, and given two upgrades: the replacement of the CW seed laser with a commercially available state-of-the-art, solid-state tunable CW laser at 589 nm, and simultaneous three-direction operation.

Through some forty years of my research pursuits on different problems, a common theme emerges: They all provided opportunities for me to learn from collaborators (i.e., colleagues, visitors, and students who have worked closely with me). These learning processes have motivated me to write this book with Jonathan. I have tried to summarize my experience, and I take this opportunity to name most of my collaborators and express a word of heartfelt thanks and appreciation.

Chiao-Yao (Joe) She
San Francisco, California
February 2021

Index

Printed in the United States
by Baker & Taylor Publisher Services